OPTIMIZATION AND CONTROL WITH APPLICATIONS

Applied Optimization

VOLUME 96

Series Editors:

Panos M. Pardalos
University of Florida, U.S.A.

Donald W. Hearn
University of Florida, U.S.A.

OPTIMIZATION AND CONTROL WITH APPLICATIONS

Edited by

LIQUN QI
The Hong Kong Polytechnic University, Hong Kong

KOKLAY TEO
The Hong Kong Polytechnic University, Hong Kong

XIAOQI YANG
The Hong Kong Polytechnic University, Hong Kong

 Springer

Library of Congress Cataloging-in-Publication Data

A C.I.P. record for this book is available from the Library of Congress.

ISBN 0-387-24254-6   e-ISBN 0-387-24255-4   Printed on acid-free paper.

Printed in the United States of America.

9 8 7 6 5 4 3 2 1   SPIN 11367154

springeronline.com

Contents

Part III   OPTIMAL CONTROL

Preface

The 34th Workshop of The International School of Mathematics G. Stampacchia, The International Workshop on Optimization and Control with Applications, was held during July 9-17, 2001 in Ettore Majorana Centre for Scientific Culture, Erice, Sicily, Italy. The Workshop was supported by Italian Ministry of University and Scientific Research, Sicilian Regional Government, The Hong Kong Polytechnic University and The National Cheng Kung University. The Director of The International School of Mathematics G. Stampacchia is Franco Giannessi. The Directors of the Workshop are Liqun Qi and Kok Lay Teo. They jointly organized the Workshop. About 90 scholars from as many as 26 countries and regions attended the Workshop. It consisted of 21 45-minute invited lectures, 45 30-minute contributed talks and 11 15-minute short communications.

This book contains 28 papers emitted from the Workshop. All papers were refereed. A special issue of Journal of Global Optimization containing 6 papers from the Workshop is also published.

The 28 papers are divided into four parts: Part I, Duality and Optimality Conditions, has 6 papers, Part II, Optimization Algorithms, has 9 papers, Part III, Optimal Control, has 8 papers, Part IV, Variational Inequality and Equilibrium Problems, has five papers.

One of the invited lecturers of the Workshop was Professor Elijah (Lucien) Polak, who is a legend in our community. As a survivor of the Holocaust, Lucien has made many significant contributions to optimization and control and their applications. Lucien was close to his 70th birthday during the Workshop. We decided to dedicate this edited volume to him.

We wish to take this opportunity to express our gratitude to Professor Franco Giannessi, the staff of Ettore Majorana Centre for Scientific Culture, and all speakers and participants for their contribution to the success of the Workshop. We would also like to thank Eva Yiu for the clerical work she provided for the workshop. We greatly appreciate the support from the referees of all submissions to this Special Issue for their reviews.

We are very happy to see that this Workshop has become a new conference series. This Workshop is now regarded as OCA 2001. During August 18-22, 2002, The Second International Conference on Optimization and Control with Applications (OCA2002) was successfully held in Tunxi, China. The Third International Conference on Optimization and Control with Applications (OCA2003) will be held in Chongqing-Chengdu, China, during July 1-7, 2003. Liqun Qi and Kok Lay Teo have continued to be the Directors of OCA 2002 and OCA 2003. We hope that OCA Series will continue to provide a forum for international researchers and practitioners working in optimization, optimal control and their applications to exchange information and ideas on the latest development in these fields.

Liqun Qi, Kok Lay Teo and Xiaoqi Yang
The Hong Kong Polytechnic University

Professor Elijah Polak

Elijah (Lucien) Polak was born August 11, 1931 in Bialystok, Poland. He is a holocaust surviver and a veteran of the death camps at Dachau, Auschwitz, Gros Rosen, and Buchenwald. His father perished in the camps, but his mother survived. After the War, he worked as an apprentice blacksmith in Poland and a clothes salesman in France. In 1949, he and his mother migrated to Australia, where, after an eight year interruption, he resumed his education, while working various part time jobs.

Elijah Polak received the B.S. degree in Electrical Engineering, from the University of Melbourne, Australia, in 1957 and the M.S. and Ph.D. degrees, both in Electrical Engineering, from the University of California, Berkeley, in 1959 and 1961, respectively.

In 1961 he married Ginette with whom he had a son and a daughter. At present, they have 5 grandchildren, of which two are beginning to show an interest in mathematics.

From 1957 to 1958 he was employed as an Instrument Engineer by Imperial Chemical Industries, Australia and New Zealand, Ltd., in Melbourne, Australia. He spent the summers of 1959 and 1960 as a Summer Student, with I.B.M. Research Laboratories, San Jose, California, and the Fall Semester of 1964 as a Visiting Assistant Professor, at the Massachusetts Institute of Technology. Since 1961, he has been on the faculty of the University of California, Berkeley, where he is at present Professor Emeritus of Electrical Engineering and Computer Sciences, as well as Professor in the Graduate School.

He was a Guggenheim Fellow in 1968 - 1969, at the Institut Blaise Pascal, in Paris, France, and a United Kingdom Science Research Council Senior Post Doctoral Fellow, at Imperial College, London, England, in 1972, in 1976, in 1979, in 1982, 1985, 1988, and in 1990.

His research interests lie in the development of optimization algorithms for computer-aided design, with applications to electronic circuit design, control system design, and structural design, as well as algorithms for optimal control and nonsmooth optimization.

He is the author or co-author of over 250 papers as well as of four books: Theory of Mathematical Programming and Optimal Control (with M. Canon and C. Cullum, 1970), Notes of a First Course on Linear Systems (with E. Wong, 1970), Computational Methods in Optimization (1971), and Optimization: Algorithms and Consistent Approximations (1997). In addition, with L. A. Zadeh, he co-edited System Theory (1969), and translated from the Russian Absolute Stability of Regulator Systems, by M. A. Aizerman and F. R. Gantmacher.

He is a Life Fellow of the Institute of Electrical and Electronic Engineers, a member of the Society of Industrial and Applied Mathematics and a member of the Mathematical Programming Society.

He is an Associate Editor the Journal of Optimization Theory and Applications, and of the Journal of Computational Optimization and Applications.

A. BOOKS

1. E. Polak, *"Absolute stability of regulator systems"*, (Translated from Russian) (M. A. Aizerman and F. R. Gantmacher) Holden-Day, 1964.

2. L. A. Zadeh and E. Polak eds., *Systems Theory*, McGraw-Hill, 521 pages, 1969.

3. Canon, M. D., C. D. Cullum and E. Polak, *Theory of Optimal Control and Mathematical Programming*, McGraw-Hill Co., New York, 285 pages, 1970.

4. E. Polak and E. Wong, *Notes for a First Course on Linear Systems*, Van Nostrand Reinhold Co. New York, 169 pages, 1970.

5. E. Polak, *Computational Methods in Optimization: A Unified Approach*, Academic Press, 329 pages, 1971.

6. E. Polak, *Optimization: Algorithms and Consistent Approximations*, Springer, New Yort, 800 pages, 1997

B. PAPERS AND REPORTS

1. E. Polak, "Stability and Graphical Analysis of First-Order Pulse-Width-Modulated Sampled-Data Regulator Systems," *IRE Trans. on Automatic Control*, Vol. AC-6, No. 3, pp. 276-282, 1961.

2. E. Polak, "Minimum Time Control of Second Order Pulse-Width-Modulated Sampled-Data Systems," *ASME Trans. Journal of Basic Engineering*, Vol. 84, Series D, No. 1, pp. 101-110, 1962.

3. E. Polak and C. A. Desoer, "A Note on Lumped linear Time Invariant Systems," *IRE Trans. on Circuit Theory*, pp. 282-283, 1962.

4. E. Polak, "Minimal Time Control of a Discrete System with a Nonlinear Plant," *IEEE Trans. on Automatic Control*, Vol. AC-8, No. 1, pp. 49-56, 1963.

5. E. Polak, "Exploratory Design of a Hydraulic Position Servo," *Instrument Society of American Trans.*, Vol. 2, Issue 3, pp. 207-215, 1963.

6. E. Polak, "On the Equivalence of Discrete Systems in Time-Optimal Control," *ASME Trans. Journal of Basic Engineering,* Series D, pp. 204-210, 1963.

7. E. Polak, C. A. Desoer, and J. Wing, "Theory of Minimum Time Discrete Regulators," *Proc. Second IFAC Congress,* Paper No. 406, Basle, 1963.

8. E. Polak, "An Application of Discrete Optimal Control Theory," *J. Franklin Inst.*, Vol. 276, No. 2, pp. 118-127, 1963.

9. H. Kwakernaak and E. Polak, "On the Reduction of the System $\dot{x} = Ax + Bu$, $y = c'x$ to its Minimal Equivalent," *IEEE Trans. on Circuit Theory,* Vol. CT-10, No. 4, 1963.

10. E. Polak, "A Note on D-Decomposition Theory," *IEEE Trans. on Control,* Vol. AC-9, No. 1, January 1964.

11. E. Polak, "Equivalence and Optimal Strategies for some Minimum Fuel Discrete Systems," *J. of the Franklin Inst.,* Vol. 277, No. 2, pp. 150-162, February 1964.

12. E. Polak, "On the Evaluation of Optimal and Non-Optimal Control Strategies," *IEEE Trans. on Automatic Control,* Vol. AC-9, No. 2, pp. 175-176, 1964.

13. M. D. Canon and E. Polak, "Analog Circuits for Energy and Fuel Optimal Control of Linear Discrete Systems," *University of California, Berkeley, Electronics Research Laboratory,* Tech. Memo. M-95, August 1964.

14. E. Polak and B. W. Jordan, "Theory of a Class of Discrete Optimal Control Systems," *Journal of Electronics and Control*, Vol. 17, No. 6, pp. 697-711, 1964.

15. E. Polak and B. W. Jordan, "Optimal Control of Aperiodic Discrete-Time Systems," *J. SIAM Control*, Vol. 2, pp. 332-346, 1965.

16. E. Polak, "Fundamentals of the Theory of Optimal Control," *Mathematical Review,* Vol. 29, No. 2, pp. 404-405, 1965.

17. E. Polak, "An Algorithm for Reducing a Linear Time-Invariant Differential System to State Form," *IEEE Trans. on Automatic Control*, Vol. AC-11, No. 3, pp. 577-580, 1966.

18. E. Polak, H. Halkin, B. W. Jordan, and J. B. Rosen, "Theory of Optimum Discrete Time Systems," *Proc. 3rd IFAC Congress*, Paper No. 28B, London, 1966.

19. E. Polak and A. Larsen, Jr., "Some Sufficient Conditions for Continuous Linear Programming Problems," *Int'l J. Eng. Science*, Vol. 4, No. 5, pp. 583-603, 1966.

20. E. Polak and C. D. Cullum, "Equivalence Relations for the Classification and Solution of Optimal Control Problems," *J. SIAM Control*, Vol. 4, No. 3, pp. 403-420, 1966.

21. E. Polak, M. D. Canon, and C. D. Cullum, "Constrained Minimization Problems in Finite Dimensional Spaces," *J. SIAM Control*, Vol. 4, No. 3, pp. 528-547, 1966.

22. E. Polak and J. P. Jacob, "On the Inverse of the Operator $O(\cdot) = A(\cdot) + (\cdot)B$," *American Mathematical Monthly*, Vol. 73, No. 4, Part I, pp. 388-390, April 1966.

23. E. Polak and N. O. Da Cunha, "Constrained Minimization Under Vector Valued-Criteria in Finite Dimensional Spaces," *J. Mathematical Analysis & Applications*, Vol. 19, No. 1, pp. 103-124, 1967.

24. E. Polak and N. O. Da Cunha, "Constrained Minimization Under Vector-Valued Criteria in Linear Topological Spaces," *Proc. Conference on Mathematical Theory of Control*, Los Angeles, February 1967.

25. E. Polak and K. Y. Wong, "Identification of Linear Discrete Time Systems Using the Instrumental Variable Method," *IEEE Trans. on Automatic Control*, Vol. AC-12, No. 6, pp. 707-718, 1967.

26. E. Polak, "Necessary Conditions of Optimality in Control and Programming," *Proc. AMS Summer Seminar on the Math. of the Decision Sciences*, Stanford University, July-August 1967.

27. E. Polak, "An Algorithm for Computing the Jordan Canonical Form of a Matrix," *University of California, Berkeley, Electronics Research Laboratory,* Memo. M-223, September 1967.

28. E. Polak and J. P. Jacob, "On a Class of Pursuit-Evasion Problems," *IEEE Trans. on Automatic Control,* Vol. AC-12, No. 4, pp.752-755, 1967.

29. E. Polak and J. P. Jacob, "On Finite Dimensional Approximations to a Class of Games," *J. Mathematical Analysis & Applications,* Vol. 21, No. 2, pp. 287-303, 1968.

30. P. L. Falb and E. Polak, "Conditions for optimality," in L. A. Zadeh and E. Polak, eds., *Systems Theory,* McGraw-Hill, 1969.

31. E. Polak, "Linear Time Invariant Systems," in L. A. Zadeh and E. Polak, eds., *Systems Theory,* McGraw-Hill, 1969.

32. E. Polak, "On the Removal of Ill Conditioning Effects in the Computation of Optimal Controls," *Automatica,* Vol. 5, pp. 607-614, 1969.

33. E. Polak and E. J. Messerli, "On Second Order Necessary Conditions of Optimality," *SIAM J. Control,* Vol. 7, No. 2, 272-291, 1969.

34. E. Polak, "On primal and Dual Methods for Solving Discrete Optimal Control Problems," *Proc. 2nd International Conference on Computing Methods in Optimization Problems,* San Remo, Italy, September 9-13, 1968. Published as: *Computing Methods in Optimization Problems -2,* L. A. Zadeh, L. W. Neustadt and A. V. Balakrishnan, eds., pp. 317-331, Academic Press, 1969.

35. E. Polak, "On the Convergence of Optimization Algorithms," *Revue Francaise d'Informatique et de Recherche Operationelle, Serie Rouge,* No. 16, pp. 17-34, 1969.

36. E. Polak and G. Ribiere, "Note sur la Convergence de Methodes de Directions Conjuguees," *Revue Francaise d'Informatique et de Recherche Operationelle, Serie Rouge,* No. 16, 1969.

37. E. Polak and M. Deparis, "An Algorithm for Minimum Energy," *IEEE Trans. on Automatic Control,* Vol. AC-14, No. 4, pp. 367-378, 1969.

38. E. Polak, "On the Implementation of Conceptual Algorithms," *Proc. Nonlinear Programming Symposium,* University of Wisconsin, Madison, Wisconsin, May 4-6, 1970.

39. E. Polak and G. Meyer, "A Decomposition Algorithm for Solving a Class of Optimal Control Problems," *J. Mathematical Analysis & Applications,* Vol. 3, No. 1, pp. 118-140, 1970.

40. E. Polak, "On the use of models in the Synthesis of Optimization Algorithms," *Differential Games and Related Topics* (Proceedings of the International Summer School on Mathematical Models of Action and Reaction, Varenna, Italy, June 15-27, 1970), H. Kuhn and G. Szego eds., North Holland, Amsterdam, pp. 263-279, 1971.

41. E. Polak, H. Mukai and O. Pironneau, "Methods of Centers and of Feasible Directions for the Solution of Optimal Control Problems," *Proc. 1971 IEEE Conference on Decision and Control,* Miami Beach, Fla., Dec. 15-17, 1971.

42. G. G. L. Meyer and E. Polak, "Abstract Models for the Synthesis of Optimization Algorithms," *SIAM J. Control,* Vol. 9, No. 4, pp 547, 560, 1971.

43. O. Pironneau and E. Polak, "On the Rate of Convergence of Certain Methods of Centers," *Mathematical Programming,* Vol. 2, No. 2, pp. 230-258, 1972.

44. R. Klessig and E. Polak, "Efficient Implementations of the Polak-Ribiere Conjugate Gradient Algorithm," *SIAM J. Control,* Vol. 10, No. 3, pp. 524-549, 1972.

45. E. Polak, "On a Class of Numerical Methods with an Adaptive Integration Subprocedure for Optimal Control Problems," Proc. *Fourth IFIP Colloquium on Optimization,* Santa Monica, Calif. Oct. 19-22, 1971. Published as: *Techniques of Optimization,* A. V. Balakrishnan, ed., Academic Press, pp. 89-105, 1972.

46. E. Polak, "A Survey of Methods of Feasible Directions for the Solution of Optimal Control Problems," *IEEE Transactions on Automatic Control,* Vol. AC-17, No. 5, pp. 591-597, 1972.

47. E. Polak, "A Modified Secant Method for Unconstrained Minimization," *Proc. VIII International Symposium on Mathematical Programming,* Stanford University, Aug. 27-31, 1973.

48. O. Pironneau and E. Polak, " A Dual Method for Optimal Control Problems with Initial and Final Boundary Constraints," *SIAM J. Control,* Vol. 11, No. 3, pp. 534-549, 1973.

49. R. Klessig and E. Polak, "An Adaptive Algorithm for Unconstrained Optimization with Applications to Optimal Control," *SIAM J. Control,* Vol. 11, No. 1, pp. 80-94, 1973.

50. O. Pironneau and E. Polak, "Rate of Convergence of a Class of Methods of Feasible Directions," *SIAM J. Numerical Analysis,* Vol. 10, No. 1, pp. 161-174, 1973.

51. R. Klessig and E. Polak, "A Method of Feasible Directions Using Function Approximations with Applications to Min Max Problems," *J. Math. Analysis and Applications,* Vol. 41, No. 3, pp. 583-602, 1973.

52. E. Polak, "On the Use of Optimization Algorithms in the Design of Linear Systems," *University of California, Berkeley, Electronics Research Lab.* Memo. No. M377, 1973.

53. E. Polak, "A Historical Survey of Computational Methods in Optimal Control," *SIAM Review,* Vol. 15, No. 2, Part 2, pp. 553-584, 1973.

54. L. J. Williamson and E. Polak, "Convergence Properties of Optimal Control Algorithms," *Proc. 1973 IEEE Conference on Decision and Control,* Dec. 5-7, 1973.

55. E. Polak, "Survey of Secant Methods for Optimization," *Proc. 1973 IEEE Conference on Decision and Control,* Dec. 5-7, 1973.

56. E. Polak "A Modified Secant Method for Unconstrained Optimization," *Mathematical Programming,* Vol. 6, No. 3, pp. 264-280, 1974.

57. E. Polak "A Globally Convergent Secant Method with Applications to Boundary Value Problems," *SIAM J. Numerical Analysis,* Vol. 11, No. 3, pp. 529-537, 1974.

58. H. J. Payne, E. Polak, D. C. Collins and S. Meisel, "An Algorithm for Multicriteria Optimization Based on the Sensitivity Function," *Proc. 1974 IEEE Conference on Decision and Control,* 1974.

59. E. Polak, R. W. H.Sargent and D. J. Sebastian, "On the Convergence of Sequential Minimization Algorithms," *J. Optimization Theory and Applications,* Vol. 14, No. 4, pp. 439-442, 1974.

60. H. Mukai and E. Polak, "Approximation Techniques in Gradient Projection Algorithms," *Proc. IEEE 1974 Allerton Conference on Circuits and Systems,* Univ. of Illinois, October, 1974.

61. E. Polak and Teodoru, I., "Newton Derived Methods for Nonlinear Equations and Inequalities," *Nonlinear Programming,* O. L. Mangasarian, R. R. Meyer and S. M. Robinson eds., Academic Press, N. Y., pp. 255-277, 1975.

62. E. Polak, "Computational Methods in Optimal Control," *Proc. Conference on Energy Related Modelling and Data Base Management,* Brookhaven National Laboratories, May 12-14, 1975.

63. R. Klessig and E. Polak, "A Survey of Convergence Theorems," *Proc. Joint National Meeting, ORSA-TIMS,* Las Vegas Nevada, Nov. 17-19, 1975.

64. H. Mukai and E. Polak, "A Quadratically Convergent Primal-Dual Algorithm with Global Convergence Properties for solving optimization with equality constraints," *Mathematical Programming,* Vol. 9, No. 3, pp. 336-350, 1975.

65. E. Polak, K. S. Pister and D. Ray, "Optimal Design of Framed Structures Subjected to Earthquakes," *Proc. Symposium on Optimization and Engineering Design in Conjunction with the 47th national Meeting of ORSA-TIMS,* Chicago, Ill. April 30 - May 2, 1975.

66. D. Q. Mayne and E. Polak, "First Order, Strong Variations Algorithms for Optimal Control," *J. Optimization Theory and Applications,* Vol. 16, No. 3/4, pp. 277-301, 1975.

67. E. Polak and D. Q. Mayne, "First Order, Strong Variations Algorithms for Optimal Control Problems with Terminal Inequality Constraints," *J.*

Optimization Theory and Applications, Vol. 16, No. 3/4, pp. 303-325, 1975.

68. E. Polak, "On the Approximation of Solutions to Multiple Criteria Decision Making Problems," *Proc. XXII International Meeting TIMS,* Kyoto, Japan, July, 1975.

69. E. Polak, "On the Global Stabilization of Locally Convergent Algorithms for Optimization and Root Finding," *Proc. 6th triannaual IFAC Congress,* Boston Mass., Aug. 24-30, 1975.

70. H. J. Payne, E. Polak, D. C. Collins and S. Meisel, "An Algorithm for Multicriteria Optimization Based on the Sensitivity Function," *IEEE Transactions on Automatic Control,* Vol. AC-20, No. 4, pp. 546-548, 1975.

71. E. Polak and D. Q. Mayne, "An Algorithm for Optimization Problems with Functional Inequality Constraints," *IEEE Transactions on Automatic Control,* Vol. AC-21, No. 2, 1976.

72. E. Polak and R. Trahan, "An Algorithm for Computer Aided Design of Control Systems," *Proc. IEEE Conference on Decision and Control,* 1976.

73. E. Polak and A. N. Payne, "On Multicriteria Optimization," *Proc. Conference on Directions in Decentralized Control, Many Person Games and Large Scale Systems,* Cambridge, Mass, Sept 1-3, 1975. Published as: *Directions in Large Scale Systems,* Y. C. Ho and K. S. Mitter, eds., Plenum Press, N.Y., pp. 77-94, 1976.

74. L. J. Williamson and E. Polak, "Relaxed Controls and the Convergence of Optimal Control Algorithms," *SIAM J. Control,* Vol. 14, No. 4, pp. 737-757, 1976.

75. E. Polak, "On the Approximation of Solutions to Multiple Criteria Decision Making Problems," *Multiple Criteria Decision Making: Kyoto 1975,* M. Zeleny Ed., Springer Verlag, New York, pp. 271-182, 1976.

76. E. Polak, "On the Global Stabilization of Locally Convergent Algorithms for Optimization and Root Finding," *Automatica,* Vol. 12, pp. 337-342, 1976.

77. H. Mukai and E. Polak, "On the Implementation of Reduced Gradient Methods," *Proc. 7th IFIP Conference on Optimization Techniques*, Nice, France, Sept. 8-18, 1975. Published as: *Optimization Techniques: Modeling and Optimization in the Service of Man*, Jean Cea, ed., Springer Verlag, Berlin, N.Y., Vol. 2, pp. 426-437, 1976.

78. E. Polak, K. S. Pister and D. Ray, "Optimal Design of Framed Structures Subjected to Earthquakes," *Engineering Optimization*, Vol. 2, pp. 65-71, 1976.

79. D. Q. Mayne and E. Polak, "Feasible Directions Algorithms for Optimization Problems with Equality and Inequality Constraints," *Mathematical Programming*, Vol. 11, pp. 67-80, 1976.

80. D. Q. Mayne and E. Polak, "A Feasible Directions Algorithm for Optimal Control Problems with Terminal Inequality Constraints," *IEEE Transactions on Automatic Control*, Vol. AC-22, No. 5, pp. 741-751, 1977.

81. I. Teodoru Gross and E. Polak, "On the Global Stabilization of Quasi-Newton Methods," *Proc. ORSA/TIMS National Meeting*, San Francisco, May 9-11, 1977.

82. D. Ray, K. S. Pister and E. Polak, "Sensitivity Analysis for Hysteretic Dynamical Systems: Theory and Applications," *Comp. Meth. in Applied Mechanics and Engineering*, Vol. 14, pp. 179-208, 1978.

83. E. Polak, "On a Class of Computer-Aided-Design Problems," *Proc. 7th IFAC World Congress*, Helsinki, Finland, June 1978.

84. H. Mukai and E. Polak, "On the Use of Approximations in Algorithms for Optimization Problems with Equality and Inequality Constraints," *SIAM J. Numerical Analysis*, Vol. 1, No. 4, pp. 674-693, 1978.

85. E. Polak and A. Sangiovanni Vincentelli, "An Algorithm for Design Centering, Tolerancing and Tuning," *Proc. European Conference on Circuit Theory and Design*, Lausanne, Switzerland, Sept. 1978.

86. A. N. Payne and E. Polak, "An Efficient Interactive Optimization Method for Multi-objective Design Problems," *Proc. 16th Allerton Conference on*

Communications, Control and Computing, Univ. of Illinois, October 4-6, 1978.

87. H. Mukai and E. Polak, "A Second Order Algorithm for Unconstrained Optimization," *J. Optimization Theory and Applications*, Vol. 26, No. 4, 1978.

88. H. Mukai and E. Polak, "A second Order Algorithm for the General Nonlinear Programming problem," *J. Optimization Theory and Applications*, Vol. 26, No. 4, 1978.

89. A. N. Payne and E. Polak, "An Interactive Method for Bi-Objective Decision Making," *Proc. Second Lawrence Symposium on Systems and Decision Sciences*, Berkeley, Ca. Oct. 1978.

90. E. Polak and A. Sangiovanni Vincentelli, "On Optimization Algorithms for Engineering Design Problems with Distributed Constraints, Tolerances and Tuning," *Proc. 1978 Joint Automatic Control Conference*, October 18, 1978.

91. M. A. Bhatti, K. S. Pister and E. Polak, "Optimal Design of an Earthquake Isolation System," *Earthquake Engineering Research Center, University of California, Berkeley*, Report No. UCB/EERC-78/22, October, 1978.

92. T. Glad and E. Polak, "A Multiplier Method with Automatic Limitation of Penalty Growth," *Mathematical Programming*, Vol. 17, No. 2, pp. 140-156, 1979.

93. E. Polak, D. Q. Mayne and R. Trahan, "An Outer Approximations Algorithm for Computer Aided Design Problems," *J. Optimization Theory and Applications*, Vol. 28, No. 3, pp. 331-352, 1979.

94. E. Polak and A. Sangiovanni Vincentelli, "Theoretical and Computational Aspects of the Optimal Design Centering, Tolerancing and Tuning Problem," *IEEE Trans. on Circuits and Systems*, Vol. CAS-26, No. 9, pp. 795-813, 1979.

95. A. N. Payne and E. Polak, "An Interactive Rectangle Elimination Method for Multi-Objective Decision Making," *IEEE Trans. on Automatic Control*, Vol. AC-25, No. 3, 1979.

96. S. Tishyadhigama, E. Polak and R. Klessig, "A Comparative Study of Several General Convergence Conditions for Algorithms Modeled by Point to Set Maps," *Mathematical Programming Study 10*, pp. 172-190, 1979.

97. E. Polak and D. Q. Mayne, "On the Finite Solution of Nonlinear Inequalities," *IEEE Trans. on Automatic Control*, Vol. AC-24, No. 3, pp. 443-445, 1979.

98. E. Polak, R. Trahan and D. Q. Mayne, "Combined Phase I - Phase II Methods of Feasible Directions," *Mathematical Programming*, Vol. 17, No. 1, pp. 32-61, 1979.

99. C. Gonzaga and E. Polak, "On Constraint Dropping Schemes and Optimality Functions for a Class of Outer Approximations Algorithms," *SIAM J. Control and Optimization*, Vol. 17, No. 4, pp. 477-493, 1979.

100. R. Trahan and E. Polak, "A Derivative Free Algorithm for a Class of Infinitely Constrained Problems," *IEEE Trans. on Automatic Control*, Vol. AC-25, No. 1, pp. 54-62, 1979.

101. C. Gonzaga, E. Polak and R. Trahan, "An Improved Algorithm for Optimization Problems with Functional Inequality Constraints," *IEEE Trans. on Automatic Control*, Vol. AC-25, No. 1, pp. 49-54 1979.

102. E. Polak and D. Q. Mayne, "Algorithms for Computer Aided Design of Control Systems by the Method of Inequalities," *Proc. 18th IEEE Conference on Decision and Control*, Fort Lauderdale, Florida, Dec. 12-14, 1979.

103. E. Polak and S. Tishyadhighama, "New Convergence Theorems for a Class of Feasible Directions Algorithms," *Proc. 18th IEEE Conference on Decision and Control*, Fort Lauderdale, Florida, Dec. 12-14, 1979.

104. E. Polak and A. Sangiovanni Vincentelli "Theoretical and Computational Aspects of the Optimal Design Centering, Tolerancing and Tuning Problem," *Proc. Fourth International Symposium on Mathematical Theory of Networks and Systems*, Delft University of Technology, Delft, Holland, July 3-6, 1979.

105. E. Polak, "On the Nature of Optimization Problems in Engineering Design," *Proc. 10th International Symposium on Mathematical Programming*, Montreal, Canada, Aug 27-30, 1979.

106. D. Q. Mayne and E. Polak, "A Superlinearly Convergent Algorithm for Constrained Optimization Problems," *Electronics Research Laboratory, University of California, Berkeley,* Memo. No. UCB/ERL M79/13, Jan. 1979; (presented at the *10th International Symposium on Mathematical Programming*, Montreal, Canada, Aug 27-30, 1979. Revised 15/1/1980, Publication No. 78/52, Department of Computing and Control, Imperial College, London.)

107. E. Polak "Algorithms for a Class of Computer Aided Design Problems: a Review," *Automatica*, Vol. 15, pp. 531-538, 1979.

108. M. A. Bhatti, E. Polak, K. S. Pister, "OPTDYN - A General Purpose Program for Optimization Problems with and without Dynamic Constraints," *University of California, Berkeley, Eathquake Engineering Research Center*, Report No. UCB/EERC-79/16, July, 1979.

109. M. A. Bhatti, E. Polak, K. S. Pister, "Optimization of Control Devices in Base isolation Systems for Aseismic Design," *Proc. International IUTAM Symposium on Structural Control*, University of Waterloo, Ontario, Canada, North Holland Pub. Co., Amsterdam, pp. 127-138, 1980.

110. E. Polak, "An Implementable Algorithm for the Optimal Design Centering, Tolerancing and Tuning Problem," *Proc. Fourth International Symposium on Computing Methods in Applied Sciences and Engineering*, Versailles, France, Dec. 10-14, 1979. Published as: *Computing Methods in Applied Science and Engineering*, R. Glowinski, J. L. Lions, ed., North Holland, Amsterdam, pp. 499-517, 1980.

111. D. Q. Mayne and E. Polak "An Exact Penalty Function Algorithm for Optimal Control Problems with Control and Terminal Equality Constraints, Part 1," *J. Optimization Theory and Applications*, Vol. 32 No. 2, pp. 211-246, 1980.

112. D. Q. Mayne and E. Polak "An Exact Penalty Function Algorithm for Optimal Control Problems with Control and Terminal Equality Constraints,

Part 2," *J. Optimization Theory and Applications*, Vol. 32 No. 3, pp. 345-363, 1980.

113. E. Polak and A. Tits, "A Globally Convergent Implementable Multiplier Method with Automatic Penalty Limitation," *J. Appl. Math. and Optimization*, Vol. 6, pp. 335-360, 1980.

114. M. A. Bhatti, K. S. Pister, and E. Polak, "Interactive Optimal Design of Dynamically Loaded Structures," *Proc. ASCE conf. on Structural Optimization*, Florida, 27-31 Oct. 1980.

115. M. A. Bhatti, K. S. Pister and E. Polak, "An Implementable Algorithm for Computer-Aided Design Problems with or without Dynamic Constraints," *Proc. ASME Century 2, International Computer Technology Conference*, Aug. 12-15, 1980, San Francisco. Published as: *Advances in Computer Technology - 1980*, A. Sierig, ed., ASME, New York, Vol. 1., pp. 392-400, Aug. 1980.

116. E. Polak and D. Q. Mayne, "On the Solution of Singular Value Inequalities" *Proc. 20th IEEE Conference on Decision and Control*, Albuqurque, N.M., Dec. 10-12, 1980.

117. E. Polak and D. Q. Mayne, "Design of Nonlinear Feedback Controllers," *Proc. 20th IEEE Conference on Decision and Control*, Albuqurque, N.M., Dec. 10-12, 1980.

118. D. Q. Mayne, E. Polak and A. Voreadis, "A Cut Map Algorithm for Design Problems with Tolerances" *Proc. 20th IEEE Conference on Decision and Control*, Albuqurque, N.M., Dec. 10-12, 1980.

119. M. A. Bhatti, K. S. Pister, and E. Polak, "Interactive Optimal Design of Dynamically Loaded Structures," *Proc. 1980 ASCE National Convention*, Florida, Dec. 1980.

120. D. Q. Mayne, E. Polak and A. Sangiovanni Vincentelli, "Computer Aided Design via Optimization," *Proc. IFAC Workshop on Control Applications of Nonlinear Programming*, Denver, Colorado, June 21, 1979. Published as: *Control Applications of Nonlinear Programming*, H. E. Rauch, ed., Pergamon Press, Oxford and New York, pp. 85-91, 1980.

121. M. A. Bhatti, K. S. Pister and E. Polak, "Optimization of Control Devices in Base Isolation Systems for Aseismic Design," *Structural Control,* H. H. E. Leipholz (ed), North Holland Pub. Co. SM Publications, pp. 127-138, 1980.

122. E. Polak, "Optimization-Based Computer-Aided Design of Engineering Systems," *FOREFRONT: Research in the College of Engineering,* University of California, Berkeley, 1979/80.

123. E. Polak and D. Q. Mayne, "A Robust Secant Method for Optimization Problems with Inequality Constraints," *J. Optimization Theory and Applications,* Vol. 33, No. 4, pp. 463-467, 1981.

124. E. Polak and D. Q. Mayne, "On the Solution of Singular Value Inequalities over a Continuum of Frequencies" *IEEE Transactions on Automatic Control,* Vol. AC-26, No. 3, pp. 690-695, 1981.

125. E. Polak and D. Q. Mayne, "Design of Nonlinear Feedback Controllers," *IEEE Transactions on Automatic Control,* Vol. AC-26, No. 3, pp. 730-733, 1981.

126. W. T. Nye, E. Polak, A. Sangiovanni-Vincentelli and A. Tits, "DELIGHT: an Optimization-Based Computer-Aided-Design System," *Proc. IEEE Int. Symp. on Circuits and Systems,* Chicago, Ill, April 24-27, 1981.

127. E. Polak and A. Tits, " On globally Stabilized Quasi-Newton Methods for Inequality Constrained Optimization Problems," *Proc. 10th IFIP Conference on System Modeling and Optimization,* New York, August 31-September 4, 1981.

128. E. Polak, "Optimization-Based Computer-Aided-Design of Control Systems," *Proc. Joint Automatic Control Conference,* University of Virginia, Charlottesville, Virginia, June 17-19, 1981.

129. E. Polak, "Algorithms for Optimal Design," *Proc. NATO Advanced Study Institute,* Univ. of Iowa, Iowa City, Ia, May 1980. Published as: *Optimization of Distributed Parameter Structures: Vol. 1,* E. J. Haug and J. Cea eds., Sijthoff & Noordhoff, pp. 586-602, 1981.

130. M. A. Bhatti, T. Essebo, W. Nye, K. S. Pister, E. Polak, A. Sangio-vanni Vincentelli and A. Tits, "A Software System for Optimization-Based Computer-Aided Design" Proc. IEEE International Symp. on Circuits and Systems, Houston, Tx, April 28-30, 1980 and also *Proc. NATO Advanced Study Institute,* Univ. of Iowa, Iowa City, Ia, May 1980. Published as: *Optimization of Distributed Parameter Structures: Vol. 1,* E. J. Haug and J. Cea eds., Sijthoff & Noordhoff, pp. 602-620, 1981.

131. E. Polak and Y. Wardi, "A Nondifferentiable Optimization Algorithm for the Design of Control Systems Subject to Singular Value Inequalities over a Frequency Range," *Proceedings IFAC/81 World Congress*, Kyoto, Japan, August 24-28, 1981.

132. E. Polak, "An Implementable Algorithm for the Design Centering, Toler-ancing and Tuning Problem," *J. Optimization Theory and Applications*, Vol. 35, No. 3, 1981.

133. D. Q. Mayne, E. Polak and A. J. Heunis, "Solving Nonlinear Inequalities in a Finite Number of Iterations," *J. Optimization Theory and Applications*, Vol. 33, No. 2, pp. 207-221, 1981.

134. M. A. Bhatti, V. Ciampi, K. S. Pister and E. Polak, "OPTNSR an Interac-tive Software System for Optimal Design of Statically Loaded Structures with Nonlinear Response," *University of California, Berkeley, Earthquake Engineering Research Center,* Report No. UCB/EERC-81/02, 1981.

135. R. J. Balling, K. S. Pister and E. Polak, "DELIGHT.STRUCT A Computer-Aided Design Environment for Structural Engineering," *University of California, Berkeley, Earthquake Engineering Research Center* Report No. UCB/EERC-81/19, Dec. 1981.

136. R. J. Balling, V. Ciampi, K. S. Pister and E. Polak, "Optimal Design of Seismic-Resistant Planar Steel Frames," *University of California, Berke-ley, Earthquake Engineering Research Center* Report No. UCB/EERC-81/20, Dec. 1981.

137. E. Polak, "Interactive Software for Computer-Aided-Design of Control Systems via Optimization," *Proc. 20th IEEE Conference of Decision and Control*, San Diego, Ca., pp. 408-411, Dec. 16-18, 1981.

138. M. A. Bhatti, K.S. Pister and E. Polak, "Package for Optimization-Based, Interactive CAD," *J. of the Structural Division of the A.S.C.E.*, Vol. 107, No.ST11, pp. 2271-2284, 1981.

139. E. Polak, K. J. Astrom and D. Q. Mayne, "INTEROPTDYN-SISO: a Tutorial," *University of California, Electronics Research Laboratory*, Memo UCB/ERL No. M81/99, Dec. 15, 1981.

140. D. Q. Mayne and E. Polak, "Algorithms for the Design of Control Systems Subject to Singular Value Inequalities," *Mathematical Programming Studies*, Vol. 18, pp. 112-134, 1982.

141. E. Polak and S. Tishyadhigama, "New Convergence Theorems for a Class of Feasible Directions Algorithms," *J. Optimization Theory and Applications*, Vol. 37, No. 1, pp. 33-44, 1982.

142. M. A. Bhatti, V. Ciampi, K. S. Pister and E. Polak, "An Interactive Software System for Optimal Design of Statically and Dynamically Loaded Structures with Nonlinear Response," *Proc. of International Symposium on Optimum Structural Design*, Tucson Arizona, October 19-22, 1981. Published as: *Optimum Structural Design*, R. H. Gallagher et al, eds., John Wiley and Sons, Chichester, England, 1982.

143. D. Q. Mayne, E. Polak and A. Voreadis, "A Cut Map Algorithm for Design Problems with Tolerances," *IEEE Trans. on Circuits and Systems*, Vol. CAS-29 No. 1, pp. 35-46, 1982.

144. D. Q. Mayne, E. Polak and A. Sangiovanni Vincentelli, "Computer Aided Design via Optimization: a Review," *Automatica*, Vol. 18, No. 2, pp. 147-154, 1982.

145. D. Q. Mayne and E. Polak, "A Superlinearly Convergent Algorithm for Constrained Optimization Problems," *Mathematical Programming Study 16*, pp. 45-61, 1982.

146. E. Polak and Y. Wardi, "A Nondifferentiable Optimization Algorithm for the Design of Control Systems Subject to Singular Value Inequalities over a Frequency Range," *Automatica*, Vol. 18, No. 3, pp. 267-283, 1982.

147. D. Q. Mayne and E. Polak, "Algorithms for the Design of Control Systems Subject to Singular Value Inequalities," *Mathematical Programming Study 18, Algorithms and Theory in Filtering and Control,* D. C. Sorensen and R. J.-B. Wets, ed., North Holland, New York, pp. 112-135, 1982.

148. E. Polak and A. Tits, " A Recursive Quadratic Programming Algorithm for Semi-Infinite Optimization Problems," *J. Appl. Math. and Optimization,* Vol. 8, pp. 325-349, 1982.

149. Balling, R. J., Ciampi, V., Pister K. S. Polak, E, Sangiovanni Vincentelli, A., Tits, A., " DELIGHT.STRUCTURE: An Interactive Software System for Optimization-Based Computer-Aided Design of Dynamically Loaded Structures with Nonlinear Response," *Proc. ASCE Convention,* Las Vegas, 1982.

150. Balling, R. J., Ciampi, V., Pister K. S. Polak, E., "Optimal Design of Structures Subjected to Earthquake Loading," *Proc. ASCE Convention,* Las Vegas, 1982.

151. E. Polak, P. Siegel and T. Wuu, W. T. Nye and D. Q. Mayne, "DELIGHT-MIMO an Interactive, Optimization based Multivariable Control System Design Package," *IEEE Control Systems Magazine,* Vol. 2, No. 4, pp. 9-14, 1982.

152. D. Q. Mayne and E. Polak, "A Quadratically Convergent Algorithm for Solving Infinite Dimensional Inequalities," *J. of Appl. Math. and Optimization,* Vol. 9., pp. 25-40, 1982.

153. E. Polak, D.Q. Mayne and Y. Wardi, "On the Extension of Constrained Optimization Algorithms from Differentiable to Nondifferentiable Problems," *SIAM J. Control and Optimization,* Vol. 21, No. 2, pp. 179-204, 1983.

154. Y.Y. Wardi and E. Polak, "A Nondifferentiable Optimization Algorithm for Structural Problems with Eigenvalue Constraints," *Journal of Structural Mechanics,* Vol. 11, No. 4, 1983.

155. E. Polak and D. Q. Mayne, "On Three Approaches to the Construction of Nondifferentiable Optimization Algorithms," *Proc. 11th IFIP Conference*

on System Modelling and Optimization, Copenhagen, Denmark, July 25-29, 1983.

156. Polak, E. and Stimler, D. M. "Optimization-Based Design of SISO Control Systems with Uncertain Plant: Problem Formulation," *University of California, Berkeley, Electronics Research Laboratory* Memo No. UCB/ERL M83/16, 1983.

157. Polak, E. and Stimler, D. M. "Optimization-based Design of SISO Control Systems with Uncertain Plant," *Proc. IFAC Symp. on Applications of Nonlinear Programming to Optimization and Control,* San Francisco, June 20-21, 1983.

158. Polak, E. and Stimler, D. M. "Complexity Reduction in Optimization-Based Design of Control Systems with Uncertain Plant," *Proc. American Automatic Control Conference,* San Francisco, June 22-24, 1983.

159. E. Polak, " Semi-Infinite Optimization in Engineering Design," *Lecture Notes in Economics and Mathematical Systems, Vol. 215: Semi-Infinite Programming and Applications,* A. V. Fiacco and K. O. Kortanek, eds., Springer-Verlag, Berlin, New York, Tokyo, 1983.

160. E. Polak and Y. Y. Wardi, "A Study of Minimizing Sequences," *Proc. IEEE Conference on Decision and Control,* San Antonio, Tx., pp. 923-928, Dec. 1983.

161. Polak and D. Q. Mayne, "Algorithm Models for Nondifferentiable Optimization," *Proc. IEEE Conference on Decision and Control,* San Antonio, Tx., pp. 934-939, Dec. 1983.

162. R. J. Balling, K. S. Pister and E. Polak, "DELIGHT.STRUCT A Computer-Aided Design Environment for Structural Engineering," *Computer Methods in Applied Mechanics and Engineering,* 38, pp. 237-251, 1983.

163. E. Polak, "A Modified Nyquist Stability Criterion for Use in Computer-Aided Design," *IEEE Trans. on Automatic Control,* Vol. AC-29, No. 1, pp. 91-93, 1984.

164. E. Polak and D.M. Stimler, "On the Design of Linear Control Systems with Plant Uncertainty via Nondifferentiable Optimization," *Proc. IX. Triennial IFAC World Congress,* Budapest, July 2-6, 1984.

165. D. Q. Mayne and E. Polak, "Outer Approximations Algorithm for Non-differentiable Optimization Problems," *J. Optimization Theory and Applications*, Vol. 42, No. 1, pp. 19-30, 1984.

166. E. Polak, "Notes on the Mathematical Foundations of Nondifferentiable Optimization in Engineering Design," *University of California, Electronics Research Laboratory*, Memo UCB/ERL M84/15, 2 Feb. 1984.

167. E. Polak and D. Q. Mayne, "Theoretical and Software Aspects of Optimization Based Control System Design," *Proceedings of the Sixth International Conference Analysis and Optimization of Systems*, Nice, France, June 19-22, 1984.

168. E. Polak, "A Perspective on the Use of Semi-Infinite Optimization in Control System Design," *Proc. 1984 Automatic Control Conference*, San Diego, June 1984.

169. D. Q. Mayne and E. Polak, "Nondifferentiable Optimization via Adaptive Smoothing," *J. Optimization Theory and Applications*, Vol. 43, No. 4, pp. 601-614, 1984.

170. E. Polak and Y. Y. Wardi, "A Study of Minimizing Sequences," *SIAM J. Control and Optimization*, Vol. 22, No. 4, pp. 599-609, 1984.

171. M. A. Bhatti, V. Ciampi, K. S. Pister and E. Polak, "An Interactive Software System for Optimal Design with Nonlinear Response," *New Directions in Optimum Structural Design*, E. Atrek, R. H. Gallagher, K. M. Ragsdell and O. C. Zienkewicz eds., pp. 633-663, John Wiley and Sons, New York, N.Y. 1984,

172. E. Polak, D. Q. Mayne and D. M. Stimler, "Control System Design via Semi-Infinite Optimization," *Proceedings of the IEEE*, pp. 1777-1795, December 1984.

173. E. Polak, S. Salcudean and D. Q. Mayne, "A Rationale for the Sequential Optimal Redesign of Control Systems," *Proc. 1985 ISCAS*, pp. 835-838, Kyoto, Japan, June 1985.

174. E. Polak, S. Salcudean and D. Q. Mayne, "A Sequential Optimal Redesign Procedure for Linear Feedback Systems," *University of California,*

Berkeley, Electronics Research laboratory Memo No. UCB/ERL M85/15, Feb.28, 1985.

175. D. Q. Mayne and E. Polak "Algorithms for Optimization Problems with Exclusion Constraints," *Proc. 1985 IEEE Conference on Decision and Control,* Fort Lauderdale, Florida, Dec. 1985.

176. E. Polak and D. M. Stimler, "On the Efficient Formulation of the Optimal Worst Case Control System Design Problem," *University of California, Electronics Research Laboratory* Memo No. UCB/ERL M85/71, 21 August 1985.

177. E. Polak and D. Q. Mayne, "Algorithm Models for Nondifferentiable Optimization," *SIAM J. Control and Optimization,* Vol. 23, No. 3, 1985.

178. T. L. Wuu, R. G. Becker and E. Polak, "A Diagonalization Technique for the Computation of Sensitivity Functions of Linear Time Invariant Systems," *IEEE Trans. on Automatic Control,* Vol. AC-31 No. 12, pp. 1141-1143, 1986.

179. E. Polak and D. M. Stimler, "Majorization: a Computational Complexity Reduction Technique in Control System Design," *Proceedings of the Seventh International Conference Analysis and Optimization of Systems,* Antibes, France, June, 1986.

180. D. M. Stimler and E. Polak , "Nondifferentiable Optimization in Worst Case Control Systems Design: a Computational Example," *Proc. IEEE Control Systems Society 3rd Symposium on CACSD,* Arlington, Va., September 24-26, 1986.

181. D. Q. Mayne and E. Polak "Algorithms for Optimization Problems with Exclusion Constraints," *J. Optimization Theory and Applications,* Vol. 51, No. 3, pp. 453-474, 1986

182. E. Polak, "A Perspective on Control System Design by Means of Semi-Infinite Optimization Algorithms," *Proc. IFIP Working Conference on Optimization Techniques,* Santiago, Chile, Aug. 1984. Springer Verlag. 1987

183. E. Polak and D. Q. Mayne, " Design of Multivariable Control Systems via Semi-Infinite Optimization," *Systems and Control Encyclopaedia,* M. G. Singh, editor, Pergamon Press, N.Y. 1987.

184. D. Q. Mayne and E. Polak "An Exact Penalty Function Algorithm for Control Problems with State and Control Constraints," *IEEE Trans. on Control,* Vol. AC-32, No. 5, pp. 380-388, 1987.

185. E. Polak, S. Salcudean and D. Q. Mayne, "Adaptive Control of ARMA Plants Using Worst Case Design by Semi-Infinite Optimization, *IEEE Trans. on Automatic Control,* Vol. AC-32, No. 5, pp. 388-397, 1987.

186. E. Polak, "On the Mathematical Foundations of Nondifferentiable Optimization in Engineering Design," *SIAM Review,* Vol.29, No.1 pp. 21-91, March 1987.

187. E. Polak, T. E. Baker, T-L. Wuu and Y-P. Harn "Optimization-Based Design of Control Systems for Flexible Structures," *Proc. 4-th Annual NASA SCOLE Workshop,* Colorado Springs, December 1987.

188. S. Daijavad, E. Polak, and R-S Tsay, "A Combined Deterministic and Random Optimization Algorithm for the Placement of Macro-Cells," *Proc. MCNC International Workshop on Placement and Routing,* Research Triangle Park, NC, May 10-13, 1988.

189. E. Polak and S. E. Salcudean, "Algorithms for Optimal Feedback Design," *Proc. International Symposium on the Mathematical Theory of Networks and Systems (MTNS/87),* Phoenix, Arizona, June 15-19, 1987: .br C. I. Byrnes, C. F. Martin, and R. E. Saeks eds., *Linear Circuits, Systems and Signal Processing: Theory and Applications,* Elsevier Science Pub. B.V. (North Holland), 1988.

190. T. E. Baker and E. Polak, "Computational Experiments in the Optimal Slewing of Flexible Structures," *Proc. Second NASA/Air Force Symposium on Recent Advances in Multidisciplinary Analysis and Optimization,* Hampton, Va., Sept. 28-30, 1988.

191. E. Polak, "Minimax Algorithms for Structural Optimization," *Proc. IU-TAM Symposium on Structural Optimization,* Melbourne, Australia, Feb. 9 - 13, 1988.

192. E. Polak and E. J. Wiest, "Domain Rescaling Techniques for the Solution of Affinely Parametrized Nondifferentiable Optimal Design Problems," *Proc. 27th IEEE Conference on Decision and Control*, Austin, Tx., Dec. 7-9 1988. Dec. 1988.

193. E. Polak and D. M. Stimler, "Majorization: a Computational Complexity Reduction Technique in Control System Design," *IEEE Trans. on Automatic Control*, Vol. 33, No.11, pp 1010-1022, 1988.

194. E. Polak and S. E. Salcudean, "On The Design of Linear Multivariable Feedback Systems via Constrained Nondifferentiable Optimization in *H supin f* Spaces," *IEEE Trans on Automatic Control*, Vol. 34, No.3, pp 268-276, 1989.

195. E. Polak and S. Wuu, "On the Design of Stabilizing Compensators via Semi-Infinite Optimization," *EEE Trans. on Control*, Vol. 34, No.2, pp 196-200, 1989.

196. E. Polak, "Nonsmooth Optimization Algorithms for the Design of Controlled Flexible Structures," *Proc. AMS-SIAM-IMS Joint Summer Research Conf. on Dynamics and Control of Multibody Systems* , July 30- August 5, 1988, Bowdoin College, Brunswick, Maine. *Contemporary Mathematics* Vol. 97, pp 337-371, J. E. Marsden, P. S. Krishnaprasad, and J. C. Simo eds., American Math Soc., Providence RI, 1989.

197. E. Polak, "Basics of Minimax Algorithms," *Proc. Fourth Course of the International School of Mathematics on Nonsmooth Optimization and Related Topics* Erice, Italy, June 19 - July 8 1988. Published as (pp 343-367): *Nonsmooth Optimization and Related Topics*, F. H. Clarke, V. F. Dem'yanov and F. Giannessi eds., Plenum Press, New York, 1989.

198. L. He and E. Polak, "Effective Discretization Strategies in Optimal Design," Proceedings *28th IEEE Conference on Decision and Control*, Tampa, FL., December 12-14, 1989.

199. Y-P. Harn and E. Polak, "On the Design of Finite Dimensional Controllers for Infinite Dimensional Feedback-Systems via Semi-Infinite Optimization," Proc. *27th IEEE Conference on Dec. and Contr.*, Austin, Tx., Dec. 7-9 1988. *IEEE Trans. on Automatic Control*, Vol. 35, No. 10, pp. 1135-1140, 1990

200. J. E. Higgins and E. Polak, "Minimizing Pseudo-Convex Functions on Convex Compact Sets," *J. Optimization Theory and Applications*, Vol.65, No.1, pp 1-28, 1990.

201. E. Polak and E. J. Wiest, "A Variable Metric Technique for the Solution of Affinely Parametrized Nondifferentiable Optimal Design Problems," *J. Optimization Theory and Applications*, Vol. 66, No. 3, pp 391-414, 1990.

202. L. He and E. Polak, "An Optimal Diagonalization Strategy for the Solution of a Class of Optimal design Problems," *IEEE, Trans. on Automatic Control*, Vol. 35, No.3, pp 258-267, 1990.

203. T. E. Baker and E. Polak, "An Algorithm for Optimal Slewing of Flexible Structures," *University of California, Electronics Research Laboratory*, Memo UCB/ERL M89/37, 11 April 1989, Revised, 4 June 1990.

204. E. Polak, T. Yang, and D. Q. Mayne, "A Method of Centers Based on Barrier Functions for Solving Optimal Control Problems with Continuum State and Control Constraints," *Proc. 29-th IEEE Conf. on Decision and Control*, Honolulu, Hawaii, Dec. 5-7, 1990.

205. L. He and E. Polak, "Effective Diagonalization Strategies for the Solution of a Class of Optimal Design Problems," *IEEE Trans. on Automatic Control*, Vol. 35, No.3, pp 258-267, 1990.

206. D. Q. Mayne, H. Michalska and E. Polak, "An Efficient Outer Approximations Algorithm for Solving Infinite Sets of Inequalities," Proc. *29-th IEEE Conf. on Dec. and Control*, Honolulu, Hawaii, Dec. 5-7, 1990.

207. Y-P. Harn and E. Polak, "On the Design of Finite Dimensional Controllers for Infinite Dimensional Feedback-Systems via Semi-Infinite Optimization," *IEEE Trans. on Automatic Control*, Vol. 35, No. 10, pp. 1135-1140, 1990.

208. E. Polak, D. Q. Mayne and J. Higgins, "A Superlinearly Convergent Algorithm for Min-Max Problems," *J. Optimization Theory and Applications* Vol. 69, No.3, pp 407-439, 1991.

209. Y-P. Harn and E. Polak, "Proportional-Plus-Multiintegral Stabilizing Compensators for a Class of MIMO Feedback Systems with Infinite-

Dimensional Plants," *IEEE Trans. on Automatic Control*, Vol. 36, No. 2, pp. 207-213, 1991.

210. E. Polak and L. He, "A Unified Phase I Phase II Method of Feasible Directions for Semi-infinite Optimization," *J. Optimization Theory and Applications*, Vol. 69, No.1, pp 83-107, 1991.

211. J. Higgins and E. Polak, "An ϵ-active Barrier Function Method for Solving Minimax Problems," *J. Applied Mathematics and Optimization*, Vol. 23, pp 275-297, 1991.

212. E. J. Wiest and E. Polak, "On the Rate of Convergence of Two Minimax Algorithms," *J. Optimization Theory and Applications* Vol. 71 No.1, pp 1-30, 1991.

213. E. Polak and L. He, "Finite-Termination Schemes for Solving Semi-Infinite Satisfycing Problems," *J. Optimization Theory and Applications*, Vol. 70, No. 3, pp 429-466, 1991.

214. L. He and E. Polak, "Multistart Method with Estimation Scheme for Global Satisfycing Problems," *Proc. European Control Conference*, Grenoble, July 2-5, 1991.

215. C. Kirjner Neto and E. Polak, "A Secant Method Based on Cubic Interpolation for Solving One Dimensional Optimization Problems," *University of California, Berkeley, Electronics Research Laboratory* Memo No. UCB/ERL M91/91, 15 October 1991.

216. E. Polak, J. Higgins and D. Q. Mayne, "A Barrier Function Method for Minimax Problems," *Mathematical Programming*, Vol. 54, No.2, pp. 155-176, 1992.

217. E. Polak, D. Q. Mayne, and J. Higgins, "On the Extension of Newton's Method to Semi-Infinite Minimax Problems," *SIAM J. Control and Optimization*, Vol. 30, No.2, pp. 376-389, 1992.

218. E. Polak and L. He, "Rate Preserving Discretization Strategies for Semi-infinite Programming and Optimal Control," *SIAM J. Control and Optimization*, Vol. 30, No. 3, pp 548-572, 1992

219. E. J. Wiest and E. Polak, "A Generalized Quadratic Programming-Based Phase I Phase II Method for Inequality Constrained Optimization," *J. Appl. Mathematics and Optimization*, Vol. 26, pp 223-252, 1992.

220. T. H. Yang and E. Polak, "Moving Horizon Control of Nonlinear Systems with Input Saturation, Disturbances, and Plant Uncertainty," *International Journal on Control* Vol. 58. No.4, pp. 875-903, 1993.

221. E. Polak and T. H. Yang, "Moving Horizon Control of Linear Systems with Input Saturation, and Plant Uncertainty, Part 2. Disturbance Rejection and Tracking," *International Journal on Control*, Vol. 68, No. 3, pp. 639-663, 1993.

222. E. Polak and T. H. Yang, "Moving Horizon Control of Linear Systems with Input Saturation, and Plant Uncertainty, Part 1. Robustness," *International Journal on Control*, Vol. 68, No. 3, pp. 613-638, 1993.

223. E. Polak, "On the Use of Consistent Approximations in the Solution of Semi-Infinite Optimization and Optimal Control Problems," *Mathematical Programming*, Series B, Vol. 62, No.2, pp 385-414, 1993.

224. L. He and E. Polak, "Multistart Method with Estimation Scheme for Global Satisfycing Problems," *J. Global Optimization*, No.3, pp 139-156, 1993.

225. E. Polak, T. Yang, and D. Q. Mayne, "A Method of Centers Based on Barrier Functions for Solving Optimal Control Problems with Continuum State and Control Constraints," *Siam J. Control and Optimization*, Vol.31, pp 159-179, 1993.

226. D. Q. Mayne and E. Polak, "Optimization Based Design and Control," plenary address, Proc. IFAC Congress, July 1993, Sydney, Australia.

227. T. E. Baker and E. Polak, "On the Optimal Control of Systems Described by Evolution Equations," *SIAM J. Control and Optimization*, Vol. 32, No. 1, pp 224-260, 1994

228. E. Polak, G. Meeker, K. Yamada and N. Kurata, "Evaluation of an Active Variable-Damping-Structure," *Earthquake Engineering and Structural Dynamics*, Vol. 23, pp 1259-1274, 1994.

229. D. Q. Mayne, H. Michalska and E. Polak, "An Efficient Algorithm for Solving Semi-Infinite Inequality Problems with Box Constraints," *J. Applied Mathematics and Optimization*, Vol. 30 No.2, pp. 135-157, 1994.

230. C. Kirjner-Neto, E. Polak, and A. Der Kiureghian, "Algorithms for Reliability-Based Optimal Design," Proc. *IFIP Working Group 7.5 Working Conference on Reliability and Optimization of Structural Systems*, Assisi (Perugia), Sept 7-9, 1994.

231. C. Kirjner Neto and E. Polak, "On the Use of Consistent Approximations for the Optimal Design of Beams," *SIAM Journal on Control and Optimization*, Vol. 34, No. 6, pp. 1891-1913, 1996.

232. A. Schwartz and E. Polak, "Consistent Approximations for Optimal Control Problems Based on Runge-Kutta Integration," *SIAM Journal on Control and Optimization*, Vol. 34., No.4, pp. 1235-69, 1996.

233. A. Schwartz and E. Polak, "A Family of Projected Descent Methods for Optimization Problems with Simple Bounds," *J. Optimization Theory and Applications*, Vol. 92, No.1, pp.1-32, 1997.

234. E. Polak, C. Kirjner-Neto, and A. Der Kiureghian, "Structural optimization with reliability constraints," *Proc. 7th IFIP WG 7.5 Conference on Reliability and Optimization of Structural Systems Boulder, Colorado, USA. 2-4 April, 1996.* Published as *Reliability and Optimization of Structural Systems*, D. M. Frangopol, R. Corotis, and R. Rackwitz eds, Pergamon, 1997.

235. I. S. Khalil-Bustany, C. J. Diederich, E. Polak, and A. W. Dutton, "A Three Dimensional Minimax Optimization-Based Inverse Treatment Planning Approach for Interstitial Thermal Therapy Using Multi-Element Applicators," Proc. *16th Annual Meeting North American Hyperthermia Soc.*, Rhode Island, 1997.

236. I. S. Khalil, C. J. Dietrich, E. Polak, and A. W. Dutton, "Three Dimensional Minimax Optimization Based Inverse Treatment Planning for Interstitial Thermal Therapy Using Multi-Element Applicators," *Proc. North American Hyperthermia Society - 16th Annual Meeting*, Providence RI, April 1997.

237. E. Polak and L. Qi, "A Globally and Superlinearly Convergent Scheme for Minimizing a Normal Merit Function", AMR 96/17, Applied Mathematics Report, University of New South Wales, 1996, and SIAM J. on Optimization, Vol. 36, No. 3, p.1005-19, 1998.

238. C. Kirjner and E. Polak, "On the Conversion of Optimization Problems with MaxMin Constraints to Standard Optimization Problems," SIAM J. Optimization, Vol. 8, No. 4, pp 887-915, 1998.

239. C. Kirjner-Neto, E. Polak and A. Der Kiureghian, "An Outer Approximations Approach to Reliability-Based Optimal Design of Structures," *J. Optimization Theory and Applications*, Vol. 98, No.1, pp. 1-17, July 1998.

240. I. S. Khalil-Bustany, C. J. Diederich, C. Kirjner-Neto, and E. Polak, "A Minimax Optimization-Based Inverse Treatment Planning Approach for Interstitial Thermal Therapy," *International Journal of Hyperthermia* Vol. 14, No. 4 pp 331-346, 1998.

241. E. Polak and L. Qi, "Some Optimality Conditions for Minimax Problems and Nonlinear Programs," *Applied Mathematics Report* AMR 98/4, University of New South Wales, 1998.

242. N. di Cesare, O. Pironneau, E. Polak. "Consistent Approximations for an Optimal Design Problem," LAN-UPMC report 98005, Universite Pierre et Marie Curie, Paris, France, January 1998. Paris, Jussieu, March 1998.

243. A. der Kiureghian and E. Polak, "Reliability-Based Optimal Structural Design: a Decoupled Approach," *Proc. 8th IFIP WG 7.5 Conference on Reliability and Optimization of Structural Systems*, Krakow, Poland, 11-13 May, 1998.

244. L. Davis, R. Evans, and E. Polak, "Maximum Likelyhood Estimation of Positive Definite hermitian Toeplitz Matrices Using Outer Approximations," *Proc. 9th IEEE Signal Processing Workshop on Statistical Signal and Array Processing*, Portland, Oregon, September 14-16, pp. 49-52, 1998.

245. J. S. Maltz, E. Polak, and T. F. Budinger, "Multistart Optimisation Algorithm for Joint Spatial and Kinetic Parameter Estimation in Dynamic

ECT," *1998 IEEE Nuclear Science Symposium and Medical Imaging Conference Record*, Toronto, Canada, November 9-14, 1998.

246. E. Polak, L. Qi, and D. Sun, "First-Order Algorithms for Generalized Finite and Semi-Infinite Min-Max Problems," *Computational Optimization and Applications*, Vol.13, No.1-3, Kluwer Academic Publishers, p.137-61, 1999.

247. E. Polak, R. J-B. Wets, and A. der Kiureghian, "On an Approach to Optimization Problems with a Probabilistic Cost and or Constraints," in *Nonlinear Optimization and Related Topics*, pp. 299-316, G.Di Pillo and F.Giannessi, Editors, Kluwer Academic Publishers B.V., 2000.

248. Geraldine Lemarchand, Olivier Pironneau, and Elijah Polak, "A Mesh Refinement Method for Optimization with DDM," *Proc. 13th International Conference on Domain Decomposition Methods*, Champfleuri, Lyons, France, October 9-12, 2000.

249. J. Royset, A. der Kiureghian, and E. Polak, "Reliability-based Optimal Design with Probabilistic Cost and Constraint," in *Proceeding of the 9th IFIP Working Conference on Optimization and Reliability of Structrual Systems*, Ann Arbor, Michigan, Sep 2000. A.S. Nowak and M.M. Szerszen (Eds.), Univ. of Michigan, Ann Arbor, pp. 209-216, 2000.

250. E. Polak, "First-Order Algorithms for Optimization Problems With a Maximum Eigenvalue/Singular Value Cost and or Constraints," M. A. Goberna and M. A. Lopez, eds, *Semi-Infinite Programming: Recent Advances*, Kluwer Academic Publishers, pp. 197-220, 2001.

251. J. Royset, A. der Kiureghian, and E. Polak, "Reliability-based Optimal Structural Design by the Decoupling Approach," Journal of Reliability Engineering and System Safety, Elsevier Science, Vol. 73, No. 3, p. 213-221, 2001.

252. J. Royset, A. der Kiureghian, and E. Polak, "Reliability-based Optimal Design of Series Structural Systems," *Journal of Engineering Mechanics*, 127, 6, p. 607-614, 2001.

253. E. Polak, L. Qi and D. Sun, "Second-Order Algorithms for Generalized Finite and Semi-Infinite Min-Max Problems," *SIAM Journal on Optimization*, Vol.11, 218 (no.4), p.937-61, 2001.

254. E. Polak and M. Wetter, "Generalized Pattern Search Algorithms with Adaptive Precision Function Evaluations," University of California ERL Memo No UCB/ERL M01/30, 7 September 2001.

255. E. Polak, "Smoothing Techniques for the Solution of Finite and Semi-Infinite Min-Max-Min Problems," *High Performance Algorithms and Software for Nonlinear Optimization*, G. Di Pillo and A. Murli, Editors, Kluwer Academic Publishers B.V., 2002

256. O. Pironneau and E. Polak, "On a Consistent Approximations Approach to Optimal Control Problems with Two Numerical Precision Parameters," *High Performance Algorithms and Software for Nonlinear Optimization*, G. Di Pillo and A. Murli, Editors, Kluwer Academic Publishers B.V., 2002.

257. O. Pironneau and E. Polak, "Consistent Approximations and Approximate Functions and Gradients in Optimal Control," *J. SIAM Control and Optimization*, Vol. 41, pp. 487-510, 2002.

258. J.O. Royset, E. Polak and A. Der Kiureghian, "FORM Analysis USIng Consistent Approximations," *Proceedings of the 15th ASCE Engineering Mechanics Conference*, New York, NY, 2002.

259. E. Polak and J.O. Royset, "Algorithms with Adaptive Smoothing for Finite Min-Max Problems," J. Optimization Theory and Applications, submitted 2002.

260. E. Polak and J.O. Royset, "Algorithms for Finite and Semi-Infinite Min-Max-Min Problems Using Adaptive Smoothing Techniques," J. Optimization Theory and Applications, submitted 2002.

261. J.O. Royset, E. Polak and A. Der Kiureghian, "Adaptive Approximations and Exact Penalization for the Solution of Generalized Semi-Infinite Min-Max Problems," to appear in *SIAM J. Optimization*,

262. A. Brockwell, E. Polak, R. Evans, and D. Ralph, "Dual-Sampling-Rate Moving Horizon Control of a Class of Linear Systems with Input Saturation and Plant Uncertainty," Carnegie Mellon, Dept. of Statistics Report No. 733, 11/00, 2000. To appear in *J. Optimization Theory and Applications*.

I DUALITY AND OPTIMALITY CONDITIONS

1   ON MINIMIZATION OF MAX-MIN FUNCTIONS

A.M. Bagirov and A.M. Rubinov

Centre for Informatics and Applied Optimization,
School of Information Technology and Mathematical Sciences,
University of Ballarat, Victoria, 3353, Australia
Email: a.bagirov@ballarat.edu.au, a.rubinov@ballarat.edu.au

Abstract: In this paper different classes of unconstrained and constrained minimization problems with max-min objective and/or constraint functions are studied. First we consider simple problems with the max-min objective function, where the explicit description of all local minimizers is given. Then we discuss the application of the cutting angle and discrete gradient methods as well as a special penalization method for solving constrained problems. We report the results of preliminary numerical experiments.

Key words: Max-min function, cutting angle method, discrete gradient method, quasidifferential.

1 INTRODUCTION

Max-min functions form one of the interesting and important for applications classes of nonconvex and nonsmooth functions. There are many practical tasks where the objective function and/or constraints belong to this class. For example, optimization problems with max-min constraints arise in different branches of engineering such as the design of electronic circuits subject to a tolerancing and tuning provision (see Bandler, et al (1976); Liu et al (1992); Muller (1976); Polak and Sangiovanni Vincentelli (1979); Polak (1981)), the design of paths for robots in the presence of obstacles (Gilbert and Johnson (1985)), the design of heat exchangers (Grossman and Sargent (1978); Halemane and Grossman (1983); Ostrovsky et al (1994)) and chemical reactors (Halemane and Grossman (1983); Ostrovsky et al (1994)), in the layout design of VLSI circuits (Cheng et al (1992); Hochbaum (1993)) etc. Optimization problems with max-min objective and constraint functions also arise when one tries to design systems under uncertainty (see Bracken and McGill (1974); Grossman and Sargent (1978); Halemane and Grossman (1983); Ierapetritou and Pistikopoulos (1994); Ostrovsky et al (1994)).

A mathematical formalization of one of the main problems of cluster analysis leads to the following optimization problem: for a given set of points $a^i \in \mathbb{R}^n$, $i = 1, \ldots, m$ find a collection $\bar{x} = \{\bar{x}^1, \ldots, \bar{x}^p\}$ of p n-dimensional vectors, which is a solution of the following problem:

$$\text{minimize } f(x^1, \ldots, x^p) = \max_{i=1,\ldots,m} \min_{l=1,\ldots,p} \|x^l - a^i\| \text{ subject to } x^l \in D, \, l = 1, \ldots, p$$

where $D \subset \mathbb{R}^n$ is a compact set.

It is known that piecewise linear functions can be represented as max-min of certain linear functions (see Bartels et al (1995)). Thus the minimization of piecewise linear functions can be reduced to the minimization of functions represented as max-min of linear functions.

In the theory of abstract convexity an arbitrary Lipschitz continuous function on the unit simplex is underestimated by certain max-min functions (see Rubinov (2000) for details). Thus, the global minimization of Lipschitz functions can be reduced to the sequence of global minimization problems with special max-min objective functions (see Bagirov and Rubinov (2000); Rubinov (2000)).

In the paper (Kirjner-Neto and Polak (1998)), the authors consider optimization problems with twice continuously differentiable objective functions and max-min constraint functions. They convert this problem to a certain problem of smooth optimization.

In this paper we consider different classes of unconstrained and constrained minimization problems with max-min objective and/or constraint functions. The paper consists of four parts. First, we investigate a special simple class of max-min functions. We give a explicit description of all local minima and show that even in such a simple case the number of local minimizers is very large.

We discuss the applicability of the discrete gradient method (see, for example Bagirov (1999a); Bagirov (1999b)) for finding a local minimizer of discrete max-min functions without constraints in the second part.

The constrained minimization problems with max-min constraints are examined in the third part. We use a special penalization approach (see Rubinov et al (2002)) for this purpose.

The unconstrained global minimization of some continuous maximum functions by the cutting angle method (see, for example, Rubinov (2000)) are studied in the fourth part. If the number of internal variables in the continuous maximum functions is large enough (more than 5) then the minimization problem of these functions cannot be solved by using traditional discretization. Application of the cutting angle method allows us to solve such problems with small number of external and large enough number of internal variables (up to 15 internal variables).

We provide results of numerical experiments, which allow us to conclude that even for simple max-min functions the number of local minima can be very large. It means that the problem of minimization of such functions is quite complicated. However, results of numerical experiments show that methods considered in this paper allow us to solve different kinds of problems of minimization of max-min functions up to 10 variables.

The paper is arranged as follows. In Section 2 we study the problem of minimization of special max-min functions over the unit simplex. Section 3 is devoted to the problem of minimization of the discrete max-min functions. Minimization problems with max-min constraints are studied in Section 4. The problem of global minimization of the continuous maximum functions is discussed in Section 5. Section 6 concludes the paper.

2 SPECIAL CLASSES OF MAX-MIN OBJECTIVE FUNCTIONS

In this section we propose an algorithm for solving special classes of minimization problems with the max-min objective functions. This algorithm allow us to calculate the set of all local minimizers.

2.1 ICAR and IPH functions

Let $I = \{1, \ldots, n\}$ and $\mathbb{R}^n$ be an n-dimensional vector space. For each $x \in \mathbb{R}^n$ we denote by x_i the i-th coordinate of the vector x. We use the following notation:

$$x \geq y \iff (x_i \geq y_i \ \forall\, i \in I); \qquad x \gg y \iff (x_i > y_i \ \forall\, i \in I).$$

Let $\mathbb{R}^n_+ = \{x \in \mathbb{R}^n : x \geq 0\}$ be the cone of vectors with nonnegative coordinates. For a vector $l \in \mathbb{R}^n_+$ consider the sets

$$I(l) = \{i \in I : l_i > 0\}, \qquad I_0(l) = \{i \in I : l_i = 0\}.$$

If $l \geq 0$ then $I(l) = \emptyset$ if and only if $l = 0$. We also have $I(l) \cup I_0(l) = I$. For each $l \in \mathbb{R}^n_+ \setminus \{0\}$ consider the function, which we denote by the same symbol l:

$$l(x) = \min_{i \in I(l)} l_i x_i, \quad x \in \mathbb{R}^n_+. \tag{2.1}$$

A function of the form (2.1) is called the *min-type* function generated by l (or simply a min-type function).

Let f be a function defined on $\mathbb{R}^n_+$. A function f is called increasing if $x \geq y$ implies that $f(x) \geq f(y)$. The restriction of the function f to a ray $R_y = \{\alpha y : \alpha > 0\}$ starting from zero and passing through y is the function of one variable

$$f_y(\alpha) = f(\alpha y). \tag{2.2}$$

Definition 2.1 *A function $f : \mathbb{R}^n_+ \to \mathbb{R}$ is called ICAR (increasing convex-along-rays) function if f is increasing and the restriction of f to each ray R_y, $y \in \mathbb{R}^n_+ \setminus \{0\}$ is a convex function.*

The class of ICAR functions is very broad. It contains all increasing convex functions and all polynomials with nonnegative coefficients and also some concave functions.

Let L be the set of all min-type functions. A function $l \in L$ is called an L-subgradient of f at a point $y \in \mathbb{R}^n_+$ if $l(x) - l(y) \leq f(x) - f(y)$ for all x. The corresponding vector l is called a support vector for f at y. The set $\partial_L f(y)$ of all subgradients of f at y is called the L-subdifferential of f at y. If $l \in \partial_L f(y)$ then the function $\tilde{\lambda}(x) = l(x) - l(y) + f(y)$ enjoys the following properties: $\tilde{\lambda}(x) \leq f(x)$ for all $x \in \mathbb{R}^n_+$ and $\tilde{\lambda}(y) = f(y)$. Let $x \in \mathbb{R}^n_+, x \neq 0$ and $u \geq 0$. We shall use the following notation:

$$\left(\frac{u}{x}\right)_i = \begin{cases} \frac{u}{x_i} & \text{if } i \in I(x), \\ 0 & \text{if } i \in I_0(x). \end{cases} \tag{2.3}$$

The following result holds:

Theorem 2.1 *(see Rubinov (2000)) Let f be a finite ICAR function defined on $\mathbb{R}^n_+$. Then for each $y \in \mathbb{R}^n_+ \setminus \{0\}$ the L subdifferential $\partial_L f(y)$ is not empty and contains all vectors of the form u/x where $u \in \partial f_y(1)$ and $\partial f_y(1)$ is the subdifferential (in the sense of convex analysis) of the convex function f_y defined by (2.2), at the point $\alpha = 1$. In particular, if f is strictly increasing at the point y $(x \leq y, x \neq y$ implies $f(x) < f(y))$ and differentiable at this point, then*

$$\partial_L f(y) = \left\{ \frac{(\nabla f(y), y)}{y} \right\},$$

where $(\cdot, \cdot)$ is a usual inner product.

The following result is based on Theorem 2.1. (See Rubinov (2000) for details.)

Proposition 2.1 *Let f be an ICAR function and let $X \subset int\ \mathbb{R}^n_+$ be a compact set. Then for each $\varepsilon > 0$ there exists a finite set $\{x^1, \ldots, x^j\} \subset X$, such that the function*

$$h(x) = \max_{k \leq j}\ \min_{i \in I(l^k)} (l^k_i x_i - c_k), \tag{2.4}$$

where l^k is an arbitrary element of $\partial_L f(x^k)$ and $c_k = f(x^k) - f'(x^k, x^k)$, $k = 1, \ldots, j$, curries out a uniform lower approximation of f, that is,

$$0 \leq f(x) - h(x) \leq \varepsilon\ for all\ x \in X.$$

It is easy to see that the function h is ICAR (see Rubinov (2000)).

An important subclass of the set of all ICAR functions consists of increasing positively homogeneous of degree one(IPH) functions. Recall that a function f

defined on $\mathbb{R}_+^n$ is called positively homogeneous of degree one if

$$f(\lambda x) = \lambda f(x), \quad x \in \mathbb{R}_+^n, \lambda > 0. \tag{2.5}$$

The function f is IPH if and only if f is positively homogeneous ICAR function. It follows from (2.5) that an ICAR function is IPH if the function f_y defined by (2.2) is linear for all $y \in \mathbb{R}_+^n$, $y \neq 0$.

The following result holds (see Rubinov (2000)).

Theorem 2.2   *1) A finite function defined on $\mathbb{R}_+^n$ is IPH if and only if*

$$f(x) = \max\{l(x) : \ l \in L, \ l \leq f\},$$

where $l(x)$ is defined by (2.1):

$$l(x) = \min_{i \in I(l)} l_i x_i;$$

2) Let $x^0 \in \mathbb{R}_+^n$ be a vector such that $f(x^0) > 0$ and $l = f(x^0)/x^0$. Then $l(x) \leq f(x)$ for all $x \in \mathbb{R}_+^n$ and $l(x^0) = f(x^0)$.

The following proposition is a special case of Proposition 2.1.

Proposition 2.2 *Let f be an IPH function and let $X \subset \operatorname{int} \mathbb{R}_+^n$ be a compact set. Then for each $\varepsilon > 0$ there exists a finite set $\{x^1, \ldots, x^j\} \subset X$, such that the function*

$$h(x) = \max_{k \leq j} \min_{i \in I(l^k)} l_i^k x_i \tag{2.6}$$

with $l^k = f(x^k)/x^k$ curries out a uniform lower approximation of f, that is,

$$0 \leq f(x) - h(x) \leq \varepsilon \text{ for all } x \in X.$$

Now we consider the following global optimization problem:

$$\text{minimize} \quad f(x) \quad \text{subject to} \quad x \in X \tag{2.7}$$

where f is an ICAR function and $X \subset \mathbb{R}_+^n$ is a compact set. The cutting angle method for solving problem (2.7) has been proposed and studied in Rubinov (2000) (p. 420). This method reduces the problem (2.7) to a sequence of auxiliary problems:

$$\text{minimize} \quad h(x) \quad \text{subject to} \quad x \in X \tag{2.8}$$

where the objective function h is defined by (2.4).

The cutting angle method can be applied for minimization of an arbitrary Lipschitz function if the set X coincides with the unit simplex S:

$$S = \left\{ x \in \mathbb{R}^n_+ : \sum_{i=1}^n x_i = 1 \right\}.$$

It follows from the following result (see Rubinov (2000) and references therein)

Theorem 2.3 *Let f be a Lipschitz positive function defined on S and g be a function defined on $\mathbb{R}^n_+$ by*

$$g(x) = \begin{cases} f\left(x/\|x\|_1\right)\|x\|_1^p & if \ \ x \neq 0, \\ 0 & if \ \ x = 0. \end{cases} \tag{2.9}$$

Then $g(y) = f(y)$ for all $y \in S$ and g is an ICAR function if $p \geq (2K)/m$ where $m = \min_{y \in S} f(y)$ and K is the Lipschitz constant of f in $\| \cdot \|_1$ norm:

$$K = \sup_{x \neq y, \ x,y \in S} \frac{|f(x) - f(y)|}{\|x - y\|_1}.$$

Let f be a Lipschitz function over S. Consider the function $f_d(x) = f(x) + d$, where $d > 0$. Note that the Lipschitz constant of f coincides with the Lipschitz constant of f_d. Hence, for each $p > 0$ there exists $d > 0$ such that the extension of f_d given by (2.9), is an ICAR function. However if d is a very large number the function g is "almost flat" and its minimization is a difficult task. On the other side, if d is small enough, then p is large and some computation difficulties can appear. Thus, an appropriate choice of a number d is an important problem.

We shall consider the cutting angle method only for minimization of ICAR functions over S, then it can be applied also for minimization of Lipschitz functions defined on S. The main idea behind this method is to approximate the objective function g by a sequence of its saw-tooth underestimates h_j of the form (2.4). Assume that $h_j(x^k) = g(x^k)$ for some points x^k, $k = 1, \dots, j$ and h_j uniformly converges to g as $j \to +\infty$. Then a global minimizer of g can be approximated by a solution of the problem:

$$\text{minimize} \quad h_j(x) \quad \text{subject to} \quad x \in S. \tag{2.10}$$

Note that objective function of (2.10) is a max-min function (this is one of the simplest max-min functions.) The detailed presentation of a theory of the cutting angle method can be found in Rubinov (2000). Here we discuss only methods for solution of the auxiliary problem (2.10). Some of them can be found in Rubinov (2000). The most applicable method can be given if the objective function g of the initial problem is IPH. In this case constant c_k in (2.4) are equal to zero, hence h_j is a positively homogeneous function. Using this property of h_j it is possible to prove that all local minimizers of h_j over S are interior points of S and then to give an explicit expression of local minima of g over S, which are interior points of S. (see Bagirov and Rubinov (2000) and also Rubinov (2000)). If g is an extension defined by (2.9) of a Lipschitz function f defined on S, then g is IPH if $p = 1$, so we need to choose a large number d in order to apply the cutting angle method. The question arises is it possible to extend results obtained for minimization of homogeneous functions of the form (2.6) for general non-homogeneous functions of the form (2.4). In the case of success we can consider more flexible versions of the cutting angle method, which can lead to its better implementation.

2.2   The simplest max-min functions

Thus in this subsection we shall study max-min functions of the form:

$$h(x) = \max_{k \leq j} \ \min_{i \in I(l^k)} \left(l_i^k x_i - c_k \right), \tag{2.11}$$

where $l^k = (l_1^k, \ldots, l_n^k) \in \mathbb{R}_+^n$ and $c_k \in \mathbb{R}^n$, $k \leq j$. Sometimes we shall denote the function h in (2.11) by h_j in order to emphasize the number of min-type functions in (2.11). The function h defined by (2.11) is the simplest (in a certain sense) non-trivial max-min function.

We consider the following minimization problem:

$$\text{minimize} \ \ h(x) \ \ \text{subject to} \ \ x \in S. \tag{2.12}$$

Our goal is to describe all local minimizers of h over the unit simplex S. We shall also indicate the structure of such minimizers. Let

$$\Phi_k(x) = \min_{i \in I(l^k)} l_i^k x_i - c_k, \ \ k = 1, \ldots, j.$$

Then

$$h(x) = \max_{k \leq j} \Phi_k(x).$$

We set

$$R(x) = \{k \in \{1, \ldots, j\} : \Phi_k(x) = h(x)\}, \tag{2.13}$$

$$Q_k(x) = \{i \in I(l^k) : l_i^k x_i - c_k = \Phi_k(x)\}. \tag{2.14}$$

The functions Φ_k, $k = 1, \ldots, j$ and h are directionally differentiable and

$$\Phi_k'(x, u) = \min_{i \in Q_k(x)} l_i^k u_i, \ u \in \mathbb{R}^n, \tag{2.15}$$

$$h'(x, u) = \max_{k \in R(x)} \Phi_k'(x, u) = \max_{k \in R(x)} \min_{i \in Q_k(x)} l_i^k u_i, \ u \in \mathbb{R}^n. \tag{2.16}$$

Recall the definition of the cone $K(x, S)$ of feasible directions at a point $x \in S$ with respect to the set S. By definition

$$K(x, S) = \{u \in \mathbb{R}^n : \ \exists \, \alpha_0 > 0 : \ x + \alpha u \in S \text{ for all } \alpha \in (0, \alpha_0)\}.$$

Recall the well-known necessary and sufficient conditions for a local minimizer of a directionally differentiable function f over the set S (see, for example, Demyanov and Rubinov (1995)).

Proposition 2.3 *Let f be a directionally differentiable and locally Lipschitz function defined on S and $x \in S$. Then*

1) *if x is a local minimizer of f over S then $f'(x, u) \geq 0$ for all $u \in K(x, S)$;*

2) *if $f'(x, u) > 0$ for all $u \in K(x, S) \setminus \{0\}$ then x is a local minimizer of f over S.*

Now we describe all local minimizers of the function h defined by (2.11). The function h is bounded from below on S. Adding a sufficiently large positive constant we can assume without loss of generality that $h(x) > 0$ for all $x \in S$. First we consider minimizers, which belong to the relative interior $\mathrm{ri}\, S$ of the set S:

$$\mathrm{ri}\, S = \{x = (x_1, \ldots, x_n) \in S : x_i > 0, \ i \in I\}.$$

Proposition 2.4 *Let $j \geq n$ and $x \in \mathrm{ri}\, S$ be a local minimizer of the function h over the set $\mathrm{ri}\, S$ such that $h(x) > 0$. Then there exists a subset $\{l^{k_1}, \ldots, l^{k_n}\}$ of the set $\{l^1, \ldots, l^j\}$ such that $l_i^{k_i} > 0$, $i = 1, \ldots, n$ and*

1)

$$x = \left(\frac{d + c_{k_1}}{l_1^{k_1}}, \ldots, \frac{d + c_{k_n}}{l_n^{k_n}} \right),$$

where

$$d = h(x) = \left(\sum_{i=1}^{n} \frac{1}{l_i^{k_i}} \right)^{-1} \left(1 - \sum_{i=1}^{n} \frac{c_{k_i}}{l_i^{k_i}} \right);$$

2)

$$\max_{k \leq j} \min_{i \in I(l^k)} \left(\frac{l_i^k}{l_i^{k_i}} - \frac{h(x) + c_k}{h(x) + c_{k_i}} \right) = 0;$$

3)

$$\frac{l_i^{k_m}}{l_i^{k_i}} > \frac{h(x) + c_{k_m}}{h(x) + c_{k_i}} \quad \text{for all} \quad i \in I(l^{k_m}), \quad i \neq m.$$

Proof. Let $x \in S$ be a local minimizer. It is easy to check that

$$K(x, S) = \left\{ u = (u_1, \ldots, u_n) \in \mathbb{R}^n : \sum_{i \in I} u_i = 0, \, u_i \geq 0 \, (i \in I_0(x)) \right\}. \quad (2.17)$$

It follows from Proposition 2.3 that $h'(x, u) \geq 0$ for all $u \in K(x, S)$. Applying (2.16) we conclude that for each $u \in K(x, S)$ there exists $k \in R(x)$ such that

$$\Phi_k'(x, u) \geq 0.$$

Let $m \in I$. Consider the following direction $u = (u_1, \ldots, u_n)$:

$$u_i = \begin{cases} 1 & \text{if } i = m, \\ -\lambda_i & \text{if } i \neq m. \end{cases}$$

Here $\lambda_i > 0$, $i \in I \setminus m$ and

$$\sum_{i \neq m} \lambda_i = 1.$$

It follows from (2.17) that $u \in K(x, S)$, hence there exists $k \in R(x)$ such that

$$\Phi_k'(x, u) \geq 0. \quad (2.18)$$

Let $Q_k(x)$ be the set defined by (2.14). We shall prove that $Q_k(x) = \{m\}$. Assume that $Q_k(x) \neq \{m\}$. Then due to (2.15) we have $\Phi_k'(x, u) < 0$ which contradicts (2.18). Thus $Q_k(x) = \{m\}$, so for any $m \in I$ there exists at least one $k \in R(x)$ such that $Q_k(x) = \{m\}$. Let $m_1, m_2 \in I$, $m_1 \neq m_2$ and k_{m_i} be an index such that $Q_{k_i}(x) = \{m_i\}$, $i = 1, 2$. Since $m_1 \neq m_2$, it follows that $k_{m_1} \neq k_{m_2}$.

Let $m \in I$ and $k_m \in R(x)$ be an arbitrary index such that $Q_{k_m}(x) = \{m\}$. Then

$$h(x) = \Phi_{k_m}(x) = l_m^{k_m} x_m - c_{k_m}. \tag{2.19}$$

It follows from (2.19) that

$$x_m = \frac{h(x) + c_{k_m}}{l_m^{k_m}}.$$

Thus we have

$$x = \left(\frac{h(x) + c_{k_1}}{l_1^{k_1}}, \ldots, \frac{h(x) + c_{k_n}}{l_n^{k_n}} \right). \tag{2.20}$$

Due to the equality $\sum_i x_i = 1$, we have

$$h(x) = \left(\sum_{i=1}^{n} \frac{1}{l_i^{k_i}} \right)^{-1} \left(1 - \sum_{i=1}^{n} \frac{c_{k_i}}{l_i^{k_i}} \right).$$

Since

$$h(x) = \max_{k \leq j} \left(\min_{i \in I(l^k)} l_i^k x_i - c_k \right) \tag{2.21}$$

we conclude that

$$h(x) \geq \min_{i \in I(l^k)} l_i^k x_i - c_k \quad \text{for all} \ \ k \leq j. \tag{2.22}$$

It follows from (2.22) that there exists $p \in I(l^k)$ such that $l_p^k x_p - c_k \leq h(x)$. Combining this inequality with (2.20) we deduce that

$$\frac{l_p^k}{l_p^{k_p}} \leq \frac{h(x) + c_k}{h(x) + c_{k_p}},$$

hence

$$\min_{i \in I(l^k)} \left(\frac{l_i^k}{l_i^{k_i}} - \frac{h(x) + c_k}{h(x) + c_{k_i}} \right) \leq 0$$

for all $k \leq j$. It follows from (2.21) that there exist indices $k \in R(x)$ and $p \in Q_k(x)$ such that $h(x) = l_p^k x_p - c_k$. Using again (2.20) we conclude that

$$\frac{l_p^k}{l_p^{k_p}} - \frac{h(x) + c_k}{h(x) + c_{k_p}} = 0,$$

so

$$\max_{k \leq j} \min_{i \in I(l^k)} \left(\frac{l_i^k}{l_i^{k_i}} - \frac{h(x) + c_k}{h(x) + c_{k_i}} \right) = 0.$$

Take now any $m \in I$ and an index k_m such that $k_m \in R(x)$ and $Q_{k_m}(x) = \{m\}$. Then we have

$$h(x) = \Phi_{k_m}(x) = l_m^{k_m} x_m - c_{k_m}$$

and

$$l_i^{k_m} x_i - c_{k_m} > \Phi_{k_m}(x) = h(x) \ \text{ for all } i \in I(l^{k_m}), \ i \neq m.$$

Hence

$$l_i^{k_m} \frac{h(x) + c_{k_i}}{l_i^{k_i}} - c_{k_m} > h(x)$$

or

$$\frac{l_i^{k_m}}{l_i^{k_i}} > \frac{h(x) + c_{k_m}}{h(x) + c_{k_i}} \ \text{ for all } \ i \in I(l^{k_m}), \ i \neq m.$$

$$\triangle$$

We again consider the function (2.11) and Problem (2.12). In next proposition we will establish a sufficient condition for a local minimum in Problem (2.12). We assume that the function (2.11) possesses the following property:

$$\Phi_k(x) := \min_{i \in I(l^k)} l_i^k x_i - c_k > 0 \ \text{ for all } \ k \leq j \text{ and } \ x \in S.$$

It can be achieved by adding the same large number to all functions Φ_k, $k \leq j$.

Proposition 2.5 *Let a subset $\{l^{k_1}, \ldots, l^{k_n}\}$ of the set $\{l^1, \ldots, l^j\}$ enjoys the following properties:*

1)

$$l_i^{k_i} > 0, \ i = 1, \ldots, n,$$

2)

$$\max_{k \leq j} \min_{i \in I(l^k)} \left(\frac{l_i^k}{l_i^{k_i}} - \frac{d + c_k}{d + c_{k_i}} \right) = 0, \tag{2.23}$$

3)

$$\frac{l_i^{k_m}}{l_i^{k_i}} > \frac{d + c_{k_m}}{d + c_{k_i}} \text{ for all } \ i \in I(l^{k_m}), i \neq m \tag{2.24}$$

where

$$d = \left(\sum_{i=1}^{n} \frac{1}{l_i^{k_i}} \right)^{-1} \left(1 - \sum_{i=1}^{n} \frac{c_{k_i}}{l_i^{k_i}} \right).$$

If

$$\bar{x} = \left(\frac{d + c_{k_1}}{l_1^{k_1}}, \ldots, \frac{d + c_{k_n}}{l_n^{k_n}} \right) \in S, \qquad (2.25)$$

then $\bar{x}$ is a local minimum of the function h over S and $h(\bar{x}) = d$.

Proof. It is sufficient to prove that $h'(\bar{x}, u) > 0$ for all $u \in K(\bar{x}, S) \setminus \{0\}$. By assumption $\Phi_k(x) > 0$ for all $x \in S$ and $k \leq j$. Then for the point $\tilde{x} = (\tilde{x}_1, \ldots, \tilde{x}_n) \in S$, where $\tilde{x}_i = 1/n$, $i = 1, \ldots, n,$, we have

$$\frac{1}{n} \min_{i \in I(l^k)} l_i^k - c_k > 0, \ k \leq j$$

which implies that

$$\frac{l_i^{k_i}}{n} - c_{k_i} > 0, \ i = 1, \ldots, n. \qquad (2.26)$$

It follows from (2.26) that

$$1 - \sum_{i=1}^{n} \frac{c_{k_i}}{l_i^{k_i}} > 0.$$

Thus $d > 0$.

Since $x \in S$ it follows that $d + c_{k_i} \geq 0$ for all $i = 1, \ldots, n$. Due to (2.23) we have

$$\min_{i \in I(l^k)} \left(\frac{l_i^k}{l_i^{k_i}} - \frac{d + c_k}{d + c_{k_i}} \right) \leq 0$$

for all $k \leq j$. The latter means that for any $k \leq j$ there exists $i = i_k \in I(l^k)$ such that

$$\frac{l_i^k}{l_i^{k_i}} - \frac{d + c_k}{d + c_{k_i}} \leq 0.$$

Combining this inequality with (2.25) we get

$$l_i^k \bar{x}_i - c_k \leq d \ \text{ for all } \ k \leq j.$$

Then

$$h(\bar{x}) = \max_{k \leq j} \min_{i \in I(l^k)} (l_i^k \bar{x}_i - c_k) \leq d.$$

Finally, it follows from (2.23) and (2.24) that

$$\min_{i \in I(l^{k_p})} l_i^{k_p} \bar{x}_i - c_{k_p} = d \ \text{ for all } \ p = 1, \ldots, n.$$

Thus $h(\bar{x}) = d$. Clearly

$$\overline{R} \equiv \{k_1, \ldots, k_n\} \subseteq R(\bar{x}).$$

Moreover it follows from (2.24) that $Q_{k_m}(\bar{x}) = \{m\}$. Let $u \in K(\bar{x}, S)$, $u \neq 0$. Due to (2.17) we conclude that there exists $m \in \{1, \ldots, n\}$ such that $u_m > 0$. We have that

$$
\begin{aligned}
h'(\bar{x}, u) &= \max_{k \in R(\bar{x})} \min_{i \in Q_k(\bar{x})} l_i^k u_i \\
&\geq \max_{k \in \overline{R}} \min_{i \in Q_k(\bar{x})} l_i^k u_i \\
&\geq \min_{i \in Q_{k_m}(\bar{x})} l_i^{k_m} u_i \\
&= l_m^{k_m} u_m > 0.
\end{aligned}
$$

Thus

$$
h'(\bar{x}, u) > 0 \quad \text{for all} \ \ u \in K(\bar{x}, S) \setminus \{0\},
$$

which implies that $\bar{x} \in S$ is a local minimum of the function h. $\triangle$

It is easy to see that a function

$$
h(x) = \max_{k \leq j} \min_{i \in I(l^k)} (l_i^k x_i - c_k)
$$

with $c_k \geq 0$ is positively homogeneous if and only if $c_k = 0$ for all $k \leq j$, that is

$$
h(x) = \max_{k \leq j} \min_{i \in I(l^k)} l_i^k x_i. \tag{2.27}
$$

Thus analogous results can be obtained for IPH functions from Propositions 2.4 and 2.5 when $c_k = 0$ for all $k \geq 0$. This case has been studied in Bagirov and Rubinov (2000).

Remark 2.1 For homogeneous max-min functions we have $d > 0$ and in this case we do not need assumption $\Phi_k(x) := \min_{i \in I(l^k)} l_i^k x_i > 0$ for all $k \leq j$ and $x \in S$.

The previous results and the following Proposition 2.6 allow one to describe all local minimizers of a homogeneous max-min function over the unit simplex:

Proposition 2.6 *(see Bagirov and Rubinov (2000)) Let $j > n$, $l^k = l_k^k e^k$, $l_k^k > 0$, $k = 1, \ldots, n$, $|I(l^k)| = n$, $k = n+1, \ldots, j$ where e^k is the kth orth vector and $|I(l)|$ is the cardinality of the set $I(l)$. Then each local minimizer of the function h defined by (2.27) over the simplex S is a strictly positive vector.*

Remark 2.2 Proposition 2.6 remains true if $|I(l^k)| \geq 2$ for $k = n+1, \ldots, j$.

Unfortunately the analogue of Proposition 2.6 for non-homogeneous functions of the form (2.11) does not hold. The following example confirms this assertion.

Example 2.1 Consider the following max-min function:

$$h(x) = \max_{k=1,\ldots,5} \min_{i \in I(l^k)} \left(l_i^k x_i - c_k \right), \quad x \in S \subset \mathbb{R}^4,$$

where

$$l^1 = (2,0,0,0), \quad l^2 = (0,4,0,0), \quad l^3 = (0,0,6,0), \quad l^4 = (0,0,0,8),$$

$$l^5 = \left(\frac{52}{25}, \frac{104}{25}, \frac{156}{25}, \frac{208}{25} \right),$$

$$c_1 = -5, \quad c_2 = -4, \quad c_3 = -3, \quad c_4 = -2, \quad c_5 = -10.$$

Let

$$\Phi_k(x) = \min_{i \in I(l^k)} l_i^k x_i - c_k, \quad x \in S, \quad k = 1, \ldots, 5.$$

We have

$$\Phi_1(x) = 2x_1 + 5, \quad \Phi_2(x) = 4x_2 + 4, \quad \Phi_3(x) = 6x_3 + 3, \quad \Phi_4(x) = 8x_4 + 2,$$

$$\Phi_5(x) = \frac{1}{25} \min \left(52x_1, 104x_2, 156x_3, 208x_4 \right) + 10.$$

Consider the boundary point of S:

$$x^1 = \left(\frac{1}{4}, \frac{3}{8}, \frac{3}{8}, 0 \right)$$

We have

$$\Phi_1(x^1) = 5.5, \quad \Phi_2(x^1) = 5.5, \quad \Phi_3(x^1) = 5.25,$$

$$\Phi_4(x^1) = 2, \quad \Phi_5(x^1) = 10$$

Since

$$h(x^1) = \max_{k=1,\ldots,5} \Phi_k(x^1)$$

it follows that

$$h(x^1) = \Phi_5(x^1) = 10 > \max_{i=1,2,3,4} \Phi_k(x^1).$$

It is clear that x^1 is the global minimum of the function Φ_5 over the set S. Moreover

$$\Phi_5(x) > \Phi_5(x^1)$$

for all $x \in \mathrm{ri}\, S$ and $\Phi_5(x) = \Phi_5(x^1)$ for all boundary points x. Since all the functions $\Phi_k, k = 1, \ldots, 5$ are continuous there exist $\epsilon > 0$ such that

$$h(x) = \Phi_5(x)$$

for all $x \in \{y \in S : \|y - x^1\| < \epsilon\}$. Hence x^1 is a local minimizer of the function h.

This example demonstrates that non-homogeneous max-min functions of the form (2.11) can attain their local minimum on the boundary of the set S. We now present a description of local minima, which are placed on the boundary of S.

Proposition 2.7 *Let $x \in S$ be a local minimizer of the function h defined by (2.11) over the simplex S such that $h(x) > 0$ and $r = |I(x)| < n$. Then there exists a subset $\{l^{k_1}, \ldots, l^{k_r}\}$ of the set $\{l^1, \ldots, l^j\}$ such that*

1)

$$l_i^{k_i} > 0, \quad i = 1, \ldots, r,$$

2)

$$x_i = \frac{d + c_{k_i}}{l_i^{k_i}}, \quad i \in I(x),$$

where

$$d = \left(\sum_{i \in I(x)} \frac{1}{l_i^{k_i}} \right)^{-1} \left(1 - \sum_{i \in I(x)} \frac{c_{k_i}}{l_i^{k_i}} \right) \quad \text{and } h(x) = d,$$

3)

$$\max_{k \leq j} \min_{i \in I(l^k) \cap I(x)} \left(\frac{l_i^k}{l_i^{k_i}} - \frac{h(x) + c_k}{h(x) + c_{k_i}} \right) = 0,$$

4)

$$\frac{l_i^{k_m}}{l_i^{k_i}} > \frac{h(x) + c_{k_m}}{h(x) + c_{k_i}} \quad \text{for all } i \in I(l^{k_m}) \cap I(x), \ i \neq m.$$

Proof. Consider the space $\mathbb{R}^r$ and the simplex

$$S' = \left\{ x = (x_i)_{i \in I(x)} : x \geq 0, \sum_{i \in I(x)} x_i = 1 \right\} \subset \mathbb{R}^r.$$

Let h' be the restriction of the function h to $\mathbb{R}^r_+$. Clearly h' is a function of the form (2.11) and x is a local minimizer of this function. Since $x \in \operatorname{ri} S'$, we can apply Proposition 2.4. The result follows directly from this proposition. $\triangle$

Propositions 2.4 and 2.5 allows us to propose the following algorithm for the computation the set of local minimizers of the function h defined by (2.11) over $\operatorname{ri} S$.

An algorithm for the calculation of local minimizers

Step 0. (Initialization) Let $\epsilon > 0$ be a tolerance. Set $t = 0$, $m = 0$, $p(k) = 0$, $k = 1, \ldots, n$.

Step 1. Set $m = m + 1$. If $m > n$ go to Step 5.

Step 2. Set $p(m) = p(m) + 1$ and $i = p(m)$. If $i > j$ then go to Step 3. Otherwise take a vector l^i and go to Step 4.

Step 3. Set $p(m) = 0$ and $m = m - 1$. If $m = 0$ then go to Step 9, otherwise go to Step 2.

Step 4. (checking the vector l^i). If $l^i_m < \epsilon$ go to Step 2, otherwise set $k_m = i$ and go to Step 1.

Step 5. If

$$\frac{l_i^{k_t}}{l_i^{k_i}} > \frac{d + c_{k_t}}{d + c_{k_i}} \quad \text{for all} \ \ t = 1, \ldots, n, \ i \in I(l^{k_t}), \ i \neq t$$

then go to Step 6, otherwise set $m = n$ and go to Step 2.

Step 6. Calculate the number

$$d = \left(\sum_{i \in I} \frac{1}{l_i^{k_i}} \right)^{-1} \left(1 - \sum_{i \in I} \frac{c_{k_i}}{l_i^{k_i}} \right).$$

and the point x with coordinates

$$x_i = \frac{d + c_{k_i}}{l_i^{k_i}}, \quad i \in I.$$

Step 7. If

$$\max_{k \leq j} \ \min_{i \in I(l^k)} \left(\frac{l_i^k}{l_i^{k_i}} - \frac{d + c_k}{d + c_{k_i}} \right) = 0$$

and $x \in S$ then go to Step 8, otherwise set $m = n$ and go to Step 2.

Step 8. Set $t = t+1$, $G(t) = x$, $m = n$ and go to Step 2.

Step 9. End.

Note that G is the set of local minimizers of the function h over the set ri S.

2.3   *Results of Numerical Experiments*

In this subsection we describe results of numerical experiments with minimization of max-min functions of the form (2.11), both homogeneous and non-homogeneous, over the unit simplex. First consider non-homogeneous functions of the form (2.11). We consider max-min functions, which appear as approximations of the following ICAR function:

$$f(x) = \sum_{i=1}^{n} \sum_{j=1}^{n} a_{ij} x_i x_j \qquad (2.28)$$

where

$$a_{ij} = \begin{cases} 12 + n/i & \text{if } i = j, \\ 0 & \text{if } i = j+1, \\ 0 & \text{if } j = i+2, \\ 15/(i+0.1j) & \text{otherwise.} \end{cases} \qquad (2.29)$$

The following points on the unit simplex S were used for the construction of min-type functions:

$$x_k^k = 1, \ x_i^k = 0, \ i, k = 1, \ldots, n, \ i \neq k, \qquad (2.30)$$

$$x_i^k = \frac{y_i^k}{\sum_{p=1}^{n} y_p^k}, \ y_i^k = i|\sin(i+k)|, \ k \geq n+1, \ i = 1, \ldots, n. \qquad (2.31)$$

Due to Proposition 2.1 and Theorem 2.1, the approximation function has the form:

$$h(x) = \max_{k \leq j} \min_{i \in I(x^k)} \left(\frac{(\nabla f(x^k), x^k)}{x_i^k} - (f(x^k) - (\nabla f(x^k)x^k, x^k)) \right), \qquad (2.32)$$

where f defined by (2.28) with a_{ij} defined by (2.29).

The algorithm for calculation of local minimizers, which was proposed in previous subsection, has been applied for the calculation of local minimizers of the function (2.32) over ri S.

The code has been written in Fortran 90 and numerical experiments have been carried out in PC IBM Pentium III with CPU 800 MHz.

For the description of the results of numerical experiments we use the following notation:

- n is the number of variables;

- j is the number of min-type functions;

- n_{lm} is the number of local minimizers;

- t_1 is the CPU time for the calculation of the set of all local minimizers;

- t_2 is the average CPU time for calculation of one local minimizer;

- n_f is the average number of the objective function evaluations for the calculation of one local minimizer.

Results of numerical experiments are presented in Table 2.1.

Table 2.1 demonstrates that even so simple max-min function as (2.11) has a large number of local minimizers. Results of numerical experiments presented in this table show that the proposed algorithm calculates the set of local minimizers quickly enough for small size problems (up to 5 variables). The number of local minimizers increases sharply as the number of min-type functions increases. The proposed algorithm can calculate one local minimizer very quickly even for large number of variables and min-type functions.

Consider now minimization of homogeneous max-min functions of the form (2.27):

$$h(x) = \max_{k \leq j} \min_{i \in I(l^k)} l_i^k x_i.$$

We consider max-min functions, which appear as approximations of the following IPH functions (see Bagirov and Rubinov (2001a)):

Example 2.2

$$f^1(x) = \left(\sum_{i=1}^{n} \sum_{j=1}^{n} a_{ij} x_i x_j \right)^{1/2},$$

where a_{ij} are numbers defined by (2.29).

Example 2.3

$$f^2(x) = \max_{1 \leq i \leq 20} \min_{1 \leq j \leq n} (a^{ij}, x),$$

Table 2.1 Results of numerical experiments for non-homogeneous max-min functions

n	3				4			
j	n_{lm}	t_1	n_f	t_2	n_{lm}	t_1	n_f	t_2
30	36	0.06	63	0.0017	85	0.17	99	0.0020
35	50	0.00	73	0.0000	129	0.33	122	0.0026
40	53	0.11	99	0.0021	173	0.49	152	0.0028
45	64	0.05	119	0.0008	252	0.82	175	0.0033
50	76	0.11	135	0.0015	286	1.27	229	0.0044

n	5				6			
30	266	0.60	83	0.0023	861	3.19	118	0.0037
35	346	1.15	113	0.0033	1719	8.78	162	0.0051
40	553	2.59	163	0.0047	2895	19.67	239	0.0068
45	687	4.22	228	0.0061	4077	38.83	323	0.0095
50	885	6.98	282	0.0079	5907	67.34	371	0.0114

n	7				8			
30	1065	6.10	174	0.0057	5090	26.64	146	0.0052
35	2251	21.36	319	0.0095	11068	91.73	234	0.0083
40	4965	65.91	456	0.0133	26230	332.02	357	0.0127
45	6489	118.53	591	0.0183	47843	745.18	435	0.0156
50	8984	230.41	770	0.0257	79629	1705.38	591	0.0214

$$a_k^{ij} = \frac{10j}{k(1 + |k - j|)} |\cos(i - 1)|, \quad i = 1, \ldots, 20, \quad j = 1, \ldots, n, \quad k = 1, \ldots, n.$$

We use points x^k given in (2.30) and (2.31) for the construction of min-type functions. Thus we describe all local minimizers of the following functions over the set S:

$$h^q(x) = \max_{k \leq j} \min_{i \in I(x^k)} \frac{f^q(x^k)}{x^k}, \qquad q = 1, 2.$$

Results of numerical experiments with the functions h^1 and h^2 are presented in Tables 2.2 and 2.3, respectively.

Table 2.2   Results of numerical experiments for the function h^1

n	3				4			
j	n_{lm}	t_1	n_f	t_2	n_{lm}	t_1	n_f	t_2
30	55	0.00	34	0.0000	131	0.05	40	0.0004
35	65	0.05	46	0.0008	167	0.17	56	0.0010
40	75	0.00	56	0.0000	194	0.27	77	0.0014
45	85	0.06	69	0.0007	229	0.39	105	0.0017
50	95	0.11	84	0.0012	269	0.55	134	0.0020

n	5				6			
30	231	0.27	42	0.0012	457	0.71	64	0.0016
35	285	0.44	59	0.0015	606	1.70	112	0.0028
40	339	0.77	93	0.0023	761	3.30	177	0.0043
45	401	1.26	125	0.0031	884	5.33	257	0.0060
50	477	1.93	159	0.0041	1042	8.67	347	0.0083

n	7				8			
30	867	1.54	61	0.0018	1509	3.90	86	0.0026
35	1169	3.85	126	0.0033	1954	9.23	166	0.0047
40	1521	8.62	214	0.0057	2810	22.96	299	0.0082
45	1855	14.88	292	0.0080	3661	49.05	468	0.0134
50	2271	25.16	397	0.0111	4526	99.19	750	0.0219

n	9				10			
30	2917	8.79	96	0.0030	4737	12.85	73	0.0027
35	4623	30.98	223	0.0067	7251	46.47	190	0.0064
40	6589	76.95	375	0.0117	11023	121.93	317	0.0111
45	9587	208.27	681	0.0217	14955	299.12	574	0.0200
50	11703	445.18	1232	0.0380	19224	734.14	1093	0.0382

Results presented in Tables 2.2 and 2.3 show that the number of local minimizers strongly depends on the original IPH function. The proposed algorithm

Table 2.3 Results of numerical experiments for the function h^2

n		3				4		
j	n_{lm}	t_1	n_f	t_2	n_{lm}	t_1	n_f	t_2
30	55	0.00	31	0.0000	130	0.05	30	0.0004
35	65	0.06	41	0.0009	161	0.16	40	0.0010
40	75	0.05	54	0.0007	200	0.22	57	0.0011
45	85	0.06	62	0.0007	228	0.39	67	0.0017
50	95	0.11	74	0.0011	271	0.44	79	0.0016

n		5				6		
30	225	0.22	29	0.0010	358	0.44	29	0.0012
35	283	0.44	41	0.0016	454	0.88	41	0.0019
40	351	0.66	58	0.0019	566	1.75	60	0.0031
45	399	1.04	67	0.0026	662	2.64	71	0.0040
50	485	1.54	80	0.0032	798	4.12	87	0.0052

n		7				8		
30	701	0.93	28	0.0013	974	1.54	27	0.0016
35	923	1.81	41	0.0020	1324	3.35	39	0.0025
40	1215	3.51	61	0.0029	1785	7.36	59	0.0041
45	1461	5.88	75	0.0040	2125	11.37	71	0.0054
50	1835	9.61	92	0.0052	2678	19.50	92	0.0073

n		9				10		
30	1339	2.80	28	0.0021	2275	5.22	33	0.0023
35	1823	6.37	42	0.0035	3240	13.73	59	0.0042
40	2483	14.23	64	0.0057	4393	33.28	97	0.0076
45	3183	26.69	82	0.0084	5504	63.50	132	0.0115
50	3987	44.87	104	0.0113	7233	118.31	176	0.0164

calculates the set of local minimizers quickly enough for problems up to 10 variables.

3 DISCRETE MAX-MIN FUNCTIONS

In this section we will consider the following minimization problem:

$$\text{minimize } f(x) \quad \text{subject to } x \in \mathbb{R}^n \tag{3.1}$$

where

$$f(x) = \max_{i \in I} \min_{j \in J} f_{ij}(x)$$

and I, J are finite set of indices and functions f_{ij} are convex. The function f can be represented as the difference of two convex functions:

$$f(x) = f_1(x) - f_2(x)$$

where

$$f_1(x) = \max_{i \in I} \left[\sum_{j \in J} f_{ij}(x) + \sum_{p \in I, p \neq i} \max_{k \in J} \sum_{q \neq k} f_{pq}(x) \right],$$

$$f_2(x) = \sum_{i \in I} \max_{k \in J} \sum_{j \neq k} f_{ij}(x).$$

Therefore f is a quasidifferentiable in the sense by Demyanov and Rubinov (see Demyanov and Rubinov (1986); Demyanov and Rubinov (1995)) and its quasidifferential $[\underline{\partial} f(x), \bar{\partial} f(x)]$ can be calculated by using methods of quasidifferential calculus. Moreover this function is semismooth (see Mifflin (1977)).

Some differential properties of this function have been studied in Kirjner-Neto and Polak (1998); Polak (1997); Polak (2003). The calculation of the Clarke subdifferential or the Demyanov-Rubinov quasidifferential of this function is complicated. Therefore here we suggest the discrete gradient method to solve Problem (3.1). This method is derivative-free. Its description can be found, for example, in Bagirov (1999a); Bagirov (1999b). For the calculation of the objective function an algorithm described in Evtushenko (1972) is used. This algorithm allows one to significantly accelerate the calculation of the objective function. The usage of this algorithm allows us to conclude that the calculation of the Clarke subdifferential or Demyanov-Rubinov quasidifferential of the discrete max-min functions is much more expensive than the calculation of values these functions. This means that methods, which use only values of the max-min functions, are more appropriate for their minimization. The discrete gradient method in conjunction with this algorithm allows us to solve minimization problems with discrete max-min objective functions. This local method sometimes can find even a global minimizer of a max-min function.

3.1 Results of numerical experiments

In this subsection results of numerical experiments by using the discrete gradient method are given. For the description of the test problems and the results of numerical experiments we use the following notation:

- $f = f(x)$ is the objective function;

- x^0 is the starting point;

- x^* is the local minimizer, $f_* = f(x^*)$;

- n is the number of variables;

- *iter* is the number of iterations;

- n_f is the number of the objective function evaluations;

- n_d is the number of the discrete gradient evaluations;

- *time* is the CPU time.

Numerical experiments have been carried out in PC IBM Pentium III with CPU 800 MHz. All problems have been solved with the precision $\delta = 10^{-4}$ that is at last point x^k:

$$f(x^k) - f_* < 10^{-4}.$$

To carry out numerical experiments unconstrained minimization problems with the following max-min objective functions were considered:

Problem 1

$$f(x) = \max_{i \in I} \min_{j \in J} \left| (a^{ij}, x) \right|,$$

$a_k^{ij} = 1/(i + j + k - 2)$, $i \in I$, $j \in J$, $k = 1, \ldots, n$, $I = \{1, \ldots, 90\}$, $J = \{1, \ldots, 10\}$, $x^0 = (10, \ldots, 10)$, $x^* = (0, \ldots, 0)$, $f_* = 0$.

Problem 2

$$f(x) = \max_{i \in I} \min_{j \in J} \left| \sum_{k=1}^{n} a_k^{ij} x_k^2 + (a^{ij}, x) \right|,$$

$a_k^{ij} = 1/(i + j + k - 2)$, $i \in I$, $j \in J$, $k = 1, \ldots, n$, $I = \{1, \ldots, 90\}$, $J = \{1, \ldots, 10\}$, $x^0 = (10, \ldots, 10)$, $x^* = (0, \ldots, 0)$, $f_* = 0$.

Results of numerical experiments are presented in Table 3.1.

Table 3.1   Results of numerical experiments for discrete max-min functions

Problem	n	$iter$	n_f	n_d	$time$
Problem 1	5	24	516	89	0.05
	10	32	1498	141	0.17
	15	39	3720	240	0.60
	20	76	10172	498	1.87
	35	135	25060	706	6.81
	50	230	55177	1089	20.70
Problem 2	5	53	1272	230	0.17
	10	98	4682	442	0.60
	15	218	21464	1388	5.33
	20	221	26908	1314	6.70
	35	409	72968	2047	31.09
	50	316	111840	2217	90.24

Results presented in Table 3.1 show that the discrete gradient method can be applied to solve minimization problems with the discrete max-min functions. This method allowed one to find global solution to the problems under consideration using reasonable computational efforts.

4   OPTIMIZATION PROBLEMS WITH MAX-MIN CONSTRAINTS

Consider the following constrained optimization problem (P):

$$\text{minimize } f(x)$$

$$\text{subject to } x \in X \subset \mathbb{R}^n,$$

$$g_i(x) \le 0, i = 1, \ldots, m, g_i(x) = 0, i = m + 1, \ldots, m + k.$$

We use a certain version of penalization for solving this problem. First we convolute all constraints g_i $(i = 1, \ldots, m \ldots, m + k)$ to a single constraint f_1:

$$f_1(x) = \max \left(g_1(x), \ldots, g_m(x), |g_{m+1}(x)|, \ldots, |g_{m+k}(x)| \right).$$

Then the problem (P) is equivalent to the following problem (P_1):

$$\text{minimize } f(x) \text{ subject to } x \in X, \ f_1(x) \le 0.$$

The classical penalization of this function leads to the unconstrained minimization of the following penalty function:

$$L^+(x; d) = f(x) + df_1^+(x),$$

where $f_1^+(x) = \max(f_1(x), 0)$ and d is a large enough number. A number d, such that the optimal value of problem (P_1) coincides with $\min_{x \in X} L^+(x, d)$, is called the exact penalty parameter. The exact parameter d, if it exists, can be very large, so the problem

$$\text{minimize } L^+(x, d) \text{ subject to } x \in X,$$

can be ill-conditioned. In order to reduce the exact parameter, we replace the objective function f with the objective function f_c, where $f_c(x) = \sqrt{f(x) + c}$ with a large enough number c. (It is assumed that f is bounded from below on X and c is a number such that $f(x) + c \geq 0$ for all $x \in X$. There are some ways to omit this assumptions, however we do not discuss them here.) This approach has been studied in Rubinov et al (2002). It was shown there that under some natural assumptions it holds:

$$\lim_{c \to +\infty} \bar{d}_c = 0,$$

where $\bar{d}_c$ is the least exact penalty parameter for the problem (P_c):

$$\text{minimize } f_c(x) \text{ subject to } x \in X, \ f_1(x) \leq 0.$$

Thus instead of the initial problem (P) we consider the problem (P_c) with a certain number c and the penalty function $L^+(x, d_c)$ corresponding to (P_c) with a certain penalty parameter d_c. This approach allows one to use two large number c and d_c instead of one "very large" number d (a penalty parameter for the problem (P_1)).

The described approach has been used for solving two test problems, which can be found in Kirjner-Neto and Polak (1998).

Problem 3 The well-known Rosenbrock function is minimized subject to max-min constraints. The description of constraint functions are given in Kirjner-Neto and Polak (1998). Here we present only results of experiments. The described penalty method has been applied to solve this problem and then the discrete gradient method has been used to solve obtained unconstrained minimization problem. Solving this problem with the starting point $x^0 = (-1, 1)$

we obtained the local minimizer $\bar{x} = (0.552992, 0.306732)$ (up to six digits), which corresponds to the objective function (in the original problem) value of 0.1999038. This minimum was reached for 34 iterations and 274 objective function (in the unconstrained minimization problem) evaluations. If we instead set $x^0 = (1.5, 1.5)$, after 54 iterations using 499 the objective function evaluations in the unconstrained optimization problem we obtained local minimizer $\bar{x} = (1.199906, 1.440341)$ which corresponds to the objective function value of 0.039994 in the original problem.

Problem 4 In this problem the well-known Beale function is minimized subject to the same max-min constraints as in Problem 3. The described penalty method has been applied to solve this problem and then the discrete gradient method has been used to solve obtained unconstrained minimization problem. Solving this problem with the starting point $x^0 = (1, 1)$ we obtained the local minimizer $\bar{x} = (2.868929, 0.466064)$ (up to six digits), which corresponds to the objective function (in the original problem) value of 0.003194. This minimum was reached for 32 iterations and 172 objective function (in the unconstrained minimization problem) evaluations. For the starting point $x^0 = (1, -1)$, after 43 iterations using 225 the objective function evaluations in the unconstrained optimization problem we obtained the same local minimizer.

5   MINIMIZATION OF CONTINUOUS MAXIMUM FUNCTIONS

The cutting angle method requires to evaluate only one value of the objective function at each iteration. This property is very beneficial, if the evaluation of values of the objective function is time-consuming. Due to this property we can use the cutting angle method for solving some continuous min-max problems. Some details of this approach can be found in Bagirov and Rubinov (2001b).

Let $Y \subset \mathbb{R}^m$ be a compact set and let $\varphi(x, y)$ be a continuous function defined on the set $\mathbb{R}^n \times Y$. Consider the following continuous min-max problem: find points $\bar{x} \in \mathbb{R}^n$ and $\bar{y} \in Y$ such that

$$\varphi(\bar{x}, \bar{y}) = \min_{x \in \mathbb{R}^n} \max_{y \in Y} \varphi(x, y). \tag{5.1}$$

We reduce (5.1) to the following unconstrained minimization problem

$$\text{minimize } f(x) \quad \text{subject to } x \in \mathbb{R}^n \tag{5.2}$$

where

$$f(x) = \max_{y \in Y} \varphi(x, y).$$

We assume that the function $y \mapsto \varphi(x, y)$ is concave for each y, then the evaluation of the values of function f can be easily done by methods of convex programming. The cutting angle method can be applied for problems with a small amount of external variables x, however the number of internal variables y can be large enough. Numerical experiments with such kind of functions were produced in Bagirov and Rubinov (2001b) and we do not repeat them here.

6 CONCLUDING REMARKS

In this paper we have considered different optimization problems with max-min objective and/or constraint functions. We have presented numerical methods which can be used for solving some of these problems. In particular, terminating algorithm for minimization of the simplest max-min functions over the unit simplex have been proposed. This algorithm allows one to describe the set of local minima of problems under consideration. We demonstrated that the discrete gradient method can be used to solve unconstrained minimization of the discrete max-min functions and a special penalty function method can be applied to solve optimization problems with max-min constraint functions. Finally, we have discussed possible applications of the cutting angle method to global minimization of the continuous maximum functions when the number of external variables is not large. Results of numerical experiments presented in this paper show that the considered methods are effective for solving some problems under consideration.

Acknowledgments

The authors are very thankful to Professor E. Polak for kindly discussions of our results during his visit to Australia and providing us his recent papers. We are also grateful to an anonymous referee for his helpful comments. This research was supported by the Victorian Partnership for Advanced Computing.

References

Bagirov, A.M. (1999a). Minimization methods for one class of nonsmooth functions and calculation of semi-equilibrium prices, In: *Progress in Optimization: Contributions from Australasia*, Eberhard, A. et al. (eds.), *Applied Optimization*, **30,** Kluwer Academic Publishers, Dordrecht, 147-175.

Bagirov, A.M. (1999b). Derivative-free methods for unconstrained nonsmooth optimization and its numerical analysis, *Investigacao Operacional*, **19,** 75-93.

Bagirov, A.M. and Rubinov, A.M. (2000). Global minimization of increasing positively homogeneous functions over unit simplex, *Annals of Operation Research*, **98,** pp. 171-189.

Bagirov, A.M. and Rubinov, A.M. (2001a). Modified versions of the cutting angle method, In: N. Hadjisavvas and P.M.Pardalos (eds.) *Advances in Convex Analysis and Global Optimization*, Kluwer Academic Publishers, Dordrecht, 245-268.

Bagirov, A.M. and Rubinov, A.M. (2001b). Global optimization of marginal functions with applications to economic equilibrium, *Journal of Global Optimization*, **20,** 215-237.

Bandler, J.W., Liu, P.C. and Tromp, H. (1976). A nonlinear programming approach to optimal design centering, tolerancing and tuning, *IEEE Trans. Circuits Systems*, **CAS-23,** 155-165.

Bartels, S.G., Kunz, L. and Sholtes, S. (1995). Continuous selections of linear functions and nonsmooth critical point theory, *Nonlinear Analysis, TMA*, **24,** 385-407.

Bracken, J. and McGill, J.T. (1974). Defence applications of mathematical programs with optimization problems in the constraints, *Oper. Res.*, **22,** 1086-1096.

Cheng, C.K., Deng, X., Liao, Y.Z. and Yao, S.Z. (1992). Symbolic layout compaction under conditional design rules, *IEEE Trans. Comput. Aided Design*, **11,** 475-486.

Demyanov, V.F. and Rubinov, A.M. (1986). *Quasidifferential Calculus*, Optimization Software, New York.

Demyanov, V.F. and Rubinov, A.M. (1995). *Constructive Nonsmooth Analysis*, Peter Lang, Frankfurt am Main.

Evtushenko, Yu. (1972). A numerical method for finding best quaranteed estimates, *USSR Computational Mathematics and Mathematical Physics*, **12**, 109-128.

Gilbert, E.G. and Johnson, D.W. (1985). Distance functions and their application to robot path planning in the presence of obstacles, *IEEE J. Robotics Automat., RA-1*, 21-30.

Grossman, I.E. and Sargent, R.W. (1978). Optimal design of chemical plants with uncertain parameters, *AIChE J.*, **24**, 1-7.

Halemane, K.P. and Grossman, I.E. (1983). Optimal process design under uncertainty, *AIChE J.*, **29**, 425-433.

Hochbaum, D. (1993). Complexity and algorithms for logical constraints with applications to VLSI layout, compaction and clocking, *Studies in Locational Analysis, ISOLD VI Proceedings*, 159-164.

Ierapetritou, M.G. and Pistikopoulos, E.N. (1994). Simultaneous incorporation of flexibility and economic risk in operational planning under uncertainty, *Comput. Chem. Engrg.*, **18**, 163-189.

Kirjner-Neto, C. and Polak, E. (1998). On the conversion of oiptimization problems with max-min constraints to standard optimization problems, *SIAM J. on Optimization*, **8(4)**, 887-915.

Liu, P.C., Chung, V.W. and Li, K.C. (1992). Circuit design with post-fabrication tuning, in: *Proc. 35th Midwest Symposium on Circuits and Systems*, Washington, DC, IEEE, NY, 344-347.

Mifflin, R. (1976). Semismooth and semiconvex functions in constrained optimization. *SIAM Journal on Control and Optimization.*

Muller, G. (1976). On computer-aided tuning of microwave filters, in: *IEEE Proc. International Symposium on Circuits and Systems*, Munich, IEEE Computer Society Press, Los Alamos, CA, 722-725.

Ostrovsky, G.M., Volin, Y.M., Barit, E.I. and Senyavin, M.M. (1994). Flexibility analysis and optimization of chemical plants with uncertain parameters, *Comput. Chem. Engrg.*, **18**, 755-767.

Polak, E. (1981). An implementable algorithm for the design centering, tolerancing and tuning problem, *J. Optim. Theory Appl.*, **35**, 45-67.

Polak, E. (1997). *Optimization. Algorithms and Consistent Approximations.* Springer Verlag, New York.

Polak, E. (2003). Smoothing techniques for the solution of finite and semi-infinite min-max-min problems, *High Performance Algorithms and Software for Nonlinear Optimization*, G. Di Pillo and A. Murli (eds.), Kluwer Academic Publishers, to appear.

Polak, E. and Sangiovanni Vincentelli, A. (1979). Theoretical and computational aspects of the optimal design centering, tolerancing and tuning problem, *IEEE Trans. Circuits and Systems, CAS-26*, 795-813.

Rubinov, A.M. (2000). *Abstract Convexity and Global Optimization*, Kluwer Academic Publishers, Dordrecht.

Rubinov, A.M., Yang, X.Q. and Bagirov, A.M. (2002). Nonlinear penalty functions with a small penalty parameter, *Optimization Methods and Software*, **17(5)**, 931-964.

2   A COMPARISON OF TWO APPROACHES TO SECOND-ORDER SUBDIFFERENTIABILITY CONCEPTS WITH APPLICATION TO OPTIMALITY CONDITIONS

A. Eberhard

Department of Mathematics,
RMIT University, GPO Box 2476V,
Melbourne, Australia

and C. E. M. Pearce

Department of Mathematics,
University of Adelaide, North Terrace,
Adelaide, Australia

Abstract: The graphical derivative and the coderivative when applied to the proximal subdifferential are in general not generated by a set of linear operators Nevertheless we find that in directions at which the subjet (or subhessian) is supported, in a rank–1 sense, we have these supported operators interpolating the contingent cone. Thus under a prox-regularity assumption we are able to make a selection from the contingent graphical derivative in certain directions, using the exposed facets of a convex set of symmetric matrices. This allows us to make a comparison between some optimality conditions. A nonsmooth formulation of a standard smooth mathematical programming problem is used to derive a novel set of sufficient optimality conditions.

Key words: Subhessians, coderivative inclusions, rank–1 representers, optimality conditions.

1 INTRODUCTION

In this paper we shall concern ourselves only with second-order subderivatives that arise from one of two constructions. First, there is the use of the contingent tangent cone to the graph of the proximal subdifferential and its polar cone to generate the graph of the contingent graphical derivative and contingent coderivative. The second one is the use of sub-Taylor expansions to construct a set of symmetric matrices as replacements for Hessians, the so called subjet (or the subhessian of Penot (1994/1)) of viscosity solution theory of partial differential equations (see Crandall *et al.* (1992)). To each of these constructions may be associated a limiting counterpart which will also be considered at times. As the first of the above constructions produces a possibly nonconvex set of vectors in $\mathbb{R}^n$ and the latter a convex set of symmetric operators, it is not clear how to compare them. One of the purposes of this paper is to begin a comparative study of these notions for the class of prox–regular functions of Poliquin and Rockafellar (1996). A second objective is to consider what this study tells us regarding certain optimality conditions that can be framed using these alternative notions. We do not take up discussion of the second-order tangent cones and the associated second-order parabolic derivatives of functions, which is beyond our present scope. We refer the reader to Bonnans, Cominetti *et al.* (1999), Rockafellar and Wets (1998) and Penot (1994/2) and the bibliographies therein for recent accounts of the use of these concepts in optimality conditions. Important earlier works include Ben-Tal (1980), Ben-Tal *et al.* (1982), Ioffe (1990) and Ioffe (1991) Some results relating parabolic derivatives to subjet-like notions may be found in Eberhard (2000).

Where possible we adhere to the terminology and notation of the book 'Variational Analysis' by Rockafellar and Wets (1998), with the notable exception of the proximal subdifferential. Henceforth we assume $f : \mathbb{R}^n \to \bar{\mathbb{R}}$ to be lower semi–continuous, minorized by a quadratic function, and $\bar{x} \in \operatorname{dom} f$.

Definition 1.1 *A vector $y \in \mathbb{R}^n$ is called a proximal sub-gradient to f at x if, for some $c > 0$,*

$$f(x') \geq f(x) + \langle y, x' - x \rangle - \frac{c}{2}\|x' - x\|^2 \tag{1.1}$$

in a neighbourhood of x. The set of all proximal sub-gradients to f at x will be denoted by $\partial_p f(x)$.

This in turn defines the basic subdifferential *via* limiting processes by

$$\partial f(x) = \limsup_{x' \to^f x} \partial_p f(x').$$

Here $x' \to^f x$ means $x' \to x$ and $f(x') \to f(x)$. In general the proximal subdifferential is only contained in the subdifferential. Next we define one concept that forces equality.

Definition 1.2 *The function f is prox–regular at $\bar{x}$ for $\bar{y}$ ($\in \partial f(\bar{x})$) with respect to ε and $c > 0$ if f satisfies (1.1) whenever $\|x' - \bar{x}\| < \varepsilon$ and $\|x - \bar{x}\| < \varepsilon$ and $|f(x) - f(\bar{x})| < \varepsilon$, while $\|y - \bar{y}\| < \varepsilon$ with $y \in \partial f(x)$.*

As noted in many papers (see, for example, Poliquin and Rockafellar (1996), Poliquin *et al.* (1996) and Rockafellar and Wets (1998)) prox–regular functions constitute an important class of nonsmooth functions in the applications of nonsmooth analysis to optimization and thus are currently undergoing intense study. The proximal subdifferential when applied to the indicator function $\delta_S(x)$ of a set S (defined to be zero if $x \in S$ and $+\infty$ otherwise) gives rise to the proximal normal cone $N_S^p(\bar{x}) := \partial_p \delta_S(\bar{x})$. This may in turn be used *via* limiting processes to define the normal cone for which $\partial \delta_S(x) := N_S(x)$. For an arbitrary set–valued mapping $F : \mathbb{R}^n \rightrightarrows \mathbb{R}^m$, the normal cone to its graph at a point $y \in F(x)$ gives rise to the coderivative mapping of Mordukhovich (1976)–Mordukhovich (1994), Kruger *et al.* (1980) and later studies by Ioffe (1986)–Ioffe (1989). We consider only the finite–dimensional case in this paper.

Definition 1.3 *Suppose $F : X \rightrightarrows Y$ is a multifunction, $(x, y) \in \operatorname{Graph} F$ and $w \in X$. The Mordukhovich–Ioffe coderivative is defined as*

$$D^* F(x, y)(w) := \{p \in Y \mid (p, -w) \in \partial \delta_{\operatorname{Graph} F}(x, y) := N_{\operatorname{Graph} F}(x, y)\}.$$

This multifunction is not necessarily convex–valued and generally has a very complicated structure. Another nonconvex set–valued derivative constructed from cones is the contingent graphical derivative.

Definition 1.4 (Aubin) *The graphical derivative (or contingent derivative) mapping of a multifunction $F : X \rightrightarrows Y$ at $(x, y) \in \operatorname{Graph} F$ in the direction $w \in X$ is defined by*

$$DF(x, y)(w) := \{z \mid (w, z) \in T_{\operatorname{Graph} F}(x, y)\}.$$

Here $T_{\mathrm{Graph}\,F}(x,y) := \limsup_{t\downarrow 0} \frac{\mathrm{Graph}\,F-(x,y)}{t}$ is the contingent tangent cone.

The generality of this construction enables one to apply this to various subdifferential multifunctions and hence arrive at a "second–order" theory of subdifferentiation, that is, for $y \in \partial_p f(x)$ or $y \in \partial f(x)$ we may construct $D^*(\partial f)(x,y)(w)$ (see Mordukhovich (1998), Mordukhovich (2001) and Mordukhovich (2002) for recent results regarding the calculus of this object). A price is paid for going down this path in that the coderivative of the basic subdifferential is, in general, no longer related to operators that generalize Hessians in any obvious manner, except for cases when the functions involved are a composition or sum of sufficiently regular or smooth functions that admit a classical Hessian. The use of calculus rules to derive explicit expression for $D^*(\partial f)(x,y)(w)$ (the coderivative to the basic subdifferential) when f involves such a composition or sum has been undertaken in Mordukhovich (1984) and Mordukhovich (1994). Such results help to characterise the structure of these set–valued mappings in specific cases. We concern ourselves here with a more generic way of using certain symmetric operators to generate elements within the graphical derivative and coderivative to the proximal subdifferential. The method used will be to establish certain inclusions which augment those of Rockafellar *et al.* (1997). Thus the intent, methods and results of this paper are quite distinct in character to those of Mordukhovich (1984) and Mordukhovich (1994).

Another path may be taken to arrive at a second–order theory of the subdifferential. The strategy here is to extend the inequality (1.1) into a sub-Taylor expansion

$$f(x) \geq f(\overline{x}) + \langle y, x - \overline{x} \rangle + \frac{1}{2}\langle Q(x - \overline{x}), (x - \overline{x}) \rangle + o(\|x - \overline{x}\|^2) \qquad (1.2)$$

of order two, where Q is chosen to be a symmetric $n \times n$ matrix and $o(\|x-\overline{x}\|^2)$ is the usual Landau small–order notation. The subhessians are the set of all such operators and denoted $\partial^{2,-} f(x,y)$. This set is always unbounded. Indeed, the subjet contains many "redundant" operators, since all the negative semi–definite operators (denoted by $-\mathcal{P}(n)$) are in the recession cone of $\partial^{2,-} f(x,y)$ when viewed as a convex set of operators in the linear space of all symmetric operators (denoted by $\mathcal{S}(n)$). This cone induces a natural ordering in $\mathcal{S}(n)$. Once again the application of ideas from set–valued analysis may be used to define limiting versions of the subjet. Details are provided in the next section.

At this juncture the natural question arises as to whether there is any relationship between $\partial^{2,-}f(x,y)$, $D(\partial_p f)(\bar{x},y)$ and $D^*(\partial_p f)(x,y)$. The answer to this question is far from obvious as on first inspection we note that $\partial^{2,-}f(x,y)$ is a set of operators while $D(\partial_p f)(\bar{x},y)$ and $D^*(\partial_p f)(x,y)$ are multifunctions whose images are contained in $\mathbb{R}^n$. When $f \in \mathcal{C}^2(\mathbb{R}^n)$ all notions coincide in that

$$\begin{aligned} D^*(\nabla f(\bar{x}))(x,y)(w) &= \{\nabla^2 f(\bar{x})w\} \\ &= D(\nabla f(\bar{x}))(x,y)(w) = \mathcal{E}_{\mathcal{P}(n)}\partial^{2,-}f(x,\nabla f(\bar{x})). \end{aligned}$$

Here the equality for the coderivative utilizes the symmetry of the operator $\nabla^2 f(\bar{x})$ and $\mathcal{E}_{\mathcal{P}(n)}\partial^{2,-}f(x,\nabla f(\bar{x}))$ denotes the Pareto efficient subset with respect to the partial order induced by $\mathcal{P}(n)$. Thus a possible relationship to consider is whether $Q \in \partial^{2,-}f(x,y)$ is to have $Qw \in D^*(\partial_p f)(x,y)(w)$ or $Qw \in D(\partial_p f)(x,y)(w)$. In Rockafellar et $al.$ (1997) conditions are given (which include prox–regularity of f at x, y) under which we have the inclusion

$$D(\partial_p f)(x,y)(w) \subseteq D^*(\partial_p f)(x,y)(w) \ . \tag{1.3}$$

When the polarity of the associated tangent and normal cones is taken into account, one can see that this inclusion does imply a kind of symmetry property for the elements of the contingent graphical derivative. This prompts one to conjecture that some elements of $\{Qw \mid Q \in \partial^{2,-}f(x,\nabla f(\bar{x}))\}$ may also be contained in $D(\partial_p f)(x,y)(w)$.

In this paper we extend the inclusion (1.3) by considering the relationship of $\partial^{2,-}f(x,y)$ to $D(\partial_p f)(x,y)(w)$. To do so we need to extract the correct operators from $\partial^{2,-}f(x,y)$. The geometry of $\partial^{2,-}f(x,y)$ is of critical importance in this development. Let us now introduce some elements of this geometry necessary to state our results. The Frobenius inner product $\langle Q, B \rangle = \operatorname{tr} B^t Q$ induces the quadratic form $\langle Q, hh^t \rangle = h^t Q h = \langle Qh, h \rangle$ when applied to a rank–one operator $hh^t (\equiv h \otimes h$ using tensor product notation). We call

$$\mathcal{A} := \{Q \in \mathcal{S}(n) \mid q(\mathcal{A})(h) \geq \langle Q, hh^t \rangle \text{ for all } h\}$$

a symmetric rank–1 representer. The function $q(\mathcal{A})(h) := \sup\{\langle Q, hh^t \rangle \mid Q \in \mathcal{A}\} : \mathbb{R}^n \mapsto \bar{\mathbb{R}}$ is called the symmetric rank–1 support function. We say that a rank–one representer $\mathcal{A}$ is exposed by the rank–one support in the direction w when

$$E(\mathcal{A}, w) := \{Q \in \mathcal{A} \mid q(\mathcal{A})(w) = \langle Q, ww^t \rangle\} \neq \emptyset.$$

It was established in Eberhard, Nyblom and Ralph (1998) that $E(\mathcal{A}, w)$ may be empty in some given directions. Despite this, we show in this paper that for an arbitrary rank–one representer we have $E(\mathcal{A}, w) \neq \emptyset$ for a dense subset in $b_s^1(\mathcal{A}) := \{w \mid q(\mathcal{A})(w) < +\infty\}$.

A multifunction $\Gamma : X \rightrightarrows Y$ is generated by a set of operators $\mathcal{A} \subseteq \mathcal{S}(n)$ if $\Gamma(w) = \mathcal{A}w = \{Qw \mid Q \in \mathcal{A}\}$. In Ioffe (1981) it is established that fans and hence many derivative–like objects are, in general, not generated by a set of linear operators. In this paper we investigate an alternative notion.

Definition 1.5 *We say that a multifunction $\Gamma : X \rightrightarrows Y$ is rank–one generated by a set $\mathcal{A}$ of bounded linear operators $A : X \rightrightarrows Y$ if*

$$\Gamma(w) = \{y = Aw \mid A \in E(\mathcal{A}, w)\}.$$

Here we are concerned with the question of when a multifunction $\Omega : X \rightrightarrows Y$ admits a rank–1 selection. That is, an operator $\Gamma : X \rightrightarrows Y$ defined as above with $\Gamma(w) \subseteq \Omega(w)$ for at least a dense set of points in $\operatorname{dom}\Omega$. We can now state the inclusions proved in this paper. Suppose f is prox–regular at x with respect to y. Suppose in addition that $Q \in E(\partial^{2,-} f(x,y), w)$ and choose w such that

$$
\begin{aligned}
f_-''(\bar{x}, y, w) &:= \liminf_{\substack{w' \to w \\ t \downarrow 0}} \frac{2}{t^2}\left(f(\bar{x} + tw') - f(\bar{x}) - t\langle y, w'\rangle\right) \\
&= f_s''(x, y, w) := \min\{f_-''(\bar{x}, y, w), f_-''(\bar{x}, y, -w)\}.
\end{aligned}
$$

Then $(w, Qw) \in T_{\operatorname{Graph} \partial_p f}(x, y)$, the contingent cone to the proximal subdifferential, that is, there exists a rank–1 selection of the graphical derivative

$$\{Qw \mid Q \in E(\partial^{2,-} f(x,y), w)\} \subseteq D(\partial_p f)(x, y)(w). \tag{1.4}$$

If we posit the same assumptions as in Rockafellar *et al.* (1997), then we have immediately

$$\{Qw \mid Q \in E(\partial^{2,-} f(\bar{x}, \bar{y}), w)\} \subseteq D^*(\partial f)(\bar{x}, \bar{y})(w). \tag{1.5}$$

Thus we are able to make a selection from these non–convex set–valued derivatives using certain rank–1 exposed facets of a convex set of symmetric matrices. The construction of such matrices is often possible and provides an alternative path to the generation of elements within $D(\partial_p f)(x, y)(w)$ and $D^*(\partial f)(\bar{x}, \bar{y})(w)$. This just the kind of selection which must be calculated if numerical optimization procedures are to be developed using such second–order information. We give an example for suprema of smooth functions.

The paper is structured as follows. Section 2 details various definitions and concepts. Section 3 discusses the geometry of rank–1 representers that is necessary for later proofs and consequently is partly a survey of results, although some new results are derived. Section 4 surveys the necessary material we use from the field of generalized convexity. Again some new results appear. Sections 5 and 6 are devoted to the development of the promised comparisons and inclusions, and contains no material previously published.

Finally, in Section 7, we explore some consequences of these results in the study of sufficient conditions for isolated, strict local minima of nonsmooth functions. A number of new observations are made. The inclusions developed allow us to compare various second–order optimality conditions based on different second–order subdifferentials. To study the particular case of a standard mathematical programming problem, we make use of the penalized Lagrangian method of Andromonov (2001). We derive a sufficient optimality condition for the so called "strict local minimum of order two" of Auslender (1984), Studniarski (1986) and Ward (1995)–Ward (1994). These conditions are of a novel character for three reasons. First is that no standard regularity condition is assumed. Secondly, one does not necessarily have to demand that $\nabla L(\bar{x}, \bar{d}) = 0$, where $\nabla L(\bar{x}, \bar{d})$ denote the usual Lagrangian function of the problem and $\bar{d}$ the optimal Lagrange multipliers. Thirdly, the second–order conditions involve some interaction of the geometry of the constraint set at the optimal point $\bar{x}$ via the curvature of the active constraints in directions h for which $\langle \nabla L(\bar{x}, \bar{d}), h \rangle \leq 0$.

2 PRELIMINARIES

In this section we define and discuss a number of concepts used throughout the paper. We assume a working knowledge of variational derivatives and the associated notion of convergence of sets taken from set–valued analysis (see Rockafellar and Wets (1998) and Aubin et $al.$ (1990)). The convergence notation used is standard and as we are working in finite dimensions there is no confusion regarding which of the many possible convergence notions is being used. One may always assume we are using Kuratowski–Painlevé convergence notions (see Rockafellar and Wets (1998) and Attouch (1984)). For a family of sets $\{C^v \mid v \in W\}$, $\limsup_{v \to w} C^v$ is defined as the set consisting of all cluster points of sequences $\{u^n\}$ with $u^n \in C^{v_n}$ (for n sufficiently large) and some $v_n \to w$ as $n \to \infty$, while $\liminf_{v \to w} C^v$ consists of points u for which, for any given sequence $v_n \to w$, there exists a convergent sequence u^n, with $u^n \in C^{v_n}$ (for n sufficiently large) and with $u = \lim_{n \to \infty} u^n$. Clearly $\liminf_{v \to w} C^v \subseteq \limsup_{v \to w} C^v$. When these coincide we say that $\{C^v \mid v \in W\}$ converges to C and we write $C = \lim_{v \to w} C^v$.

Denote by $\mathcal{S}(n)$ the set of all real $n \times n$ symmetric matrices and by $\mathbb{R}_+$ (respectively $\overline{\mathbb{R}}$) the real intervals $[0, +\infty)$ (respectively $(-\infty, +\infty]$). When C is a convex set in a vector space X, denote the recession directions of C by $0^+C = \{x \in X \mid C + x \subseteq C\}$. When C is not convex, we denote the horizontal directions by $C^\infty := \{x \in X \mid \exists \mu^n \downarrow 0 \text{ and } c^n \in C \text{ such that } x = \lim_{n \to \infty} \mu^n c^n\}$.

Definition 2.1 *Let $\{f, f^v : \mathbb{R} \to \bar{\mathbb{R}}, v \in W\}$ be a family of proper extended-real-valued functions, where W is a neighbourhood of w in some topological space. Then the lower epi-limit $e\text{-li}_{v \to w} f^v$ is the function having as its epigraph the outer limit of the sequence of sets $\operatorname{epi} f^v$:*

$$\operatorname{epi}\left(e\text{-li}_{v \to w} f^v\right) := \limsup_v (\operatorname{epi} f^v).$$

The upper epi-limit $e\text{-ls}_{v \to w} f^v$ is the function having as its epigraph the inner limit of sets $\operatorname{epi} f^v$:

$$\operatorname{epi}\left(e\text{-ls}_{v \to w} f^v\right) := \liminf_v (\operatorname{epi} f^v).$$

When these two are equal, the epi-limit $e\text{--}\lim_{v \to w} f^v$ is said to exist. In this case, $\{f^v\}_{v \in W}$ is said to epi-converge to f.

As $e\text{-li}_{v \to w} f^v(x) \leq e\text{-ls}_{v \to w} f^v(x)$, we have epi-convergence of f^v occurring when $e\text{-ls}_{v \to w} f^v(x) \leq f(x)$ and $f(x) \leq e\text{-li}_{v \to w} f^v(x)$ for all x. The upper and lower epi-limits of the sequence f^v may also be defined *via* composite limits (see Rockafellar and Wets (1998)). In particular

$$
\begin{aligned}
e\text{-ls}_{v \to w} f^v(x) &= \sup_{\delta > 0} \limsup_{v \to w} \inf_{x' \in B_\delta(x)} f^v(x') \text{ and} \\
e\text{-li}_{v \to w} f^v(x) &= \sup_{\delta > 0} \liminf_{v \to w} \inf_{x' \in B_\delta(x)} f^v(x')
\end{aligned}
$$

In the introduction we defined the notions of proximal subdifferential and prox-regularity. Let us now define a number of subderivative concepts arising in nonsmooth analysis.

Definition 2.2 *Let $f : \mathbb{R}^n \mapsto \overline{\mathbb{R}}$ be lower semi–continuous, $\bar{x} \in \operatorname{dom} f$ and $\bar{x}$, z, $h \in \mathbb{R}^n$. Put $\Delta_2 f(x, t, p, u) := 2\frac{1}{t^2}[f(x + tu) - f(x) - t\langle p, u\rangle]$.*

1. *The basic subdifferential is given by*

$$\partial f(x) = \limsup_{x' \to^f x} \partial_p f(x') := \{\, \lim_{v \to \infty} z_v \mid \text{for some } z_v \in \partial_p f(x_v),\ x_v \to^f x\}.$$

2. *The lower second–order epi–derivative at $\bar{x}$ with respect to z and h is given by*

$$f''_-(\bar{x}, z, h) := \liminf_{\substack{t\downarrow 0 \\ h' \to h}} \Delta_2 f(\bar{x}, t, z, h') = \text{e–li}_{\,t\downarrow 0}\Delta_2 f(\bar{x}, t, z, h).$$

3. *A function $f : X \to \overline{\mathbb{R}}$ is said to be twice sub–differentiable (or possess a subjet) at x if the set*

$$\partial^{2,-} f(x) \;=\; \{(\nabla\varphi(x), \nabla^2\varphi(x)) \mid \varphi \in C^2(\mathbb{R}^n) \text{ and}$$
$$f - \varphi \text{ has a local minimum at } x\}$$

is nonempty. The subhessians are denoted by

$$\partial^{2,-} f(x, p) := \{Q \in \mathcal{S}(n) \mid (p, Q) \in \partial^{2,-} f(x)\}\,.$$

4. *The limiting subjet of f at x is defined to be*

$$\underline{\partial}^2 f(x) = \limsup_{u \to^f x} \partial^{2,-} f(u).$$

5. *The set $\underline{\partial}^2 f(x, p) = \{Q \in \mathcal{S}(n) \mid (p, Q) \in \underline{\partial} f(x)\}$ is called the limiting subhessians of f.*

As $-\partial^{2,-} f(x, -p) = \partial^{2,+} f(x, p)$ we study only the subjet. One may define corresponding superjet quantities by reversing the inequality. It must be stressed that these quantities may not exists everywhere but $\partial^{2,-} f(x)$ and $\partial_p f(x)$ are defined densely. If $f''_-(\bar{x}, p, h)$ is finite, then $f'_-(\bar{x}, h) = \langle p, h\rangle$, where the first–order lower epi–derivative is defined by

$$f'_-(\bar{x}, h) = \liminf_{t\downarrow 0,\ h' \to h} \frac{1}{t}\left(f(x + th') - f(x)\right).$$

Another commonly used concept is that of the upper second–order epi–derivative at $\bar{x}$ with respect to z and h, which is given by

$$f''_+(\bar{x}, z, h) := \limsup_{t\downarrow 0} \inf_{h' \to h} \Delta_2 f(\bar{x}, t, z, h') := \text{e–ls}_{\,t\downarrow 0}\Delta_2 f(\bar{x}, t, z, h).$$

When $f''_-(\bar{x}, p, h) = f''_+(\bar{x}, p, h) := f''_e(\bar{x}, p, h)$, we say f possesses a second–order epi–derivative.

We recall that there is a one–to–one correspondence between subsets of $X \times Y$ and the graphs $\operatorname{Graph} F := \{(x, y) \mid y \in F(x)\}$ of multifunctions $F : X \rightrightarrows Y$. The next concept has been extensively studied (see, for example, Rockafellar (1989) and Rockafellar (1988)).

Definition 2.3 *The intermediate cone is given by*

$$T^i_{\operatorname{Graph} F}(x, y) := \liminf_{t \downarrow 0} \frac{\operatorname{Graph} F - (x, y)}{t}.$$

A multifunction is said to be proto–differentiable if this cone coincides with the contingent cone, that is,

$$T_{\operatorname{Graph} F}(x, y) = T^i_{\operatorname{Graph} F}(x, y).$$

Using this tangent cone and the contingent tangent cone one may define many derivative like concepts using the multifunction $F(x) := f(x) + [0, +\infty)$ with graph $\operatorname{Graph} F = \operatorname{epi} f$. It is well–known that the contingent cone to $\operatorname{epi} f$ corresponds to the epigraph of the function $h \mapsto f'_-(x, h)$ (see Aubin *et al.* (1990), Rockafellar (1989) and Rockafellar (1988)). Also proto–differentiability of $\operatorname{epi} f$ corresponds to first–order epi–differentiability of f.

The normal cone to a set S is given by

$$N_S(\bar{x}) = \limsup_{x(\in S) \to \bar{x}} (T_S(x))^0.$$

The contingent coderivative of a multifunction $F : X \rightrightarrows Y$ at $(x, y) \in \operatorname{Graph} F$ in the direction $w \in X$ is defined by

$$\hat{D}^* F(x, y)(w) := \{p \in Y \mid \langle p, h \rangle \leq \langle w, z \rangle; \, \forall (h, z) \in T_{\operatorname{Graph} F}(x, y)\} \qquad (2.1)$$

The Ioffe–Mordukhovich coderivative is generated by the contingent coderivative through

$$D^* F(x, y)(w) = \limsup_{\substack{(x', y')(\in \operatorname{Graph} F) \to (x, y) \\ w' \to w}} \hat{D}^* F(x', y')(w'). \qquad (2.2)$$

Note that (2.2) is consistent with Definition 1.3 in that

$$\hat{D}^* F(x, y)(w) = \{p \in Y \mid (p, -w) \in (T_{\operatorname{Graph} F}(x, y))^0\}, \text{ where}$$

$$(T_{\operatorname{Graph} F}(x, y))^0 := \{(r, s) \mid \langle (r, s), (h, z) \rangle \leq 0; \forall (h, z) \in T_{\operatorname{Graph} F}(x, y)\}.$$

In finite dimensions the subjet concept is closely related to the proximal differential. Denote by $\mathcal{P}(n)$ the cone in $\mathcal{S}(n)$ of all positive semi–definite symmetric matrices. It is well–known (see Penot (1994/1)) that $\partial^{2,-} f(\bar{x}) \neq \emptyset$ is equivalent to $\partial_p f(\bar{x}) \neq \emptyset$ which is in turn equivalent to the existence of $(p, X) \in \mathbb{R}^n \times \mathcal{S}(n)$ such that

$$f''_-(\bar{x}, p, h) \geq \langle X, hh^t \rangle \text{ for all } h.$$

This prompts the following definitions (see Eberhard, Nyblom and Ralph (1998)).

Definition 2.4 *Denote by $\mathcal{M}(n, m)$ the class of real $n \times m$ matrices.*

1. *The rank–one hull of a set $\mathcal{A} \subseteq \mathcal{M}(n, m)$ is given by*

$$\mathcal{A}^1 = \{A \in \mathcal{M}(n, m) \mid \langle A, vu^t \rangle \leq q(\mathcal{A})(u, v) \text{ for all } v \in \mathbb{R}^m, u \in \mathbb{R}^n\},$$

   where $q(\mathcal{A})(u, v) := \sup\{\langle Q, vu^t \rangle \mid Q \in \mathcal{A}\}$.

2. *A set $\mathcal{A}$ is said to be a rank–one representer if $\mathcal{A}^1 = \mathcal{A}$.*

3. *When $\mathcal{A} \subseteq \mathcal{S}(n)$, the set of real symmetric matrices, we denote the symmetric rank–1 support by $q(\mathcal{A})(u) := \sup\{\langle Q, uu^t \rangle \mid Q \in \mathcal{A}\}$ and the symmetric rank–1 hull by*

$$\mathcal{A}^1_s = \{A \in \mathcal{S}(n) \mid \langle A, uu^t \rangle \leq q(\mathcal{A})(u) \text{ for all } u \in \mathbb{R}^n\}.$$

4. *The rank–1 (resp. ε–rank–1) supported points in the direction u are given respectively by*

$$E(\mathcal{A}, u) := \{A \in \mathcal{A} \mid \langle A, uu^t \rangle = q(\mathcal{A})(u)\} \text{ and}$$
$$E_\varepsilon(\mathcal{A}, u) := \{A \in \mathcal{A} \mid \langle A, uu^t \rangle \geq q(\mathcal{A})(u) - \varepsilon\} \, .$$

5. *When $-\mathcal{P}(n) \subseteq 0^+\mathcal{A} \subseteq \mathcal{S}(n)$ we define the symmetric rank–one barrier cone as $b^1_s(\mathcal{A}) := \{u \in \mathbb{R}^n \mid q(\mathcal{A})(u) < \infty\}$.*

Remark 2.1 *If we restrict attention to $\mathcal{S}(n)$ and sets $\mathcal{A}$ such that $-\mathcal{P}(n) \subseteq 0^+\mathcal{A}$, then $q(\mathcal{A})(u, v) = +\infty$ unless $v = u$. Thus in this case we need only consider the symmetric supports $q(\mathcal{A})(u)$ and $\mathcal{A}^1_s = \mathcal{A}^1$. Indeed we always have $-\mathcal{P}(n) \subseteq 0^+\{A \in \mathcal{S}(n) \mid \langle A, uu^t \rangle \leq q(\mathcal{A})(u) \text{ for all } u \in \mathbb{R}^n\}$ and $q(\mathcal{A} - \mathcal{P}(n))(u) = q(\mathcal{A})(u)$. Even $E(\mathcal{A}, u)$ contains recession directions in $-\mathcal{P}(n)$.*

The subjet is always a closed convex set of matrices while $\underline{\partial}^2 f(\bar{x}, p)$ may not be convex, just as $\partial_p f(\bar{x})$ is convex while $\partial f(\bar{x})$ often is not. Eberhard, Nyblom and Ralph (1998) observed that

$$
\begin{aligned}
q\left(\partial^{2,-} f(\bar{x}, p)\right)(u) &= \min\{f''_-(\bar{x}, p, u), f''_-(\bar{x}, p, -u)\} \\
&= \liminf_{\substack{t \to 0 \\ h' \to h}} \left(\frac{2}{t^2}\right) (f(\bar{x} + th') - f(\bar{x}) - t\langle p, h'\rangle) \\
&:= f''_s(\bar{x}, p, u).
\end{aligned}
$$

Hence if we work with subjets we are in effect dealing with objects dual to the lower, symmetric, second-order epi-derivative. The study of second–order directional derivatives as dual objects to some kind of second order subdifferential was begun in J.–B. Hiriart–Urruty (1986), Seeger (1986) and Hiriart–Urruty *et al.* (1989). These works studied the particular case where f is a convex function.

Definition 2.5 *The function f is subdifferentially continuous at $\bar{x}$ for $\bar{v}$, where $\bar{v} \in \partial f(\bar{x})$, if for every $\delta > 0$ there exists $\varepsilon > 0$ such that $|f(x) - f(\bar{x})| \le \delta$ whenever $|x - \bar{x}| \le \varepsilon$ and $|v - \bar{v}| \le \varepsilon$ with $v \in \partial f(x)$.*

In general we have for all h (see Ioffe and Penot (1997)) that

$$
\begin{aligned}
q\left(\underline{\partial}^2 f(\bar{x}, p)\right)(h) &= \sup\{\langle Q, hh^t\rangle \mid Q \in \underline{\partial}^2 f(\bar{x}, p)\} \le f^{\uparrow\uparrow}(\bar{x}, p, h) \\
&:= \limsup_{(x', p') \to_{S_p(f)} (\bar{x}, p), \, t \downarrow 0} \inf_{u' \to h} \Delta_2 f(x', t, p', u'),
\end{aligned}
$$

where $(x', p') \to_{S_p(f)} (\bar{x}, p)$ means $x' \to \bar{x}$, $f(x') \to f(\bar{x})$, $z' \in \partial_p f(x')$ and $p' \to p$. Equality holds when f is prox–regular and subdifferentially continuous (see Corollary 6.1 of Eberhard (2000)).

3   CHARACTERIZATION OF SUPPORTED OPERATORS

In this section we outline some important facts regarding the geometry of rank–1 sets which are relevant to later sections. We show that there is a dense set of directions $h \in b^1_s(\mathcal{A})$ for which a rank–1 representer $\mathcal{A}$ has $E(\mathcal{A}, h) \ne \emptyset$. We discuss also some results from previous papers relevant to the characterization of rank–1 exposed operators, as these figure strongly in the later development of the paper. One can completely characterise symmetric rank–one supports. The following characterization may be found in Eberhard, Nyblom and Ralph

(1998). This result generalizes those of J.-B. Hiriart–Urruty (1986) and Seeger (1986) which treats only the case when the symmetric rank–1 support happens to be a convex function. The results of J.-B. Hiriart–Urruty (1986) and Seeger (1986) are sufficient to study the second–order derivative notion for convex functions (see, for example, Seeger (1992), Hiriart–Urruty *et al.* (1989) and Seeger (1994)), while Theorem 3.1 below is applicable to a more general study of second–order epi–derivatives. Important as these earlier works are, they do not figure in the present development.

Theorem 3.1 *Let $p : \mathbb{R}^n \mapsto \bar{\mathbb{R}}$ be proper (that is, $p(u) \neq -\infty$ anywhere). For $u, v \in \mathbb{R}^n$, define $q(u, v) = \infty$ if u is not a positive scalar multiple of v or vice versa, and $q(\alpha u, u) = q(u, \alpha u) = \alpha p(u)$ for $\alpha \geq 0$.*

Then q is a symmetric rank–one support of a set $\mathcal{A} \subseteq \mathcal{S}(n)$ with $-\mathcal{P}(n) \subseteq 0^+ \mathcal{A}$ if and only if

1. p is positively homogeneous of degree two;

2. p is lower semicontinuous;

3. $p(-u) = p(u)$.

Even in finite dimensions there may not exist, in every direction, an operator in a given rank–1 representer achieving the symmetric rank–one support. Thus some caution is required in the subsequent development when using rank–1 exposed facets and operators.

As noted earlier we endow $\mathcal{S}(n)$ with the Frobenius inner product $\langle Q, A \rangle = \operatorname{tr} A^t Q$. The natural norm induced by this inner product is the so–called projective norm $\|A\|_{\text{proj}}$, given by

$$\|A\|_{\text{proj}} = \sup_{\|Q\|_{\text{op}} \leq 1} \langle Q, A \rangle = \inf \left\{ \sum_i \|v_i\| \|u_i\| \mid \sum_i v_i u_i^t = A \right\}.$$

Here $\|Q\|_{\text{op}} = \sup_{\|x\|_2 \leq 1} \|Qx\|_2$ denotes the usual operator norm induced by the Euclidean norm. One may use the symmetric rank–1 support to reconstruct the support of the symmetric rank–1 representer as a convex subset of $\mathcal{S}(n)$, confirming that all relevant data for this problem is contained in such rank–one facets. We quote the following result from Eberhard (2000).

Theorem 3.2 *Let* $\dim \mathcal{S}(n) = m \ (= \frac{n}{2}(n+1))$. *Then if* $\mathcal{A}$ *is a rank–one representer which has* $-\mathcal{P}(n) \subseteq 0^+\mathcal{A}$, *we have for any* $Q \in \mathcal{P}(n)$ *that*

$$S(\mathcal{A}, Q) \ := \ \sup\{\langle A, Q\rangle \mid A \in \mathcal{A}\}$$

$$= \ \min\left\{\sum_{i=1}^{l} q(\mathcal{A})(u_i) \mid \sum_{i=1}^{l} u_i u_i^t = Q \text{ for some } l \le m+1\right\} \ (3.1)$$

On the space $\mathcal{S}(n)$ we may define the barrier cone to $\mathcal{A}$ (as a convex subset) by $b(\mathcal{A}) := \{P \in \mathcal{S}(n) \mid S(\mathcal{A}, P) < \infty\}$.

Recall that an exposed point of $\mathcal{A}$ is the unique maximizer Q in $\mathcal{A}$ of a linear function $\langle \cdot, P\rangle$. Recall also that $\mathcal{P}(n)$ induces a natural ordering in the space $\mathcal{S}(n)$ with respect to which we may define a Pareto (or undominated) subset of the rank–1 representer $\mathcal{A}$. We quote next another result from Eberhard (2000) that we require in this section.

Corollary 3.1 *Any point* $Q \in \mathcal{A}$ *which is supported by a linear functional* $\langle \cdot, P\rangle$ *on* $\mathcal{S}(n)$ *also lies on at least as many rank–one supporting hyperplanes as the rank of* P. *In particular if* Q *is an exposed point of* $\mathcal{A}$, *then it is a Pareto maximizer of* $E(\mathcal{A}, h)$ *for some* h.

In convex analysis the concept of supporting points plays an important role. Unfortunately even when $b_s^1(\mathcal{A}) = \mathbb{R}^n$ we are not assured of the existence of a supporting point for the rank–one support. See Eberhard, Nyblom and Ralph (1998) for an example of a rank–one representer with $E(\mathcal{A}, u) = \emptyset$ for a given direction $u \in b_s^1(\mathcal{A}) = \mathbb{R}^n$. Thus it is not immediately clear that rank–one supported operators are sufficiently numerous to provide a description of the rank–one representer. The best that can be hoped for is $E(\mathcal{A}, u) \ne \emptyset$ for a dense set in $b_s^1(\mathcal{A})$.

Denote by $\partial S(\mathcal{A}, Q)$ the subdifferential (in the sense of ordinary convex analysis) of the convex support $Q' \mapsto S(\mathcal{A})(Q')$ at a given point $Q \in \mathcal{S}(n)$. It is a standard result in convex analysis that $\partial S(\mathcal{A}, Q) = \{A \in \mathcal{A} \mid S(\mathcal{A}, Q) = \langle A, Q\rangle\}$. It is well–known that $\partial S(\mathcal{A}, Q)$ reduces to a singleton almost every-where in $\operatorname{int} b(\mathcal{A})$ at which $\nabla S(\mathcal{A}, Q) \in \operatorname{bd} \mathcal{A}$ is the unique supporting point, that is, the intersection of the supporting hyperplane and the set $\mathcal{A}$). Even when $\operatorname{int} b(\mathcal{A}) = \emptyset$, we know from Section 20D of Holmes (1975) that any lower semi–continuous convex function (on a Banach space) still has $\partial S(\mathcal{A}, Q) \ne \emptyset$ densely on its effective domain. We need the following lemma.

Lemma 3.1 *Suppose that $Q = \sum_{i \in F} u_i u_i^t \in B_{\delta^2}(uu^t) \cap \mathcal{P}(n)$, where $u \neq 0$ and F is a finite index set. Then there exist α_i such that $|\sum_{i \in F} \alpha_i^2 - 1| \leq \frac{\delta^2}{\|u\|^2}$ and for $i \in F$ we have $\|u_i - \alpha_i u\| \leq \delta$.*

Proof We have $Q = \sum_{i \in F} u_i u_i^t = uu^\dagger + Q_\delta$, where $Q_\delta \in B_{\delta^2}(0)$. It follows that for all $v \in (u)^\perp$ with $\|v\| = 1$, we have $\sum_{i \in F} \langle u_i, v \rangle^2 = \langle Q_\delta, vv^t \rangle$ and so $\sum_{i \in F} \langle u_i, v \rangle^2 \leq \delta^2$, as $\langle Q_\delta, vv^t \rangle \leq \|Q_\delta\| \|v\|^2 < \delta^2$. By a theorem of Hörmander (see Holmes (1975)) we have for $M = \{\alpha u \mid \alpha \in \mathbb{R}\}$ that there exists $v \in (u)^\perp \cap B_1(0)$ such that

$$d(u', M) = \sup_{w \in B_1(0)} \{\langle u', w \rangle - S(M, w)\} = \langle u', v \rangle. \tag{3.2}$$

The supremum is attained as $B_1(0)$ is compact and $S(M, \cdot)$ is lower semi-continuous.

Let α_i attain the minimum in $\min_\alpha \|u_i - \alpha u\|$. Using (3.2) for each i, we have $v_i \in (u)^\perp \cap B_1(0)$ such that $d(u_i, M) = \langle u_i, v_i \rangle$ and so

$$\|u_i - \alpha_i u\|^2 = \langle u_i, v_i \rangle^2 \leq \sum_{j \in F} \langle u_j, v_i \rangle^2 \leq \delta^2.$$

Finally take zz^t with $z = \nabla \|u\|$ and note that $\langle u_i, z \rangle = \alpha_i u$. We compose zz^t with $Q = \sum_{i \in F} u_i u_i^t = uu^t + Q_\delta$ to get

$$\sum_{i \in F} \langle u_i, z \rangle^2 = \|u\|^2 + \langle Q_\delta, zz^t \rangle$$

and so $\delta^2 \geq |\sum_{i \in F} \alpha_i^2 \|u\|^2 - \|u\|^2|$, which implies $\frac{\delta^2}{\|u\|^2} \geq |\sum_{i \in F} \alpha_i^2 - 1|$.

We may prove the promised density of directions in $b_s^1(\mathcal{A})$ that expose $\mathcal{A}$ in a symmetric rank–1 sense.

Theorem 3.3 *Suppose that $\mathcal{A}$ is a rank-one representer with $-\mathcal{P}(n) \subseteq 0^+ \mathcal{A}$ and with $b^1(\mathcal{A}) \neq \emptyset$. Then the set of directions u for which $E(\mathcal{A}, u) \neq \emptyset$ is dense in $b^1(\mathcal{A})$.*

Proof Take an arbitrary nonzero $u \in b(\mathcal{A})$. As $\partial S(\mathcal{A}, P) \neq \emptyset$ densely there exists a sequence $P_m \to uu^t (\neq 0)$, where $P_m \in \mathcal{P}(n)$ and

$$A_m \in \partial S(\mathcal{A}, P_m) = \{A \in \mathcal{A} \mid S(\mathcal{A}, Q) = \langle A, Q \rangle\}.$$

By Lemma 3.1 we have for any representation $P_m = \sum_{i=1}^l u_i^m (u_i^m)^t$ (with $l \leq \dim \mathcal{S}(n) + 1$) that $u_i^m \to \alpha_i u$ (taking a subsequence and renaming if necessary).

Here $\sum_{i=1}^{l} \alpha_i^2 = 1$ are the accumulation points of the α's in Lemma 3.1. Thus for any $\varepsilon > 0$ we have for $m \geq m_0$ that $\|u_i^m - \alpha_i u\| \leq \alpha_i \varepsilon$ for all i with $\alpha_i \neq 0$ and $\|u_i^m\| \leq \varepsilon$ for all i with $\alpha_i = 0$. For each m, the supporting point $A_m \in \mathcal{A}$ has $S(\mathcal{A}, P_m) = \langle A_m, P_m \rangle$. By Corollary 3.1 there exists a representation $P_m = \sum_{i=1}^{l} u_i^m (u_i^m)^t$, containing at least as many linearly independent u_i^m as the rank of P_m but no more than $\dim \mathcal{S}(n) + 1$, such that for each i

$$q(\mathcal{A}) \left(\frac{u_i^m}{\alpha_i} \right) = \left\langle A_m, \left(\frac{u_i^m}{\alpha_i} \right) \left(\frac{u_i^m}{\alpha_i} \right)^t \right\rangle,$$

where $\| \frac{u_i^m}{\alpha_i} - u \| \leq \varepsilon$ whenever $\alpha_i \neq 0$. In particular we have $A_m \in E(\mathcal{A}, \frac{u_i^m}{\alpha_i})$. It remains only to establish that at least one $\alpha_i \neq 0$. But if we assume to the contrary that all $\alpha_i = 0$, then for all i we have $u_i^m \to 0$ and so $P_m = \sum_{i=1}^{l} u_i^m (u_i^m)^t \to 0$. This contradicts $u \neq 0$, so we are done.

Remark 3.1 *In Eberhard (2000), Corollary 6.2 it is shown that if f is prox-regular at $\bar{x}$ for $p \in \partial f(\bar{x})$ and in addition possesses a second–order epi-derivative at $\bar{x}$, then $E\left(\partial^{2,-} f(\bar{x}, p), h\right) \neq \emptyset$ for all directions in the relative interior of the set $b^1\left(\partial^{2,-} f(\bar{x}, p)\right)$.*

As noted earlier the identification of rank–1 exposed points is of importance in the development of later results. With regard to the supported points of a rank–1 representer, we have the following result (see Eberhard, Nyblom and Ralph (1998), Theorem 4 and Eberhard (2000), Theorem 4.2).

Theorem 3.4 *Suppose that $\mathcal{A} \subseteq \mathcal{S}(X)$ is a rank–one representer with $-\mathcal{P}(X) \subseteq 0^+ \mathcal{A}$ and $u \in b(\mathcal{A})$. Then $(2Au, 2A) \in \partial^{2,-} S(\mathcal{A})(u)$ and $A \in \mathcal{A}$ if and only if $A \in E(\mathcal{A}, u)$.*

It should be noted that Eberhard, Nyblom and Ralph (1998) contains a counterexample to the possible conjecture that $(2Au, 2A) \in \partial^{2,-} S(\mathcal{A})(u)$ implies $A \in \mathcal{A}$.

We finish this section by consider what effect the infimal convolution (or Moreau envelope) has on the rank–1 support of a rank–1 representer and its rank–1 exposed operators. Recall that the infimal convolution (Moreau envelope) of a function f is defined *via*

$$f_\lambda(x) := \inf_u \left\{ f(u) + \frac{1}{2\lambda} \|x - u\|^2 \right\} \leq f(x) .$$

This is finite when the function f is proper, lower semi–continuous and minorized by a quadratically function $\gamma - \frac{1}{2r}\| \cdot -y\|^2$ with $r > 0$ (see Rockafellar and Wets (1998) Example 1.44). In this case each f_λ is proper and lower semi–continuous. The supremum of all parameters r is called the prox–threshold of f. It is well–known that f_λ is pointwise nondecreasing with decreasing λ and epi–convergent to f as $\lambda \to 0$ (see Rockafellar and Wets (1998) Example 1.44 and Proposition 7.4 (d)). Note that even when $\operatorname{int} \operatorname{dom} f = \emptyset$ we have f_λ finite–valued everywhere (for λ sufficiently small) as long as f is quadratically minorized (or prox–bounded). Thus $\operatorname{dom} f_\lambda = \mathbb{R}^n$. Moreover f_λ can be shown to be locally Lipschitz continuous. In Eberhard, Nyblom and Ralph (1998) we may find the following series of results.

Definition 3.1 *Suppose that $\mathcal{A}$ is a rank–1 representer. Put*

$$\mathcal{A}_\lambda = \left\{ Q \in \mathcal{S}(n) \mid \langle Q, uu^t \rangle \leq 2q_\lambda(\frac{1}{2}\mathcal{A})(u) \text{ for all } u \in \mathbb{R}^n \right\},$$

where

$$
\begin{aligned}
2q_\lambda\left(\frac{1}{2}\mathcal{A}\right)(u) &= 2\inf_{w \in \mathbb{R}^n}\left\{ q\left(\frac{1}{2}\mathcal{A}\right)(w) + \frac{1}{2\lambda}\|w - u\|^2 \right\} \\
&= \inf_{w \in \mathbb{R}^n}\left\{ q(\mathcal{A})(w) + \frac{1}{\lambda}\|w - u\|^2 \right\} = q_{\lambda/2}(\mathcal{A}) .
\end{aligned}
$$

Write $\partial_\lambda^{2,-} f(x,0) = (\partial^{2,-} f(x,0))_\lambda$. Not surprisingly we have the following (see Eberhard, Nyblom and Ralph (1998)).

Theorem 3.5 *Let $\mathcal{A}$ be a rank–one representer with $-\mathcal{P}(n) \subseteq 0^+ \mathcal{A}$. Then the infimal convolution of the support $u \mapsto q(\mathcal{A})(u)$ is also the support of a rank–1 representer. That is, there exists a rank–1 representer $\mathcal{A}_\lambda$ for which $-\mathcal{P}(n) = 0^+ \mathcal{A}_\lambda$ and*

$$q_{\lambda/2}(\mathcal{A})(u) = q(\mathcal{A}_\lambda)(u).$$

Note that, since $q(\mathcal{A}_\lambda)(u) = q_\lambda(\mathcal{A})(u) \leq q(\mathcal{A})(u)$ for all u, we have $\mathcal{A}_\lambda \subseteq \mathcal{A}$. We now quote a couple of results from Eberhard, Nyblom and Ralph (1998) detailing the approximation properties of the infimal convolution which will be used in subsequent proofs. In particular we have the following.

Theorem 3.6 *Suppose that $f : \mathbb{R}^n \mapsto \overline{\mathbb{R}}$ is a lower semi-continuous, prox-bounded function. We have for all $\lambda > 0$ sufficiently small that*

$$\left(\frac{1}{2}f_s''(x,0,\cdot)\right)_\lambda(h) = \frac{1}{2}(f_\lambda)_s''(x,0,h)$$

and so

$$\partial^{2,-} f_\lambda(x,0) = \partial_\lambda^{2,-} f(x,0) . \tag{3.3}$$

Some related observations may be found in Poliquin and Rockafellar (1996). The limiting behaviour is also good (see Eberhard, Nyblom and Ralph (1998)).

Theorem 3.7 *Suppose that* $f : \mathbb{R}^n \mapsto \overline{\mathbb{R}}$ *is a lower semi-continuous function which is prox–bounded. Then for all* $\lambda > 0$ *sufficiently small*

$$\partial^{2,-} f_\lambda(\bar{x},0) \subseteq \partial^{2,-} f(\bar{x},0) \quad \text{and} \quad \limsup_{\lambda \to 0} \partial^{2,-} f_\lambda(\bar{x},0) = \partial^{2,-} f(\bar{x},0).$$

Such smoothing may be viewed alternatively in terms of infimal convolution smoothings of the associated quadratic forms rather than the smoothing of the rank–one support (see Eberhard, Nyblom and Ralph (1998), Proposition 7). An alternative approach to these results follows from the following observation communicated to us by A. Seeger. Suppose star denotes the convex conjugate. The form $q_Q(u) := \langle Q, uu^t \rangle$ has an infimal convolution characterised by

$$(q_Q)_{\lambda/2}(h) = 2\left(q_{\frac{1}{2}Q}\right)_\lambda(h) = \lambda^{-1}\|h\|^2 - 2\lambda q^*_{I+\lambda Q}(h), \text{ where} \tag{3.4}$$

$$q^*_{I+\lambda Q}(h) = \begin{cases} -\infty & \text{if } I + \lambda Q \notin \mathcal{P}(n) \\ q_{(I+\lambda Q)^+}(h) & \text{if } I + \lambda Q \in \mathcal{P}(n)\setminus(\text{int } \mathcal{P}(n)), \, h \in \text{Im } (I + \lambda Q) \\ q_{(I+\lambda Q)^{-1}}(h) & \text{if } I + \lambda Q \in \text{int } \mathcal{P}(n) \\ +\infty & \text{if } h \notin \text{Im } (I + \lambda Q). \end{cases}$$

Here Im $(I + \lambda Q)$ denotes the image or range and $(I+\lambda Q)^+$ the Moore-Penrose inverse. Thus $I + \lambda Q \in \text{int } \mathcal{P}(n)$ is a necessary and sufficient condition for $(q_Q)_{\lambda/2} = q_{Q_\lambda}$, where

$$Q_\lambda := Q(I + \lambda Q)^{-1} = (I + \lambda Q)^{-1} Q = \frac{1}{\lambda}\left(I - (I + \lambda Q)^{-1}\right) . \tag{3.5}$$

Such results have a long history when one notes that Q_λ is constructed *via* a 'parallel sum' (see Mazure (1996), Anderson *et al.* (1969) and Seeger (1991)). The following result is very useful in subsequent proofs (see Eberhard, Nyblom and Ralph (1998)).

Proposition 3.1 *Suppose that* $\mathcal{A}$ *is a rank–one representer with* $-\mathcal{P}(n) \subseteq 0^+\mathcal{A}$. *Then for* $\lambda > 0$ *sufficiently small we have*

$$(\mathcal{A})_\lambda = \overline{\{Q_\lambda \mid Q \in \mathcal{A} \text{ and } Q_\lambda \text{ is a quadratic form}\}} - \mathcal{P}(n).$$

The next result allows us to study $E(\mathcal{A}, \cdot)$ *via* the more regular $E(\mathcal{A}_\lambda, \cdot)$.

Proposition 3.2 *Suppose $\mathcal{A}$ is a non–empty rank–one representer. If $Q \in E(\mathcal{A}, h)$, then for all $\lambda > 0$ sufficiently small we have $Q_\lambda \in E(\mathcal{A}_\lambda, h_\lambda)$, where $h_\lambda = (I + \lambda Q)\, h \to h$ as $\lambda \to 0$ and $q(\mathcal{A}_\lambda)(h_\lambda) = q(\mathcal{A})(h) + \lambda \|Qh\|^2$.*

Proof Since $\langle Q, hh^t \rangle \leq q(\mathcal{A})(h)$ for all h, application of the infimal convolution (with parameter λ^{-1}) to both sides of this inequality provides $\langle Q_\lambda, uu^t \rangle \leq q_\lambda(\mathcal{A})(u) = q(\mathcal{A}_\lambda)(u)$ for all u. Thus $Q_\lambda \in \mathcal{A}_\lambda$. For $\lambda^{-1} > |\min\{\mu \mid \mu$ is an eigenvalue of $Q\}|$, the matrix $\langle Q, \eta\eta^t \rangle + \lambda^{-1}\|\eta\|^2$ is positive definite and so the problem $\inf_\eta \left\{ \langle Q, \eta\eta^t \rangle + \frac{1}{\lambda}\|h - \eta\|^2 \right\}$ has a unique solution at $\eta = (I + \lambda Q)^{-1}\, h$. In particular, for any fixed $h \in \mathbb{R}^n$, we have for each $\lambda^{-1} > \|Q\|_2$ that $h_\lambda := (I + \lambda Q)\, h$ has $h = (I + \lambda Q)^{-1}\, h_\lambda$. Thus

$$
\begin{aligned}
\langle Q_\lambda, h_\lambda h_\lambda^t \rangle &= \langle Q, hh^t \rangle + \frac{1}{\lambda}\|h_\lambda - h\|^2 \\
&= \langle Q, hh^t \rangle + \frac{1}{\lambda}\| (I - (I + \lambda Q))\, h \|^2 = \langle Q, hh^t \rangle + \lambda \|Qh\|^2.
\end{aligned}
$$

When $Q \in E(\mathcal{A}, h)$ we have $q(\mathcal{A})(h) = \langle Q, hh^t \rangle$ and so

$$
\begin{aligned}
\langle Q_\lambda, h_\lambda h_\lambda^t \rangle &= \langle Q, hh^t \rangle + \lambda \|Qh\|^2 \\
&= q(\mathcal{A})(h) + \frac{1}{\lambda}\|h_\lambda - h\|^2 \geq q(\mathcal{A}_\lambda)(h_\lambda),
\end{aligned}
$$

which implies $Q_\lambda \in E(\mathcal{A}_\lambda, h_\lambda)$.

4   GENERALIZED CONVEXITY AND PROXIMAL SUBDERIVATIVES

We shall need to use the notion of a generalized subgradient with respect to certain generalized convexity generating classes of functions Φ. The following concept was probably first introduce in Ky Fan (1963).

Definition 4.1 *Let Φ be subset of the mappings $\varphi : \mathbb{R}^n \to \overline{\mathbb{R}}$. A function f is called Φ–convex if $f(x) = \sup_{\varphi \in \Phi'} \varphi(x)$ for some subclass $\Phi' \subseteq \Phi$.*

It is widely recognised that the differential information extracted from the functions in such Φ–subdifferentials provided information regarding certain subdifferentials of nonsmooth analysis (see Rockafellar and Wets (1998)). This section concerns itself with the problem of quantifying this relationship more formally.

    Notions of abstract convexity were introduced in Janin (1973) and developed later in Balder (1977) and Dolecki *et al.* (1978). This approach has a long history. Many earlier papers were concerned with the case of paraconvex/paraconcave functions (or strong/weak convexity) (see Penot and Volle (1988) and Vial (1983)). Essentially this corresponds to taking the class Φ to consist of quadratics with a fixed maximum negative curvature. This restriction is dispensed with in Dolecki *et al.* (1978) and greatly generalized *via* the use of abstract 'dualities' (see Martinez–Legaz (1988) and Martinez–Legaz and Singer (1995)). This approach is detailed in the text of Singer (1997). This approach is developed in a different direction and applied to many optimization problems in Rubinov (2000).

The approach discussed above is more general than required here as we require only the use of abstract conjugations in the spirit of Pallaschke and Rolewicz (1998). This approach has been exploited in Eberhard, Nyblom and Ralph (1998)–Eberhard (2000) to study the approximate subdifferential and consequently the basic subdifferential along with certain second–order derivative concepts. It has long been recognised that abstract convexity gives information about the subdifferentials of nonsmooth analysis. One of the contributions of Eberhard and Nyblom (1998) was to show that, in finite dimensions, the study of the Φ_2–subdifferential was equivalent to the study of the proximal subdifferential for the class of lower semi–continuous, prox–bounded functions. To establish this one must show how to extend the local inequalities (1.1) and (1.2) to ones which hold globally. The reason for doing this is that generalized conjugates require global suprema rather than local ones. We present a result of this kind in this section but defer the long proof to an appendix.

As we are mainly concerned with sub-Taylor expansions, the class

$$
J_2(\bar{x}) := \left\{ \varphi(x) := \alpha + \langle p, x - \bar{x} \rangle + \frac{1}{2} \langle Q, (x - \bar{x})(x - \bar{x})^t \rangle - r(\|\bar{x} - x\|)\|\bar{x} - x\|^2 \right.
$$

$$
\left. \mid \text{with } \varphi \in C^2(\mathbb{R}^n) \text{ and } r : R_+ \to \mathbb{R} \text{ and } \lim_{t \downarrow 0} r(t) = 0 \right\} .
$$

will be of importance in subsequent proofs. Clearly it follows that $\varphi \in \partial_{J_2(\bar{x})} f(\bar{x})$ implies $(\nabla \varphi(\bar{x}), \nabla^2 \varphi(\bar{x})) \in \partial^{2,-} f(\bar{x})$.

Lemma 4.1 *Suppose there exists a function* $w(\cdot) : \mathbb{R}_+ \mapsto \mathbb{R}$ *with* $\lim_{t \downarrow 0} w(t) = 0$ *such that*

$$
f(y) - f(\bar{x}) \geq \langle p, y - \bar{x} \rangle + \frac{1}{2} \langle Q, (y - \bar{x})(y - \bar{x})^t \rangle - w(\|y - \bar{x}\|)\|y - \bar{x}\|^2. \quad (4.1)
$$

Then there exists a function $\varepsilon(\cdot) : \mathbb{R}_+ \mapsto \mathbb{R}_+$ *with* $\varepsilon \geq w$, $\lim_{t\downarrow 0} \varepsilon(t) = 0 = \varepsilon(0)$ *and a function* $r : y \mapsto \varepsilon(\|y - \bar{x}\|)\|y - \bar{x}\|^2 \in \mathcal{C}^2(\mathbb{R}^n)$ *satisfying* $\square(r, \bar{x}) :=$ $(\nabla r(\bar{x}), \nabla^2 r(\bar{x})) = (0, 0)$ *and (4.1) is satisfied with* ε *in place of* w.

The proof is deferred to the appendix. We shall provide a proper introduction to the operator $\square$ in Definition 4.4.

Corollary 4.1 *Suppose* $f : \mathbb{R}^n \to \mathbb{R}$ *is prox–bounded (or* Φ_2 *bounded). If* $(p, X) \in \partial^{2,-} f(\bar{x})$, *then there exists* $\bar{r} : \mathbb{R}_+ \to \mathbb{R}_+$ *such that*

$$\phi(x) = \langle p, x - \bar{x} \rangle + \frac{1}{2}\langle X, (x - \bar{x})(x - \bar{x})^t \rangle - \bar{r}(\|\bar{x} - x\|)\|\bar{x} - x\|^2 \qquad (4.2)$$

belongs to $\partial_{J_2(\bar{x})} f(\bar{x})$.

Proof It follows from Lemma 4.1 that this may be achieved locally around $\bar{x}$. To extend outside this neighbourhood we follow the proof of Eberhard, Nyblom and Ralph (1998) Proposition 6, noting that we may begin with equation (4.1) of Eberhard, Nyblom and Ralph (1998) replaced by (4.1) locally about $\bar{x}$, that is, with $-r(\|\bar{x} - x\|)\|\bar{x} - x\|^2 \in \mathcal{C}^2(\mathbb{R}^n)$ replacing the term $-\lambda\|x - \bar{x}\|^2$. The argument of Eberhard, Nyblom and Ralph (1998) Proposition 6 now applies and establishes the result.

Thus any $\varphi \in J_2(\bar{x})$ that satisfies the subgradient inequality locally around $\bar{x}$ may be suitably modified outside a neighbourhood of $\bar{x}$ to arrive at a sub-gradient inequality that holds globally. Thus for all u

$$f(u) - \varphi(u) \geq \min_u \left(f - \varphi\right)(u) \quad = \quad f(\bar{x}) - \varphi(\bar{x}),$$
$$\text{which implies } \max_u \left(\varphi - f\right)(u) + f(\bar{x}) \quad \geq \quad \varphi(\bar{x}).$$

This generalized Fenchel inequality may be used to define a generalized derivative.

Definition 4.2 *Let* $f : X \mapsto \overline{\mathbb{R}}$ *be a* Φ*–convex function. Then*

1. *the* Φ*–conjugate is given by* $f^c(\varphi) := \sup_{u \in X} (\varphi(u) - f(u))$;

2. *the Fenchel inequality is* $f(x) + f^c(\varphi) \geq \varphi(x)$;

3. *the* Φ*–subdifferential is* $\partial_\Phi f(x) := \{\varphi \in \Phi \mid f(x) + f^c(\varphi) = \varphi(x)\}$;

Consider the class   .

$$\Phi_2 := \left\{ \varphi(x) = \alpha - \frac{c}{2}\|x - y\|^2 \mid (\alpha, c, y) \in \mathbb{R}^{n+2} \right\}.$$

In Eberhard, Nyblom and Ralph (1998) and Eberhard and Nyblom (1998) it is shown that the proximal subdifferential of Rockafellar, Murdukhovich and Ioffe, denoted by $\partial_p f(x)$, is equivalently characterised via $\partial_p f(x) = \{\nabla\varphi(u)|_{u=x} \mid \varphi \in \partial_{\Phi_2} f(x)\} := \nabla\partial_{\Phi_2} f(x)$. This is a set of elements from X^* rather than a set of nonlinear functions defined on X. One may see that $\partial_p f(x) \subseteq \nabla\partial_\Phi f(x)$ for any class $\Phi_2 \subseteq \Phi \subseteq C^2(\mathbb{R}^n)$. It should be noted that such Φ–convex functions are simply those lower semi–continuous functions which are bounded below by some $\varphi \in \Phi$. Traditionally this has been termed Φ–bounded (see Dolecki et $al.$ (1978)). If $\varphi \in \partial_{C^2} f(\bar{x})$, then by taking $c > \rho(\nabla^2\varphi(\bar{x}))$ (the spectral radius) we have $\psi \in \partial_{\Phi_2} f(\bar{x})$ with $\psi(x) := \varphi(\bar{x}) + \langle\nabla\varphi(x), (x - \bar{x})\rangle - \frac{c}{2}\|x - \bar{x}\|^2$, since locally around $\bar{x}$

$$f(x) - f(\bar{x}) \geq \varphi(x) - \varphi(\bar{x}) \geq \psi(x) - \psi(\bar{x}).$$

We may "globalize" this inequality using Proposition 2.2 of Eberhard and Nyblom (1998). This globalization property for Φ_2–bounded functions may be generalized to the class $C^2(\mathbb{R}^n)$ as shown by Proposition 6 in Eberhard, Nyblom and Ralph (1998). Thus we always have $\partial_p f(x) = \nabla\partial_\Phi f(x)$ for any class $\Phi_2 \subseteq \Phi \subseteq C^2(\mathbb{R}^n)$.

If a function is not $-\infty$ anywhere and $f \not\equiv +\infty$ (that is, f is proper), then to be a supremum of Φ_2 functions it must be at least bounded below by one such function, that is, Φ_2–bounded. When one is interested only in the local differentiability properties of the function, this assumption may be dropped by setting $f(x) = +\infty$ for $x \notin B_\delta(\bar{x})$. If f is lower semi–continuous then $\delta > 0$ may be chosen so that the resultant function is actually bounded below by a constant. This will not affect the local differentiability properties of f at $\bar{x}$.

Definition 4.3 *Given a function* $f : \mathbb{R}^n \mapsto \overline{\mathbb{R}}$, *a generalized convexity generating family* $\Phi \subseteq C^2(\mathbb{R}^n)$ *and an abstract convex cone*

$$\Xi \subseteq \{\varepsilon(\cdot) : \mathbb{R}_+ \mapsto \mathbb{R}_+ \text{ with } \lim_{t\downarrow 0}\varepsilon(t) = 0\},$$

for any given $\varepsilon \in \Xi$ *we define the* ε–*subdifferential of* f *at* x *by*

$$\partial_\Phi^\varepsilon f(x) := \{\varphi \in \Phi \mid f(y) - f(x) \geq \varphi(y) - \varphi(x) - \varepsilon(\|y - x\|) \text{ for all } y\}.$$

We use the notation $\varepsilon > 0$ to mean that $\varepsilon(t) > 0$ for all $t > 0$. As is usual $(\varepsilon + \lambda)(\cdot) := \varepsilon(\cdot) + \lambda(\cdot)$ for any ε and $\lambda \in \Xi$. We have deliberately left the precise choice of Ξ open for time being but note that in Eberhard and Nyblom

(1998) it is shown that for the choice $\Xi = \{\varepsilon t \mid \varepsilon > 0\}$ the quantities $\partial^\varepsilon_{\Phi_2} f(x)$ approximate the proximal subdifferential in that for all $\delta > 0$ and any $\varepsilon \geq 0$ we have (slightly abusing notation in suppressing the t for $\varepsilon t \in \Xi$)

$$\nabla \partial^\varepsilon_{\Phi_2} f(\bar{x}) := \{\nabla \varphi(\bar{x}) \mid \varphi \in \partial^\varepsilon_{\Phi_2} f(\bar{x})\}$$

$$\subseteq \quad \partial^-_\varepsilon f(\bar{x}) := \{z \mid \langle z, y \rangle \leq f'_-(\bar{x}; y) + \varepsilon \|y\|, \ \forall y\} \quad (4.3)$$

$$\subseteq \quad \nabla \partial^{\varepsilon + \delta}_{\Phi_2} f(\bar{x}),$$

$$\text{where } f'_-(\bar{x}; y) \quad = \quad \liminf_{t \downarrow 0, \, y' \to y} \frac{f(x + ty') - f(x)}{t}. \quad (4.4)$$

In particular this implies that $\nabla \partial^\varepsilon_{\Phi_2} f(x)$ in effect estimates the closed sets $\partial^-_\varepsilon f(\bar{x}) = \partial^- f(\bar{x}) + \varepsilon \overline{B}_1(0)$, where $\partial^- f(\bar{x}) = \{z \mid \langle z, y \rangle \leq f'_-(\bar{x}; y) \text{ for all } y\}$ is the lower Dini subdifferential. As $\varepsilon \downarrow 0$ both will approximate the closure of the proximal subdifferential $\overline{\partial_p f(x)} = \nabla \partial_{\Phi_2} f(x)$. In order to drop the closure operation we need the following concept.

Definition 4.4 *We say a function f is ε–proximally regular at $\bar{x}$ if $\nabla \partial^\varepsilon_{\Phi_2} f(\bar{x})|_{\bar{x}}$ (the set of points generated by evaluating these derivatives at $\bar{x}$) is closed for all $\varepsilon t \in \Xi := \{\varepsilon t \mid \varepsilon > 0\}$.*

Corollary 4.2 *Let $f : \mathbb{R}^n \mapsto \overline{\mathbb{R}}$ be lower semi–continuous and Φ_2–bounded. Then*

$$\overline{\nabla \partial^\varepsilon_\Phi f(\bar{x})} = \overline{\nabla \partial^\varepsilon_{\Phi_2} f(\bar{x})} = \partial^-_\varepsilon f(\bar{x})$$

for any class $\Phi_2 \subseteq \Phi \subseteq C^2(\mathbb{R}^n)$. In particular for all $\delta > 0$ we have $\nabla \partial^\varepsilon_\Phi f(\bar{x}) \subseteq \nabla \partial^{\varepsilon + \delta}_{\Phi_2} f(\bar{x})$ and when $\nabla \partial^\varepsilon_{\Phi_2} f(\bar{x})$ is closed for all $\varepsilon > 0$ (that is, is ε–proximally regular) we have $\nabla \partial^\varepsilon_\Phi f(\bar{x}) = \nabla \partial^\varepsilon_{\Phi_2} f(\bar{x})$.

Proof The result will follow immediately on showing $\nabla \partial^\varepsilon_\Phi f(\bar{x}) \subseteq \nabla \partial^{\varepsilon + \delta}_{\Phi_2} f(\bar{x})$ for all $\delta > 0$. If $z \in \nabla \partial^\varepsilon_\Phi f(\bar{x})$, then by definition $z = \nabla \varphi(\bar{x})$ for some $\varphi \in \partial^\varepsilon_\Phi f(\bar{x})$. Put $c = \rho(\nabla^2 \varphi(\bar{x}))$. Then for all $\delta > 0$ we have

$$
\begin{aligned}
f(y) - f(\bar{x}) \quad &\geq \quad \varphi(y) - \varphi(\bar{x}) - \varepsilon \|y - \bar{x}\| \\
&= \quad \langle \nabla \varphi(\bar{x}), y - \bar{x} \rangle + \frac{1}{2} \langle \nabla^2 \varphi(\bar{x}), (y - \bar{x})(y - \bar{x})^t \rangle \\
&\qquad\qquad\qquad + o(\|y - \bar{x}\|^2) - \varepsilon \|y - \bar{x}\| \\
&\geq \quad \langle z, y - \bar{x} \rangle - \frac{1}{2} \langle cI(x - \bar{x}), x - \bar{x} \rangle - (\varepsilon + \delta)\|x - \bar{x}\| \\
&= \quad v(x) - v(\bar{x}) - (\varepsilon + \delta)\|x - \bar{x}\|,
\end{aligned}
$$

58   OPTIMIZATION AND CONTROL WITH APPLICATIONS

where $v(x) = -\frac{c}{2}\|x - \bar{y}\|^2 \in \Phi_2 \subseteq \Phi$, $\bar{y} = \bar{x} + \frac{z}{c}$ and $z = \nabla v(\bar{x})$. By Theorem 5.3 part 1.3 of Eberhard and Nyblom (1998) we have

$$z \in \bigcap_{\delta > 0} \overline{\nabla \partial_{\Phi_2}^{\varepsilon + \delta} f(\bar{x})} \;\; = \;\; \overline{\nabla \partial_{\Phi_2}^{\varepsilon} f(\bar{x})} \subseteq \overline{\nabla \partial_{\Phi}^{\varepsilon} f(\bar{x})},$$

$$\text{which implies} \quad \nabla \partial_{\Phi}^{\varepsilon} f(\bar{x}) \;\; \subseteq \;\; \overline{\nabla \partial_{\Phi_2}^{\varepsilon} f(\bar{x})} \subseteq \overline{\nabla \partial_{\Phi}^{\varepsilon} f(\bar{x})}$$

and so $\overline{\nabla \partial_{\Phi}^{\varepsilon} f(\bar{x})} = \overline{\nabla \partial_{\Phi_2}^{\varepsilon} f(\bar{x})} = \partial_{\varepsilon}^{-} f(\bar{x})$. If $\nabla \partial_{\Phi_2}^{\varepsilon + \delta} f(\bar{x})$ is closed for all $\delta > 0$, then as noted in Theorem 5.1 part 1 (c) of Eberhard and Nyblom (1998) the closures may be dropped.

We note that $J_2(\bar{x})$ is indeed a convex conic subspace of $\mathcal{C}^2(\mathbb{R}^n)$ which clearly contains Φ_2 (for any $\bar{x}$). Put $D_{J_2}(f) := \left\{ x \mid \partial_{J_2(x)}^{\varepsilon} f(x) \neq \emptyset \text{ for all } \varepsilon > 0 \right\}$.

Definition 4.5 *Define an operator* $\square : \mathcal{C}^2(\mathbb{R}^n) \times \mathbb{R}^n \mapsto \mathbb{R}^n \times \mathcal{S}(n)$ *given by* $\square(\varphi, \bar{x}) := (\nabla \varphi(\bar{x}), \nabla^2 \varphi(\bar{x}))$ *and let* $\varepsilon \in \Xi := \{\frac{\epsilon}{2} t^2 \mid \epsilon > 0\}$. *Put*

1. $\partial_{\varepsilon}^{2,-} f(\bar{x}, p) := \{Q \mid f''_{-}(\bar{x}, p, h) + \varepsilon\|h\|^2 \geq \langle Q, hh^t \rangle \text{ for all } h\}$;

2. $\square \partial_{J_2(\bar{x})}^{\varepsilon} f(\bar{x}) := \{\square(\varphi, \bar{x}) \mid \varphi \in \partial_{J_2(\bar{x})}^{\varepsilon} f(\bar{x})\}$ *and*

3. $\nabla^2 \partial_{J_2(\bar{x})}^{\varepsilon} f(\bar{x}, p) := \{\nabla^2 \varphi(\bar{x}) \mid \varphi \in \partial_{J_2(\bar{x})}^{\varepsilon} f(\bar{x}) \text{ and } \nabla \varphi(\bar{x}) = p\}$

When $\varepsilon = 0$ we have $\partial_{\varepsilon}^{2,-} f(\bar{x}, p) = \{Q \mid f''_{-}(\bar{x}, p, h) \geq \langle Q, hh^t \rangle \text{ for all } h\} = \partial^{2,-} f(\bar{x}, p)$ is a closed subset of $\mathcal{S}(n)$ (with the operator norm). We must now consider the relationship between $\nabla^2 \partial_{J_2(\bar{x})}^{\varepsilon} f(\bar{x}, p)$ and $\partial_{\varepsilon}^{2,-} f(\bar{x}, p)$. Clearly $\partial_{\varepsilon}^{2,-} f(\bar{x}, p)$ is closed for any $\varepsilon \geq 0$ and it follows from definitions that $\nabla^2 \partial_{J_2(\bar{x})}^{\varepsilon} f(\bar{x}, p) \subseteq \partial_{\varepsilon}^{2,-} f(\bar{x}, p)$. We wish to show that equality holds. To do so we need the following slight generalization of Lemma 5.2 in Eberhard and Nyblom (1998). The proof is easily extended to this more general context.

Lemma 4.2 *Let* $g : \mathbb{R}^n \to \overline{\mathbb{R}}$ *be a proper lower semi–continuous Φ-bounded function and* $\varphi \in \Phi$. *Then for all* $\varepsilon \geq 0$

$$\partial_{\Phi}^{\varepsilon}(g - \varphi)(x) \;\; = \;\; (\partial_{\Phi}^{\varepsilon} g(x) - \varphi) \cap \Phi,$$

$$\partial_{\Phi}^{\varepsilon} g(x) + \varphi \;\; = \;\; \partial_{\Phi}^{\varepsilon}(g + \varphi)(x) \cap (\Phi + \varphi). \tag{4.5}$$

With this in hand it is easy to extend Eberhard and Nyblom (1998) Proposition 5.1 to the form we state next without proof.

Proposition 4.1 *Let $g : \mathbb{R}^n \to \overline{\mathbb{R}}$ be a proper lower semi–continuous Φ_2-bounded function and $\varphi \in \Phi$, where $\Phi_2 \subseteq \Phi \subseteq C^2(\mathbb{R}^n)$. Then for all $\varepsilon \in \Xi$ and $x \in \operatorname{dom} g$, where $\nabla \partial_\Phi^\varepsilon g(\bar{x}) \neq \emptyset$, we have*

$$\nabla \partial_\Phi^\varepsilon (g + \varphi)(\bar{x}) = \nabla \partial_\Phi^\varepsilon g(\bar{x}) + \nabla \varphi(\bar{x}).$$

We can now show the equivalence of two notions.

Lemma 4.3 *Suppose $f : \mathbb{R}^n \mapsto \overline{\mathbb{R}}$ is lower semi–continuous and proper. Then $\nabla^2 \partial_{J_2(\bar{x})}^\varepsilon f(\bar{x}, p) = \partial_\varepsilon^{2,-} f(\bar{x}, p)$ and so $\nabla^2 \partial_{J_2(\bar{x})}^\varepsilon f(\bar{x}, p)$ is closed.*

Proof To show that $\nabla^2 \partial_{J_2(\bar{x})} f(\bar{x}, p) \supseteq \partial^{2,-} f(\bar{x}, p)$, take $Q \in \partial^{2,-} f(\bar{x}, p)$. By (1.2) we have

$$f(x) \geq f(\bar{x}) + \langle p, x - \bar{x} \rangle + \frac{1}{2}\langle X, (x - \bar{x})(x - \bar{x})^t \rangle - w(\|x - \bar{x}\|)\|x - \bar{x}\|^2,$$

where $-w(\|x - \bar{x}\|) := \frac{o(\|x - \bar{x}\|^2)}{\|x - \bar{x}\|^2}$ induces a mapping $w(\cdot) : \mathbb{R}_+ \mapsto \mathbb{R}$. Now apply Lemma 4.1 to obtain $r(\cdot) : \mathbb{R}_+ \mapsto \mathbb{R}_+$ with $r(\cdot)\| \cdot - \bar{x}\|^2 \in \mathcal{C}^2(\mathbb{R})$ such that $\varphi(x) := \langle p, x - \bar{x} \rangle + \frac{1}{2}\langle X, (x - \bar{x})(x - \bar{x})^t \rangle - r(\|x - \bar{x}\|)\|x - \bar{x}\|^2 \in \partial_{J_2(\bar{x})} f(\bar{x})$. Then $(p, Q) \in \Box \partial_{J_2(\bar{x})} f(\bar{x})$ as required. As the reverse inequality is always true the result $\nabla^2 \partial_{J_2(\bar{x})} f(\bar{x}, p) = \partial^{2,-} f(\bar{x}, p)$ follows. Now apply this result to the function $x \mapsto f(x) + \varepsilon\|x - \bar{x}\|^2$, noting that $\varepsilon\| \cdot - \bar{x}\|^2 \in J_2(\bar{x})$. By Lemma 4.2 we have $\partial_{J_2(\bar{x})}(f + \varepsilon\| \cdot - \bar{x}\|^2)(\bar{x}, p) = \partial_{J_2(\bar{x})} f(\bar{x}, p) + \varepsilon\| \cdot - \bar{x}\|^2$. We have *via* an elementary argument that $\partial_\varepsilon^{2,-}(f + \varepsilon\| \cdot - \bar{x}\|^2)(\bar{x}, p) = \partial_\varepsilon^{2,-} f(\bar{x}, p) + 2\varepsilon I$. This gives

$$\begin{aligned} \nabla^2 \partial_{J_2(\bar{x})}(f + \varepsilon\| \cdot - \bar{x}\|^2)(\bar{x}, p) &= \partial_\varepsilon^{2,-}(f + \varepsilon\| \cdot - \bar{x}\|^2)(\bar{x}, p) \\ \text{and so} \quad \nabla^2 \partial_{J_2(\bar{x})}^\varepsilon f(\bar{x}, p) &= \partial_\varepsilon^{2,-} f(\bar{x}, p). \end{aligned}$$

Because the closure is required to relate $\nabla \partial_{C^2} f(x)$ and $\partial_p f(x)$, we can equate $\Box \partial_{J_2(\bar{x})}^\varepsilon f(\bar{x})$ and $\partial^{2,-} f(x)$ only under the assumption of ε–proximally regular. The next result is taken from Eberhard and Nyblom (1998).

Lemma 4.4 *Let f be locally Lipschitz in a neighbourhood of $\bar{x}$. Then f is ε – proximally regular at $\bar{x}$ if and only if for every $\varepsilon > 0$ there exists a $c_\varepsilon > 0$ such that for each $z \in \nabla \partial_{\Phi_2}^\varepsilon f(\bar{x})$ (with $\Xi := \{\varepsilon t \mid \varepsilon > 0\}$) there exists a $\varphi \in \partial_{\Phi_2}^\varepsilon f(\bar{x})$ such that $\nabla \varphi(\bar{x}) = z$ and $\nabla^2 \varphi(\bar{x}) \geq -c_\varepsilon I$.*

Indeed prox–regular functions are locally ε–proximally regular.

Theorem 4.1 *Suppose the function f is prox–regular at $\bar{x}$ for $\bar{v}$ with respect to η and r, where $\bar{v} \in \partial f(\bar{x})$. Then f is locally $\varepsilon-$ proximally regular for all x with $\|x - \bar{x}\| < \eta$ and $|f(x) - f(\bar{x})| < \eta$ and all $p \in \partial_p f(x)$ with $\|p - \bar{v}\| < \eta$.*

Proof We sketch the proof leaving the details to the reader. First note that when f is prox–regular at $\bar{x}$ it is also locally prox–regular and so is $f(\cdot)+\varepsilon\|\cdot-x\|$. The addition of $\varepsilon\|\cdot-x\|$ will only help to ensure that T, the f attentive $\eta-$ localization of $\partial(f(\cdot) + \varepsilon\|\cdot-x\|)(x)$ (see Poliquin and Rockafellar (1996) or Rockafellar and Wets (1998) for definitions), will have $T + rI$ monotone (use Poliquin and Rockafellar (1996) Theorem 3.2 and subdifferential calculus on the sum). In addition note that an f attentive $\eta-$neighbourhood of $\bar{x}$ is contained in an $f(\cdot) + \varepsilon\|\cdot-x\|$ attentive $(\eta + \varepsilon\eta)-$neighbourhood of $\bar{x}$. Thus in an f attentive neighbourhood (that is, $\|x - \bar{x}\| < \eta$ and $\|f(x) - f(\bar{x})\| < \eta$) of $\bar{x}$ we have $\partial(f(\cdot) + \varepsilon\|\cdot-x\|)(x) = \partial_p(f(\cdot) + \varepsilon\|\cdot-x\|)(x)$ and so

$$\begin{aligned}
\nabla\partial^{\varepsilon}_{\Phi_2} f(x) &= \partial_p(f(\cdot) + \varepsilon\|\cdot-x\|)(x) \\
&\subseteq \partial^-_\varepsilon f(x) \subseteq \partial(f(\cdot) + \varepsilon\|\cdot-x\|)(x) = \partial_p(f(\cdot) + \varepsilon\|\cdot-x\|)(x).
\end{aligned}$$

This implies $\nabla\partial^{\varepsilon}_{\Phi_2} f(x) = \partial^-_\varepsilon f(x)$ is closed. Hence f is locally $\varepsilon-$ proximally regular.

5 GENERALIZED CONVEXITY AND SUBJETS

As we have shown, the derivative information supplied by $\varphi \in \partial_{C^2} f(\bar{x})$ is closely related to proximal subdifferentials and subjets. We pursue this approach further here in order to study the graphical derivative of the proximal subdifferential. For the rest of this section let $\Phi_2 \subseteq \Phi \subseteq C^2(\mathbb{R}^n)$ with Φ convex and conic (using the usual operations scalar multiplication and addition of functions). For $\varphi \in \partial_\Phi f(x)$ we define

$$\Delta_2 f(x, t, \varphi, h) := \left(\frac{2}{t^2}\right)\left(f(x + th) - f(x) - t\left(\frac{\varphi(x + th) - \varphi(x)}{t}\right)\right).$$

Theorem 5.1 *Suppose that f is $\Phi-$convex and proper and Φ is a convex conic subset of mappings from X to $\overline{\mathbb{R}}$. Then for any $v : X \mapsto \mathbb{R}$ with $\frac{1}{t}(v(x+t(\cdot)) - v(x)) \in \Phi$ and $\varphi \in \partial_\Phi f(x)$ we have*

$$\left(\frac{1}{2}\Delta_2 f(x, t, \varphi, \cdot)\right)^c\left(\frac{1}{t}(v(x + t(\cdot)) - v(x))\right) = \frac{1}{2}\Delta_2 f^c(\varphi, t, x, v).$$

As a consequence the following are equivalent:

1. $\frac{1}{t}(v(x+t(\cdot)) - v(x)) \in \partial_\Phi \left(\frac{1}{2}\Delta_2 f(x,t,\varphi,\cdot)\right)(h)$;

2. $(h,v) \in \dfrac{\text{Graph } \partial_\Phi f - (x,\varphi)}{t}$.

Proof We apply the Fenchel equality using the facts that Φ is a convex conic subset and the evaluation mapping $\psi \mapsto x(\psi) := \psi(x)$ is linear for $\psi \in \Phi$. For $\varphi \in \partial_\Phi f(x)$ take the conjugate $(\frac{1}{2}\Delta_2 f(x,t,\varphi,\cdot))^c(\phi) = \sup_h(\phi(h) - \frac{1}{2}\Delta_2 f(x,t,\varphi,h))$, where $\phi(h) := \frac{1}{t}(v(x+th) - v(x))$. On using $f^c(\varphi) + f(x) = \varphi(x)$ we obtain that

$$\left(\frac{1}{2}\Delta_2 f(x,t,\varphi,\cdot)\right)^c \left(\frac{1}{t}(v(x+t(\cdot)) - v(x))\right)$$

$$= \sup_h \left(\left(\frac{1}{t}\right)(v(x+th) - v(x)) - \frac{f(x+th) - f(x) - (\varphi(x+th) - \varphi(x))}{t^2}\right)$$

$$= \left(\frac{1}{t^2}\right)\sup_h (\varphi(x+th) + tv(x+th) - f(x+th) + f(x) - \varphi(x) - tv(x))$$

$$= \left(\frac{1}{t^2}\right)\sup_h ((\varphi+tv)(x+th) - f(x+th) - f^c(\varphi) - tv(x))$$

$$= \left(\frac{1}{t^2}\right)\left(f^c(\varphi+tv) - f^c(\varphi) - t\left(\frac{x(\varphi+tv) - x(\varphi)}{t}\right)\right) = \frac{1}{2}\Delta_2 f^c(\varphi,t,x,v).$$

Thus we have $\frac{1}{t}(v(x+t(\cdot)) - v(x)) \in \partial_\Phi \left(\frac{1}{2}\Delta_2 f(x,t,\varphi,\cdot)\right)(h)$ if and only if

$$\left(\frac{1}{2}\Delta_2 f(x,t,\varphi,\cdot)\right)^c \left(\frac{1}{t}(v(x+t(\cdot)) - v(x))\right) + \frac{1}{2}\Delta_2 f(x,t,\varphi,h) = \frac{1}{t}(v(x+th) - v(x))$$

or when

$$\left(\frac{1}{t^2}\right)(f^c(\varphi+tv) \quad - \quad f^c(\varphi) - t(\frac{x(\varphi+tv) - x(\varphi)}{t}))$$

$$+ \quad \left(\frac{1}{t^2}\right)(f(x+th) - f(x) - t(\frac{\varphi(x+th) - \varphi(x)}{t}))$$

$$= \quad \frac{1}{t}(v(x+th) - v(x))$$

Substituting $f^c(\varphi) + f(x) = \varphi(x)$ again and cancelling gives

$$\left(\frac{1}{t^2}\right)(f^c(\varphi+tv) + f(x+th) - \varphi(x+th)) - \frac{v(x)}{t} = \frac{v(x+th) - v(x)}{t}$$

which is equivalent to $f^c(\varphi+tv) + f(x+th) = (\varphi+tv)(x+th)$ or $\varphi+tv \in \partial_\Phi f(x+th)$. This is true if and only if $(x+th, \varphi+tv) \in \text{Graph } \partial f$ or $(h,v) \in \dfrac{\text{Graph } \partial f - (x,\varphi)}{t}$.

These imply some familiar structures.

Corollary 5.1 *Suppose f is a Φ–convex function, $v \in \Phi$ and $\varphi \in \partial_\Phi f(x)$. Then*

$$(h, v) \in \frac{\text{Graph}\, \partial_\Phi f - (x, \varphi)}{t}$$

implies

$$(h, \nabla^2 \varphi(x)h + \nabla v(x)) \in \frac{\text{Graph}\, \partial_p f - (x, \nabla\varphi(x))}{t} + o(1)B_1,$$

where $\partial_p f(x) := \{\nabla\varphi(u)|_{u=x} \mid \varphi \in \partial_\Phi f(x)\}$.

Proof We use $\nabla\partial f(x + th) := \{\nabla\phi(x + th) \mid \phi \in \partial_\Phi f(x + th)\} = \partial_p f(x + th)$ along with $\nabla\varphi(x+th) = \nabla\varphi(x)+t\nabla^2\varphi(x)h+o(t)$ and $\nabla v(x+th) = \nabla v(x)+o(1)$. In particular whenever (2) of Theorem 5.1 hold we have equivalently

$$\varphi(x + th) + tv(x + th) \in \partial_\Phi f(x + th)$$

and so $\nabla\varphi(x + th) + t\nabla v(x + th) \in \nabla\partial_\Phi f(x + th)$. This implies

$$\nabla\varphi(x) + t\nabla^2\varphi(x)h + o(t) + t(\nabla v(x) + o(1)) \quad \in \quad \partial_p f(x + th)$$
$$\text{or } \nabla^2\varphi(x)h + \nabla v(x) + \frac{o(t)}{t} \quad \in \quad \frac{\partial_p f(x + th) - \nabla\varphi(x)}{t},$$

which in turn implies

$$(h, \nabla^2\varphi(x)h + \nabla v(x)) \quad \in \quad \frac{\text{Graph}\, \partial_p f - (x, \nabla\varphi(x))}{t} + \frac{o(t)}{t}B_1. \quad (5.1)$$

Note that for $\varphi \in C^2(X)$ we have

$$\Delta_2 f(x, t, \varphi, h)$$
$$= \left(\frac{2}{t^2}\right)\left(f(x + th) - f(x) - t\langle\nabla\varphi(x), h\rangle - t^2\langle\nabla^2\varphi(x), hh^t\rangle\right) + o(1).$$

We thus define $\Delta_2 f(x, t, \nabla\varphi, h) := \left(\frac{2}{t^2}\right)\left(f(x + th) - f(x) - t\langle\nabla\varphi(x), h\rangle\right)$ and obtain

$$\Delta_2 f(x, t, \varphi, h) = \Delta_2 f(x, t, \nabla\varphi, h) - \langle\nabla^2\varphi(x), hh^t\rangle + o(1).$$

Recall that $A \in E(\mathcal{A}, u)$ if and only if $(2Au, 2A) \in \partial^{2,-}q(\mathcal{A})(u)$ and $A \in \mathcal{A}$. This situation is rather more natural than first may appear.

Corollary 5.2 *Suppose f is a Φ–convex function, $\varphi \in \partial_\Phi f(x)$; $h \in \mathbb{R}^n$ and $v : X \mapsto \overline{\mathbb{R}}$ is differentiable at x with $v \in \Psi \supseteq \Phi$.*

1. *The condition*

$$\frac{1}{t}(v(x+t(\cdot))-v(x)) \in \partial_\Psi \left(\frac{1}{2}\Delta_2 f(x,t,\varphi,\cdot)\right)(h) \qquad (5.2)$$

for all $t > 0$ sufficiently small implies for all $\rho > 0$ and $w \in X$ that

$$\left(\frac{1}{\rho}\right)(f''_-(x,\nabla\varphi(x),h+\rho w')-f''_s(x,\nabla\varphi(x),h))$$

$$\geq 2\langle\nabla v(x),w\rangle + \left(\frac{1}{\rho}\right)(\langle\nabla^2\varphi(x),(h+\rho w)(h+\rho w)^t\rangle - \langle\nabla^2\varphi(x),hh^t\rangle).$$

2. *If $\nabla v(x) = 0$, then there exists h such that*

$$f''_-(x,\nabla\varphi(x),h) = \min\{f''_-(x,\nabla\varphi(x),h),f''_-(x,\nabla\varphi(x),-h)\}$$

and for this h we have $\nabla^2\varphi(x) \in E(\partial^{2,-}f(x,\nabla\varphi(x)),h)$.

Proof The subgradient inequality for (5.2), namely $\frac{1}{t}(v(x+t(\cdot))-v(x)) \in \partial_\Psi\left(\frac{1}{2}\Delta_2 f(x,t,\varphi,\cdot)\right)(h)$, gives that for all h' and w'

$$\frac{1}{2}\Delta_2 f(x,t,\varphi,h'+\rho w') - \frac{1}{2}\Delta_2 f(x,t,\varphi,h)$$

$$\geq \frac{1}{t}(v(x+t(h'+\rho w'))-v(x)) - \frac{1}{t}(v(x+th)-v(x)).$$

If $X \in \partial^{2,-}f(x,\nabla\varphi(x))$ then we have $\Delta_2 f(x,t,\varphi,h) \geq \langle X - \nabla^2\varphi(x),hh^\dagger\rangle + o_t(1)$. On taking the limit infimum as $t \downarrow 0$, $w' \to \bar{w}$ and $h' \to \bar{h}$ we obtain

$$f''_-(x,\nabla\varphi(x),\bar{h}+\rho\bar{w}) - \langle\nabla^2\varphi(x),(\bar{h}+\rho\bar{w})(\bar{h}+\rho\bar{w})^t\rangle \qquad (5.3)$$

$$-\langle X - \nabla^2\varphi(x),hh^t\rangle \geq 2(\langle\nabla v(x),\bar{h}+\rho\bar{w}\rangle - \langle\nabla v(x),h\rangle)$$

for arbitrary $\bar{h}$ and $\bar{w}$. On choosing $\bar{h} = h$, $\rho = \rho_n$ and $\bar{w} = w_n$, it follows that for $\rho_n \downarrow 0$ and $w_n \to w$

$$f''_-(x,\nabla\varphi(x),h+\rho_n w_n) - \langle X,hh^t\rangle \geq 2\rho_n\langle\nabla v(x),w_n\rangle$$

$$+\langle\nabla^2\varphi(x),(h+\rho_n w_n)(h+\rho_n w_n)^t\rangle - \langle\nabla^2\varphi(x),hh^t\rangle.$$

This implies

$$f''_-(x,\nabla\varphi(x),h+\rho_n w_n) - \sup\{\langle X,hh^t\rangle \mid X \in \partial^{2,-}f(x,\nabla\varphi(x))\}$$

$$\geq 2\rho_n\langle\nabla v(x),w_n\rangle + \langle\nabla^2\varphi(x),(h+\rho_n w_n)(h+\rho_n w_n)^t\rangle - \langle\nabla^2\varphi(x),hh^t\rangle$$

or

$$f_-''(x, \nabla\varphi(x), h + \rho_n w_n) - f_s''(x, \nabla\varphi(x), h)$$
$$\geq 2\rho_n \langle \nabla v(x), w_n \rangle + \langle \nabla^2\varphi(x), (h + \rho_n w_n)(h + \rho_n w_n)^t \rangle - \langle \nabla^2\varphi(x), hh^t \rangle$$
$$= 2\langle \nabla v(x), w_n \rangle + \rho_n \langle 2\nabla^2\varphi(x)h, w_n \rangle + \frac{\rho_n^2}{2}\langle 2\nabla^2\varphi(x), w_n(w_n)^t \rangle. \tag{5.4}$$

We observe that for any h and w' we have

$$f_s''(x, \nabla\varphi(x), h) = \min\{f_-''(x, \nabla\varphi(x), h), f_-''(x, \nabla\varphi(x), -h)\}$$
$$= f_-''(x, \nabla\varphi(x), h) \quad \text{and}$$
$$f_s''(x, \nabla\varphi(x), h + \rho w') = \min\{f_-''(x, \nabla\varphi(x), h + \rho w'), f_-''(x, \nabla\varphi(x), -(h + \rho w'))\}.$$

We separate out two cases, when

$$f_-''(x, \nabla\varphi(x), -h) > f_-''(x, \nabla\varphi(x), h)$$

and when $f_-''(x, \nabla\varphi(x), -h) = f_-''(x, \nabla\varphi(x), h)$.

Consider first the former. As $w' \mapsto f_-''(x, \nabla\varphi(x), h + \rho w')$ is lower semi-continuous and by assumption $f_-''(x, \nabla\varphi(x), -h) > f_-''(x, \nabla\varphi(x), h)$, it follows that for all w

$$\liminf_{\rho\downarrow 0,\ w'\to w} f_-''(x, \nabla\varphi(x), -(h + \rho w')) \geq f_-''(x, \nabla\varphi(x), -h) > f_-''(x, \nabla\varphi(x), h).$$

Thus there exists a $\delta > 0$ such that for $\rho < \delta$ and $w' \in B_\delta(w)$ we have $f_-''(x, \nabla\varphi(x), -(h + \rho w')) \geq f_-''(x, \nabla\varphi(x), h) + \delta$. In particular this implies

$$\frac{1}{\rho}\left(f_-''(x, \nabla\varphi(x), -(h + \rho w')) - f_s''(x, \nabla\varphi(x), h))\right) \geq \frac{\delta}{\rho} \to +\infty$$

as $\rho \downarrow 0$. Thus if $\rho_n \downarrow 0$, $w_n \to w$ results in a finite limit in

$$\frac{1}{\rho_n}\left(f_s''(x, \nabla\varphi(x), h + \rho_n w_n) - f_s''(x, \nabla\varphi(x), h))\right), \tag{5.5}$$

then for n sufficiently large we have

$$f_s''(x, \nabla\varphi(x), h + \rho_n w_n) = f_-''(x, \nabla\varphi(x), h + \rho_n w_n). \tag{5.6}$$

Indeed in this case there exists a $\delta > 0$ such that for all $w \in B_\delta(w)$ and $\rho < \delta$ we have

$$f_s''(x, \nabla\varphi(x), h + \rho w) = f_-''(x, \nabla\varphi(x), h + \rho w). \tag{5.7}$$

We now consider the second case. Let $\rho_n \downarrow 0$, $w_n \to w$ attain the limit infimum in

$$\liminf_{\rho \downarrow 0,\; w' \to w} \frac{1}{\rho}\, (f_s''(x, \nabla\varphi(x), h + \rho w') - f_s''(x, \nabla\varphi(x), h))) \tag{5.8}$$

and assume once again this limit infimum is finite. We address three subcases. The first two are when either $f_s''(x, \nabla\varphi(x), h + \rho_n w_n) = f_-''(x, \nabla\varphi(x), h + \rho_n w_n)$ for n sufficiently large or $f_s''(x, \nabla\varphi(x), h + \rho_n w_n) = f_-''(x, \nabla\varphi(x), -(h + \rho_n w_n))$ for all n sufficiently large. With either we have *via* the finiteness of (5.8) that

$$
\begin{aligned}
f_s''(x, \nabla\varphi(x), h + \rho_n w_n) &= f_-''(x, \nabla\varphi(x), h + \rho_n w_n) \\
&\to f_s''(x, \nabla\varphi(x), h) = f_-''(x, \nabla\varphi(x), h) \\
\text{or}\quad f_s''(x, \nabla\varphi(x), h + \rho_n w_n) &= f_-''(x, \nabla\varphi(x), -(h + \rho_n w_n)) \\
&\to f_s''(x, \nabla\varphi(x), h).
\end{aligned}
$$

In the former case (5.6) holds for all n. In the latter, after renaming h as $-h$ and w as $-w$, (5.6) then holds for n sufficiently large. The third subcase is when $f_s''(x, \nabla\varphi(x), h + \rho_n w_n) = f_-''(x, \nabla\varphi(x), \pm(h + \rho_n w_n))$ infinitely often. We may reduce this to (5.6) on taking a subsequence along which a positive sign is always chosen.

The only case remaining occurs when the limit infimum is not finite in which case any arbitrarily chosen sequence will attain the limit infimum. Thus we may assume that the limit infimum in (5.5) is attained by a subsequence along which (5.6) holds for all n.

Now consider $\rho_n \downarrow 0$ and $w_n \to w$ achieving the limit infimum

$$\liminf_{\rho \downarrow 0,\, w' \to w} \left(\frac{2}{\rho^2}\right) (f_s''(x, \nabla\varphi(x), h + \rho w') - f_s''(x, \nabla\varphi(x), h) - \rho\langle 2\nabla^2\varphi(x)h, w'\rangle). \tag{5.9}$$

When this is infinite we have (5.10) holding trivially, since (5.4) implies the quotient in (5.9) is bounded below by $\langle 2\nabla^2\varphi(x), ww^t\rangle$. Finiteness of the limit

$$\lim_{n \to \infty} \left(\frac{2}{\rho_n^2}\right) (f_s''(x, \nabla\varphi(x), h + \rho_n w_n) - f_s''(x, \nabla\varphi(x), h) - \rho_n\langle 2\nabla^2\varphi(x)h, w_n\rangle)$$

implies finiteness of the limit (5.5). Arguing as before we may once again assume that (5.6) holds along this sequence. Then when $\nabla v(x) = 0$ we have by (5.7) and (5.4) that for arbitrary w

$$\liminf_{\rho \downarrow 0,\, w' \to w} \left(\frac{2}{\rho^2}\right) (f_s''(x, \nabla\varphi(x), h + \rho w') - f_s''(x, \nabla\varphi(x), h) - \rho\langle 2\nabla^2\varphi(x)h, w\rangle)$$

$$\geq \langle 2\nabla^2\varphi(x), ww^t\rangle. \tag{5.10}$$

This implies $(2\nabla^2\varphi(x)h, 2\nabla^2\varphi(x)) \in \partial_h^{2,-} f_s''(x, \nabla\varphi(x), h)$. Since

$$\nabla^2\varphi(x) \in \partial^{2,-} f(x, \nabla\varphi(x))$$

we have, by Theorem 3.4, that $\nabla^2\varphi(x) \in E(\partial^{2,-} f(x, \nabla\varphi(x)), h)$.

6 SUBJET, CONTINGENT CONE INCLUSIONS

This section explores the relationship between subjets and coderivatives. This is achieved by first obtaining results relating the subjet and the contingent graphical derivative and then appealing to the results of Rockafellar *et al.* (1997). This will allow us to explore the connection between the symmetric operators used in subjets and the symmetry notions introduced in Rockafellar *et al.* (1997) for coderivatives. We find once again that the crucial operators are those exposed by rank–one supports. In Dolecki *et al.* (1978) it was shown that a necessary and sufficient condition for a function $f : \mathbb{R}^n \mapsto \overline{\mathbb{R}}$ to be Φ–convex (for $\Phi_2 \subseteq \Phi \subseteq C^2(\mathbb{R}^n)$) is for f to be lower semi–continuous and minorized by at least one element of Φ (that is, Φ_2–bounded or alternatively prox–bounded).

Recall that a function f belongs to $C^{1,1}(\mathbb{R}^n)$ when its gradient exists everywhere and the gradient itself is a locally Lipschitz function.

Proposition 6.1 *Suppose that $f : \mathbb{R}^n \mapsto \overline{\mathbb{R}}$ is Φ–convex with $\Phi_2 \subseteq \Phi \subseteq C^2(\mathbb{R}^n)$ and $\varphi \in \partial_\Phi f(x)$. Suppose $\nabla^2\varphi(x) \in E(\partial^{2,-} f(x, \nabla\varphi(x)), h)$ and choose h such that $f_-''(x, \nabla\varphi(x), h) = \min\{f_-''(x, \nabla\varphi(x), h), f_-''(x, \nabla\varphi(x), -h)\}$. Then there exist sequences $\{t_n\} \downarrow 0$ and $\{h_n\} \to h$ such that for all y*

$$\left(\frac{1}{2}\Delta_2 f(x, t, \nabla\varphi(x), y) - \frac{1}{2}\langle\nabla^2\varphi(x), yy^t\rangle\right)$$

$$- \left(\frac{1}{2}\Delta_2 f(x, t_n, \nabla\varphi(x), h_n) - \frac{1}{2}\langle\nabla^2\varphi(x), h_n(h_n)^t\rangle\right)$$

$$\geq \frac{1}{t}(v(x + ty) - v(x)) - \frac{1}{t_n}(v(x + t_n h_n) - v(x)) \quad (6.1)$$

where $v(x + ty) := v(x) - t\varepsilon(ty)\|y\|^2$ (with $v(x)$ taken as an arbitrary fixed value) is differentiable at x with $\nabla v(x) = 0$ and $\varepsilon(\cdot) : \mathbb{R}^n \to \mathbb{R}$ is as given Lemma 4.1. If in addition we assume that f is locally Lipschitz and v may be chosen so that it is strictly differentiable at x, which may be achieved if f is $C^{1,1}(\mathbb{R}^n)$, then

$$(h_n, \nabla^2\varphi(x)h_n) \in \frac{\text{Graph}\,\partial f - (x, \nabla\varphi(x))}{t_n} + o(1)B_1 \qquad (6.2)$$

and so $(h, \nabla^2\varphi(x)h) \in T_{\mathrm{Graph}\,\partial f}(x, \nabla\varphi(x))$, *the contingent tangent cone to* Graph ∂f.

Proof Suppose $\nabla^2\varphi(x) \in E(\partial^{2,-}f(x, \nabla(x)), h)$ and h is chosen as above. Then $f''_-(x, \nabla\varphi(x), h) = \langle \nabla^2\varphi(x), hh^t \rangle$ and so there exists $\{t_n\} \downarrow 0$ and $\{h_n\} \to h$ and an error term $\varepsilon(y)\|y\|^2 = o(\|y\|^2)$ such that

$$f(x + t_n h_n) - f(x) = t_n\langle \nabla\varphi(x), h_n \rangle +$$
$$\frac{1}{2}t_n^2\langle \nabla^2\varphi(x), h_n(h_n)^t \rangle - \varepsilon(t_n h_n)\|t_n h_n\|^2$$

and

$$f(x + ty) - f(x) \geq t\langle \nabla\varphi(x), y \rangle + \frac{1}{2}t^2\langle \nabla^2\varphi(x), yy^t \rangle - \varepsilon(ty)\|ty\|^2 \quad (6.3)$$

for all y. Here $\varphi \in \partial_\Phi f(x) \subseteq C^2(\mathbb{R}^n)$ and $\varepsilon : \mathbb{R}^n \mapsto \mathbb{R}$ with $\lim_{y\to 0}\varepsilon(y) = 0 := \varepsilon(0)$. When f is $C^{1,1}(\mathbb{R}^n)$ we may assume that $y \mapsto \varepsilon(ty)\|y\|^2$ is strictly differentiable at the origin with derivative zero. Indeed we could put

$$\varepsilon(ty)\|y\|^2 = \frac{2}{t^2}\left(f(x + ty) - f(x) - t\langle \nabla\varphi(x), y \rangle\right).$$

We have on rearranging and subtracting (6.3) that

$$\left(\frac{1}{2}\Delta_2 f(x, t, \nabla\varphi(x), y)\right) - \frac{1}{2}\langle \nabla^2\varphi(x), yy^t \rangle)$$
$$- \left(\frac{1}{2}\Delta_2 f(x, t_n, \nabla\varphi(x), h_n)\right) - \frac{1}{2}\langle \nabla^2\varphi(x), h_n(h_n)^t \rangle)$$
$$\geq \varepsilon(t_n h_n)\|h_n\|^2 - \varepsilon(ty)\|y\|^2.$$

If $v(x + ty) := v(x) - t\varepsilon(ty)\|y\|^2$ (for $v(x) := 0$), it follows that

$$\frac{1}{t}(v(x + ty) - v(x)) = -\varepsilon(ty)\|y\|^2 \to 0 = \nabla v(x) \text{ as } t \to 0$$

along with (6.1). Define $g(y) := \varepsilon(ty)\|y\|^2$ and note that $v(x+ty) = v(x) - tg(y)$.

When f is $C^{1,1}$ we have f twice differentiable almost everywhere with $\|\nabla^2 f(x + ty)\| \leq L$, where L is the Lipschitz constant of ∇f. Thus $y \mapsto v(x + ty)$ is twice differentiable almost everywhere with value $\nabla_y^2 v(x + ty) = -t\nabla^2 g(y)$, where

$$\nabla^2 g(y) = \frac{2}{t^2}\left(\nabla^2 f(x + ty)t^2\right) = 2\nabla^2 f(x + ty).$$

Thus almost everywhere we have $\|\nabla_y^2 v(x + ty)\| \leq t\left(2\|\nabla^2 f(x + ty)\|\right) := t2L$ and so for fixed t we have $y \mapsto v(x + ty)$ is $C^{1,1}$ with a Lipschitz constant $t2L$.

Next observe that for fixed $t > 0$ we have by the chain rule that $\nabla_y v(x + ty) = t\nabla_u v(u)\,|_{u=x+ty} = t\nabla v(x + ty)$. Thus

$$\|\nabla v(x + ty) - \nabla v(x)\| \;=\; \frac{1}{t}\|\nabla_y v(x + ty) - \nabla_y v(x + ty)|_{y=0}\|$$
$$\leq\; \frac{1}{t}t2L\,\|x + ty - x\| = 2Lt\|y\|,$$

which implies that $\nabla v(\cdot)$ is strictly differentiable at x when f is $C^{1,1}$.

We show next that when v is strictly differentiable at zero we have

$$\limsup_{y\to th'}\left|\frac{\varepsilon(y)\|y\|^2 - \varepsilon(th')\|th'\|^2}{\|y - th'\|}\right| = \limsup_{y'\to h'}\left|\frac{t^2\,(g(y') - g(h'))}{\|ty' - th'\|}\right|$$

$$= \limsup_{y\to h'}\left|\frac{tg(y) - tg(h')}{\|y - h'\|}\right| = o(t) \text{ as } t \to 0. \tag{6.4}$$

To prove (6.4) consider the Clarke subdifferential of the function $y \mapsto tg(y)$. As $\frac{1}{t}(v(x + ty) - v(x)) = -g(y)$ we have $tg(y) = -(v(x + ty) - v(x))$. The chain rule implies $-t\partial v(x + th') = \partial\,(tg)\,(h')$. As v is strictly differentiable at x, we have $\partial v(x + th') \to \nabla v(x) = 0$ as $t \to 0$. Thus

$$\operatorname{diam}\partial\,(tg)\,(h') = t\operatorname{diam}\partial v(x + th') = o(t)$$

for h' in a bounded set. By the Lebourg mean–value Theorem we have

$$\left|\frac{tg(y) - tg(h')}{\|y - h'\|}\right| = \left|\left\langle z(t), \frac{y - h'}{\|y - h'\|}\right\rangle\right|,$$

where $z(t) \in \partial\,(tg)\,(y')$ for some y' in the interval between y and h'. Thus by the upper semi–continuity of the Clarke subgradient

$$\limsup_{y\to th'}\left|\frac{\varepsilon(y)\|y\|^2 - \varepsilon(th')\|th'\|^2}{\|y - th'\|}\right| = \left|\frac{tg(y) - tg(h')}{\|y - h'\|}\right|$$

$$\leq \sup\{\|z\| \mid z \in \partial\,(tg)\,(y')$$

$$\text{for } y' = \lambda y + (1 - \lambda)th' \text{ and } \lambda \in (0, 1)\} = o(t)$$

as $t \to 0$ and $h' \to h$ as claimed in (6.4).

Consequently for each t there is a small neighbourhood $B_{\delta_t}(th')$ within which we have

$$\left|\varepsilon(y)\|y\|^2 - \varepsilon(th')\|th'\|^2\right| \leq o(t)\|y - th'\|.$$

Thus for every $y \in B_{\delta_n}(t_n h_n)$ we have

$$f(x + t_n h_n) - f(x) = t_n\langle\nabla\varphi(x), h_n\rangle + \frac{1}{2}t_n^2\langle\nabla^2\varphi(x), h_n(h_n)^t\rangle - \varepsilon(t_n h_n)\|t_n h_n\|^2$$

and $f(x + y) - f(x) \geq \langle\nabla\varphi(x), y\rangle + \frac{1}{2}\langle\nabla^2\varphi(x), yy^t\rangle - \varepsilon(y)\|y\|^2$ for all y.

Hence

$$f(x+y) - f(x+t_n h_n)$$
$$= f(x+y) - f(x) + f(x) - f(x+t_n h_n)$$
$$\geq \langle \nabla\varphi(x), y\rangle + \frac{1}{2}\langle \nabla^2\varphi(x), yy^t\rangle$$
$$- \left(t_n\langle\nabla\varphi(x), h_n\rangle + \frac{1}{2}t_n^2\langle\nabla^2\varphi(x), h_n(h_n)^t\rangle \right) + \varepsilon(t_n h_n)\|t_n h_n\|^2 - \varepsilon(y)\|y\|^2$$
$$= \langle\nabla\varphi(x), y - th\rangle + \frac{1}{2}\langle\nabla^2\varphi(x), yy^t - t_n^2 h_n(h_n)^t\rangle + \varepsilon(t_n h_n)\|t_n h_n\|^2 - \varepsilon(y)\|y\|^2$$
$$\geq \langle\nabla\varphi(x), y - t_n h_n\rangle + \frac{1}{2}\langle\nabla^2\varphi(x), yy^t - t_n^2 h_n(h_n)^t\rangle - o(t_n)\|y - t_n h_n\|, \quad (6.5)$$

where the little "o" notation is taken to mean that implied by (6.4). Define the function $k(y) := f(x+y)$ and put $\psi(y) := \langle\nabla\varphi(x), y - t_n h_n\rangle + \frac{1}{2}\langle\nabla^2\varphi(x), yy^t - t_n^2 h_n(h_n)^t\rangle \in C^2(\mathbb{R}^n)$. Then by (6.5) we have

$$\psi(y) - \psi(t_n h_n) - o(t_n)\|y - t_n h_n\| \leq k(y) - k(t_n h_n), \quad \text{which implies}$$
$$\psi(y) - \psi(t_n h_n) \leq (k(\cdot) + o(t_n)\|\cdot - t_n h_n\|)(y) - (k(\cdot) + o(t_n)\|\cdot - t_n h_n\|)(t_n h_n)$$

for all $y \in B_{\delta_n}(t_n h_n)$. Since the function $o(t_n) \geq 0$ the Φ_2–boundedness of k is transferred to $k(\cdot) + o(t_n)\|\cdot - t_n h_n\|$ and so we may use Corollary 4.1 to globalize this inequality and thus obtain $\psi \in \partial_{C^2}(k(\cdot) + o(t_n)\|\cdot - t_n h_n\|)(t_n h_n)$. This implies, using the subdifferential calculus for locally Lipschitz functions and the basic subdifferential, that

$$\begin{aligned}
\nabla_y\psi(y)|_{y=t_n h_n} &= \nabla\varphi(x) + t_n\nabla^2\varphi(x)h_n\\
&\in \nabla\partial_{C^2}(k + o(t_n)\|\cdot - t_n h_n\|)(t_n h_n)|_{t_n h_n}\\
&= \partial_p(k + o(t_n)\|\cdot - t_n h_n\|)(t_n h_n)\\
&\subseteq \partial k(t_n h_n) + o(t_n)\overline{B}_1(0)\\
&= \partial f(x + t_n h_n) + o(t_n)\overline{B}_1(0)
\end{aligned}$$
$$\text{and so } (h, \nabla^2\varphi(x)h) \in \frac{\mathrm{Graph}\,\partial f - (x, \nabla\varphi(x))}{t_n} + o(1)\overline{B}_1.$$

This immediately implies $(h, \nabla^2\varphi(x)h) \in T_{\mathrm{Graph}\,\partial f}(x, \nabla\varphi(x))$.

Remark 6.1 *The construct of Lemma 4.1 ensures that we can always find a strictly differentiable remainder term for the second–order subjet expansion. Unfortunately this construction does not preserve the first equation in (6.3) and so we are unable to use it here.*

The following is a rephrasing of the result A5 of Crandall *et al.* (1992) (see Eberhard, Nyblom and Ralph (1998) for the following version), under only the assumption of Φ_2–boundedness and lower semi–continuity, that we have

$$(p, Q) \in \partial^{2,-} f_\lambda(\bar{x}) \text{ implies } \qquad (p, Q) \in \partial^{2,-} f(\bar{x} - \lambda p) \text{ and}$$

$$f(x - \lambda p) \quad = \quad f_\lambda(x) - (\lambda/2)\|p\|^2. \tag{6.6}$$

Set $L_\lambda(x, p) := (x + \lambda p, p) : \mathbb{R}^{2n} \to \mathbb{R}^{2n}$, which is clearly linear and invertible. The following is a corollary.

Lemma 6.1 *Suppose f is prox–bounded and lower semi–continuous. Then for all $\lambda > 0$ sufficiently small we have L_λ^{-1} (Graph $\partial_p f_\lambda$) $\subseteq$ Graph $\partial_p f$ and $T_{\mathrm{Graph}\,\partial_p f_\lambda}(\bar{x}, 0) \subseteq L_\lambda\left(T_{\mathrm{Graph}\,\partial_p f}(\bar{x}, 0)\right)$.*

Proof We know that there exists a Q such that $(p, Q) \in \partial^{2,-} f_\lambda(\bar{x})$ if and only if $(\bar{x}, p) \in \mathrm{Graph}\,\partial_p f_\lambda$. This implies $(p, Q) \in \partial^{2,-} f(\bar{x} - \lambda p)$, for some Q, which is equivalent to saying that $(\bar{x} - \lambda p, p) = L_\lambda^{-1}(\bar{x}, p) \in \mathrm{Graph}\,\partial_p f$ and so L_λ^{-1} (Graph $\partial_p f_\lambda$) $\subseteq$ Graph $\partial_p f$. Finally we note that if $p = 0$, then $L_\lambda^{-1}(\bar{x}, 0) = (\bar{x}, 0)$, giving

$$L_\lambda^{-1}\left(\frac{\mathrm{Graph}\,\partial_p f_\lambda - (\bar{x}, 0)}{t}\right) \subseteq \frac{\mathrm{Graph}\,\partial_p f - (\bar{x}, 0)}{t}.$$

The result follows on taking the limit supremum.

For the set of vectors $D^*(\partial f)(\bar{x}, \bar{v})(w)$ we put

$$S(D^*(\partial f)(\bar{x}, \bar{v})(w))(w) = \sup\{\langle z, w\rangle \mid z \in S(D^*(\partial f)(\bar{x}, \bar{v})(w), w)\}$$

and similarly for $S(D(\partial f)(\bar{x}, \bar{v})(w))(w)$. Recall that

$$f_s''(x, \nabla\varphi(x), h) = \min\{f_-''(x, \nabla\varphi(x), h), f_-''(x, \nabla\varphi(x), -h)\}.$$

Corollary 6.1 *Suppose f is prox–regular at x with respect to v. Suppose in addition that $Q \in E(\partial^{2,-} f(x, v), h)$ and choose h such that $f''(x, v, h) = f_s''(x, v, h)$ (or $f_-''(x, v, h) \leq f_-''(x, v, -h)$). Then*

$$(h, Qh) \in T_{\mathrm{Graph}\,\partial_p f}(x, v), \tag{6.7}$$

the contingent cone to the proximal subdifferential. It follows that

$$f_-''(x, y, h) \leq S(D(\partial_p f)(x, y)(h))(h). \tag{6.8}$$

Proof We may translate the graph of ∂f so that $x = 0$ and $v = \nabla\varphi(x) = 0$ and add an indicator of a neighbourhood of $(0,0)$ to f so that the resultant function is bounded below, since f is lower semi–continuous. Next note that prox–regularity ensures that locally we have $\partial f(x) = \partial_p f(x)$. We extend the result from functions that are $C^{1,1}(\mathbb{R}^n)$ to those that are prox–regular. We use the results in Poliquin and Rockafellar (1996) and Poliquin *et al.* (1996) to deduce that f_λ is $C^{1,1}(\mathbb{R}^n)$ locally around $\bar{x}$ (see Poliquin and Rockafellar (1996) Theorem 4.4) with $\nabla f_\lambda(0) = 0$. Thus we may apply Proposition 6.1 and Lemma 6.1 to deduce that for all $\lambda > 0$ and any h_λ with $(f_\lambda)''_s(0,0,h_\lambda) = (f_\lambda)''_-(0,0,h_\lambda)$ and $Q_\lambda \in E(\partial^{2,-} f_\lambda(0,0), h_\lambda)$ we have

$$(h_\lambda, Q_\lambda h_\lambda) \in T_{\mathrm{Graph}\,\nabla f_\lambda}(0,0) \subseteq L_\lambda\left(T_{\mathrm{Graph}\,\partial_p f}(0,0)\right). \tag{6.9}$$

Now use Proposition 3.2. If $Q \in E(\partial^{2,-} f(0,0), h)$, then for all $\lambda > 0$ sufficiently small we have $Q_\lambda \in E(\partial^{2,-}_\lambda f(0,0), h^Q_\lambda)$, where $h^Q_\lambda = (I + \lambda Q)h \to h$ as $\lambda \to 0$. Observe that Theorem 3.6 implies $\partial^{2,-}_\lambda f(0,0) = \partial^{2,-} f_\lambda(0,0)$ and so $Q_\lambda \in E(\partial^{2,-} f_\lambda(0,0), h^Q_\lambda)$. For this Q_λ to satisfy the inclusion (6.9). we need only to show that $(f_\lambda)''_s(0,0,h^Q_\lambda) = (f_\lambda)''_-(0,0,h^Q_\lambda)$. First we observe that $(f_\lambda)''_s(0,0,h^Q_\lambda) \le (f_\lambda)''_-(0,0,h^Q_\lambda)$ always holds. To establish the reverse inequality, use Proposition 3.2 again with $\mathcal{A} = \partial^{2,-} f_\lambda(0,0)$ to establish

$$
\begin{aligned}
(f_\lambda)''_s(0,0,h^Q_\lambda) &= q(\partial^{2,-} f_\lambda(0,0))(h^Q_\lambda) = q(\partial^{2,-} f(0,0))(h) + \lambda\|Qh\|^2 \\
&= f''_s(0,0,h) + \frac{1}{\lambda}\|h_\lambda - h\|^2 = f''_-(0,0,h) + \frac{1}{\lambda}\|h_\lambda - h\|^2 \\
&\ge (f''_-(0,0,\cdot))_\lambda (h^Q_\lambda) \ge (f_\lambda)''_-(0,0,h^Q_\lambda).
\end{aligned}
$$

The last inequality follows *via* direct calculation as follows. From (6.6) we have $f_\lambda(0) = f(0)$ and

$$
\begin{aligned}
(\tfrac{1}{2}\, f''_-(0,0,\cdot))_\lambda(h^Q_\lambda) &= \inf_{u\in\mathbb{R}^n}\left\{\liminf_{\substack{u'\to u\\ t\downarrow 0}}\left(\frac{1}{2}2\frac{f(0+tu')-f(0)}{t^2} + \frac{1}{2\lambda}\|u' - h^Q_\lambda\|^2\right)\right\} \\
&= \inf_{u\in\mathbb{R}^n}\left\{\sup_{\substack{\delta>0\\ t\in(0,\delta]}}\inf_{u'\in B_\delta(u)}\left(\frac{f(0+tu') + \frac{1}{2\lambda}\|(0+tu') - (0+th^Q_\lambda)\|^2 - f(0)}{t^2}\right)\right\} \\
&\ge \sup_{\substack{\delta>0\\ t\in(0,\delta]}}\inf_{u\in\mathbb{R}^n}\inf_{u'\in B_\delta(u)}\left(\frac{f(0+tu') + \frac{1}{2\lambda}\|(0+tu') - (0+th^Q_\lambda)\|^2 - f(0)}{t^2}\right) \\
&= \sup_{\substack{\delta>0\\ t\in(0,\delta]}}\inf\left(\frac{(\inf_{u\in\mathbb{R}^n} f(u) + \frac{1}{2\lambda}\|u - (0+th^Q_\lambda)\|^2) - f(0)}{t^2}\right)
\end{aligned}
$$

$$= \liminf_{t\downarrow 0}\left(\frac{f_\lambda(0+th_\lambda^Q) - f_\lambda(0)}{t^2}\right)$$

$$\geq \frac{1}{2}\liminf_{\substack{h'\to h_\lambda^Q \\ t\downarrow 0}} 2\left(\frac{f_\lambda(0+th') - f_\lambda(0)}{t^2}\right) = \frac{1}{2}(f_\lambda)''_-(0,0,h_\lambda^Q).$$

Now use (3.5) and a Neumann series (see for example Anderson *et al.* (1969)) to deduce that for $\lambda > 0$ sufficiently small

$$Q_\lambda = \frac{1}{\lambda}\left(I - (I+\lambda Q)^{-1}\right) = Q - \lambda\,(I+\lambda Q)^{-1} \to Q$$

as $\lambda \to 0$. Thus on taking the limit as $\lambda \to 0$ through (6.9) we obtain $(h, Qh) \in T_{\text{Graph}\,\partial_p f}(x, 0)$.

This suggests that $\{Qw \mid Q \in E(\partial^{2,-} f(\bar{x},\bar{v}), w)\}$ may indeed provide a better description of the function than the coderivative $D^*\,(\partial f)\,(\bar{x},\bar{v})(w)$ for some functions!

Remark 6.2 *Under the assumptions of Corollary 6.1, we have for all h that*

$$\{Qh \mid Q \in E(\partial^{2,-} f(x,y), h)\} \subseteq D(\partial_p f)(x,y)(h) \cup [-D(\partial_p f)(x,y)(-h)]$$
$$= D_s\,(\partial_p f)\,(x,y)(h).$$

R. T. Rockafellar and D. Zagrodny in Rockafellar *et al.* (1997) proved the following related result (see also Rockafellar and Wets (1998)).

Theorem 6.1 *Suppose that $f : \mathbb{R}^n \mapsto \overline{\mathbb{R}}$ is prox–regular and subdifferentially continuous at $\bar{x}$ for $\bar{v} \in \partial f(\bar{x})$, and is proto–differentiable at $\bar{x}$ for $\bar{v}$. Then*

$$D\,(\partial f)\,(\bar{x},\bar{v})(w) \subseteq D^*\,(\partial f)\,(\bar{x},\bar{v})(w).$$

In fact for all w we have

$$D\,(\partial f)\,(\bar{x},\bar{v})(w) \quad \cup \quad [-D\,(\partial f)\,(\bar{x},\bar{v})(-w)] \tag{6.10}$$
$$\subseteq \quad D^*\,(\partial f)\,(\bar{x},\bar{v})(w) \cap [-D^*\,(\partial f)\,(\bar{x},\bar{v})(-w)]. \tag{6.11}$$

In the following we use the convention $\sup \emptyset = -\infty$ and $\inf \emptyset = +\infty$. Put

$$b_D(f,x,v) := \{w \in b^1\left(\partial^{2,-} f(x,v)\right) \mid \text{ such that } E(\partial^{2,-} f(x,v), w) \neq \emptyset$$
$$\text{and } f''_-(x,v,w) = f''_s(x,v,w)\}.$$

Corollary 6.2 *Suppose that* $f : \mathbb{R}^n \mapsto \overline{\mathbb{R}}$ *is prox–regular, subdifferentially continuous at* $\bar{x}$ *for* $\bar{v} \in \partial f(\bar{x})$ *and proto–differentiable at* $\bar{x}$ *for* $\bar{v}$. *Then for all* $w \in b_D(f, \bar{x}, \bar{v})$, *we have*

$$\{Qw \mid Q \in E(\partial^{2,-} f(\bar{x}, \bar{v}), w)\} \subseteq D(\partial f)(\bar{x}, \bar{v})(w) \subseteq D^*(\partial f)(\bar{x}, \bar{v})(w) \quad (6.12)$$

and

$$f''_-(\bar{x}, \bar{v}, w) \leq S(D(\partial f)(\bar{x}, \bar{v})(w))(w) \leq S(D^*(\partial f)(\bar{x}, \bar{v})(w))(w).$$

Proof The containment (6.12) follows immediately from Theorem 6.1 and Corollary 6.1, noting that $\partial_p f = \partial f$ locally. The inequality follows from $w^t Q w = f''_s(\bar{x}, \bar{v}, w) = f''_-(\bar{x}, \bar{v}, w)$ for all $Q \in E(\partial^{2,-} f(\bar{x}, \bar{v}), w)$.

In Poliquin and Rockafellar (1996) and Poliquin *et al.* (1996) it is observed that the proto–differentiability of ∂f at $(\bar{x}, \bar{v})$ implies the second–order epi-differentiability of f at $(\bar{x}, \bar{v})$.

Whenever we have $(h, Qh) \in T_{\text{Graph}\, \partial_p f}(x, p)$, then by definition for all $x^* \in \hat{D}^*(\partial_p f(x, p))(y^*)$ we have

$$\langle x^*, h \rangle \leq \langle y^*, Qh \rangle \text{ which implies } S(\hat{D}^* \partial_p f(x, p)(y^*), h) \leq \langle Q, h(y^*)^t \rangle.$$

If $\partial^{2,-} f(x, \nabla\varphi(x)) = \{Q\}$ then we have for all h that $0 \leq \langle y^* Q - x^*, h \rangle$ and so $x^* = y^* Q$ as one would expect.

Proposition 6.2 *Suppose f is prox–regular at x with respect to v. Then for every $w \in b_D(f, x, v)$ we have*

1.

$$S(\hat{D}^*(\partial_p f)(x, v)(w), w) \leq f''_-(x, v, w) ; \quad (6.13)$$

2.

$$\hat{D}^*(\partial_p f)(x, v)(w) \subseteq \{p \in \mathbb{R}^n \mid \langle p, w \rangle \leq f''_-(x, v, w)\}. \quad (6.14)$$

3. Suppose in addition we assume f is subdifferentially continuous at $\bar{x}$ for $\bar{v} \in \partial f(\bar{x})$, proto–differentiable at $\bar{x}$ for $\bar{v}$ and $b^1(\partial^{2,-} f(x, v))$ is a polyhedral convex set. Then (6.13) holds for all $w \in b^1(\partial^{2,-} f(x, v))$ with $f''_-(x, v, w) = f''_s(x, v, w)$.

Proof We use (2.1) and (6.7). Hence for $w \in \mathbb{R}^n$

$$\hat{D}^* \left(\partial_p f\right)(x, v)(w)$$

$$:= \{p \in \mathbb{R}^n \mid \langle p, h \rangle \le \langle w, z \rangle; \forall (h, z) \in T_{\text{Graph}\, \partial_p f}(x, v)\} \qquad (6.15)$$

$$\subseteq \{p \in \mathbb{R}^n \mid \langle p, h \rangle \le \langle w, Qh \rangle; \forall Q \in E(\partial^{2,-} f(x, v), h), h \in b_D(f, x, v)\}.$$

On using the symmetry of Q we have $p \in \hat{D}^* \left(\partial_p f\right)(x, v)(w)$, which implies

$$\langle w^t Q, h \rangle \ge \langle p, h \rangle \text{ for all } Q \in E(\partial^{2,-} f(x, v), h) \qquad (6.16)$$

for all $h \in b_D(f, x, v)$.

Let $w \in b_D(f, x, v)$. We may use (6.16) with $w = h$ to get $w^t Q w \ge \langle p, w \rangle$ for all $Q \in E(\partial^{2,-} f(x, v), w)$. This implies

$$f_-''(x, v, w) = f_s''(x, v, w) = w^t Q w \ge \langle p, w \rangle$$

for all $p \in \hat{D}^* \left(\partial_p f\right)(x, v)(w)$. If in addition we make the assumption of 3, then by Remark 3.1 $E(\partial^{2,-} f(x, v), w) \ne \emptyset$ for all $w \in \text{rel-int}\, b^1 \left(\partial^{2,-} f(x, v)\right)$.

Now consider $\bar{w} \in b^1 \left(\partial^{2,-} f(x, v)\right) \setminus \text{rel-int}\, b^1 \left(\partial^{2,-} f(x, v)\right)$ with $f_-''(x, v, \bar{w}) = f_s''(x, v, \bar{w})$. Using Corollary 6.1 of Eberhard (2000) we find that

$$w \mapsto f_-''(\bar{x}, 0, w) + r\|w\|^2 \qquad (6.17)$$

is a convex function. We wish to take $Q_n \in E(\partial^{2,-} f(x, v), w_n)$ with

$$w_n \in \text{rel-int}\, b^1 \left(\partial^{2,-} f(x, v)\right)$$

for all n and such that $\pm w_n \in \text{dom}\, f_-''(x, v, \cdot)$; $\pm w_n \to \pm \bar{w}$ and for either the plus or minus sign we have $f_s''(x, v, \pm w_n) = f_-''(x, v, \pm w_n) \to f_-''(x, v, \pm \bar{w}) = f_s''(x, v, \bar{w})$. We may need to take the minus sign if $f_-''(x, v, \bar{w}) = f_-''(x, v, -\bar{w}) = f_s''(x, v, \bar{w})$. By relabeling the vectors we can assume without loss of generality that we may use $w_n \to \bar{w}$. The existence of such a sequence may be established by invoking Theorem 10.2 of Rockafellar (1970) on a simplicial convex subset S of $\text{rel-int}\, \text{dom}\, f_-''(x, v, \cdot) \cup \{\bar{w}\} \subseteq b^1 \left(\partial^{2,-} f(x, v)\right)$ containing $\bar{w}$ as a vertex. Also as $b^1 \left(\partial^{2,-} f(x, v)\right)$ is a polyhedral convex set we may contain any sequence $w_n \in b^1 \left(\partial^{2,-} f(x, v)\right)$ with $w_n \to \bar{w}$ in such a simplex. The Theorem 10.2 of the cited reference states that the convex function $f_-''(x, v, \cdot) + r\| \cdot \|^2$ is upper semi–continuous relative to S. As the function is by construction lower semi–continuous on all of S, it must also be continuous at $\pm \bar{w}$. It follows that

$f''_-(x, v, w_n) \geq \langle p, w_n \rangle$ for all $p \in \hat{D}^* (\partial_p f) (x, v)(w_n)$. Thus for any convergent sequence $(p_n, -w_n) \in \left(T_{\mathrm{Graph}\, \partial_p f}(x, v) \right)^o$, converging to $(p, -\bar{w})$, we have

$$\lim_{n \to \infty} f''_-(x, v, w_n) = f''_-(x, v, \bar{w}) = f''_s(x, v, \bar{w}) \geq \langle p, \bar{w} \rangle.$$

As the graph of $\hat{D}^* (\partial_p f) (x, v)(\cdot)$ equals $\left(T_{\mathrm{Graph}\, \partial_p f}(x, v) \right)^o$, a closed convex cone, p may be taken as an arbitrary element of $\hat{D}^* (\partial_p f) (x, v)(\bar{w})$. Thus (6.13) holds for all $w \in b^1 \left(\partial^{2,-} f(x, v) \right)$ with $f''_-(x, v, w) = f''_s(x, v, w)$.

Under the assumption of 3 we always have $b^1 \left(\partial^{2,-} f(x, v) \right)$ convex (see Eberhard (2000) Corollary 4.1). This set is actually a polyhedral set when f is "fully amenable" (see Rockafellar and Wets (1998) Theorem 13.67).

Corollary 6.3 *Suppose that $f : \mathbb{R}^n \mapsto \overline{\mathbb{R}}$ is prox–regular at $\bar{x}$ for $\bar{v} \in \partial f(\bar{x})$ with respect to ε and r and subdifferentially continuous and proto–differentiable at $\bar{x}$ for $\bar{v}$. Then for all w we have*

$$f''_-(\bar{x}, \bar{v}, w) = S(D (\partial f) (\bar{x}, \bar{v})(w))(w) \leq S(D^* (\partial f) (\bar{x}, \bar{v})(w), w). \qquad (6.18)$$

Proof The first equality in (6.18) follows from an application of Corollary 6.2 of Poliquin and Rockafellar (1996) that states that for all w we have

$$D (\partial_p f) (\bar{x}, \bar{v})(w) = \partial_p \left(\frac{1}{2} f''_-(\bar{x}, \bar{v}, \cdot) \right) (w) . \qquad (6.19)$$

Also without loss of generality we may translate $\bar{v}$ to zero and consider the inequality for the function $g(\cdot) := f(\cdot) - \langle \bar{v}, \cdot \rangle + \frac{r}{2} \| \cdot - \bar{x} \|^2$. By the results of Rockafellar and Wets (1998) and Poliquin and Rockafellar (1996)) (see also Corollary 6.1 of Eberhard (2000)) under the current assumptions $h \mapsto g''_-(\bar{x}, 0, h)$ is convex. Hence (6.19) implies

$$D (\partial_p g) (\bar{x}, 0)(w) = \partial_p \left(\frac{1}{2} g''_-(\bar{x}, 0, \cdot) \right) (w) , \qquad (6.20)$$

where the subgradient ∂_p coincides with ∂ the usual one from convex analysis. Now use the fact that support function of the convex subdifferential $\partial \left(\frac{1}{2} g''_-(\bar{x}, 0, \cdot) \right) (w)$ in the direction w is equal to the one sided radial directional derivative (see Rockafellar (1970))

$$\lim_{t \downarrow 0} \frac{1}{t} \left(\frac{1}{2} g''_-(\bar{x}, 0, w + tw) - \frac{1}{2} g''_-(\bar{x}, 0, w) \right)$$

$$= \lim_{t \downarrow 0} \frac{1}{t} \left(\frac{1}{2}(1 + t)^2 - 1 \right) g''_-(\bar{x}, 0, w) = g''_-(\bar{x}, 0, w) . \qquad (6.21)$$

Thus

$$D\left(\partial_p g\right)(\bar{x},0)(w) = g_-''(\bar{x},0,w)$$

and on removing the translations (using some basic calculus) we obtain the equality in (6.18) holding for all w. The final inequality is true for all w as shown by R. T. Rockafellar and D. Zagrondny in Rockafellar *et al.* (1997) and stated in Theorem 6.1.

7   SOME CONSEQUENCES FOR OPTIMALITY CONDITIONS

In this section we give a few examples showing that formulæ (6.12) and (6.15) can be used to obtain estimates of graphical derivatives and coderivatives. These estimates provide a connection between classical optimality concepts based on derivatives of smooth functions and those using graphical derivatives. Another approach which is successful in achieving this goal is the study of optimality conditions for $C^{1,1}$ functions (see Yang and Jeyakumar (1992)) and for convex composite functions (see Yang (1998)) both of which are particular examples of prox–regular functions.

We shall apply these ideas to a nonsmooth penalization of the Lagrangian associated with a standard smoothly constrained mathematical programming problem. Indeed we do not have to assume *a priori* any regularity of the constraint set but allow a condition to arise out of the construction of the rank–1 exposed facet of the subjet of the penalized Lagrangian. In this way, what appears to be a new and in some ways a more refined second–order sufficiency condition for a strict local minimum is derived.

For unconstrained nonsmooth functions the optimality conditions we investigate, when using various second–order subdifferential objects, are as follows.

Definition 7.1 *Let* $f : \mathbb{R}^n \to \overline{\mathbb{R}}$ *and assume the first–order condition* $0 \in \partial_p f(\bar{x})$ *holds.*

The necessary (sufficient) condition of the first kind holds at $\bar{x}$ *when we have*

$$f_-''(\bar{x},0,h) \geq (>)0 \quad \text{for all } h \neq 0 \text{ with } f_-'(\bar{x},h) \leq 0.$$

The necessary (sufficient) condition of the second kind holds at $\bar{x}$ *when we have*

$$\forall h \in \operatorname{dom} D\left(\partial_p f\right)(\bar{x},0)(\cdot) \text{ with } h \neq 0$$
$$\exists p \in D\left(\partial_p f\right)(\bar{x},0)(h) \text{ such that } \langle p,h \rangle \geq (>)0.$$

Definition 7.2 *Let $f : \mathbb{R}^n \to \overline{\mathbb{R}}$ and assume the first–order condition $0 \in \partial f(\bar{x})$ holds.*

The necessary (sufficient) condition of the third kind holds at $\bar{x}$ when we have

$$\forall h \neq 0, \, p \in D^* \left(\partial f \right) (\bar{x}, 0)(h) \text{ such that } \langle p, h \rangle \geq (>)0.$$

The necessary (sufficient) condition of the forth kind holds at $\bar{x}$ when we have

$$\forall h \neq 0, \, Q \in E \left(\underline{\partial}^2 f(\bar{x}, 0), h \right) \text{ we have } \langle Qh, h \rangle \geq (>) \, 0.$$

We shall say nothing here about conditions of the fourth kind, leaving this to a later paper. The conditions of the first kind are easier to study. This was first done by Auslender (1984), Studniarski (1986) and later by Ward (1995), Ward (1994). Some related results may be found in Eberhard (2000). In this context the sufficient optimality condition of the first kind is equivalent to the concept of a strict local minimum of order two (see Studniarski (1986) and Ward (1995)).

Definition 7.3 *We say $\bar{x} \in C$ is a strict local minimizer of order 2 for $f : \mathbb{R}^n \to \overline{\mathbb{R}}$ if there exist $\beta > 0$ and $\delta > 0$ such that*

$$f(x) \geq f(\bar{x}) + \beta \|x - \bar{x}\|^2 \tag{7.1}$$

for all $x \in B_\delta(\bar{x})$.

As we assume we have extended real–valued functions, constrained problems may easily be included *via* the use of indicator functions. That is, $\bar{x}$ is a strict local minimizer of order two for the problem $\inf \{f(x) \mid x \in C\}$ if it is also one for the function $f(x) + \delta_C(x)$, where $\delta_C(x) = +\infty$ when $x \notin C$ and $\delta_C(x) = 0$ when $x \in C$.

Remark 7.1 *In Eberhard (2000) it is noted that $0 \in \partial_p f(\bar{x})$ and*

$$f_s''(\bar{x}, 0, h) = q \left(\partial^{2,-} f(\bar{x}, 0) \right) (h) > 0 \text{ for all } h \neq 0 \text{ with } f_-'(\bar{x}, h) \leq 0 \tag{7.2}$$

is necessary and sufficient for a strict local minimizer of order two. Proposition 3.3 of Ward (1995) states that $\bar{x}$ is strict local minimizer of order two if and only if

$$f_-''(\bar{x}, 0, h) > 0 \text{ for all } h \neq 0 \text{ with } f_-'(\bar{x}, h) \leq 0. \tag{7.3}$$

Clearly (7.2) implies (7.3), since $f_s''(\bar{x}, 0, h) = \min \{f_-''(\bar{x}, 0, h), f_-''(\bar{x}, 0, -h)\}$ and hence implies $\bar{x}$ is a strict local minimizer of order two. For the converse, it

*is immediate from definitions that when (7.1) holds we have $f_s''(\bar{x}, 0, h) \geq \beta > 0$
in all directions h.*

Remark 7.2 *If f is not prox–regular, then in order to frame a sufficient optimality condition in terms of the subjet we need to use a condition postulating
$0 \in \partial f(\bar{x})$ and the existence of a $\varepsilon > 0$ such that when $f_-'(\bar{x}, h) \leq 0$ for*

$$h \in b^1\left(\partial^{2,-} f(\bar{x}, 0)\right), \text{ then } \exists Q \in E_\varepsilon\left(\partial^{2,-} f(\bar{x}, 0), h\right) \text{ with } \langle Q, hh^t \rangle > 0. \quad (7.4)$$

Remark 7.3 *When $\bar{x}$ is a strict local minimum order two, then by Remark 7.1
we have the sufficient conditions of the first kind holding. Consider the case
when $f_-'(\bar{x}, h) \leq 0$ and $f_-'(\bar{x}, -h) > 0$ with f prox-regular and subdifferentially
continuous. Now $f_-'(\bar{x}, -h) > 0$ implies the existence of a $\delta > 0$ such that
$f(\bar{x} + t(-h)) - f(\bar{x}) \geq \delta t$ for t small, so we have*

$$\frac{2}{t^2}\left(f(\bar{x} + t(-h)) - f(\bar{x})\right) \geq \frac{2\delta}{t} \rightarrow_{t\downarrow 0} +\infty$$

Hence we always have

$$f_-''(\bar{x}, 0, h) < +\infty \Rightarrow f_-'(\bar{x}, h) \leq 0.$$

*Thus $f_-''(\bar{x}, 0, -h) = +\infty$ and $f_s''(\bar{x}, 0, h) = f_-''(\bar{x}, 0, h)$. Now invoke Corollary
6.1 to obtain*

$$0 < f_-''(\bar{x}, 0, h) \leq S(D\left(\partial f\right)(\bar{x}, 0)(h))(h).$$

*Thus there exists a $p \in D\left(\partial f\right)(\bar{x}, 0)(h)$ such that $\langle p, h \rangle > 0$. This is precisely
the sufficient condition of the second kind.*

The sufficient conditions of the third kind were studied (for the prox-regular
subdifferentially continuous function) in Poliquin *et al.* (1998) in association
with the concept of the tilt stable local minimum.

Definition 7.4 *A point $\bar{x}$ is said to give a tilt local minimum of the function
$f : \mathbb{R}^n \to \overline{\mathbb{R}}$ if $f(\bar{x})$ is finite and there exists $\delta > 0$ such that the mapping*

$$M : v \mapsto \arg\min_{|x - \bar{x}| \leq \delta} \{f(x) - f(\bar{x}) - \langle v, x - \bar{x} \rangle\}$$

*is single–valued and Lipschitz continuous on some neighbourhood of $v = 0$ with
$M(0) = \bar{x}$.*

Remark 7.4 *In Poliquin et al. (1998) it is shown that if f is prox-regular and subdifferentially continuous at $\bar{x}$ for $\bar{v} = 0$ then a tilt stable local minimum exists at $\bar{x}$ if and only if the sufficient conditions of the third kind hold at $\bar{x}$.*

Lemma 7.1 *Suppose that $f : \mathbb{R}^n \to \overline{\mathbb{R}}$ is prox–regular and subdifferentially continuous at $\bar{x}$ for $\bar{v} = 0 \in \partial f(\bar{x})$, and ∂f is proto–differentiable at $\bar{x}$ for $\bar{v}$. Then*

$$\operatorname{dom} D\left(\partial_p f\right)(\bar{x}, 0)(\cdot) := \{h \mid D\left(\partial_p f\right)(\bar{x}, 0)(h) \neq \emptyset\} = \operatorname{dom} \partial_p \left(\frac{1}{2} f_-''(\bar{x}, 0, \cdot)\right)(\cdot)$$

$$= \{h \mid f_-''(\bar{x}, 0, h) < +\infty\}. \tag{7.5}$$

In particular $\{h \mid f_-'(\bar{x}, h) \leq 0\} \supseteq \{h \mid f_-''(\bar{x}, 0, h) < +\infty\} = \operatorname{dom} D\left(\partial_p f\right)(\bar{x}, 0)(\cdot)$ and

$$b^1\left(\partial^{2,-} f(\bar{x}, 0)\right) = \operatorname{dom} D\left(\partial_p f\right)(\bar{x}, 0)(\cdot) \cup \left(-\operatorname{dom} D\left(\partial_p f\right)(\bar{x}, 0)(\cdot)\right). \tag{7.6}$$

Proof Using (6.19) there exists a $p \in D\left(\partial_p f\right)(\bar{x}, 0)(h')$ if and only if $\partial_p \left(\frac{1}{2} f_-''(\bar{x}, 0, \cdot)\right)(h') \neq \emptyset$ which is only possible if $f_-''(\bar{x}, 0, h') < +\infty$. Indeed by Corollary 6.1 of Eberhard (2000) (see also the results of Rockafellar and Wets (1998) and Poliquin and Rockafellar (1996)) that under the current assumptions $h \mapsto f_-''(\bar{x}, 0, h) + r\|h\|^2 = g_-''(\bar{x}, 0, h)$ (where $g(\cdot) := f(\cdot) + \frac{r}{2}\| \cdot -\bar{x}\|^2$ as in (6.21)) is convex and proper (see Rockafellar and Wets (1998) Theorem 13.40). Thus $-\infty < f_-''(\bar{x}, 0, \cdot)$ and consequently the directional derivative (6.21) is also never equal to $-\infty$ in any direction. Invoking Theorem 23.3 of Rockafellar (1970) we have convex analysis subdifferential $\partial \left(\frac{1}{2} g_-''(\bar{x}, 0, \cdot)\right)(h) \neq \emptyset$ and consequently $\partial_p \left(\frac{1}{2} f_-''(\bar{x}, 0, \cdot)\right)(h') \neq \emptyset$. Thus by (6.19) we have $h' \in \operatorname{dom} D\left(\partial_p f\right)(\bar{x}, 0)(\cdot) := \{h \mid D\left(\partial_p f\right)(\bar{x}, 0)(h) \neq \emptyset\}$ and (7.5) holding.

As $f_s''(\bar{x}, 0, h) = \min\{f_-''(\bar{x}, 0, h), f_-''(\bar{x}, 0, -h)\} < +\infty$ if and only of $\pm h \in \operatorname{dom} D\left(\partial_p f\right)(\bar{x}, 0)(\cdot)$ we have (7.6) holding. Finally note that as $f_-''(\bar{x}, 0, h') < +\infty$ implies $f_-'(\bar{x}, h') \leq 0$ and we have the relations preceding (7.6) holding.

The following are immediate from the results thus proved.

Theorem 7.1 *Suppose that $f : \mathbb{R}^n \mapsto \overline{\mathbb{R}}$ is prox–regular, subdifferentially continuous and possesses a second–order epi–derivative at $\bar{x}$ for $0 \in \partial f(\bar{x})$.*

1. *The necessary conditions of the second kind implies the necessary conditions the first kind while the sufficient conditions of the first kind hold if and only if the sufficient conditions of the second kind hold.*

2. *The sufficient conditions of the first and second kind hold if and only if $\bar{x}$ is strict local minimum of order two.*

3. *If $\bar{x}$ is a tilt stable local minimum, then it is also a strict local minimum of order two.*

Proof On applying Corollary 6.3 we have for all w that

$$f''_-(\bar{x}, 0, h) = S(D\,(\partial f)\,(\bar{x}, 0)(h))(h) \tag{7.7}$$

and so $0 \leq(<)f''_-(\bar{x}, 0, h)$ if and only if $0 \leq (<)S(D\,(\partial f)\,(\bar{x}, 0)(h))(h)$. Suppose the necessary (sufficient) conditions of the first kind hold. Then when $h \in \operatorname{dom} D\,(\partial f)\,(\bar{x}, 0)(\cdot)$ we have $f''_-(\bar{x}, 0, w) < +\infty$ and hence $f'_-(\bar{x}, 0) \leq 0$ implying $0 \leq(<)f''_-(\bar{x}, 0, h)$. The existence of $p \in D\,(\partial f)\,(\bar{x}, 0)(h)$ with $0 < \langle p, h \rangle$ follows from (7.7) when $0 < f''_-(\bar{x}, 0, h)$. When the necessary condition of the second kind hold then there exists a $p \in D\,(\partial f)\,(\bar{x}, 0)(h)$ with $0 \leq \langle p, h \rangle$ and (7.7) implies $0 \leq f''_-(\bar{x}, 0, h)$.

When the necessary (sufficient) conditions of the second kind hold, take h such that $f'_-(\bar{x}, 0) \leq 0$. Suppose first that $f''_-(\bar{x}, 0, w) < +\infty$ in which case $h \in \operatorname{dom} D\,(\partial f)\,(\bar{x}, 0)(\cdot)$. Then (7.7) along with the necessary (sufficient) conditions of the second kind imply $0 \leq(<)f''_-(\bar{x}, 0, h)$. Otherwise $f''_-(\bar{x}, 0, w) = +\infty > 0$ as required in the necessary (sufficient) conditions of the first kind. As the sufficient conditions of the first kind hold if and only if $\bar{x}$ is a strict local minimum order two the sufficient conditions of the second kind are also equivalent to $\bar{x}$ being a strict local minimum order two.

Finally suppose $\bar{x}$ is a tilt stable local minimum and hence the sufficient conditions of third kind hold at $\bar{x}$. Then (6.11) of Theorem 6.1 implies for all $h \in \operatorname{dom} D\,(\partial f)\,(\bar{x}, 0)(\cdot)$ and $p \in D\,(\partial f)\,(\bar{x}, 0)(h)$ that $0 \leq (<)\langle p, h \rangle$. Thus the sufficient conditions of the second kind follow and $\bar{x}$ is a strict local minimum order two.

Jets do possess a kind of monotonicity property under fairly natural conditions. Recall that the rank–1 hull of a set of symmetric operators can be much larger than its convex hull (see Eberhard *et al.* (2002)).

Lemma 7.2 *Suppose $\{f_\alpha\}_{\alpha \in \Lambda}$ is a family of lower semi–continuous functions.*

1. *Define $f(x) = \sup_{\alpha \in \Lambda} f_\alpha(x)$. Let $\Lambda(x) := \{\alpha \in \Lambda \mid f_\alpha(x) = f(x)\}$. Then*

$$\bigcup_{\alpha \in \Lambda(x)} \partial^{2,-} f_\alpha(x) \subseteq \partial^{2,-} f(x). \tag{7.8}$$

When $\Lambda(x)$ is finite the convex hull of the left hand side of (7.8) is contained in the right hand side of (7.8).

2. *Now suppose in addition that each $f_\alpha \in C^2(\mathbb{R}^n)$ for $\alpha \in \Lambda = \{1,\ldots,m\}$ and let*

$$\bar{\Lambda}(x,h) \quad := \quad \limsup_{x' \to_h x} \Lambda(x')$$
$$= \quad \{\alpha \in \Lambda(x) \mid \exists (t_n, h_n) \to (0^+, h) \text{ with } \alpha \in \Lambda(x + t_n h_n), \forall n\}.$$

Then

$$\emptyset \neq \bar{\Lambda}(x,h) \subseteq \Lambda(x,h) := \{\alpha \in \Lambda(x) \mid f'_-(x,h) = \langle \nabla f_\alpha(x), h \rangle\}.$$

If also $p \in \mathrm{co}\left\{\nabla f_\alpha(x) \mid \alpha \in \bar{\Lambda}(x,h)\right\}$, then

$$E\left(\partial^{2,-} f(x,p), h\right) \quad \supseteq \quad \left\{ \sum_{i \in \bar{\Lambda}(x,h)} \lambda_i \nabla^2 f_i(x) \mid \lambda \in \overline{\Omega}(x,h,p) \right\}^1$$

where

$$\overline{\Omega}(x,h,p) \quad = \quad \{\lambda \in \mathbb{R}^m_+ \mid \exists (t_n, h_n) \to (0^+, h) \text{ such that } \lambda_\alpha = 0,$$
$$\forall \alpha \in \{i \mid i \notin \Lambda(x + t_n h_n) \text{ for } n \text{ sufficiently large}\} \text{ and}$$
$$\sum_{i \in \bar{\Lambda}(x,h)} \lambda_i = 1 \text{ and } p = \sum_{i \in \bar{\Lambda}(x,h)} \lambda_i \nabla f_i(x) \}. \qquad (7.9)$$

3. *When $\{\nabla f_\alpha(x) \mid \alpha \in \bar{\Lambda}(x,h)\}$ are linearly independent we have $\Omega(x,h,p) = \overline{\Omega}(x,h,p)$ where*

$$\emptyset \neq \Omega(x,h,p) = \{\lambda \in \mathbb{R}^m_+ \mid \lambda_\alpha = 0 \text{ if } \alpha \notin \bar{\Lambda}(x,h) \text{ and}$$
$$\sum_{\alpha \in \bar{\Lambda}(x,h)} \lambda_\alpha = 1 \text{ and } p = \sum_{\alpha \in \bar{\Lambda}(x,h)} \lambda_\alpha \nabla f_i(x) \}. \,(7.10)$$

Proof The first containment for $f(x) = \sup_{\alpha \in \Lambda} f_\alpha(x)$ follows from definitions in that when $(p_\alpha, Q_\alpha) \in \partial^{2,-} f_\alpha(x)$ and $\alpha \in \Lambda(x)$ then we have

$$f(y) \quad \geq \quad f_\alpha(y) \geq f_\alpha(x) + \langle p_\alpha, y - x \rangle + \frac{1}{2}\langle Q_\alpha(y - x), (y - x)\rangle + o_\alpha\left(\|y - x\|^2\right)$$
$$= \quad f(x) + \langle p_\alpha, y - x \rangle + \frac{1}{2}\langle Q_\alpha(y - x), (y - x)\rangle + o_\alpha\left(\|y - x\|^2\right).$$

Thus for any $\{\lambda_\alpha\}_{\alpha\in\Lambda(x)}$ with $\sum_{\alpha\in\Lambda(x)}\lambda_\alpha = 1$ and $\lambda_\alpha \geq 0$ we have for any y

$$f(y) \geq f(x) + \langle \sum_{\alpha\in\Lambda(x)} \lambda_\alpha p_\alpha, y - x \rangle + \frac{1}{2}\langle \sum_{\alpha\in\Lambda(x)} \lambda_\alpha Q_\alpha(y-x), (y-x)\rangle$$

$$+ \sum_{\alpha\in\Lambda(x)} \lambda_\alpha o_\alpha \left(\|y - x\|^2\right)$$

where the last term is clearly of small order when $\Lambda(x)$ is countably finite. This implies $\sum_{\alpha\in\Lambda(x)}\lambda_\alpha (p_\alpha, Q_\alpha) \in \partial^{2,-} f(x)$. Next note that if $\alpha \in \bar{\Lambda}(x, h)$, then by the lower semi–continuity of f we have for $x_n = \bar{x} + t_n h_n$ that

$$f_\alpha(x) = \liminf_n f_\alpha(x_n) = \liminf_n f(x_n) \geq f(\bar{x})$$

and so $\alpha \in \Lambda(x)$. Observe that in this case, if $f_\alpha \in C^2(\mathbb{R}^n)$ and Λ is finite, then f is locally Lipschitz, prox–regular and subdifferentially continuous (see Example 2.9 of Poliquin and Rockafellar (1996)). Also f is regular (see Rockafellar (1982) regarding lower C^2 functions) and semi–smooth (see Mifflin (1977) regarding suprema of semi–smooth functions). Hence by the results of Spingarn (1981) we must have ∂f submonotone and hence directionally continuous in the sense that if $(t_n, h_n) \to (0^+, h)$ and $p_n \in \partial f(x + t_n h_n)$ with $p_n \to p$, then

$$p \in \partial f(x)_h := \{z \in \partial f(x) \mid \langle z, h\rangle = f'_-(x, h)\}.$$

In particular, if $\alpha \in \bar{\Lambda}(x, h)$ then $\exists (t_n, h_n) \to (0^+, h)$ such that $\nabla f_\alpha(x+t_n h_n) \in \partial f(x + t_n h_n)$. This implies

$$\lim_n \nabla f_\alpha(x + t_n h_n) = \nabla f_\alpha(x) \in \partial f(x)_h$$

and so $\alpha \in \Lambda(x, h)$. If $p \in \mathrm{co}\left\{\nabla f_\alpha(x) \mid \alpha \in \bar{\Lambda}(x, h)\right\}$, then for any $\lambda \in \bar{\Omega}(x, h, p)$ (which may be empty) we have $p = \sum_{\alpha\in\bar{\Lambda}(x,h)}\lambda_\alpha \nabla f_\alpha(x)$ and the existence of $(t_n, h_n) \to (0^+, h)$ such that $\lambda_\alpha = 0$ for all $\alpha \notin \Lambda(x + t_n h_n)$, for n sufficiently large. Summing across the second–order Taylor expansion of each f_α for $\alpha \in \bar{\Lambda}(x, h)$ yields

$$\sum_{\alpha\in\bar{\Lambda}(x,h)} \lambda_\alpha\, f_\alpha(x + t_n h_n)$$

$$= \sum_{\alpha\in\bar{\Lambda}(x,h)} \lambda_\alpha f_\alpha(x) + t_n \left\langle \sum_{\alpha\in\bar{\Lambda}(x,h)} \lambda_\alpha \nabla f_\alpha(x), h_n \right\rangle$$

$$+ \frac{1}{2} t_n^2 \left\langle \sum_{\alpha\in\bar{\Lambda}(x,h)} \lambda_\alpha \nabla^2 f_\alpha(x), h_n h_n^t \right\rangle + o\left(t_n^2\right),$$

where $o\left(t_n^2\right) = \sum_{\alpha \in \bar{\Lambda}(x,h)} o_\alpha\left(\|t_n h_n\|^2\right)$, a finite sum. As $f_\alpha(x) = f(x)$ and $f_\alpha(x + t_n h_n) = f(x + t_n h_n)$ for all $\lambda_\alpha > 0$ and n sufficiently large, we obtain for $Q = \sum_{\alpha \in \bar{\Lambda}(x,h)} \nabla^2 f_\alpha(x)$ that

$$f(x + t_n h_n) = f(x) + t_n \langle p, h_n \rangle + \frac{1}{2}\langle Q, h_n h_n^t \rangle + o\left(t_n^2\right)$$

or

$$\liminf_{n \to \infty} \frac{2}{t_n^2}\left(f(x + t_n h_n) - f(x) - t_n \langle p, h_n \rangle\right) = \frac{1}{2}\langle Q, hh^t \rangle$$
$$\geq \quad f''_-(x,p,h) \geq \min\left\{f''_-(x,p,h), f''_-(x,p,-h)\right\} = f''_s(\bar{x},p,h). \ (7.11)$$

In order for $Q \in E\left(\partial^{2,-} f(x)\right)(h)$ we require $\langle Q, hh^t \rangle = f''_s(\bar{x},p,h)$. By standard arguments (see Example 13.16 of Rockafellar and Wets (1998)) one may show that

$$f''_-(\bar{x},p,h) = \max\{\langle \ \sum_{\alpha \in \Lambda(x)} \ \lambda_\alpha \nabla^2 f_\alpha(x), hh^t \rangle \mid \ \lambda \in \mathbb{R}^n_+ \ \text{and}$$
$$\sum_{\alpha \in \Lambda(x)} \ \lambda_\alpha = 1 \ \text{with} \ p = \sum_{\alpha \in \Lambda(x)} \lambda_\alpha \nabla f_\alpha(x)\} \ (7.12)$$

whenever h is a direction such that $p \in \text{co }\{\nabla f_\alpha(x) \mid \alpha \in \Lambda(x,h)\}$, as is the case here. If $p \in \text{co }\{\nabla f_\alpha(x) \mid \alpha \in \Lambda(x,-h)\}$, then as (7.12) holds for h and $-h$ we have $f''_-(\bar{x},p,h) = f''_-(\bar{x},p,-h)$ (due to the symmetry of the Hessians) and so $f''_-(\bar{x},p,h) = f''_s(\bar{x},p,h)$. Otherwise by the Clarke regularity of f we have $\langle p, -h \rangle < f^\circ(x,-h) = f'(x,-h)$, implying $f''_-(x,p,-h) = +\infty$ and once again $f''_-(\bar{x},p,h) = f''_s(\bar{x},p,h)$. By (7.8), (7.12) and the fact that $\partial^{2,-} f(x,p)$ is a rank–1 representer, we have immediately that $\partial^{2,-} f(x,p)$ equals

$$\left\{\sum_{\alpha \in \Lambda(x)} \lambda_\alpha \nabla^2 f_\alpha(x) \mid \ \lambda \in \mathbb{R}^n_+ \ \text{and} \ \sum_{\alpha \in \Lambda(x)} \lambda_\alpha = 1 \ \text{with} \ p = \sum_{\alpha \in \Lambda(x)} \lambda_\alpha \nabla f_\alpha(x)\right\}^1$$
$$(7.13)$$

In particular

$$\left\{\sum_{\alpha \in \bar{\Lambda}(x,h)} \lambda_\alpha \nabla^2 f_\alpha(x) \mid \lambda \in \overline{\Omega}(x,h,p)\right\} \subseteq \partial^{2,-} f(x,p).$$

Combining this observation with (7.11) yields

$$\left\{\sum_{\alpha \in \bar{\Lambda}(x,h)} \lambda_\alpha \nabla^2 f_\alpha(x) \mid \lambda \in \overline{\Omega}(x,h,p)\right\}^1 \subseteq E\left(\partial^{2,-} f(x,p),h\right).$$

Now suppose $p \in \mathrm{co}\,\{\nabla f_\alpha(x) \mid \alpha \in \bar{\Lambda}(x,h)\}$. It is clear that $\overline{\Omega}(x,h,p) \subseteq \Omega(x,h,p)$. By the linear independence of $\{\nabla f_\lambda(x) \mid \alpha \in \overline{\Lambda}(x,h)\}$ there exists a unique $\lambda \in \Omega(x,h,p)$ such that $p = \sum_{\alpha \in \Lambda(x,h)} \lambda_\alpha \nabla f_\alpha(x)$. Thus $\Omega(x,h,p) = \{\lambda\}$ forcing equality.

Remark 7.5 *If $f_\alpha \in C^2\,(\mathbb{R}^n)$ and Λ is finite, then f is locally Lipschitz, prox-regular and subdifferentially continuous (see Example 2.9 of Poliquin and Rockafellar (1996) and we have*

$$\partial_p f(x) = \partial f(x) = \mathrm{co}\,\{\,\nabla f_\alpha(x) \mid \alpha \in \Lambda(x)\}$$

(see, exercise 8.31 of Rockafellar and Wets (1998)). In particular, when $0 \in \partial f(x)$ and $0 = \max\{\langle \nabla f_\alpha(x), h \rangle \mid i \in \Lambda(x)\}$ this implies

$$0 \in \mathrm{co}\,\{\,\nabla f_\alpha(x) \mid \alpha \in \Lambda(x,h)\}$$

and so there exists λ such that

$$0 = \sum_{i \in \Lambda(x,h)} \lambda_i \nabla f_i(x) \text{ for some } \sum_{i \in \Lambda(x,h)} \lambda_i = 1 \text{ with } \lambda_i \geq 0.$$

Corollary 7.1 *Let $f_\alpha \in C^2(\mathbb{R}^n)$ for each $\alpha \in \Lambda = \{1,\dots,m\}$ and $f = \max_{\alpha \in \Lambda} f_\alpha$ as in Lemma 7.2 part 2. Then*

$$\left\{ \sum_{i \in \bar{\Lambda}(x,h)} \lambda_i \nabla^2 f_i(x) h \mid \lambda \in \overline{\Omega}(x,h,p) \right\} \subseteq D(\partial_p f)(x,p)(h) \subseteq D^*\,(\partial f)\,(\bar{x},p)(h).$$

Proof By Example 2.9 of Poliquin and Rockafellar (1996) and the discussion of Rockafellar and Wets (1998), we have all the assumptions of Corollary 6.2 holding.

Consider a standard non–linear programming problem

$$f^* := \min\,\{f_0(x) \mid f_i(x) \leq 0 \text{ for } i = 1,\dots,m \text{ and } x \in X\}. \tag{7.14}$$

We consider a modified Lagrangian penalty which is studied in Andromonov (2001) (but in greater generality) and is given by

$$U(x,d,r) \;:=\; \max\,\{L(x,d), r_1 f_1(x),\dots, r_m f_m(x)\},$$

$$\text{where } L(x,d) \;:=\; f_0(x) + \sum_{i=1}^{m} d_i f_i(x).$$

This can be viewed as a combination of penalty and Lagrangian methods. We posit the following assumptions:

A1 all f_i are continuous for all $i = 0, \ldots, m$;

A2 f_0 is positive on the feasible set $\mathcal{F} \cap X$, where

$$\mathcal{F} := \{x \mid f_i(x) \leq 0 \text{ for } i = 1, \ldots, m\} \cap X$$

(this can easily be arranged *via* a reformulation);

A3 all the constraints f_i are bounded below on $\mathbb{R}^n$ (once again this can easily be arranged *via* a reformulation).

Then it can be shown (see Andromonov (2001), Chapter 5, but in greater generality) that the following are all true (note the absence of regularity assumptions).

1. Suppose $(\bar{x}, \bar{d}) \in X \times \mathbb{R}^m_+$ is a saddle point of $L(x, d)$, $f_i(\bar{x}) \leq 0$ for all $i = 1, \ldots, m$ and $\bar{d}_i f_i(\bar{x}) = 0$ for all $i = 1, \ldots, m$. Then $(\bar{x}, \bar{d})$ is also a saddle point of the function $U(x, d, \bar{r})$ for any fixed value of the penalty parameters $\bar{r}$, that is, for all $x \in X$ and $d \in \mathbb{R}^n_+ := \{d \mid d_i \geq 0\}$ we have

$$U(x, \bar{d}, \bar{r}) \geq U(\bar{x}, \bar{d}, \bar{r}) \geq U(\bar{x}, d, \bar{r}) \tag{7.15}$$

2. We have $f^* \geq \sup_{d \in \mathbb{R}^n_+} \inf_{x \in X} U(x, d, \bar{r})$.

Proof (Provided for completeness.) Let $x_n \in \mathcal{F} \cap X$ be a minimizing sequence i.e. $f_0(x_n) \to f^*$. Then $f_i(x_n) \leq 0$ for all i and n and as $d \in \mathbb{R}^m_+$ we have $\sum_{i=1}^m d_i f_i(x_n) \leq 0$ so $f_0(x_n) \geq L(x_n, d)$. Next note that

$$\inf_{x \in X} U(x, d, \bar{r}) \leq U(x_n, d, \bar{r}). \tag{7.16}$$

If $L(x_n, d) \leq 0$, then $U(x_n, d, \bar{r}) \leq 0 \leq f_0(x_n)$ (since $x_n \in \mathcal{F} \cap X$ and f_0 is positive there). Otherwise

$$U(x_n, d, \bar{r}) = L(x_n, d) \leq f_0(x_n).$$

In both cases we have on combining with (7.16) that

$$\inf_{x \in X} U(x, d, \bar{r}) \leq f_0(x_n)$$

for all n and so $\inf_{x \in X} U(x, d, \bar{r}) \leq f^*$ for all $d \in \mathbb{R}^m_+$.

3. Under the assumption that (for any $\varepsilon > 0$)

$$D(\varepsilon) := \{x \in X \mid f_i(x) \leq \varepsilon \, , \, \forall i = 1, \ldots, m\}$$

is upper semi–continuous for $\varepsilon \downarrow 0$, there exists $\bar{r}$ sufficiently large such that $f^* - \varepsilon \leq \sup_{d \in \mathbb{R}^n_+} \inf_{x \in X} U(x, d, \bar{r})$.

4. If $(\bar{x}, \bar{d}) \in X \times \mathbb{R}^m_+$ is a saddle point (as given in (7.15)), then $\bar{x}$ is an optimal solution for the problem (7.14).

Proof (Provided for completeness.) First we show that $\bar{x}$ is feasible. As we have a saddle point, it follows that for all $d \in \mathbb{R}^m_+$ we have

$$f^* \geq U(\bar{x}, \bar{d}, \bar{r}) \geq U(\bar{x}, d, \bar{r}) \geq f_0(\bar{x}) + \sum_{i=1}^{m} d_i f_i(\bar{x}).$$

If there is a k such that $f_k(\bar{x}) > 0$, we simply let $d_k \to \infty$ with $d_i = 0$ for $i \neq k$ to obtain a contradiction. Thus $\bar{x} \in \mathcal{F} \cap X$. Next let $d = 0$ to obtain $f^* \geq f_0(\bar{x})$ and so that $\bar{x}$ is optimal for (7.14).

5. From the work of Poliquin and Rockafellar (1996), we know that as

$$
\begin{aligned}
U(x, d, r) \quad &= \quad g \circ F(x) \\
\text{with } g(y) \quad &:= \quad \max \{y_0, \ldots, y_m\} \text{ and} \\
F(x) \quad &:= \quad \{L(x, d), r_1 f_1(x), \ldots, r_m f_m(x)\} \, ,
\end{aligned}
$$

then for any fixed (d, r) we have $x \mapsto U(x, d, r)$ prox–regular and subdifferentially continuous (see Example 2.9 of Poliquin and Rockafellar (1996)) and as observed before also Clarke regular (see (Rockafellar and Wets (1998))), semi–smooth (see (Mifflin (1977))) and hence submonotone and directionally upper semi–continuous (see Spingarn (1981) and Rockafellar (1982)). It is a very well–behaved function indeed. We now derive the necessary optimality condition associated with this Lagrangian penalty method.

Theorem 7.2 *Suppose that all $f_i \in C^2(\mathbb{R}^n)$ for $i = 0, \ldots, m$ and f_0 is bounded below on the feasible set*

$$\mathcal{F} := \{x \mid f_i(\bar{x}) \leq 0, \, \forall i \in \Lambda := \{1, \ldots, m\}\} \neq \emptyset.$$

Let $\Lambda(\bar{x}) := \{i \mid f_i(\bar{x}) = 0\}$ be the active set at $\bar{x}$ and

$$\Lambda(\bar{x}, h) := \{i \in \Lambda(\bar{x}) \mid \langle \nabla_x f_i(\bar{x}), h \rangle = 0\}.$$

Then the following are sufficient conditions for a strict minimum of order two at $\bar{x}$ for the problem (7.14) with $X = \mathbb{R}^n$. Suppose there exists $\bar{d} \in \mathbb{R}^n_+$ satisfying:

1. *$f_i(\bar{x}) \leq 0$ for all $i = 1, \ldots, m$ and $\bar{d}_i f_i(\bar{x}) = 0$ for all $i = 1, \ldots, m$;*

2. *$0 \in \mathrm{co}\left(\{\nabla_x L(\bar{x}, \bar{d})\} \cup \{\nabla_x f_i(\bar{x}) \mid i \in \Lambda(\bar{x})\}\right)$;*

3. *for each h such that $0 = \sup\left\{\langle \nabla_x L(\bar{x}, \bar{d}), h \rangle, \{\langle \nabla_x f_i(\bar{x}), h \rangle \mid i \in \Lambda(\bar{x})\}\right\}$ we have that*

    there exists $r_i > 0$ for $i = 1, \ldots, m$ such that $\Lambda(\bar{x}, h) = \bar{\Lambda}_r(\bar{x}, h)$, where

$$\bar{\Lambda}_r(\bar{x}, h)$$
$$:= \{i \in \Lambda(\bar{x}) \mid \exists (t_n, h_n) \to (0^+, h) \text{ such that} \tag{7.17}$$
$$r_i f_i(\bar{x} + t_n h_n) \geq \max\{r_1 f_1(\bar{x} + t_n h_n), \ldots, r_m f_m(\bar{x} + t_n h_n)\}$$

and there exists $(\lambda_0, \lambda_1, \ldots, \lambda_m) \geq 0$ with $\lambda_i = 0$ if $i \notin \Lambda(\bar{x}, h)$ and with $\lambda_0 = 0$ if $\langle \nabla_x L(\bar{x}, \bar{d}), h \rangle < 0$ such that

$$0 = \lambda_0 \nabla_x L(\bar{x}, \bar{d}) + \sum_{i \in \bar{\Lambda}(\bar{x}, h)} \lambda_i \nabla f_i(\bar{x}) \quad \text{and}$$

$$0 < \langle \lambda_0 \nabla_x^2 L(\bar{x}, \bar{d}) + \sum_{i \in \bar{\Lambda}(\bar{x}, h)} \lambda_i \nabla^2 f_i(\bar{x}), hh^t \rangle. \tag{7.18}$$

Remark 7.6 *We note that if (7.18) holds, then for any $r_i > 0$ there exists $\bar{\lambda}_i \geq 0$ with $\sum_{i \in \bar{\Lambda}(\bar{x}, h)} \bar{\lambda}_i = 1$ such that*

$$0 = \bar{\lambda}_0 \nabla_x L(\bar{x}, \bar{d}) + \sum_{i \in \bar{\Lambda}(\bar{x}, h)} \bar{\lambda}_i r_i \nabla f_i(\bar{x}) \quad \text{and}$$

$$0 < \langle \bar{\lambda}_0 \nabla_x^2 L(\bar{x}, \bar{d}) + \sum_{i \in \bar{\Lambda}(\bar{x}, h)} \bar{\lambda}_i r_i \nabla^2 f_i(\bar{x}), hh^t \rangle. \tag{7.19}$$

In particular this implies $0 \in \mathrm{co}\left(\{\nabla_x L(\bar{x}, \bar{d})\} \cup \{r_i \nabla_x f_i(\bar{x}) \mid i \in \Lambda(\bar{x}, h)\}\right)$. To see this, simply divide the equations in (7.18) by $\left(\sum_{i \in \bar{\Lambda}(\bar{x}, h)} \frac{\lambda_i}{r_i} + \lambda_0\right)$ and define

$$\bar{\lambda}_0 = \lambda_0 \left(\sum_{i \in \bar{\Lambda}(\bar{x}, h)} \frac{\lambda_i}{r_i} + \lambda_0\right)^{-1} \text{ and } \bar{\lambda}_i = \frac{\lambda_i}{r_i} \left(\sum_{i \in \bar{\Lambda}(\bar{x}, h)} \frac{\lambda_i}{r_i} + \lambda_0\right)^{-1}.$$

A similar calculation shows that the condition (2) implies

$$0 \in \mathrm{co}\left(\{\nabla_x L(\bar{x},\bar{d})\} \cup \{r_i \nabla_x f_i(\bar{x}) \mid i \in \Lambda(\bar{x})\}\right) \text{ for any } (r_1,\ldots,r_m) > 0.$$

Proof (Theorem 7.2). Since f_0 is bounded below on $\mathcal{F}$ by adding a sufficiently large constant to f_0 we may assume without loss of generality that f_0 is positive on $\mathcal{F}$. Consider the localized problem

$$f_0^* := \min\{f_0(x) \mid f_i(x) \le 0 \text{ for } i = 1,\ldots,m \text{ and } x \in B_\delta(\bar{x})\} \ge 0 \qquad (7.20)$$

for $X = B_\delta(\bar{x})$ for some $\delta > 0$ yet to be specified. We observe that as $f_i \in C^2(\mathbb{R}^n)$ for $i = 0,\ldots,m$ these are all bounded below for $\delta > 0$ sufficiently small. Define $U(x,d,\bar{r}) = \max\{L(x,d), \bar{r}_1 f_1(x),\ldots,\bar{r}_m f_m(x)\}$ and note that $U(\bar{x},\bar{d},\bar{r}) = f_0(\bar{x})$.

Observe that as $\bar{d}_i f_i(\bar{x}) = 0$ for all $i = 1,\ldots,m$, we have $L(\bar{x},\bar{d}) = f_0(\bar{x}) \ge 0$. Now suppose that contrary to our assertion that $\bar{x}$ is not a minimum of f_0 in $B_\delta(\bar{x})$ for any $\delta > 0$. Then for any $\bar{r} \in \mathbb{R}_+^m$ we have $U(\bar{x},\bar{d},\bar{r}) = L(\bar{x},\bar{d}) = f_0(\bar{x}) > 0$ and there exists $x' \in \mathcal{F} \cap B_\delta(\bar{x})$ such that $0 \le f_0(x') < f_0(\bar{x})$. This implies

$$U(\bar{x},\bar{d},\bar{r}) = f_0(\bar{x}) = L(\bar{x},\bar{d}) > f(x') + \sum_{i=1}^m \bar{d}_i f_i(x') = L(x',\bar{d}),$$

since $\sum_{i=1}^m \bar{d}_i f_i(x') \le 0$ for any $x' \in \mathcal{F}$. Now as all $\bar{r}_i f_i(x') \le 0$ for all $x' \in \mathcal{F}$ and any $\bar{r} \ge 0$, when $L(x',\bar{d}) > 0$ we have

$$\begin{aligned} U(x',\bar{d},\bar{r}) &= \max\{L(x',\bar{d}), \bar{r}_1 f_1(x'),\ldots,\bar{r}_m f_m(x')\} = L(x',\bar{d}) \\ &< L(\bar{x},\bar{d}) = U(\bar{x},\bar{d},\bar{r}). \end{aligned}$$

When $L(x',\bar{d}) \le 0$ we have

$$U(x',\bar{d},\bar{r}) \le 0 < L(\bar{x},\bar{d}) = U(\bar{x},\bar{d},\bar{r}).$$

In both cases we have, for $\delta > 0$, an $x' \in B_\delta(\bar{x}) \cap \mathcal{F}$ such that $U(x',\bar{d},\bar{r}) < U(\bar{x},\bar{d},\bar{r})$. If this is true for all $\delta > 0$, then there exists $x_n \to_h \bar{x}$ such that $f_0(x_n) < f_0(\bar{x})$ and

$$\begin{aligned} U'_-(\bar{x},\bar{d},r,h) &\le 0, \text{ which implies } L'_-(\bar{x},\bar{d},h) \le 0 \\ \text{since } U(x_n,\bar{d},\bar{r}) &= L(x_n,\bar{d}) \text{ for large } n \text{ since } f_0(\bar{x}) > 0. \end{aligned}$$

Now consider the function

$$\hat{U}(x,\bar{d},r) := \max\left\{ L(x,\bar{d}), r_1 f_1(x) + f_0(\bar{x}), \ldots, r_i f_i(x) + f_0(\bar{x}) \right\},$$

for which $\hat{U}(\bar{x},\bar{d},r) = f_0(\bar{x})$ and the maximum is attained for the indices $\Lambda(\bar{x}) \cup \{0\}$. By assumption 2. we have $0 \in \partial\hat{U}(x,\bar{d},r)$ and so

$$\sup\left\{ \langle \nabla_x L(\bar{x},\bar{d}), h\rangle, \{\langle r_i \nabla_x f_i(\bar{x}), h\rangle \mid i \in \Lambda(\bar{x},h)\}\right\} \geq 0 . \tag{7.21}$$

For $i \in \Lambda(\bar{x})$ we have $f_i(\bar{x}) = 0$ and so $r_i f_i(x_n) \leq r_i f_i(\bar{x}) = 0$ implying $\langle \nabla f_i(\bar{x}), h\rangle \leq 0$. As $L'_-(\bar{x},\bar{d},h) \leq 0$ we have the supremum in (7.21) equal to zero. Thus we may invoke 3. to obtain the existence of a $\lambda \geq 0$, $\sum_i \lambda_i = 1$ with $\lambda_i = 0$ if $i \notin \Lambda(\bar{x},h) \cup \{0\}$ such that $0 = \lambda_0 \nabla L(\bar{x},\bar{d}) + \sum_{i \in \Lambda(\bar{x},h)} \lambda_i r_i \nabla_i f(\bar{x})$. By (7.13) we have for any such choice of λ that

$$Q := \lambda_0 \nabla_x^2 L(\bar{x},\bar{d}) + \sum_{i \in \bar{\Lambda}(\bar{x},h)} \lambda_i r_i \nabla^2 f_i(\bar{x}) \in \partial^{2,-}\hat{U}(x,\bar{d},r,0).$$

Thus on choosing λ as given in (7.19) we have

$$\hat{U}(x_n,\bar{d},r) \geq \hat{U}(\bar{x},\bar{d},r) \;\; + \;\; \frac{1}{2}\left\langle Q, \left(\frac{x_n - \bar{x}}{\|x_n - x\|}\right)\left(\frac{x_n - \bar{x}}{\|x_n - x\|}\right)^t\right\rangle \|x_n - x\|^2$$
$$+ o(\|x_n - x\|^2).$$

This implies by (7.19) that for n sufficiently large and some $\varepsilon > 0$ we have

$$\hat{U}(x_n,\bar{d},r) \geq f(\bar{x}) + \varepsilon\|x_n - x\|^2 > f_0(\bar{x}),$$

but since we assumed $f_0(x_n) < f_0(\bar{x})$ with $x_n \in \mathcal{F}$ we have $\hat{U}(x_n,\bar{d},r) = f_0(\bar{x})$ contradicting $\varepsilon > 0$. Thus $\bar{x}$ is at least a local minimum.

Let us now fix $\delta > 0$ sufficiently small so that $\bar{x}$ is a local minimum of f_0 on $B_\delta(\bar{x})$. Let us now redefine the objective to be $f_0 - f_0^* \geq 0$ and henceforth will assume $f_0 \geq 0$ on the feasible set $\mathcal{F} \cap B_\delta(\bar{x})$. This translation does not change any the form of the optimality conditions. As $f_0(\bar{x}) = f_0^* = 0$ we have for all $r \geq 0$ that $U(\bar{x},\bar{d},r) = 0$. We now proceed to show that $\bar{x}$ is a strict local minimum order two. Note that for any $d \in \mathbb{R}_+^m$ we have

$$U(\bar{x},d,\bar{r}) \leq U(\bar{x},\bar{d},\bar{r}),$$

since $\sum_{i=1}^m d_i f_i(\bar{x}) \leq 0 = \sum_{i=1}^m \bar{d}_i f_i(\bar{x})$ as $f_i(\bar{x}) \leq 0$ for all $i = 1, \ldots, m$. Thus to verify that $\bar{x}$ is indeed a strict local minimum order two for (7.20) we need only consider if it is also one for $x \mapsto U(x, \bar{d}, \bar{r})$ on $B_\delta(\bar{x})$ for some $\delta > 0$. Indeed if $\bar{x}$ is a strict local minimum order two for $x \mapsto U(x, \bar{d}, \bar{r})$ on $B_\delta(\bar{x})$ then for all $x' \in \mathcal{F} \cap B_\delta(\bar{x})$ we have $U(x', d, \bar{r}) = \max\left\{L(x', \bar{d}), \bar{r}_1 f_1(x'), \ldots, \bar{r}_m f_m(x')\right\} = L(x', \bar{d}) \leq f_0(x')$. On recalling that $f_0(\bar{x}) = U(\bar{x}, \bar{d}, r) = 0$, we have thus that

$$f_0(x') - f_0(\bar{x}) \geq U(x', \bar{d}, \bar{r}) - U(\bar{x}, \bar{d}, \bar{r}) \geq \gamma \|x' - \bar{x}\|^2$$

and $\bar{x}$ is also a strict local minimum order two for the problem (7.14).

Next note that the first–order condition for this amounts to

$$0 \in \partial_x U(x, \bar{d}, \bar{r}) = \mathrm{co}\left(\{\nabla_x L(\bar{x}, \bar{d})\} \cup \{r_i \nabla_x f_i(\bar{x}) \mid i \in \Lambda(\bar{x})\}\right). \qquad (7.22)$$

We now use sufficient conditions of the first kind in the form of (7.4). We need to show that for all h with $U'_-(\bar{x}, \bar{d}, \bar{r}, h) \leq 0$ and $h \in b^1\left(\partial^{2,-}U(\bar{x}, \bar{d}, \bar{r}, 0)\right)$, for some $\varepsilon > 0$ we have the existence of a $Q \in E_\varepsilon\left(\partial^{2,-}U(\bar{x}, \bar{d}, \bar{r}, 0), h\right)$ with $0 < \langle Q, hh^t \rangle$. As (7.22) holds we have $U'_-(\bar{x}, \bar{d}, \bar{r}, h) \geq 0$ and hence we need to consider only those h such that $U'_-(\bar{x}, \bar{d}, \bar{r}, h) = 0$ or

$$0 = \sup\langle \nabla_x L(\bar{x}, \bar{d}), h \rangle, \{\langle r_i \nabla_x f_i(\bar{x}), h \rangle \mid i \in \Lambda(\bar{x})\}.$$

This means that $0 \in \mathrm{co}\left\{\langle \nabla_x L(\bar{x}, \bar{d}), \{r_i \nabla_x f_i(\bar{x}) \mid i \in \Lambda(\bar{x}, h)\}\right\}$, so we need to consider all $(\lambda_0, \lambda_1, \ldots, \lambda_m) \geq 0$ with $\lambda_i = 0$ if $i \notin \Lambda(\bar{x}, h)$ such that $\sum_{i \in \Lambda(\bar{x}, h)} \lambda_i r_i + \lambda_0 = 1$ and

$$0 = \lambda_0 \nabla_x^2 L(\bar{x}, \bar{d}) + \sum_{i \in \bar{\Lambda}(\bar{x}, h)} \lambda_i r_i \nabla^2 f_i(\bar{x}). \qquad (7.23)$$

We are at liberty to assume that λ satisfies (7.19). We now use (7.10) for the function $x \mapsto U_{-1}(x, \bar{d}, r) := \max\{\bar{r}_1 f_1(x), \ldots, \bar{r}_m f_m(x)\}$. Let $\mu_i := \left(\dfrac{\lambda_i}{\sum_{i \in \Lambda(\bar{x}, h)} \lambda_i}\right)$ and define

$$p_{-1} := \sum_{i \in \Lambda(\bar{x}, h)} \mu_i r_i \nabla f_i(\bar{x}) \in \partial U_{-1}(\bar{x}, \bar{d}, r).$$

Note that as $\Lambda(\bar{x}, h) = \bar{\Lambda}_r(\bar{x}, h)$, we have by (7.13) that

$$Q_{-1} := \sum_{i \in \Lambda(\bar{x}, h)} \mu_i r_i \nabla^2 f_i(\bar{x}) \in \partial^{2,-}U_{-1}(\bar{x}, \bar{d}, r; p_{-1}).$$

Since $U(\bar{x}, \bar{d}, \bar{r}) = L(\bar{x}, \bar{d}) = 0$ and $U(\bar{x}, \bar{d}, \bar{r}) = \max\{L(\bar{x}, \bar{d}), U_{-1}(\bar{x}, \bar{d}, r)\}$, we may apply (7.8) to obtain for our given $\lambda_0 \in (0, 1)$ such that

$$\lambda_0\left(\nabla L(\bar{x}, \bar{d}), \nabla^2 L(\bar{x}, \bar{d})\right) + (1 - \lambda_0)\left(p_{-1}, Q_{-1}\right) \in \partial^{2,-}U(\bar{x}, \bar{d}, r).$$

As $1 - \lambda_0 = \sum_{i \in \Lambda(\bar{x},h)} \lambda_i$, we have by (7.23) that $\lambda_0 \nabla L(\bar{x},\bar{d}) + (1 - \lambda_0) p_{-1} = 0$ and hence

$$
\begin{aligned}
Q' &:= \lambda_0 \nabla^2 L(\bar{x},\bar{d}) + (1 - \lambda_0) Q_{-1} \\
&= \lambda_0 \nabla_x^2 L(\bar{x},\bar{d}) + \sum_{i \in \bar{\Lambda}(\bar{x},h)} \lambda_i r_i \nabla^2 f_i(\bar{x}) \in \partial^{2,-} U(\bar{x},\bar{d},r,0),
\end{aligned}
$$

with Q' satisfying (7.19), that is, $\langle Q', hh^t \rangle > 0$. When $h \in b^1 \left(\partial^{2,-} U(\bar{x},\bar{d},r,0) \right)$, there always exists a $Q \in (Q' + \mathcal{P}(n)) \cap E_\varepsilon \left(\partial^{2,-} U(\bar{x},\bar{d},r,0), h \right) \neq \emptyset$ such that (7.19) holds and so we have $\langle Q, hh^t \rangle > 0$, the desired conclusion follows from (7.4).

Remark 7.7 *The standard second–order sufficiency conditions correspond to the case when $\nabla L(\bar{x},\bar{d}) = 0$ and so $0 \in \Lambda(\bar{x},h)$ for all h. If we assume also that $\langle \nabla^2 L(\bar{x},\bar{d}) hh^t \rangle > 0$ for all h satisfying $\langle \nabla L(\bar{x},\bar{d}), h \rangle = 0$, we find that (7.18) is automatically satisfied. Assumption (7.17) corresponds to a kind of well–posedness condition which replaces the usual regularity at $\bar{x}$. The correspondence between these two conditions is a topic of current research.*

8   APPENDIX

In this appendix we provide the proof of Lemma 4.1. Let Θ be a class of functions from $\mathbb{R}_+$ to $\mathbb{R}$ and define $J_2(\Theta, \bar{x})$ to be the class

$$
\left\{ \varphi \in \mathcal{C}^2(\mathbb{R}^n) \mid \text{there exists } (\alpha, p, Q) \in \mathbb{R}^{n+1} \times \mathcal{S}(n) \text{ and } r \in \Theta \text{ such that} \right.
$$
$$
\left. \varphi(x) := \alpha + \langle p, x - \bar{x} \rangle + \frac{1}{2} \langle Q, (x - \bar{x})(x - \bar{x})^t \rangle - r(\|\bar{x} - x\|) \|\bar{x} - x\|^2 \right\}.
$$

Lemma 4.1 claims that it is sufficient to consider only the class $J_2(\Theta, \bar{x})$ generated by

$$
\Theta := \{ r : \mathbb{R}_+ \mapsto \mathbb{R} \mid r(\cdot) \in \mathcal{C}^2(\mathbb{R}_+ \setminus \{0\}) \text{ with } \lim_{t \downarrow 0} r(t) = r(0) = 0 \},
$$

which we denote by $J_2(\bar{x})$.

In Penot (1994/1), the subjet is defined *via* a notion equivalent to taking $\Theta := \{ r : \mathbb{R}_+ \mapsto \mathbb{R} \mid \lim_{t \downarrow 0} r(t) = 0 \}$. Penot shows that in finite dimensions that this amounts to demanding

$$
\liminf_{|h| \to 0} \left(\frac{1}{\|h\|^2} \right) \left(f(\bar{x} + h) - f(\bar{x}) - \langle p, h \rangle - \frac{1}{2} \langle Q, hh^t \rangle \right) \geq 0. \tag{8.1}
$$

On reflection one can see that this corresponds to the second–order characterization of subjets as provided in the first–order case by Proposition 1.2 (page 341) in Deville *et al.* (1993). We extract and extend to the second–order level the relevant parts of this argument in the proof of Lemma 4.1.

Proof [**Lemma 4.1**] The only problem with w in (4.1) is that it is not necessarily $C^2(\mathbb{R}^n)$. As a $C^2(\mathbb{R}^n)$ bump function exists, the construction found in Lemma 1.3 of Deville *et al.* (1993) (page 340) leads to a function $d : \mathbb{R}^n \mapsto \mathbb{R}_+$, where

$$d(x) := \frac{2}{h(x)} \quad \text{and} \quad h(x) = \sum_{n=0}^{\infty} b(nx).$$

Here $b : \mathbb{R}^n \mapsto \mathbb{R}$ is a $C^2(\mathbb{R}^n)$ bump function such that $0 \le b \le 1$ on $\mathbb{R}^n$, $b(0) = 1$ and $b(x) = 0$ for all $\|x\| \ge 1$. Since $\|\cdot\|^2$ is $C^2(\mathbb{R}^n)$ by any one of a standard set of constructions, we may choose here $b(x) = R(\|x\|^2)$ (for some $R : \mathbb{R}_+ \mapsto \mathbb{R}_+$) for a $C^2(\mathbb{R})$ bump function on $\mathbb{R}$. Thus we may assume that b is of the form $b(x) = D(\|x\|)$ for some function D defined on the positive reals. The function d possess all the properties announced in Lemma 1.3 of Deville *et al.* (1993) as well as being $C^2(\mathbb{R}^n)$ on $\mathbb{R}^n$ (note that $h(x) \ge 1$). To see this, observe that the sum is locally finite on $\mathbb{R}^n$ and so

$$\nabla^2 d(x) \;=\; -2 \left(\sum_{n=0}^{\infty} n^2 \nabla^2 b(nx) \right) \left(\sum_{n=0}^{\infty} b(nx) \right)^{-2}$$
$$+4 \left(\sum_{n=0}^{\infty} n \nabla b(nx) \right) \left(\sum_{n=0}^{\infty} n \nabla b(nx) \right)^t \left(\sum_{n=0}^{\infty} b(nx) \right)^{-3}$$

is well–defined as $\sum_{n=0}^{\infty} b(nx) \ge 1$. Without loss of generality we may assume that $w(\cdot) \ge 0$. Now arguing as in Proposition 1.2 of Deville *et al.* (1993) (page 341) we define

$$\rho(t) := \inf\{u(h) \mid \|h\| \le t\} \quad \text{where} \quad u(h) := \sup\{-w(\|h\|)\|h\|^2, -1\}.$$

Then ρ is nonincreasing, $\rho(0) = 0$ and so $\rho \le 0$ and

$$\liminf_{\|h\| \to 0} \frac{u(h)}{\|h\|^2} \;\ge\; 0, \quad \text{implying} \quad \lim_{t \to 0} \frac{\rho(t)}{t^2} = 0,$$
$$\text{because} \quad \inf_{\|h\| \le t} \frac{u(h)}{\|h\|^2} \;\le\; \inf_{\|h\| \le t} \frac{u(h)}{t^2} = \frac{\rho(t)}{t^2} \le 0.$$

Continuing in a parallel fashion to Deville *et al.* (1993) Proposition 1.2 we define

$$\rho_1(t) := \int_t^{et} \frac{\rho(s)}{s}\, ds, \quad \rho_2(t) := \int_t^{et} \frac{\rho_1(s)}{s}\, ds \quad \text{and} \quad \rho_3(t) := \int_t^{et} \frac{\rho_2(s)}{s}\, ds.$$

As in the proof of Proposition 1.2 of Deville *et al.* (1993) (page 342) we have

$$\rho(e^3 t) \leq \rho_1(e^2 t) \leq \rho_2(et) \leq \rho_3(t) \leq \rho_2(t) \leq \rho_1(t) \leq \rho(t) \leq 0. \tag{8.2}$$

Thus we have

$$\lim_{t \to 0} \frac{\rho_3(t)}{t^2} = \lim_{t \to 0} \frac{\rho_2(t)}{t^2} = \lim_{t \to 0} \frac{\rho_1(t)}{t^2} = \lim_{t \to 0} \frac{\rho(t)}{t^2} = 0. \tag{8.3}$$

As ρ_1 is continuous, ρ_3 is C^2–smooth on $(0, +\infty)$. Put $\psi(x) := \rho_3(d(x))$ and $\psi(0) = 0$. This function is clearly Fréchet differentiable on $\mathbb{R}^n \setminus \{0\}$, as $d(x) \neq 0$ for $x \neq 0$. Since ρ is nonincreasing, it follows that each ρ_i, $i = 1, 2, 3$ are nonincreasing. Moreover as $(u - \psi)(0) = 0$ and $\|x\| \leq d(x)$ for $\|x\| \leq 1$, we have

$$(u - \psi)(h) = u(h) - \rho_3(d(h)) \geq u(h) - \rho_3(\|h\|) \geq u(h) - \rho(\|h\|) \geq 0$$

and so $u - \psi$ has a local minimum at 0. Using (4.1), we find that in some neighbourhood of $\bar{x}$ we have

$$\begin{aligned}
f(y) - f(\bar{x}) - \langle p, y - \bar{x} \rangle - \frac{1}{2}\langle Q, (y - \bar{x})(y - \bar{x})^t \rangle &\geq -w(\|y - \bar{x}\|)\|y - \bar{x}\|^2 \\
&= u(y - \bar{x}) \geq \psi(y - \bar{x}).
\end{aligned}$$

It remains only to show that ψ is twice Fréchet differentiable at 0. *via* direct calculation we have for $x \in \mathbb{R}^n \setminus \{0\}$ that

$$\begin{aligned}
\nabla\psi(h) &= \rho_3'(d(x))\nabla d(x) \quad \text{and} \\
\nabla^2\psi(h) &= \rho_3''(d(x))\left(\nabla d(x)\nabla d(x)^t\right) + \rho_3'(d(x))\nabla^2 d(x), \quad \text{where} \\
\rho_3'(t) &= \left(\frac{\rho_2(et) - \rho_2(t)}{t}\right) \quad \text{and} \\
\rho_3''(t) &= \left(\frac{\rho_1(e^2 t) - \rho_1(et)}{t^2}\right) e - \left(\frac{\rho_1(et) - \rho_1(t)}{t^2}\right) - \left(\frac{\rho_2(et) - \rho_2(t)}{t^2}\right).
\end{aligned} \tag{8.4}$$

By (8.3), $\rho_3'(t) \to 0$ along with $\rho_3''(t) \to 0$ as $t \to 0$. Since $d(x)$ is Lipschitz continuous, we have $d(x) \to 0$ as $\|x\| \to 0$. In the proof of Proposition 1.2 of Deville *et al.* (1993) it is shown that $\|\nabla\psi(x)\| \to 0$ as $x \to 0$ and so we need only show $\|\nabla^2\psi(x)\| \to 0$ as $x \to 0$. To do this we need estimates on $\nabla^2 d(x)$ for such x. From Proposition 1.2 of Deville *et al.* (1993) we know that $h(x) \leq \frac{C}{\|x\|}$

for all small $x \neq 0$ (that is, $\frac{1}{h(x)} = O(\|x\|)$. Letting M be a global bound for $\|\nabla b(x)\|$ and $\|\nabla^2 b(x)\|$, we have since that $\nabla b(x)$ and $\nabla^2 b(x)$ are zero outside the unit ball that

$$\left\| \sum_{n=0}^{\infty} n \nabla b(nx) \right\| \leq M \sum_{n=1}^{\left[\frac{1}{\|x\|}\right]} n = \frac{M}{2} \left[\frac{1}{\|x\|}\right] \left(\left[\frac{1}{\|x\|}\right] + 1\right)$$

$$\leq M'\left(2 + \frac{1}{\|x\|}\right)^2 = O\left(\frac{1}{\|x\|^2}\right).$$

Similarly,

$$\left\| \sum_{n=0}^{\infty} n^2 \nabla^2 b(nx) \right\| \leq M \sum_{n=1}^{\left[\frac{1}{\|x\|}\right]} n^2$$

$$= \frac{M}{6} \left[\frac{1}{\|x\|}\right] \left(\left[\frac{1}{\|x\|}\right] + 1\right) \left(2\left[\frac{1}{\|x\|}\right] + 1\right) = O\left(\frac{1}{\|x\|^3}\right).$$

Thus

$$\|\nabla d(x)\| = O\left(\frac{1}{\|x\|^3}\right) O\left(\|x\|\right)^2 + O\left(\frac{1}{\|x\|^2}\right)^2 O\left(\|x\|\right)^3 = O\left(\frac{1}{\|x\|}\right)$$

as $x \to 0$. From (8.4) and $\lim_{t \to 0} \frac{\rho_2(t)}{t^2} = 0$ we have

$$\frac{\rho_3'(t)}{t} = e^2 \frac{\rho_2(et)}{(et)^2} - \frac{\rho_2(t)}{t^2} \to 0$$

and since $\|x\| \leq d(x) \leq K\|x\|$ for $\|x\| \leq 1$, it follows that $\rho_3'(d(x)) = o\left(\|x\|\right)$. Hence since $\rho_3''(d(x)) \to 0$ and $\nabla d(x)$ is bounded as $x \to 0$ we have

$$\nabla^2 \psi(x) = \rho_3''(d(x)) \nabla d(x) \nabla d(x)^t + \rho_3'(d(x)) \nabla^2 d(x)$$

$$= o(1)\left(O(1)\right)^2 + o(\|x\|) O\left(\frac{1}{\|x\|}\right) = o(1).$$

By the mean–value theorem applied initially to the function ψ we find $\nabla\psi(0)$ which exists and is zero at the origin (see the proof of Proposition 1.2 of Deville et al. (1993)). Next we apply the mean–value theorem to the first derivative $\nabla\psi(x)$ to deduce that $\nabla^2\psi(0) = 0$ exists and $\nabla^2\psi$ is continuous at 0. That is, for some $\gamma_t \in (0,1)$, we have

$$\frac{1}{\|h\|}\left(\nabla\psi(x+h) - \nabla\psi(x)\right) = \nabla^2\psi(x + \gamma_t h)h \to 0$$

as $t \to 0$ and $h \to 0$. Thus ψ is twice Fréchet differentiable at 0 with $\Box\psi(\bar{x}) = (0,0)$. Finally we note the following. As d is $C^2(\mathbb{R}^n)$ it is locally Lipschitz and so Lipschitz on the (compact) unit ball. Since $d(x) = 0$ outside the unit ball, d is globally Lipschitz. Thus

$$0 \geq \liminf_{y \to \bar{x}} \frac{\psi(y - \bar{x})}{\|y - \bar{x}\|^2} = \liminf_{y \to \bar{x}} \frac{\rho_3(d(y - \bar{x}))}{d(y - \bar{x})^2} \left(\frac{d(y - \bar{x})}{\|y - \bar{x}\|} \right)^2$$

$$\geq K^2 \left(\liminf_{y \to \bar{x}} \frac{\rho_3(d(y - \bar{x}))}{d(y - \bar{x})^2} \right) = 0,$$

where K is the global Lipschitz constant of d. Putting $r(t) := -\frac{\rho_3(D(t))}{t^2} \geq 0$ and $r(0) = 0$, we derive $\psi(h) = -r(\|h\|)\|h\|^2$ is as required.

References

Anderson W. Jr. and R. J. Duffin (1969), Series and Parallel Addition of Matrices *J. Math. Anal. Appl.*, Vol. 26, pp. 576–594.

Andromonov M. (2001), *Global Minimization of Some Classes of Generalized Convex Functions*, PhD Thesis, University of Ballarat, Australia.

Attouch H. (1984) , *Variational Convergence for Functions and Operators*, Pitman Adv. Publ. Prog. Boston–London–Melbourne.

Aubin J. P. and Frankowska H. (1990), *Set Valued Analysis*, Birkhauser.

Auslender A. (1984), Stability in Mathematical Programming with Nondifferentiable Data, *SIAM J. Control and Optimization*, Vol. 22, pp 239–254.

Balder E. J. (1977), An Extension of Duality Relations to Nonconvex Optimzation Problems, SIAM J. Control and Optim., Vol. 15, pp. 329–343.

Ben-Tal A. (1980) Second–Order and Related Extremality Conditions in Nonlinear Programming, *J. Optimization Theory Applic.* Vol. 31, pp. 143–165.

Ben-Tal A. and Zowe J. (1982) *Necessary and Sufficient Conditions for a Class of Nonsmooth Minimization problems*, Mathematical Programming Study 19, pp. 39-76.

Bonnans J. F., Cominetti R. and Shapiro A. (1999) Second Order Optimality Conditions Based on Parabolic Second Order Tangent Sets, *SIAM J. Optimization*, Vol. 9, No. 2, pp. 466-492.

Crandall M., Ishii H. and Lions P.-L. (1992), User's Guide to Viscosity Solutions of Second Order Partial Differential Equations, *Bull. American Math. Soc.*, Vol. 27, No. 1, pp. 1–67.

Cominetti R. and Penot J.-P. (1995), Tangent sets to unilateral convex sets, *C. R. Acad. Sci. Ser. I Math.*, 321, pp 1631–1636.

Dolecki S. and Kurcyusz S. (1978), On Φ–Convexity in Extremal Problems, *Soc. for. Indust. and Applied Maths. (SIAM), J.of Control and Optimization,* Vol. 16, pp. 277-300.

Deville R., Godefroy G. and Zizler V. (1993), *Smoothness and Renorming in Babach Spaces,* Pitman Monographs and Surveys in Pure and Applied Mathematics 64, Longman Science and Technical–Wiley and Sons, Inc., New York.

Eberhard A., Nyblom M. and Ralph D. (1998), Applying Generalised Convexity Notions to Jets, J.–P. Crouzeix *et al..* (eds), in *Generalized Convexity, Generalized Monotonicity: Recent Results,* Kluwer, pp. 111–157.

Eberhard A. and Nyblom M. (1998), Jets, Generalized Convexity, Proximal Normality and Differences of Functions, *Nonlinear Analysis* Vol. 34, pp. 319–360.

Eberhard A. (1998), Optimality Conditions using a Generalized Second Order Derivative, in *Proceedings of ICOTA98, Optimization Tecniques and Applications,* Vol. 2, Curtin University Press, pp. 811–818.

Eberhard A. (2000), Prox–Regularity and Subjets, in *Optimization and Related Topics,* Applied Optimization Volumes, Kluwer Academic Pub., Ed. A. Rubinov, pp. 237–313.

Eberhard A. and Ralph D. (2002), Rank One Representers, to appear in the *Journal of Mathematical Sciences,* published by Kluwer/Plenum.

Holmes R. (1975),*Geometric Functional Analysis,* Springer–Verlag, New York Berlin.

Ioffe A. D. (1981), Nonsmooth Analysis: Differential Calculus of Nondifferentiable Mappings, *Trans. Amer. Math. Soc.,* Vol. 266, No. 1, pp. 1–56.

Ioffe A. D. (1984), Calculus of Dini Subdifferentials of Functions and Contingent Coderivatives of Set–Valued Mappings, *Nonlinear Analysis: Theory, Methods and Applications,* Vol. 8, pp 517–539.

Ioffe A. D. (1996), Approximate Subdifferentials and Applications 2, *Mathematika* 33, pp 111–128.

Ioffe A. D. (1989), Approximate Subdifferentials and Application 3. The Metric Theroy, *Mathematika 36,* Vol. 36, No. 3, pp 1–38.

Ioffe A. D. (1990) Composite Optimization: Second Order Conditions, Value Functions and Sensitivity, *Analysis and optimization of systems,* Lecture Notes in Control and Inform. Sci., 144 (Antibes), pp. 442–451.

Ioffe A. D. (1991) Variational Analysis of Composite Function: A Formula for the Lower Second Order Epi–Derivative, *J. Math. Anal. and Applic.*, Vol. 160, pp. 379–405.

Ioffe A. and Penot J.-P. (1997), Limiting Subhessians, Limiting Subjets and Their Calculus, *Transactions of the American Mathematics Society*, No. 2, pp. 789–807.

Janin R. (1973) Sur la dualité in programmation dynamique, *C.R. Acad. Sci. Paris A* 277 , pp. 1195-1197.

Kruger A. Ya. and Mordukhovich B. S. (1980), Extremal Points and the Euler Equation in Nonsmooth Optimzation Problems, *Dokl. Akad. Nauk BSSR*, Vol. 24, pp. 684–687.

Ky Fan (1963) On the the Krein–Milman Theorem, *Proc. of Symposia in Pure Mathematics. Vol VII*, American Math. Soc., Providence, RI, 1963, pp 211–220.

Martinez–Legas J–E (1988), Generalized Conjugation and related Topics, in 'Generalized Convexity and Fractional Programming with Economic Applications', Proceedings, Pisa, Italy, 1998 *Lecture Notes Economics and Mathematical Systems Vol. 345*, Springer–Verlag, pp. 168–179.

Martinez–Legas J–E and Singer I. (1995), Subdifferentials with respect to Dualities, *Mathematical Methods of Operations Research*, Vol. 42, pp 109–125.

Mazure M.–L. (1986), L'addition parallèle d'opérareurs Interprétée Comme Inf-convolution de Formes Quadratiques Convexes, *RAIRO Modél. Math. Anal. Numér.*, Vol. 20, No. 3, pp. 497–515.

Mifflin R. (1977), Semismooth and Semiconvex Functions in Constrained Optimization, *SIAM J. Control Optim.*, Vol. 15, pp. 957-972.

Mordukhovich B. S. (1976), Maximum Principle in the Problem of Time Optimal Response with Nonsmooth Constraints, *J. Appl. Math. Mech.*, Vol. 40, pp. 960–969; translation from Prikl. Mat. Mekh. 40, pp. 1014–1023.

Mordukhovich B. S. (1984), Nonsmooth Analysis with Nonconvex generalized Differentials and Conjugate Mappings, *Dokl. Akad. Nauk. BSSR*, Vol. 28, pp. 976–979.

Mordukhovich B. S. (1994), Generalized Differential Calculus for Nonsmooth and Set–Valued Mappings, *J. Math. Anal. and Applic.*, Vol. 183, pp. 1805–1838.

Mordukhovich B. S. and Shao Y. (1998), Mixed Coderivatives of Set–Valued Mappings in Variational Analysis, *J. Appl. Analysis*, Vol. 4, No.2, pp. 269–294.

Mordukhovich B. S. and Outrata J. V. (2001), On Second Order Subdifferentials and their Applications, *SIAM J. Optim.* , Vol. 12, No. 1, pp. 139–169.

Mordukhovich B. S. (2002), Calculus of Second–Order Subdifferentials in Infinite Dimensions, *personal communication.*

Pallaschke D. and Rolewicz S. (1998) *Foundations of Mathematical Optimization. Convex Analysis without Linearity*, Maths. and its Appl., Vol. 388, Kluwer, Dordrecht.

Penot J.–P. and Volle M. (1988), On Strongly Convex and paraconvex Dualities, in 'Generalized Convexity and Fractional Programming with Economic Applications', Proceedings, Pisa, Italy, 1998 *Lecture Notes Economics and Mathematical Systems Vol. 345*, Springer–Verlag, pp 198–218.

Penot J.-P. (1992), Second-Order Generalised Derivatives: Relationship with Convergence Notions, *Non-Smooth Optimization Methods and Applications*, Gordon and Breach Sc, Pub., Ed. F. Giannessi, pp. 303–322.

Penot J.-P.(1994), Sub–Hessians, Super–Hessians and Conjugation, *Non–linear Analysis, Theory, Methods and Applications*, Vol. 23, No. 6, pp. 689–702.

Penot J.–P. (1994) Optimality Conditions in Mathematical Programming and Composite Optimization, *Mathematical Programming* Vol. 67, pp. 225–245.

Poliquin R. A. and Rockafellar R. T. (1993), A Calculus of Epiderivatives Applicable to Optimization, *Canadian Journal of Mathematics*, Vol. 45, No. 4, pp. 879-896.

Poliquin R. A. and Rockafellar R. T. (1996), Prox-Regular Functions in Variational Analysis, *Transactions of the American Mathematical Society*, Vol. 348, No. 5, 99 pp. 1805-1838.

Poliquin R. A. and Rockafellar R. T. (1996), Generalised Hessians Properties of Regularized Nonsmooth Functions, *SIAM J Optimization*, Vol. 6, No. 4, pp. 1121-1137.

Poliquin R. A. and Rockafellar R. T. (1998), Tilt Stability of Local Minimum, *SIAM J. Optimization*, Vol. 8, No.2, pp. 287–299.

R. T. Rockafellar (1970) *Convex Analysis*, Princeton University Press, Princeton New Jersey.

Rockafellar R. T. (1982), Favorable classes of Lipschitz Continuous Functions in Subgradient Optimization, *Progress in Nondifferentiable Optimization*, E, Nurminski, ed., (IIASA, Luxenberg, Austria, 1982), pp. 125-143.

Rockafellar R. T. (1989), Proto-Differentiability of Set-Valued Mappings and its Applications in Optimization, *Analyse Non Linéaire*, (editor H. Attouch *et al.*), Gauthier-Villars, Paris, pp. 449-482.

Rockafellar R. T. (1988), First and Second Order Epi–Differentiability in Nonlinear Programming, *Transactions of the American Mathematical Society*, Vol. 307, No. 1, pp. 75-108.

Rockafellar R. T. and Zagrodny D. (1997), A Derivative–Coderivative Inclusion in Second–Order Nonsmooth Analysis, *Set–Valued Analysis*, No. 5, pp. 89–105.

Rockafellar R. T. (1990), Generalized Second Derivatives of Convex Functions and Saddle Functions, *Trans. Amer. Math. Soc.*, Vol. 320, pp. 810–822.

Rockafellar R. T. and Wets R. J-B. (1998), *Variational Analysis*, Volume 317, A series of Comprehensive Studies in Mathematics, Pub. Springer.

Rubinov A. (200), Abstract Convexity and Global Optimization., *Nonconvex Optimization and its Applications 44*, Dordrecht: Kluwer Academic Publishers.

SeegerA. (1986), *Analyse du Second Ordre de problèmes Non Différentiables*, Thèsis de l'Université Paul Sabatier, Toulouse, 1986.

Seeger A. (1991), Complément de Shur at Sous-différentriel du Second-ordre d'une Function Convexe, *Aequationes mathematicae*, Vol. 42, No.1, pp. 47–71.

Seeger A. (1992) Limiting Behaviour of the Approximate Second–Order Subdifferential of a Convex Function, *Journal of Optimization theory and Applications*, Vol. 74, No. 3, pp. 527–544.

Seeger A. (1994), Second–Order Normal Vectors to Convex Epigraph, *Bull. Austral. Math. Soc.*,Vol. 50, pp. 123–134.

Singer I., *Abstract convex analysis*, Wiley, New York.

Spingarn J.E. (1981), Sub-monotone Sub-differentials of Lipschitz Functions, *Transactions of the American Mathematicla Society*, Vol. 264, pp. 77-89.

Studniarski M. (1986), Necessary and Sufficient Conditions for Isolated Local Minima of Nonsmooth Functions, *SIAM J. Control and Optimization*, Vol. 24, No. 5, pp. 1044–1049.

Hiriart–Urruty J.–B. (1986), A New Set–Valued Second Order Derivative for Convex Functions, *Mathematics for Optimization*, Mathematical Studies 129, (North–Holland, Amsterdam, 1996), pp 157–182.

Hiriart–Urruty J.–B. and Seeger A. (1989), The Second–Order Subdifferential and the Dupin Indicators of a Non–differential Convex Function, *Proc. London Math, Soc.*, Vol. 58, No. 3, pp. 351–365.

Vial J.–P. (1983), Strong and Weak Convexity of Sets and Functions, *Mathematics of Operations Research*, Vol. 8, No. 2, pp. 231–259.

Ward D. (1995), A Comparison of Second-Order Epiderivativess: Calculus and Optimality Conditions, *Journal of Mathematical Analysis and Applications*, Vol. 193, pp. 465-482.

Ward D. (1994), Characterizations of Strict Local Minima and Necessary Conditions for Weak Sharp Minima, *Journal of Optimization Theory and Applications*, Vol.80, No. 3, pp. 551–571.

Yang, X.Q. and Jeyakumar, V. (1992), Generalized second-order directional derivatives and optimization with $C^{1,1}$ functions, *Optimization*, Vol. 26, pp. 165-185.

Yang, X.Q. (1998), Second-order global optimality conditions for convex composite optimization, *Mathematical Programming*, Vol. 81, pp. 327-347.

3   DUALITY AND EXACT PENALIZATION VIA A GENERALIZED AUGMENTED LAGRANGIAN FUNCTION

X.X. Huang

Department of Mathematics and Computer Science,
Chongqing Normal University,
Chongqing 400047, China

and X.Q. Yang

Department of Applied Mathematics,
The Hong Kong Polytechnic University,
Kowloon, Hong Kong, China

Abstract: In this paper, we introduce generalized augmented Lagrangian by relaxing the convexity assumption on the usual augmenting function. Applications are given to establish strong duality and exact peanlty representation for the problem of minizing an extended real valued function. More specifically, a strong duality result based on the generalized augmented Lagrangian is established, and a necessary and sufficient condition for the exact penalty representation in the framework of generalized augmented Lagrangian is obatined.

Key words: Extended real-valued function, generalized augmented Lagrangian, duality, exact penalty representation.

1 INTRODUCTION

It is well-known that for nonconvex optimization problems, a nonzero duality gap may exist when using ordinary Lagrangian. In order to overcome this drawback, augmented Lagrangians were introduced in, e.g., Rockafellar (1974); Rockafellar (1993) for constrained optimization problems. Recently in Rockafellar, et al (1998) a general augmented Lagrangian was introduced where the augmenting function is assumed to be convex and, under mild conditions, a zero duality gap and a necessary and sufficient condition for the exact penalization were established. Most recently, there is an increasing interest in the use of "lower order" penalty penalty functions Luo et al (1996); Luo, et al (2000); Pang (1997); Robinov, et al (1999). A special feature of these penalty functions is that they are generally neither convex composite functions nor locally Lipschitz functions. However, the conditions to guarantee the exact penalty property of these "lower order" functions are weaker than those required by the usual l_1 penalty function.

In this paper, we will introduce a generalized augmented Lagrangian for a primal problem of minimizing an extended real-valued function without the convexity requirement on the augmenting function. This generalized augmented Lagrangian includes the lower order penalty function in Luo et al (1996) as a special case. This relaxation also allows us to derive strong duality results and exact penalty representation results under weaker conditions than those of Rockafellar, et al (1998). More detailed study of this generalized augmented Lagrangian can be found in Huang et al (2003).

The outline of this paper is as follows. In section 2, we introduce a generalized augmented Lagrangian. In section 3, we address the issue of strong duality. Section 4 is devoted to exact penalization.

2 GENERALIZED AUGMENTED LAGRANGIAN

In this section, we introduce some concepts and obtain some basic properties of augmented Lagrangians.

Let $\bar{R} = R \bigcup \{+\infty, -\infty\}$ and $\varphi : R^n \to \bar{R}$ be an extended real-valued function. Consider the primal problem

$$\inf_{x \in R^n} \varphi(x). \tag{2.1}$$

A function $\bar{f} : R^n \times R^m \to \bar{R}$ is said to be a dualizing parameterization function for φ if $\varphi(x) = \bar{f}(x, 0), \quad \forall x \in R^n$.

Remark 2.1 *A standard constrained nonlinear program can be written as*

$$
\begin{aligned}
(CP) \quad & \inf \quad f(x) \\
& s.t. \quad x \in X, \\
& \qquad g_j(x) \le 0, \quad j = 1, \cdots, m_1, \\
& \qquad g_j(x) = 0, \quad j = m_1 + 1, \cdots, m,
\end{aligned}
$$

where $X \subset R^n$ is a nonempty and closed set, $f, g_j : X \to R^1, j = 1, \cdots, m_1$ are lsc and $g_j : X \to R^1, j = m_1 + 1, \cdots, m$ are continuous.

It is clear that (CP) is equivalent to the following unconstrained optimization problem

$$
(P') \quad \inf_{x \in R^n} \varphi(x),
$$

where

$$
\varphi(x) = \begin{cases} f(x), & \text{if } x \in X_0, \\ +\infty, & \text{otherwise}, \end{cases}
$$

$$
X_0 = \{x \in X : g_j(x) \le 0, j = 1, \cdots, m_1, g_j(x) = 0, j = m_1 + 1, \cdots, m\}.
$$

So considering the model (2.1) provides a unified approach to the usual constrained and unconstrained optimization problems.

A simple way to define the dualizing parametrization function for (CP) is:

$$
\bar{f}(x, u) = \begin{cases} f(x), & \text{if } x \in X_u, \\ +\infty, & \text{otherwise}, \end{cases}
$$

where

$$
X_u = \{x \in X : g_j(x) + u_j \le 0, j = 1, \cdots, m_1, g_j(x) + u_j = 0, j = m_1 + 1, \cdots, m\}.
$$

In the sequel, we will use this dualizing parametrization function for (CP).

Definition 2.1 *(Rockafellar, et al (1998)). (i) Let $X \subset R^n$ be a closed subset and $f : X \to \bar{R}$ be an extended real-valued function. The function f is said to be level-bounded on X if, for any $\alpha \in R$, the set $\{x \in X : f(x) \le \alpha\}$ is bounded.*

(ii) A function $F : R^n \times R^m \to \bar{R}$ with value $F(x, u)$ is said to be level-bounded in x locally uniform in u if, for each $\bar{u} \in R^m$ and $\alpha \in R$, there exists a neighborhood $U(\bar{u})$ of $\bar{u}$ along with a bounded set $D \subset R^n$, such that $\{x \in R^n : F(x, u) \le \alpha\} \subset D$ for any $u \in U(\bar{u})$.

Definition 2.2 *A function $\sigma : R^m \to R_+ \bigcup \{+\infty\}$ is said to be a generalized augmenting function if it is proper, lower semicontinuous (lsc, for short), level-bounded on R^m, $\mathrm{argmin}_y \sigma(y) = \{0\}$ and $\sigma(0) = 0$.*

It is worth noting that this definition of generalized augmenting function is different from that of the augmenting function given in Definition 11.55 of Rockafellar, et al (1998) in that no convexity requirement is imposed on generalized augmenting functions.

Definition 2.3 *Consider the primal problem (2.1). Let $\bar{f}$ be any dualizing parameterization function for φ, and σ be a generalized augmenting function.*

(i) The generalized augmented Lagrangian (with parameter $r > 0$) $\bar{l} : R^n \times R^m \times (0, +\infty) \to \bar{R}$ is defined by

$$\bar{l}(x, y, r) = \inf\{\bar{f}(x, u) - \langle y, u \rangle + r\sigma(u) : u \in R^m\}, \quad x \in R^n, y \in R^m, r > 0,$$

where $\langle y, u \rangle$ denotes the inner product.

(ii) The generalized augmented Lagrangian dual function is defined by

$$\bar{\psi}(y, r) = \inf\{\bar{l}(x, y, r) : x \in R^n\}, \quad y \in R^m, r > 0. \tag{2.2}$$

(iii) The generalized augmented Lagrangian dual problem is defined as

$$\sup_{(y,r) \in R^m \times (0, +\infty)} \bar{\psi}(y, r). \tag{2.3}$$

Remark 2.2 *(i) By Definition 2.2, any augmenting function used in Definition 11.55 of Rockafellar, et al (1998) is a generalized augmenting function. Thus, any augmented Lagrangian defined in Definition 11.55 of Rockafellar, et al (1998) is also a generalized augmented Lagrangian.*

(ii) If σ is an augmenting function in the sense of Rockafellar, et al (1998), then for any $\gamma > 0$, σ^γ is a generalized augmenting function. In particular,

(a) let $\sigma(u) = \|u\|_1$, then for any $\gamma > 0$, $\sigma^\gamma(u) = \|u\|_1^\gamma$ is a generalized augmenting function;

(b) take $\sigma(u) = \|u\|_\infty$, then $\sigma^\gamma(u) = \|u\|_\infty^\gamma$ is a generalized augmenting function for any $\gamma > 0$.

(c) let $\sigma(u) = \sum_{j=1}^m |u_j|^\gamma$, where $\gamma > 0$. Then $\sigma(u)$ is a generalized augmenting function.

It is clear that none of these three classes (a), (b) and (c) of generalized augmenting functions is convex when $\gamma \in (0,1)$, namely, none of them is an augmenting function.

Remark 2.3 *Consider (CP). Let the dualizing parametrization function be as in Remark 2.1. It is routine to check (see, e.g., Huang et al (2003)) that the corresponding augmented Lagrangian is*

$$\bar{l}(x,y,r) = \begin{cases} f(x) + \sum_{j=1}^{m} y_j g_j(x) + \inf_{v \geq 0}\{\sum_{j=1}^{m_1} y_j v_j + r\sigma(-g_1(x) - v_1, \cdots, \\ \qquad -g_{m_1}(x) - v_{m_1}, -g_{m_1+1}(x), \cdots, -g_m(x))\}, \\ \qquad\qquad\qquad\qquad\qquad\qquad\qquad\qquad \text{if } x \in X, \\ +\infty, \qquad\qquad\qquad\qquad\qquad\qquad\qquad\quad \text{otherwise,} \end{cases}$$

where $v = (v_1, \cdots, v_{m_1})$. In particular, if $\sigma(u) = \frac{1}{2}\|u\|_2^2$, then the augmented Lagrangian and the augmented Lagrangian dual problem are the classical augmented Lagrangian and the classical augmented Lagrangian dual problem studied in Rockafellar (1974); Rockafellar (1993), respectively; if $\sigma(u) = \|u\|_1^{\gamma}, \gamma > 0$ and $m_1 = 0$ (i.e., (CP) does not have inequality constraints), then the augmented Lagrangian for (CP) is

$$\bar{l}_{\gamma}(x,y,r) = f(x) + \sum_{j=1}^{m} y_j g_j(x) + r\left[\sum_{j=1}^{m} |g_j(x)|\right]^{\gamma}$$

and the corresponding augmented Lagrangian dual problem is

$$\sup_{(y,r)\in R^m \times (0,+\infty)} \bar{\psi}_{\gamma}(y,r),$$

where

$$\bar{\psi}_{\gamma}(y,r) = \inf\{f(x) + \sum_{j=1}^{m} y_j g_j(x) + r\left[\sum_{j=1}^{m} |g_j(x)|\right]^{\gamma} : x \in X\}, \quad y \in R^m, r > 0.$$

Define the perturbation function by

$$p(u) = \inf\{\bar{f}(x,u) : x \in R^n\}.$$

Then $p(0)$ is just the optimal value of the problem (2.1).

Remark 2.4 *Consider (CP). Let the dualizing parametrization function be as in Remark 2.1. Then the perturbation function for (CP) is*

$$p(u) = \inf\{f(x) : x \in X : g_j(x) + u_j \leq 0, j = 1, \cdots, m_1,$$

$$g_j(x) + u_j = 0, j = m_1 + 1, \cdots, m\},$$

which is the optimal value of the standard perturbed problem of (CP) (see, e.g., Clarke (1983); Rosenberg (1984)). Denote by M_{CP} the optimal value of (CP). Then we have $p(0) = M_{CP}$.

The following proposition summarizes some basic properties of the generalized augmented Lagrangian, which will be useful in the sequel. Its proof is elementary and omitted.

Proposition 2.1 *For any dualizing parameterization and any generalized augmenting function, we have*

(i) the generalized augmented Lagrangian $\bar{l}(x, y, r)$ is concave, upper semicontinuous in (y, r) and nondecreasing in r.

(ii) weak duality holds:

$$\bar{\psi}(y, r) \le p(0), \quad \forall (y, r) \in R^m \times (0, +\infty).$$

3   STRONG DUALITY

The following strong duality result generalizes and improves Theorem 11.59 of Rockafellar, et al (1998).

Theorem 3.1 *(strong duality). Consider the primal problem (2.1) and its generalized augmented Lagrangian dual problem (2.3). Assume that φ is proper, and that its dualizing parameterization function $\bar{f}(x, u)$ is proper, lsc, and level-bounded in x locally uniform in u. Suppose that there exists $(\bar{y}, \bar{r}) \in R^m \times (0, +\infty)$ such that*

$$inf\{\bar{l}(x, \bar{y}, \bar{r}) : x \in R^n\} > -\infty.$$

Then zero duality gap holds:

$$p(0) = \sup_{(y,r) \in R^m \times (0,+\infty)} \bar{\psi}(y, r).$$

Proof. By Proposition 2.1, we have

$$p(0) \ge \sup_{(y,r) \in R^m \times (0,+\infty)} \bar{\psi}(y, r).$$

Suppose to the contrary that there exists $\delta > 0$ such that

$$p(0) - \delta \geq \sup_{(y,r) \in R^m \times (0,+\infty)} \bar{\psi}(y,r).$$

Then,

$$p(0) - \delta \geq \bar{\psi}(y,r), \quad \forall y, r.$$

In particular,

$$p(0) - \delta \geq \bar{\psi}(\bar{y},r) = \inf_{x \in X} \bar{l}(x,\bar{y},r).$$

Let $0 < r_k \uparrow +\infty$. Then,

$$p(0) - \delta \geq \inf_{x \in X} \bar{l}(x,\bar{y},r_k).$$

Thus, $\exists x_k \in X$ such that

$$p(0) - \delta/2 \geq \bar{l}(x_k,\bar{y},r_k).$$

Hence, $\exists u_k \in R^m$ such that

$$p(0) - \delta/4 \geq \bar{f}(x_k,u_k) - \langle \bar{y}, u_k \rangle + r_k \sigma(u_k). \tag{3.1}$$

From the assumption of the theorem, we suppose that

$$\bar{l}(x,\bar{y},\bar{r}) \geq -m_0 > -\infty, \quad \forall x.$$

Then,

$$\bar{f}(x_k,u_k) - \langle \bar{y}, u_k \rangle + \bar{r}\sigma(u_k) \geq -m_0, \quad \forall k. \tag{3.2}$$

(3.1) and (3.2) give us

$$p(0) - \delta/4 \geq -m_0 + (r_k - \bar{r})\sigma(u_k). \tag{3.3}$$

From the level-boundedness of σ, we see that $\{u_k\}$ is bounded. Assume without loss of generality that $u_k \to \bar{u}$. Then from (3.3) we deduce

$$\sigma(\bar{u}) \leq \liminf_{k \to +\infty} \sigma(u_k) \leq 0.$$

Consequently, $\bar{u} = 0$. Moreover, from (3.1) we see that

$$p(0) - \delta/4 + \|\bar{y}\| \geq p(0) - \delta/4 + \langle \bar{y}, u_k \rangle \geq \bar{f}(x_k,u_k) \tag{3.4}$$

when k is sufficiently large. This combined with the fact that $u_k \to 0$ and the fact that $f(x,u)$ is level-bounded in x locally uniform in u implies that $\{x_k\}$

is bounded. Assume without loss of generality that $x_k \to \bar{x}$. Taking the lower limit in (3.4) as $k \to +\infty$, we obtain

$$p(0) - \delta/4 \geq \liminf_{k \to +\infty} \bar{f}(x_k, u_k) \geq \bar{f}(\bar{x}, 0) = \varphi(\bar{x}).$$

This contradicts the definition of $p(0)$.

Remark 3.1 *For the standard constrained optimization problem (CP), let the generalized augmented Lagrangian be defined as in Remark 2.3. Further assume the following conditions hold:*
   (i)

$$f(x) \geq m^*, \quad \forall x \in X, \tag{3.5}$$

for some $m^* \in R$.
   (ii)

$$\lim_{\|x\| \to +\infty, x \in X} max\{f(x), g_1(x), \cdots, g_{m_1}(x), |g_{m_1+1}(x)|, \cdots, |g_m(x)|\} = +\infty.$$

Then all the conditions of Theorem 3.1 hold. It follows that there exists no duality gap between (CP) and its generalized augmented Lagrangian dual problem.

4   EXACT PENALTY REPRESENTATION

In this section, we present exact penalty representation results in the framework of generalized augmented Lagrangian.

Definition 4.1 *(exact penalty representation) Consider the problem (2.1). Let the generalized augmented Lagrangian $\bar{l}$ be defined as in Definition 2.3. A vector $\bar{y} \in R^m$ is said to support an exact penalty representation for the problem (2.1) if there exists $\bar{r} > 0$ such that*

$$p(0) = inf_{x \in R^n} \bar{l}(x, \bar{y}, r), \quad \forall r \geq \bar{r} \tag{4.1}$$

and

$$argmin_x \varphi(x) = argmin_x \bar{l}(x, \bar{y}, r), \quad \forall r \geq \bar{r}.$$

The following result can be proved similarly to Theorem 11.61 in Rockafellar, et al (1998).

Theorem 4.1 *In the framework of the generalized augmented Lagrangian defined in Definition 2.3. The following statements are true:*

(i) If $\bar{y}$ supports an exact penalty representation for the problem (2.1), then there exist $\bar{r} > 0$ and a neighborhood W of $0 \in R^m$ such that

$$p(u) \geq p(0) + \langle \bar{y}, u \rangle - \bar{r}\sigma(u), \quad \forall u \in W.$$

(ii) The converse of (i) is true if
(a) $p(0)$ is finite;
(b) there exists $\bar{r}' > 0$ such that

$$\inf\{\bar{f}(x,u) - \langle \bar{y}, u \rangle + \bar{r}'\sigma(u) : (x,u) \in R^n \times R^m\} > -\infty;$$

(c) there exist $\tau > 0$ and $N > 0$ such that $\sigma(u) \geq \tau\|u\|$ when $\|u\| \geq N$.

Proof. Since $\bar{y}$ supports an exact penalty representation, there exists $\bar{r} > 0$ such that

$$p(0) = \inf\{\bar{l}(x,\bar{y},\bar{r}) : x \in R^n\} = \inf\{\bar{f}(x,u) - \langle \bar{y}, u \rangle + \bar{r}\sigma(u) : (x,u) \in R^n \times R^m\}.$$

Consequently,

$$p(0) \leq \bar{f}(x,u) - \langle \bar{y}, u \rangle + \bar{r}\sigma(u), \quad \forall x \in R^n, u \in R^m,$$

implying

$$p(0) \leq p(u) - \langle \bar{y}, u \rangle + \bar{r}\sigma(u), \quad \forall u \in R^m.$$

This proves (i).

It is evident from the proof of Theorem 11.61 in Rockafellar, et al (1998) that (ii) is true.

Remark 4.1 *In Rockafellar, et al (1998), σ was assumed to be proper, lsc, convex and $\mathrm{argmin}_y\sigma(y) = \{0\}$. As noted in Rockafellar, et al (1998), σ is level-coercive. It follows that this assumption implies the existence of $\tau > 0$ and $N > 0$ satisfying $\sigma(u) \geq \tau\|u\|$ when $\|u\| \geq N$.*

Remark 4.2 *Consider (CP) with $m_1 = 0$. Suppose that (3.5) holds, $X_0 \neq \emptyset$, and $\sigma(u) = \|u\|_1^\gamma, \gamma \geq 1$. Let the generalized augmented Lagrangian $\bar{l}_\gamma(x,y,r)$ for (CP) be given as in Remark 2.3. Then $\bar{y}$ supports an exact penalty representation for (CP) in the framework of its augmented Lagrangian $\bar{l}_\gamma(x,y,r)$, namely,* *there exists $\bar{r} > 0$ such that*

$$M_{CP} = \inf_{x \in X} \left\{ f(x) + \sum_{j=1}^{m} \bar{y}_j g_j(x) + r \left[\sum_{j=1}^{m} |g_j(x)| \right]^{\gamma} \right\}$$

and the solution set of (CP) is the same as that of the problem of minimizing
$f(x) + \sum_{j=1}^{m} \bar{y}_j g_j(x) + r \left[\sum_{j=1}^{m} |g_j(x)| \right]^{\gamma}$ *over $x \in X$ whenever $r \geq \bar{r}$,*
if and only if there exist $\bar{r}' > 0$ and a neighborhood W of $0 \in R^m$ such that

$$p(u) \geq M_{CP} + \langle \bar{y}, u \rangle - \bar{r}' \|u\|_1^{\gamma}, \quad \forall u \in W,$$

where p is defined as in Remark 2.4.

For the special case where $\bar{y} = 0$ supports an exact penalty representation
for the problem (2.1), we have the following result.

Theorem 4.2 *In the framework of the generalized augmented Lagrangian huángl
defined in Definition 2.3. The following statements are true:*

*(i) If $\bar{y} = 0$ supports an exact penalty representation, then there exist $\bar{r} > 0$
and a neighborhood W of $0 \in R^m$ such that*

$$p(u) \geq p(0) - \bar{r}\sigma(u), \quad \forall u \in W. \tag{4.2}$$

(ii) The converse of (i) is true if
(a) $p(0)$ is finite;
*(b) there exist $\bar{r}' > 0$ and $m^{**} \in R$ such that $\bar{f}(x, u) + \bar{r}'\sigma(u) \geq m^{**}, \forall x \in
R^n, u \in R^m$.*

Proof. (i) follows from Theorem 4.1 (i). We need only to prove (ii). Assume
that (4.2) holds.

First we prove (4.1) by contradiction. Suppose by the weak duality that
there exists $0 < r_k \to +\infty$ with

$$p(0) > \inf_{x \in R^n} \bar{l}(x, 0, r_k).$$

Then there exist $x^k \in R^n$ and $u^k \in R^m$ such that

$$
\begin{aligned}
p(0) \quad &> \quad \bar{f}(x^k, u^k) + r_k \sigma(u^k) \\
&= \quad \bar{f}(x^k, u^k) + \bar{r}'\sigma(u^k) + (r_k - \bar{r}')\sigma(u^k) \\
&\geq \quad m^{**} + (r_k - \bar{r}')\sigma(u^k). \tag{4.3}
\end{aligned}
$$

The level-boundedness of σ implies that $\{u^k\}$ is bounded. Assume, without loss of generality, that $u^k \to \bar{u}$. It follows from (4.3) that

$$\sigma(\bar{u}) \le \liminf_{k \to +\infty} \sigma(u^k) \le \lim_{k \to +\infty} \frac{p(0) - m^{**}}{r_k - \bar{r}'} = 0.$$

Thus $\bar{u} = 0$. From the first inequality in (4.3), we deduce that

$$p(0) > p(u^k) + r_k \sigma(u^k), \quad \forall k. \tag{4.4}$$

Since $u^k \to 0$, we conclude that (4.4) contradicts (4.2). As a result, there exists $\bar{r} > \bar{r}'$ such that (4.1) holds. Hence, for any $x^* \in \mathrm{argmin}_x \varphi(x)$, we have

$$\varphi(x^*) = \bar{l}(x^*, 0, r) = \inf_{x \in R^n} \bar{l}(x, 0, r), \quad r \ge \bar{r}.$$

Consequently, $x^* \in \mathrm{argmin}_x \bar{l}(x, 0, r)$. This shows that

$$\mathrm{argmin}_x \varphi(x) \subseteq \mathrm{argmin}_x \bar{l}(x, 0, r)$$

whenever $r \ge \bar{r}$. Now we show that there exists $r^* > \bar{r} + 1 > 0$ such that

$$\mathrm{argmin}_x \bar{l}(x, 0, r) \subseteq \mathrm{argmin}_x \varphi(x), \quad \forall r > r^*.$$

Suppose to the contrary that there exist $\bar{r} + 1 \le r_k \uparrow +\infty$ and $x^k \in \mathrm{argmin}_x \bar{l}(x, 0, r_k)$ such that $x^k \notin \mathrm{argmin}_x \varphi(x), \forall k$. Then

$$\varphi(x^k) > p(0), \quad \forall k. \tag{4.5}$$

For each fixed k, by the definition of $\bar{l}(x^k, 0, r_k)$, $\exists \{u^{k,i}\} \subset R^m$ with

$$\bar{f}(x^k, u^{k,i}) + r_k \sigma(u^{k,i}) \to \bar{l}(x^k, 0, r_k) \tag{4.6}$$

as $i \to +\infty$. Since $r_k \ge \bar{r} + 1$, $x^k \notin \mathrm{argmin}_x \varphi(x)$, by (4.1), we have

$$\bar{l}(x^k, 0, r_k) = p(0). \tag{4.7}$$

From (4.6) and (4.7), we deduce that

$$\bar{f}(x^k, u^{k,i}) + \bar{r}' \sigma(u^{k,i}) + (r_k - \bar{r}')\sigma(u^{k,i}) \to p(0)$$

as $i \to +\infty$. It follows that $\{(r_k - \bar{r}')\sigma(u^{k,i})\}_{i=1}^{+\infty}$ is bounded since $\bar{f}(x^k, u^{k,i}) + \bar{r}'\sigma(u^{k,i}) \ge m^{**}$. As σ is level-bounded, we know that $\{u^{k,i}\}_{i=1}^{+\infty}$ is bounded. Without loss of generality, assume that $u^{k,i} \to \bar{u}^k$. Then

$$\bar{f}(x^k, \bar{u}^k) + r_k \sigma(\bar{u}^k) \le \liminf_{i \to +\infty} \bar{f}(x^k, u^{k,i}) + r_k \sigma(u^{k,i}) = p(0). \tag{4.8}$$

Hence,

$$\bar{f}(x^k, \bar{u}^k) + \bar{r}'\sigma(\bar{u}^k) + (r_k - \bar{r}')\sigma(\bar{u}^k) \le p(0). \tag{4.9}$$

So

$$(r_k - \bar{r}')\sigma(\bar{u}^k) \le p(0) - m^{**}. \tag{4.10}$$

Again, by the level-boundedness of σ, we see that $\{\bar{u}^k\}$ is bounded. Suppose, without loss of generality, that $\bar{u}^k \to \bar{u}$. Then, from (4.10), we obtain

$$\sigma(\bar{u}) \le \liminf_{k \to +\infty} \sigma(\bar{u}^k) \le \lim_{k \to +\infty} \frac{p(0) - m^*}{r_k - \bar{r}'} = 0.$$

So we know that $\bar{u}^k \to 0$. Note from (4.5) that $\bar{u}^k \neq 0, \forall k$. Otherwise, suppose that $\exists k^*$ such that $\bar{u}^{k^*} = 0$. Then from (4.9) we have

$$\varphi(x^{k^*}) = \bar{f}(x^{k^*}, 0) \le p(0),$$

contradicting (4.5). As a result, (4.8) contradicts (4.2). The proof is complete.

Remark 4.3 *Comparing Theorems 4.1 and 4.2, the special case where $\bar{y} = 0$ supports an exact penalty representation requires weaker conditions, i.e., condition (c) of (ii) in Theorem 4.1 is not needed.*

Remark 4.4 *Consider (CP) and its generalized augmented Lagrangian defined in Remark 2.3. Suppose that $X_0 \neq \emptyset$ and (3.5) holds. Then $\bar{y} = 0$ supports an exact penalty representation for (CP) in the framework of its generalized augmented Lagrangian if and only if there exist $\bar{r} > 0$ and a neighborhood W of $0 \in R^m$ such that*

$$p(u) \ge M_{CP} - \bar{r}\sigma(u), \quad \forall u \in W,$$

where p is defined as in Remark 2.4.

Example 4.1 *Consider (CP). Let X^* denote the set of optimal solutions of (CP). Suppose that $X_0 \neq \emptyset$ and (3.5) hold. Let the generalized augmenting function be $\sigma(u) = \|u\|_1^\gamma, \gamma > 0$ and the generalized augmented Lagrangian be defined as in Remark 2.3. It is easily computed that*

$$\bar{l}(x, 0, r) = \begin{cases} f(x) + r[\sum_{j=1}^{m_1} g_j^+(x) + \sum_{j=m_1+1}^{m} |g_j(x)|]^\gamma, & \text{if } x \in X \\ +\infty, & \text{otherwise.} \end{cases}$$

This is a typical form of "lower order" penalty function when $0 < \gamma < 1$. Remark 4.4 says that the following two statements are equivalent

(i) there exists $\bar{r}' > 0$ such that

$$\inf_{x \in X} \left\{ f(x) + r[\sum_{j=1}^{m_1} g_j^+(x) + \sum_{j=m_1+1}^{m} |g_j(x)|]^\gamma \right\} = M_{CP}$$

and $X^* = X_r^*$, $r \geq \bar{r}'$, where X_r^* is the set of optimal solutions of the problem of minimizing $f(x) + r[\sum_{j=1}^{m_1} g_j^+(x) + \sum_{j=m_1+1}^{m} |g_j(x)|]^\gamma$ over $x \in X$;
(ii) there exist $\bar{r} > 0$ and a neighborhood W of $0 \in R^m$ such that

$$p(u) \geq M_{CP} - \bar{r}\|u\|_1^\gamma, \quad \forall u \in W,$$

where p is defined as in Remark 2.4.

5   CONCLUSIONS

We introduced generalized augmented Lagrangian by relaxing the convexity assumption imposed on the ordinary augmenting function. As a result, weaker conditions are required to guarantee the strong duality and exact penalization results similar to those obtained in Rockafellar, et al (1998). This work was supported by a grant from the Research Grants Council of the Hong Kong Special Administrative Region, China (Project No. PolyU 5141/01E).

References

Clarke, F.H. (1983), *Optimization and Nonsmooth Analysis*, John Wiley & Sons, New York.

Huang, X.X. and Yang, X.Q. (2003), A Unified Augmented Lagrangian Approach to Duality and Exact Penalization, *Mathematics of Operations Research*, to appear.

Luo, Z.Q., Pang, J.S. and Ralph, D. (1996), *Mathematical Programs with Equilibrium Constraints*, Cambridge University Press, New York.

Luo, Z.Q. and Pang, J.S. (eds.) (2000), Error Bounds in Mathematical Programming, *Mathematical Programming*, Ser. B., Vol. 88, No. 2.

Pang, J.S. (1997), Error bounds in mathematical programming, *Mathematical Programming*, Vol. 79, pp. 299-332.

Rockafellar, R.T. (1974), Augmented Lagrange multiplier functions and duality in nonconvex programming, *SIAM Journal on Control and Optimization*, Vol. 12, pp. 268-285, 1974.

Rockafellar, R.T. (1993), Lagrange multipliers and optimality, *SIAM Review*, Vol. 35, pp. 183-238.

Rockafellar, R.T. and Wets, R.J.-B. (1998), *Variational Analysis*, Springer-Verlag, Berlin.

Rosenberg, E. (1984), Exact penalty functions and stability in locally Lipschitz programming, *Mathematical Programming*, Vol. 30, pp. 340-356.

Rubinov, A. M., Glover, B. M. and Yang, X. Q. (1999), Decreasing functions with applications to penalization, *SIAM J. Optimization*, Vol. 10, No. 1, pp. 289-313.

4 DUALITY FOR SEMI-DEFINITE AND SEMI-INFINITE PROGRAMMING WITH EQUALITY CONSTRAINTS

S.J. Li,

Department of Information and Computer Sciences,
College of Sciences, Chongqing University,
Chongqing, 400044, China.
E-mail: 00900470r@polyu.edu.hk

X.Q. Yang and K.L. Teo

Department of Applied Mathematics,
The Hong Kong Polytechnic University,
Kowloon, Hong Kong.
E-mail: mayangxq@polyu.edu.hk
and mateokl@polyu.edu.hk

Abstract: In this paper, we study a semi-definite and semi-infinite programming problem (SDSIP) with equality constraints. We establish that a uniform duality between a homogeneous (SDSIP) problem and its Lagrangian-type dual problem is equivalent to the closedness condition of certain cone. A corresponding result for a nonhomogeneous (SDSIP) problem is also obtained by transforming it into an equivalent homogeneous (SDSIP) problem.

Key words: Duality, semi-definite program, semi-infinite program.

1   INTRODUCTION AND PRELIMINARIES

Let S^n denote the set of real symmetric $n \times n$ matrices. By $X \succeq 0$, where $X \in S^n$, we mean that the matrix X is positive semidefinite. The set $K = \{X \in S^n | X \succeq 0\}$ is called the positive semidefinite cone. For any $S \subset S^n, cl(S)$ denotes the closure of S in S^n. For the compact set B in a metric space, let

$$\Lambda_B = \{y = \{y(t)\}_{t \in B} \in R^B | (\exists \text{a finite set } F \subseteq B)(\forall t \in B \backslash F) \, y(t) = 0\}.$$

For the set $W = \{A(t) | t \in B\}$, $sp(W)$ denotes the subspace generated by W, i.e.,

$$sp(W) = \{\sum_{t \in B} y(t)A(t) | \forall y \in \Lambda_B\},$$

The standard inner product on S^n is

$$A \bullet B = \text{tr}AB = \sum_{i,j} a_{ij}b_{ij}.$$

We consider the following semi-definite and semi-infinite linear programming problem (SDSIP):

$$\begin{aligned} \inf \quad & C \bullet X \\ \text{s.t} \quad & A(t) \bullet X = b(t), \quad t \in B, \\ & X \succeq 0. \end{aligned} \qquad (1.1)$$

Here B is a compact set in R, C and $A(t)(t \in B)$ are all fixed matrices in S^n, $b(t) \in R \ (t \in B)$ and the unknown variable X also lies in S^n.

Obviously, (SDSIP) problem includes the semi-definite programming problem and the linear semi-infinite programming problem with equality constraints as special cases. See Charnes et al (1962) and Wolkowicz et al (2000).

For the (SDSIP) problem , we introduce the Lagrangian dual problem (DS-DSIP) as follows:

$$\begin{aligned} \sup \quad & \sum_{t \in B} y(t)b(t) \\ \text{s.t} \quad & \sum_{t \in B} y(t)A(t) + Z = C, \, y \in \Lambda_B, \\ & Z \succeq 0. \end{aligned} \qquad (1.2)$$

When the parameter set B is finite, Then, (SDSIP) and (DSDSIP) is a pair of primal and dual (SDP). See Vandenberghe and Boyd (1996), Ramana et al (1997) and Wolkowicz et al (2000).

Proposition 1.1 *Suppose that X and (y, Z) are feasible solutions for (SDSIP) problem and (DSDSIP) problem, respectively. Then,*

$$C \bullet X \geq \sum_{t \in B} y(t) b(t).$$

Proof Since (y, Z) is a feasible solution for (DSDSIP),

$$\sum_{t \in B} y(t) A(t) + Z = C.$$

Then, we have

$$C \bullet X = Z \bullet X + \sum_{t \in B} y(t) A(t) \bullet X.$$

Since X and Z are positive semidefinite matrices, $Z \bullet X \geq 0$. It follows that

$$C \bullet X \geq \sum_{t \in B} y(t) A(t) \bullet X.$$

Thus, the result holds.   □

(SDSIP) problem is said to be consistent if there exists $X \succeq 0$ such that (1.1) holds. It is said to be bounded in value if it is consistent and there exists a number z^* such that all feasible solutions $X \in S^n$ to (SDSIP) satisfy $C \bullet X \geq z^*$. It is said to unbounded in value if, for each integer n, there exists a feasible solution $X^{(n)}$ to (SDSIP) with $C \bullet X^{(n)} \leq -n$.

Definition 1.1 *The system of linear equalities*

$$A(t) \bullet X = b(t), \quad t \in B \tag{1.3}$$

yields duality with respect to $C \in S^n$, if exactly one of the following conditions holds:

(i) *(SDSIP) is unbounded in value and (DSDSIP) is inconsistent;*

(ii) *(DSDSIP) is unbounded in value and (SDSIP) is inconsistent;*

(iii) *Both (SDSIP) and (DSDSIP) are inconsistent;*

(iv) *Both (SDSIP) and (DSDSIP) are consistent and have the same optimal value, and the value is attained in (DSDSIP).*

We say that (SDSIP) yields uniform duality if the constraint system (1.3) yields duality for every $C \in S^n$.

In the paper, we firstly establish that a uniform duality between the homogeneous (SDSIP) and its Lagrangian-type dual problem is equivalent to the closedness condition of certain cone. With aid of the result, we also obtain a corresponding result for nonhomogeneous (SDSIP).

Detailed study of uniform duality for (SDSIP) problems with inequality constrains can be found in Li et al (2002).

2   UNIFORM DUALITY FOR HOMOGENEOUS (SDSIP)

We firstly discuss the homogeneous case in (SDSIP): $b(t) = 0, \forall\, t \in B$. Then (SDSIP) becomes the following problem (SDSIP$_h$):

$$\begin{aligned} \inf\quad & C \bullet X \\ s.t\quad & A(t) \bullet X = 0, \quad t \in B, \\ & X \succeq 0, \end{aligned} \qquad (2.1)$$

and (DSDSIP) becomes the following problem (DSDSIP$_h$):

$$\begin{aligned} \sup\quad & 0 \\ s.t\quad & \sum_{t \in B} y(t)A(t) + Z = C, \ y \in \Lambda_B, \\ & Z \succeq 0. \end{aligned} \qquad (2.2)$$

Lemma 2.1 *The problem (SDSIP$_h$) is unbounded in value if and only if there exists $X^* \succeq 0$ satisfying:*

$$A(t) \bullet X^* \ = 0, \quad t \in B \qquad (2.3)$$

$$and \quad C \bullet X^* \ < 0. \qquad (2.4)$$

Proof. Suppose that there is $X^* \succeq 0$ such that (2.3) and (2.4) hold. Without loss of generality, assume $C \bullet X^* < -1$. For each n we have $A(t) \bullet X^{(n)} \geq 0, t \in B$ and $C \bullet X^{(n)} < -n$ with $X^{(n)} = nX^*$. Hence, (SDSIP$_h$) is unbounded in value.

Conversely, by the unbounded definition, the case holds. This completes the proof. $\qquad \square$

Remark 2.1 Since $X = 0$ is a feasible solution of (SDSIP$_h$), (SDSIP$_h$) is always consistent. If the optimal value of (SDSIP$_h$) is bounded below, the optimal value of (SDSIP$_h$) is zero. Thus, (ii) and (iii) in Definition 1.1 do not happen.

Theorem 2.1 *(SDSIP$_h$) yields uniform duality if and only if $sp(W) + K$ is a closed set.*

Proof. Suppose (SDSIP$_h$) yields uniform duality. Let $C \in cl(sp(W) + K)$ and $C \notin sp(W) + K$. Then, there exists no $y \in \Lambda_B$ and $Z \succeq 0$ such that

$$\sum_{t \in B} y(t)A(t) + Z = C.$$

Thus, (DSDSIP$_h$) is inconsistent. Since (SDSIP$_h$) is consistent, the problem (SDSIP$_h$) must be unbounded in value. By Lemma 2.1, there exists $X^* \succeq 0$ satisfying (2.3) and (2.4). By (2.3), we have that

$$V \bullet X^* = 0, \quad \forall\, V \in sp(W).$$

Take any $S \in sp(W) + K$. Then, there exist $V \in sp(W)$ and $Q \in K$ such that

$$S = V + Q.$$

We have

$$S \bullet X^* = V \bullet X^* + Q \bullet X^* = Q \bullet X^*.$$

Since Q and X^* are positive semidefinite matrices, we have that

$$Q \bullet X^* \geq 0,$$

and

$$S \bullet X^* \geq 0.$$

Therefore, we get that

$$S \bullet X^* \geq 0, \quad \forall\, S \in cl(sp(W) + K),$$

and

$$C \bullet X^* \geq 0.$$

However, it follows from (2.4) that $C \bullet X^* < 0$, which is a contradiction. Hence, $sp(W) + K$ is closed.

Conversely, suppose $sp(W) + K$ is closed. Let $C \in S^n$ be arbitrary. Since (SDSIP$_h$) is consistent, either (SDSIP$_h$) is unbounded in valued or bounded in value. If (SDSIP$_h$) is unbounded in value, by Proposition 1.1, (DSDSIP$_h$) is inconsistent. If (SDSIP$_h$) is bounded in value, its value is zero by Lemma

2.1. Now we show that clause (iv) of Definition 1.1 holds. If (DSDSIP_h) is not consistent for C, $C \notin \text{sp}(W) + K = cl(\text{sp}(W) + K)$. By the definitions of $\text{sp}(W)$ and K, we have that $\text{sp}(W) + K$ is a closed and convex cone in S^n. Thus, by the separation theorem, there exists X^* in S^n such that

$$C \bullet X^* < 0 \text{ and } V \bullet X^* \geq 0, \quad \forall V \in \text{sp}(W) + K.$$

Obviously,

$$A(t) \in \text{sp}(W), \quad \forall t \in B.$$

Therefore,

$$C \bullet X^* < 0 \text{ and } A(t) \bullet X^* = 0, \quad \forall t \in B.$$

Thus, it is necessary that we prove $X^* \succeq 0$. Take any $Q \in K$ and $0 \in \text{sp}(W)$. We have

$$Q \bullet X^* \geq 0, \quad \forall Q \in K. \tag{2.5}$$

Thus, X^* is a positive semidefinite matrix. It completes this proof. $\qquad \square$

Proposition 2.1 *The constraint system*

$$\bar{A}(t) \bullet \bar{X} \geq 0, \; t \in B$$

yields duality for any $\bar{C} \in S^{n+1}$ if and only if $\text{sp}(\bar{W}) + \bar{K}$ is a closed set.

Proof. The proof is similar to that of Theorem 2.1 and is omitted. $\qquad \square$

3   UNIFORM DUALITY FOR NONHOMOGENEOUS (SDSIP)

We now establish the duality for the nonhomogeneous constraint system (1.1) of (SDSIP) by reformulating it as a form of homogeneous system (2.1) and applying Proposition 2.1. For any real number $d \in R$, we define:

$$\tilde{C} = \begin{pmatrix} C & \mathbf{0} \\ \mathbf{0}^T & -d \end{pmatrix}, \tilde{A}(t) = \begin{pmatrix} A(t) & \mathbf{0} \\ \mathbf{0}^T & -b(t) \end{pmatrix},$$

$\mathbf{0}$ is zero element in R^n and $\tilde{X}, \tilde{Z} \in S^{n+1}$.

Now, we introduce a new semi-definite and semi-infinite programming problem (SDSIP1):

$$\inf \; \tilde{C} \bullet \tilde{X}$$
$$\text{s.t} \; \tilde{A}(t) \bullet \tilde{X} = 0, \; t \in B, \tag{3.1}$$
$$\tilde{X} \succeq 0.$$

The Lagrangian dual program (DSDSIP1) of (SDSIP1) is as follows:

$$\sup \quad 0$$
$$\text{s.t.} \quad \sum_{t \in B} y(t)\tilde{A}(t) + \tilde{Z} = \tilde{C}, \quad y \in \Lambda_B, \tag{3.2}$$
$$\tilde{Z} \succeq 0.$$

which is equivalent to the program

$$\sup \quad 0$$
$$\text{s.t.} \quad \sum_{t \in B} y(t)A(t) + Z = C, \quad y \in \Lambda_B, \tag{3.3}$$
$$\sum_{t \in B} y(t)b(t) \geq d, \ y \in \Lambda_B, \tag{3.4}$$
$$Z \succeq 0. \tag{3.5}$$

Lemma 3.1 *The nonhomogeneous constraint system (1.1) yields duality with respect to $C \in S^n$ if and only if, for every $d \in R$, the constraint system (3.1) yields duality with respect to $\tilde{C} \in S^{n+1}$.*

Proof Suppose that the constraint system (1.1) yields duality with respect to $C \in S^n$ and let $d \in R$. We will show that the constraint system (3.1) yields duality with respect to $\tilde{C}$.

Since (SDSIP1) is a homogeneous system, by Remark 2.1, we need only show two cases:

Case one: if its dual problem (DSDSIP1) is consistent, (iv) in Definition 1.1 holds.

By homogeneous property of (SDSIP1) problem, we have that the optimal value of (DSDSIP1) problem is zero. It follows from Proposition 1.1 that (SDSIP1) problem is bounded in value. By Remark 2.1, we have that (iv) in Definition 1.1 holds.

Case two: if its dual problem (DSDSIP1) is inconsistent, (i) in Definition 1.1 holds. Namely, (SDSIP1) has a value of $-\infty$ with respect to $\tilde{C}$.

Assume that its dual problem (DSDSIP1) is inconsistent. Note that (3.3) and (3.5) are the constraint system of (DSDSIP). If (SDSIP) is consistent and (3.3) and (3.5) are hold, by hypothesis condition that the constraint system (1.1) yields duality with respect to $C \in S^n$, we have

$$\inf\{C \bullet X | A(t) \bullet X = b(t), \ t \in B, \ X \succeq 0\} =$$
$$\sup\{\textstyle\sum_{t \in B} y(t)b(t) | \sum_{t \in B} y(t)A(t) + Z = C, \ y \in \Lambda_B, \ Z \succeq 0\}. \tag{3.6}$$

Thus, it follows from the inconsistency of the dual problem (DSDSIP1) that at least one of two conditions holds: (i) the constraint (3.3) does not hold; (ii) the constraint (3.3) holds, but the constraint (3.4) does not hold. Namely,

$$d > \sup\{\sum_{t \in B} y(t)b(t) \mid \sum_{t \in B} y(t)A(t) + Z = C, \ y \in \Lambda_B, \ Z \succeq 0\}.$$

If (i) holds, then the problem (SDSIP) is unbounded in value by hypothesis. Thus, for any number n however large, there exists $X^{(n)} \succeq 0$ such that

$$\begin{aligned} A(t) \bullet X^{(n)} &= b(t), \quad \forall \, t \in B, \\ C \bullet X^{(n)} &< d - n. \end{aligned}$$

Set

$$\tilde{X}^{(n)} = \begin{pmatrix} X^{(n)} & \mathbf{0} \\ \mathbf{0}^T & 1 \end{pmatrix} \text{ and } \mathbf{0} \in R^n.$$

It follows from $X^{(n)}$ that $\tilde{X}^{(n)}$ is a positive semidefinite matrix. Thus, we have that

$$\begin{aligned} \tilde{A}(t) \bullet \tilde{X}^{(n)} &= 0, \ \forall t \in B, \\ \tilde{C} \bullet \tilde{X}^{(n)} &= C \bullet X^{(n)} - d < -n. \end{aligned}$$

Therefore, (SDSIP1) has value $-\infty$.

If (ii) holds and (SDSIP) is inconsistent, then we have that (SDSIP) is inconsistent and (DSDSIP) is consistent. Since (SDSIP) has duality with respect to C, for any n, there exists a solution $y \in \Lambda_B$ for (DSDSIP) with $\sum_{t \in B} y(t)b(t) \geq n$. Then, the dual problem (DSDSIP1) is consistent for any $d \in R$, which contradicts with assumption.

If (ii) holds and (SDSIP) is consistent, by (3.6), we have

$$d > \inf\{C \bullet X \mid A(t) \bullet X = b(t), \ t \in B, \ X \succeq 0\}.$$

Therefore, there exists a point $\bar{X} \succeq 0$ with

$$A(t) \bullet \bar{X} = b(t), \forall \, t \in B,$$

and

$$d > C \bullet \bar{X}.$$

Set

$$\tilde{X} = \begin{pmatrix} \bar{X} & \mathbf{0} \\ \mathbf{0}^T & 1 \end{pmatrix} \text{ and } \mathbf{0} \in R^n.$$

Then,

$$\tilde{A}(t) \bullet \tilde{X} \; = 0, \quad \forall\, t \in B,$$
$$\tilde{C} \bullet \tilde{X} \; < 0.$$

Thus, by Lemma 2.1, (SDSIP1) has value $-\infty$. We have proved the necessity of this lemma.

To prove the sufficiency of the lemma, we suppose that, for all $d \in R$ and $C \in S^n$, (SDSIP1) yields duality with respect to $\tilde{C}$. We need to prove that (SDSIP) yields duality with respect to C.

If (SDSIP) is inconsistent, then, there is only zero to solve (SDSIP1). By the definition of duality, (DSDSIP1) is consistent for any $d \in R$. Take $d = n$. Thus, there exists $y^{(n)} \in \Lambda_B$ such that

$$\sum_{t \in B} y^{(n)}(t)A(t) + Z = C, \quad Z \succeq 0, \quad y^{(n)} \in \Lambda_B,$$

and

$$\sum_{t \in B} y^{(n)}(t)b(t) \geq n, \quad y^{(n)} \in \Lambda_B.$$

Therefore, clause (ii) of the Definition 1.1 holds.

If (SDSIP) is consistent, then there are two cases:

(a) (SDSIP) is unbounded in value. If (DSDSIP) is consistent, then take any feasible solutions X and (y, Z) for (SDSIP) and (DSDSIP), respectively. Thus,

$$C \bullet X \geq Z \bullet X + \sum_{t \in B} y(t)b(t).$$

Since X and Z are positive semidefinite matrices, $Z \bullet X \geq 0$. It follows that

$$C \bullet X \geq \sum_{t \in B} y(t)b(t),$$

which contradicts unboundedness for (SDSIP). Thus, (DSDSIP) is inconsistent.

(b) (SDSIP) is bounded in value. Let $z_0 = \inf\{C \bullet X \,|\, A(t) \bullet X \geq b(t), \; t \in B\}$. We first show that (SDSIP1) for $d = z_0$ cannot be unbounded in value.

If (SDSIP1) is unbounded in value, then by Lemma 2.1, there is a solution to

$$\tilde{A}(t) \bullet \tilde{X} \; = 0, \quad t \in B, \; \tilde{X} \succeq 0,$$
$$\tilde{C} \bullet \tilde{X} \; < 0.$$

Suppose $\tilde{X} = \begin{pmatrix} X^* & x \\ x^T & x_{n+1} \end{pmatrix}$, where $x \in R^n, X^* \in S^n$ and $X^* \succeq 0$. Take $\bar{X} = \begin{pmatrix} X^* & \mathbf{0} \\ \mathbf{0}^T & x_{n+1} \end{pmatrix}$, where $\mathbf{0} \in R^n$. Obviously, $\bar{X}$ is positive semidefinite and satisfy:

$$\tilde{A}(t) \bullet \bar{X} = 0, \quad t \in B, \ \bar{X} \succeq 0,$$

$$\tilde{C} \bullet \bar{X} < 0.$$

If $x_{n+1} > 0$, then we may assume $x_{n+1} = 1$ by homogeneity, and we have

$$A(t) \bullet X^* = b(t), \quad t \in B, \ X^* \succeq 0,$$

$$C \bullet X^* < d,$$

which is a contradiction to the definition of d.

If $x_{n+1} = 0$, then, we have

$$A(t) \bullet X^* = 0, \quad t \in B, \ X^* \succeq 0,$$

$$C \bullet X^* < 0.$$

Let X_0 be a feasible solution for (SDSIP). Then, for any $\lambda \geq 0, X_0 + \lambda X^*$ is a solution of (SDSIP). Thus, we have that $C \bullet (X_0 + \lambda X^*) = C \bullet X_0 + \lambda C \bullet X^* < z_0$, for large $\lambda > 0$. This is a contradiction to the definition of z_0. Then, (SDSIP1) is bounded in value for $d = z_0$. Since (SDSIP1) has duality, there exists a solution (y, Z) satisfying (3.3) and (3.4). By (3.4), the optimal value of (SDSIP) is equal to that of (DSDSIP). So (SDSIP) yields duality with respect to C. $\quad\square$

It follows from Lemma 3.1 that we can get the following corollary.

Corollary 3.1 *(SDSIP) yields uniform duality if and only if, for any $d \in R$ and $C \in S^n$, the constraint system (3.1) yields duality with respect to $\tilde{C} \in S^{n+1}$.*

Theorem 3.1 *(SDSIP) yields uniform duality if and only if $sp(\tilde{W}) + \tilde{K}$ is a closed set, where*

$$\tilde{W} = \{\tilde{A}(t) \in S^{n+1} | t \in B\}, \quad sp(\tilde{W}) = \{\sum_{t \in B} y(t)\tilde{A}(t) | y \in \Lambda_B\},$$

and

$$\tilde{K} = \left\{ \begin{pmatrix} K & \mathbf{0} \\ \mathbf{0}^T & k \end{pmatrix} \in S^{n+1} | K \in S^n, K \succeq 0 \text{ and } k \in R^+ \right\}.$$

Proof. By Corollary 3.1, (SDSIP) yields uniform duality if and only if, for any $d \in R$ and $C \in S^n$, the constraint system (3.1) yields duality with respect to each $\tilde{C} \in S^{n+1}$. By Proposition 2.1, for any $d \in R$ and $C \in S^n$, the constraint system (3.1) yields duality with respect to each $\tilde{C} \in S^{n+1}$ if and only if $sp(\tilde{W}) + \tilde{K}$ is a closed set. Then, the conclusion follows readily. $\qquad \square$

Acknowledgments

This research is partially supported by the Research Committee of The Hong Kong Polytechnic University and the National Natural Science Foundation of China.

References

Charnes, A., Cooper, W. W. and Kortanek, K. (1962), Duality in Semi-Infinite Programs and Some Works of Haar and Caratheodory, Management Sciences, Vol.9, pp.209-229.

Duffin, R. J., Jeroslow, R. G. and Karlovitz, L. A. (1983), Duality in Semi-infinite Linear Programming, in Semi-Infinite Programming and Applications, Fiacco, A.V. and Kortanek, K. O., Eds., Lecture Notes in Economics and Mathematical Systems 215, Spring-Verlag Berlin Heidelberg New York Tokyo, pp.50-62.

Lieven Vandenberghe and Stephen Boyd (1996), Semidefinite Programming, SIAM Review, Vol.38, No.1, pp.49-95.

Madhu V. Nayakkankuppam and Michael L. Overton (1999), Conditioning of Semidefinite Programs, Mathematical Programming, Series A, Vol.85, No.3, pp.525-540.

Ramana, M. V., Tuncel, L. and Wolkowicz, H. (1997), Strong Duality for Semidefinite Programming, SIAM Journal on Optimization, Vol.7, pp.641-662.

Reemtsen, R. and Ruckmann, J.J. (1998), Semi-Infinite Programming, Kluwer Academic Publishers.

Wolkowicz, H., Saigal, R. and Vandenberghe, L. (2000), Handbook of Semidefinite Programming Theory, Algorithms, and Applications, Kluwer Academic Publishers.

Li, S. J., Yang, X. Q. and Teo, K. L. (2002), Duality and Discretization for Semi-Definite and Semi-Infinite Programming, submitted.

5 THE USE OF NONSMOOTH ANALYSIS AND OF DUALITY METHODS FOR THE STUDY OF HAMILTON-JACOBI EQUATIONS

Jean-Paul Penot

Université de Pau, Faculté des Sciences,
Laboratoire de Mathématiques Appliquées, CNRS ERS 2055
Av. de l'Université, BP 1155, 64013 PAU, France

Abstract: We consider some elements of the influence of methods from convex analysis and duality on the study of Hamilton-Jacobi equations.

Key words: Conjugacy, convexity, duality, Hamilton-Jacobi equation, subdifferential, viscosity solution.

1 INTRODUCTION

A huge literature has been devoted to Hamilton-Jacobi equations during the last decades and several monographs are devoted to them, partially or entirely (Bardi and Capuzzo-Dolcetta (1998), Barles (1994), Clarke et al (1998), Evans (1998), Lions (1982), Subbotin (1995), Vinter (2000)). The amount of methods used to study them is amazing. Here we consider them from the point of view of unilateral analysis (nonsmooth analysis, convex analysis and variational convergences).

Given a Banach space X, with dual X^* and functions $g : X \to \mathbb{R} \cup \{+\infty\}$, $H : X^* \to \mathbb{R} \cup \{+\infty\}$, the evolution Hamilton-Jacobi equation we study consists in finding solutions to the system

$$(H - J) \qquad \frac{\partial u}{\partial t}(x,t) + H(Du(x,t)) \;=\; 0 \qquad (1.1)$$

$$(B) \qquad\qquad\qquad\qquad u(x,0) \;=\; g(x) \qquad (1.2)$$

where $u : X \times \mathbb{R}_+ \to \overline{\mathbb{R}}$ is the unknown function, and Du (resp. $\frac{\partial u}{\partial t}$) denotes the derivative of u with respect to its first (resp. second) variable. Note that we accept solutions, Hamiltonians and initial value functions taking the values $+\infty$ or being discontinuous.

In the present paper we survey some questions which occurred to us while studying this equation and the papers which were available to us. Needless to say that many other questions could be considered. We refer to Alvarez et al (1999), Barron (1999), Borwein and Zhu (1996), Crandall et al (1992), Deville (1999), Frankowska (1993), Imbert (1999), Imbert and Volle (1999), Penot (2000), Penot and Volle (2000) and Volle (1997) for more complete recent developments and to the monographs quoted above for classical results and references.

2 THE INTEREST OF CONSIDERING EXTENDED REAL-VALUED FUNCTIONS

The origins of equations (H-J), (B) incite to consider such data. These equations arise when, for a given $(x,t) \in X \times \mathbb{P}$, with $\mathbb{P}$ the set of positive numbers, one considers the Bolza problem:

$(\mathcal{B})$ find $V(x,t) :=$

$$\inf \left\{ g(w(0)) + \int_{-t}^{0} L(w(s), w'(s))ds : w \in W^{1,1}([-t,0], X),\ w(-t) = x \right\}$$

where $W^{1,1}([-t,0],X)$ is the set of primitives of integrable functions on $[-t,0]$ with values in X.

For considering problems in which a target A has to be reached, it can be useful to take for g the *indicator function* ι_A of the subset A of X given by $\iota_A(x) = 0$ if $x \in A$, $+\infty$ if $x \in X\backslash A$, so that the terminal constraint $x(0) \in A$ is taken into account.

Classically, one associates to the Lagrangian L an Hamiltonian H via the partial conjugacy formula

$$H(x,p) = \sup_{v \in X}(p.v - L(x,-v)).$$

or $H(x,p) = (L_x)^*(-p)$, where $L_x(v) = L(x,v)$ and f^* denotes the Fenchel conjugate of f given by $f^*(p) := \sup\{p.x - f(x) : x \in X\}$. We observe that, even when L is everywhere finite, the function H may take the value $+\infty$. Moreover, for modeling control problems, L may take infinite values. For instance, in order to take into account a differential inclusion

$$\dot{w}\,(s) \in E(w(s)) \quad \text{a.e. } s \in [-t,0]$$

where $E : X \rightrightarrows X$ is a multimapping, one may set $L(x,\cdot) = \iota_{E(x)}(\cdot)$, the indicator function of the set $E(x)$. In such a case, $H(x,\cdot)$ is the support function of the set $E(x)$; it is finite everywhere iff $E(x)$ is bounded, a condition which is not always satisfied.

Allowing the solutions to take the value $+\infty$ brings difficulties in defining notions of solution and in questions of convergence. When H depends on the derivative of u only, one disposes of explicit formulae designed by Hopf, Lax and Oleinik: for $(x,t) \in X \times \mathbb{P} := X \times (0,+\infty)$

$$u(x,t) \quad : \quad = \inf_{y \in X}\,\sup_{p \in X^*}\,(p.(x-y) + g(y) - tH(p)), \qquad \text{(Lax-Oleinik)}$$

$$v(x,t) \quad : \quad = \sup_{p \in X^*}\,\inf_{y \in X}\,(p.(x-y) + g(y) - tH(p)) \qquad \text{(Hopf)}$$

These formulae can be interpreted with the help of the Legendre-Fenchel transform:

$$v(x,t) := (g^* + tH)^*(x) := \sup_{p \in \mathrm{dom}H \cap \mathrm{dom}g^*}(p.x - g^*(p) - tH(p)) \qquad (2.1)$$

and, using the infimal convolution $\square$ given by $(g\square h)(x) := \inf\{g(x-y)+h(y) : y \in Y\}$,

$$u(x,t) := (g\square(tH)^*)(x) := \inf_{y\in X} g(x-y) + (tH)^*(y), \qquad (2.2)$$

One can extend these functions to $X \times \mathbb{R}$ by setting $u(x,t) = v(x,t) = +\infty$ for $x \in X$, $t < 0$, $v(x,0) := (g^* + \iota_{\operatorname{dom} H})^*(x)$, $u(x,0) := (g\square h_0)(x)$ with h_0 being interpreted as $\iota_{\operatorname{dom} H}^*$, the support function of $\operatorname{dom} H$. In doing so, one gets another interpretation made in several recent contributions (Imbert (1999), Imbert and Volle (1999), Penot and Volle (2000)), under various degrees of generality, starting with the pioneering work of Plazanet Plazanet (1990) dealing with the Moreau regularization of convex functions using a convex kernel. Setting

$$F(p,r) := \iota_E(p,r), \qquad G(x,t) = g(x) + \iota_{\{0\}}(t),$$

where $E := \{(p,r) \in X^* \times \mathbb{R} : -r \geq H(p)\} = S(\operatorname{epi} H)$ and S is the symmetry $(p,r) \mapsto (p,-r)$, one has

$$\begin{aligned} u &= F^*\square G, & \text{on } X \times \mathbb{R}_+ & \qquad (2.3)\\ v &= (F + G^*)^*, & \text{on } X \times \mathbb{R}_+, & \qquad (2.4) \end{aligned}$$

where the conjugates are taken with respect to the pairs (p,r), (x,t) and where the infimal convolution is taken with respect to the variable (x,t) (the notation is unambiguous inasmuch F and G are defined on $X^* \times \mathbb{R}$ and $X \times \mathbb{R}$ respectively). In fact, for $t \in \mathbb{R}_+$ one has $F^*(\cdot,t) = (tH)^*$ and $G^* = g^* \circ p_{X^*}$, where p_{X^*} is the first projection from $X^* \times \mathbb{R}$ to X^*.

In order to avoid the trivial case in which v is the constant function $-\infty^X$, we assume that

$$\operatorname{dom} H \cap \operatorname{dom} g^* \neq \emptyset, \qquad (2.5)$$

while to avoid the case the Lax solution u is an improper function we assume the condition

$$\operatorname{dom} g \neq \emptyset, \quad \operatorname{dom} H \neq \emptyset, \quad \operatorname{dom} H^* \neq \emptyset. \qquad (2.6)$$

In Imbert and Volle (1999), Penot (2000) and Penot and Volle (2000) some criteria for the coincidence of the Hopf and the Lax solutions are presented and some consequences of this coincidence are drawn. In particular, in Penot and Volle (2000) we introduced the use of the Attouch-Brézis type condition

$$Z := \mathbb{R}_+(\operatorname{dom} g^* - \operatorname{dom} H) = -Z = \operatorname{cl}(Z) \qquad (2.7)$$

which ensures that $u = v$ when g and H are closed proper convex functions. Simple examples show that, without convexity assumptions, u and v may differ.

The interchange of inf and sup in the explicit formulae above shows that $u \geq v$. A more precise comparison can be given. Without any assumption, for $t \in \mathbb{P}$, one has

$$
\begin{aligned}
u(\cdot, t) \quad &\geq \quad u(\cdot, t)^{**} := (g \square (tH)^*)^{**} \\
&= (g^* + tH^{**})^* \geq (g^* + tH)^* = v(\cdot, t), \quad (2.8) \\
u^{**} \quad &= \quad (F^* \square G)^{**} = (F^{**} + G^*)^* \geq (F + G^*)^* = v. \quad (2.9)
\end{aligned}
$$

Under the convexity assumption

$$
H^{**} \mid \operatorname{dom} g^* = H \mid \operatorname{dom} g^*, \quad (2.10)
$$

which is milder than the condition $H^{**} = H$, one has a close connection between u and v.

Proposition 2.1 *Under assumptions (2.5), (2.10), one has $u(\cdot, t)^{**} = v(\cdot, t)$, $u^{**} = v$. If moreover g is convex, then $v = \overline{u}$, the lower semicontinuous hull of u.*

Proof. Let us prove the second equality of the first assertion, the first one being similar and simpler. In view of relation (2.9) it suffices to show that $(F^{**} + G^*)(p, q) = (F + G^*)(p, q)$ for any $(p, q) \in X \times \mathbb{R}$, or for any $(p, q) \in \operatorname{dom} g^* \times \mathbb{R}$ since both sides are $+\infty$ when $p \notin \operatorname{dom} g^*$. Now F^{**} is the indicator function of the closed convex hull $\overline{co}(E)$ of E. Since $\overline{co}(E) = \overline{co}(S(\operatorname{epi} H)) = S(\overline{co}(\operatorname{epi} H) = S(\operatorname{epi} H^{**})$, for $p \in \operatorname{dom} g^*$ we have $(F^{**} + G^*)(p, q) = g^*(p)$ iff $(p, q) \in \overline{co}(E) = S(\operatorname{epi} H^{**})$ iff $(F + G^*)(p, q) = g^*(p)$.

3   SOLUTIONS IN THE SENSE OF UNILATERAL ANALYSIS

Defining an appropriate notion of solution is part of the challenge. When considering existence and uniqueness of a solution as the crucial question, the notion of viscosity solution (or Crandall-Lions solution) is a fine concept. It is certainly preferable to the notion of generalized solution in which the derivative of u exists a.e. and satisfies equation (H-J).

Since the initial condition can also be given different interpretations, we chose in Penot and Volle (2000) to treat separately equation (H-J) and the initial condition (B). In doing so, one can detect interesting properties of functions which are good candidates for the equation but do not satisfy the initial condition in a classical sense. When a function u satisfies both (H-J) and (B) in an appropriate sense, we speak of a solution of the system (H-J)-(B).

The notion of viscosity solution (Crandall and Lions (1983), Crandall et al (1984)) which yielded existence and uniqueness results, introduced a crucial one-sided viewpoint since in this notion, equalities are replaced by inequalities. A further turn in the direction of nonsmooth analysis occurred with Barron and Jensen (1990), Frankowska (1987), Frankowska (1993) (see also Bardi and Capuzzo-Dolcetta (1998), Clarke et al (1998), Vinter (2000)) in which only subdifferentials are involved. We retain this viewpoint here and we admit the use of an arbitrary subdifferential. Although this concept is generally restricted to some natural conditions, here we adopt a loose definition which encompasses all known proposals: a *subdifferential* is just a triple $(\mathcal{X}, \mathcal{F}, \partial)$ where $\mathcal{X}$ is a class of Banach spaces, $\mathcal{F}(X)$ is a class of functions on the member X of $\mathcal{X}$ and ∂ is a mapping from $\mathcal{F}(X) \times X$ into the family of subsets of X^*, denoted by $(f, x) \mapsto \partial^? f(x)$, with empty value at (f, x) when $|f(x)| = \infty$. The *viscosity subdifferential* of f at x is the set of derivatives at x of functions φ of class C^1 such that $f - \varphi$ attains its minimum at x. For most results, it suffices to require that φ is Fréchet differentiable at x. This variant coincides with the notion of *Fréchet subdifferential* defined for $f \in \overline{\mathbb{R}}^X$, $x \in f^{-1}(\mathbb{R})$ by

$$\partial^- f(x) := \left\{ x^* \in X^* : \liminf_{\|u\| \to 0_+} \frac{1}{\|u\|} \left[f(x + u) - f(z) - \langle x^*, u \rangle \right] \geq 0 \right\}$$

in view of the following simple lemma which shows that there is no misfit between the two notions.

Lemma 3.1 *For any Banach space X, the Fréchet subdifferential of f at $x \in$ dom f coincides with the set of derivatives at x of functions φ which are Fréchet-differentiable at x and such that $f - \varphi$ attains its minimum at x. If X is reflexive (or more generally can be renormed by a norm of class C^1 on $X \backslash \{0\}$) then the Fréchet subdifferential coincides with the viscosity subdifferential.*

A similar result holds for the *Hadamard* (or contingent) subdifferential defined for $f \in \overline{\mathbb{R}}^X$, $x \in f^{-1}(\mathbb{R})$ by

$$\partial f(x) := \left\{ x^* \in X^* : \forall w \in X, \liminf_{(t,v) \to (0_+, w)} \frac{1}{t} \left[f(x + tv) - f(x) - \langle x^*, tv \rangle \right] \geq 0 \right\}$$

when the Fréchet differentiability of φ is replaced with Hadamard differentiability.

Definition 3.1 *Given a subdifferential $\partial^?$, a function $w : X \times \mathbb{R} \to \overline{\mathbb{R}}$ is a $\partial^?$-supersolution of (H-J) if for any $(x,t) \in X \times \mathbb{P}$ and any $(p,q) \in \partial^? w(x,t)$ one has $q + H(p) \geq 0$. If moreover* weak-lim inf$_{(x',t)\to(x,0_+)} w(x',t) \geq g(x)$ *for each $x \in X$, with X endowed with its weak topology, w is said to be a supersolution to (H-J)-(B).*

It is called a $\partial^?$-subsolution of (H-J) if for $(x,t) \in X \times \mathbb{P}$ and any $(p,q) \in \partial^? w(x,t)$ one has $q + H(p) \leq 0$. If moreover weak-lim inf$_{(x',t)\to(x,0_+)} w(x',t) \leq g(x)$ *for each $x \in X$, w is said to be a subsolution to (H-J)-(B). It is a $\partial^?$-solution of (H-J) if it is both a $\partial^?$-supersolution and a $\partial^?$-subsolution.*

The preceding definition is a natural one from the point of view of nonsmooth analysis or unilateral analysis (note that in Bardi and Capuzzo-Dolcetta (1998) a solution is called a bilateral solution in view of the relation $q + H(p) = 0$ for any $(p,q) \in \partial^? u(x,t)$; however, we prefer to stress the exclusive use of the subdifferential). The preceding definition has been introduced by Barron and Jensen in Barron and Jensen (1990) (along with a lower semicontinuity assumption) for the viscosity subdifferential.

We recover a more classical notion in the following definition.

Definition 3.2 *A function $u : X \times \mathbb{R} \to \overline{\mathbb{R}}$ is a Crandall-Lions $\partial^?$-subsolution if $-u$ is a supersolution of the equation associated with the Hamiltonian $p \mapsto -H(-p)$. It is a Crandall-Lions $\partial^?$-solution if it is both a $\partial^?$-supersolution and a Crandall-Lions $\partial^?$-subsolution.*

4 VALIDITY OF SOME EXPLICIT FORMULAE

Another interest of formulas (2.3)-(2.4) lies in the simplicity of the proofs of existence theorems (see Imbert (1999) and Imbert and Volle (1999)). Here we simplify the original approach of Penot and Volle (2000). Note that no assumption is needed for assertion (a).

Theorem 4.1 *(a) The Hopf solution v is a Hadamard (hence a Fréchet) supersolution of (H-J).*

(b) Under assumptions (2.5), (2.10), the Hopf solution v is a Hadamard solution of (H-J).

Proof. (a) If assumption (2.5) does not hold, one has $v \mid X \times \mathbb{P} = -\infty^{X \times \mathbb{P}}$ and there is nothing to prove. If it holds, given $t > 0$, $x \in X$ and $(p, q) \in \partial v(x, t)$, since v is convex, for each $s \in \mathbb{R}_+$, one has

$$p.x + qt - qs - v(x, t) \geq p.w - v(w, s) \qquad \forall w \in X. \qquad (4.1)$$

For $s = 0$, taking the supremum on w and using $v(\cdot, 0)^* = (g^* + \iota_{\text{dom } H})^{**} \geq g^*$, one gets

$$
\begin{aligned}
p.x + qt - (p.x - g^*(p) - tH(p)) &\geq p.x + qt - v(x, t) \\
&\geq v(\cdot, 0)^*(p) \geq g^*(p).
\end{aligned}
$$

The last inequalities show that $g^*(p) < \infty$; it follows that $q + H(p) \geq 0$.

(b) For $s > 0$, using (2.5), (2.10) to note that $g^* + sH = g^* + sH^{**}$ is closed, proper, convex, one gets

$$
\begin{aligned}
p.x + q(t - s) - v(x, t) &\geq \sup_{w \in X} \left(p.w - (g^* + sH)^*(w) \right) \\
&= (g^* + sH)^{**}(p) \\
&= (g^* + sH)(p). \qquad (4.2)
\end{aligned}
$$

Since $g^*(p) < +\infty$ and since s can be arbitrarily large, one obtains $-q \geq H(p)$.

In order to get a similar property for the Lax solution, we use a coercivity condition:

(C) $\liminf_{\|x\| \to \infty} H^*(x)/\|x\| > -\liminf_{\|x\| \to \infty} g(x)/\|x\|$.

Let us recall that an infimal convolution is said to be *exact* if the infimum is attained. Under assumption (C), exactness occurs in (2.2) when X is reflexive and g is weakly l.s.c.

Theorem 4.2 *(a) When (C) holds, the Lax solution u is a Fréchet supersolution.*

(b) If the inf-convolution in the definition of u is exact, then u is a Hadamard supersolution.

(c) If g is convex, then u is a Hadamard supersolution.

(d) If $H = H^{**}$, then u is a Hadamard subsolution.

Since the definition of u involves H through H^*, the assumption $H = H^{**}$ is sensible.

Proof Since assertions (b)-(d) are proved in Penot and Volle (2000) and elsewhere, we just prove (a). Let $(x,t) \in X \times \mathbb{P}$ be fixed and let k be given by $k(w) = g(x-w) + tH^*(t^{-1}w)$. Assumption (C) ensures that k is coercive (this fact justifies the observation preceding the statement). Let B be a bounded subset of X such that $\inf k(B) = \inf k(X)$. For each $s \in]0,t[$ let us pick $z_s \in B$ such that $k(z) < \inf k(X) + s^2 = u(x,t) + s^2$. Then, a short computation shows that

$$u(x - st^{-1}z_s, t-s) \leq u(x,t) - sH^*(t^{-1}z_s) + s^2.$$

It follows that for each $(p,q) \in \partial^- u(x,t)$ one can find a function $\varepsilon : \mathbb{R}_+ \to \mathbb{R}_+$ with limit 0 at 0 such that

$$-p.st^{-1}z_s - sq - \varepsilon(s)s \leq u(x - st^{-1}z_s, t-s) - u(x,t) \leq s^2 - sH^*(t^{-1}z_s).$$

Therefore, dividing by s, and passing to the limit inferior, we obtain

$$q + H(p) \geq \liminf_{s \to 0_+} \left(q + p.t^{-1}z_s - h(t^{-1}z_s) \right) \geq \lim_{s \to 0_+} (-s - \varepsilon(s)) = 0.$$

5   UNIQUENESS AND COMPARISON RESULTS

Let us turn to this important question which has been at the core of the viscosity method. There are several methods for such a question: partial differential equations techniques (Bardi and Capuzzo-Dolcetta (1998), Barles (1994), Lions (1982)...), invariance and viability for differential inclusions (Subbotin (1995), Frankowska (1993), Plaskacz and Quincampoix (2000), Plaskacz and Quincampoix (2000)...), nonsmooth analysis results such as the Barron-Jensens striking touching theorem (Barron (1999), Barron and Jensen (1990)), the fuzzy sum rule (Borwein and Zhu (1996), Deville (1999), El Haddad and Deville (1996)), multidirectional mean value inequalities (Imbert (1999), Imbert and Volle (1999), Penot and Volle (2000)). Let us note that the last two results are almost equivalent and are equivalent in reflexive spaces.

In order to simplify our presentation of recent uniqueness results arising from nonsmooth analysis, we assume in the sequel that X is reflexive and we use the Fréchet subdifferential. In such a case, a fuzzy sum rule is satisfied and mean value theorems are available.

Theorem 5.1 *(Penot and Volle (2000) Th. 6.2). For any l.s.c. Fréchet subsolution w of (H-J)-(B) one has $w \leq u$, the Lax solution.*

The next corollary has been obtained in Alvarez et al (1999) Th. 2.1 under the additional assumptions that X is finite dimensional, H is finite everywhere and for the subclass of solutions which are l.s.c. and bounded below by a function of linear growth. It is proved in Imbert and Volle (1999) under the additional condition that $\operatorname{dom} H^*$ is open.

Corollary 5.1 *Suppose X is reflexive, g and H are closed proper convex functions and $\operatorname{dom} g^* \subset \operatorname{dom} H$. Then the Hopf solution is the greatest l.s.c. Fréchet subsolution of (H-J)-(B).*

The use of the mean value inequality for a comparison result first appeared in Imbert (1999), Imbert and Volle (1999) Theorem 3.3 which assumes that H is convex and globally Lipschitzian and that X is a Hilbert space. Let us note that in our framework the mean value theorem is equivalent to the fuzzy sum rule; the fuzzy sum rule has been used in Borwein and Zhu (1996), Borwein and Zhu (1999), Deville (1999), El Haddad and Deville (1996) for a similar purpose.

Theorem 5.2 *Suppose X is reflexive, H is u.s.c. on $\operatorname{dom} g^* \neq \emptyset$ and such that $H(\cdot) \leq b + c \, \|\cdot\|$ for some $b, c \in \mathbb{R}$. Let $w : X \times \mathbb{R}_+ \to \mathbb{R}$ be a weakly l.s.c. Fréchet supersolution to (H-J)-(B). Then $w \geq v$, the Hopf solution.*

Thus v is the lowest Fréchet supersolution to (H-J)-(B) when H fulfils the assumptions.

Corollary 5.2 *Suppose X is reflexive and H is u.s.c. on $\operatorname{dom} g^*$ and such that $H(\cdot) \leq b + c \, \|\cdot\|$ for some $b, c \in \mathbb{R}$. Let w be a weakly l.s.c. function on $X \times \mathbb{R}_+$ which is such that $w(\cdot, 0) = g$ and is a Fréchet (or viscosity) solution to (H-J)-(B). Then $v \leq w \leq u$.*

If moreover g and H are convex, then $w = v$, the Hopf solution.

It is shown in (Alvarez et al (1999) Thms 2.1 and 2.5) that the growth condition on H can be dropped when $\dim X < +\infty$.

Well-posedness in the sense of Hadamard requires that when the data (g, H) are perturbed in a continuous way, the solution is perturbed in a continuous way. Up to now, this question seems to have been studied essentially in the sense of local uniform convergence. While this mode of convergence is well-suited to the finite dimensional case with finite data and solutions, it does not fit our framework. Thus, in Penot (2000) and in Penot and Zalinescu (2001b) this question is considered with respect to sublevel convergence and to epiconvergence (and various other related convergences). These convergences are well adapted to fonctions taking infinite values since they involve convergence of epigraphs. They have a nice behavior with respect to duality. However, the continuity of the operations involved in the explicit formulae require technical "qualification" assumptions (Penot and Zalinescu (2001a), Penot and Zalinescu (2001b)).

We have not considered here the case H depends on x; we refer to Rockafellar and Wolenski (2000a), Rockafellar and P.R Wolenski (2000b) for recent progress on this question. We also discarded the case H depends on $u(x)$. In such a case one can use operations similar to the infimal convolution $\Box$ such as the sublevel convolution $\Diamond$ and quasiconvex dualities as introduced in Penot and Volle (1987)-Penot and Volle (1990) (see also Martinez-Legaz (1988), Martinez-Legaz (1988)). The papers Barron et al (1996)-Barron et al (1997) opened the way and have been followed by Alvarez et al (1999), Barron (1999), Volle (1998), Volle (1997)). A panorama of quasiconvex dualities is given in Penot (2000) which incites to look for the use of rare dualities, reminding the role the Mendeleiev tableau played in chemistry.

References

O. Alvarez, E.N. Barron and H. Ishii (1999), Hopf-Lax formulas for semicontinuous data, Indiana Univ. Math. J. 48 (3), 993-1035.

O. Alvarez, S. Koike and I Nakayama (2000), Uniqueness of lower semicontinuous viscosity solutions for the minimum time problem, SIAM J. Control Optim. 38 (2), 470-481.

H. Attouch (1984), Variational convergence for functions and operators, Pitman, Boston.

M. Bardi and I. Capuzzo-Dolcetta, Optimal Control and Viscosity Solutions of Hamilton-Jacobi-Bellman Equations, Birkhäuser, Basel.

G. Barles (1994), Solutions de viscosité des équations de Hamilton-Jacobi, Springer, Berlin.

G. Barles and B. Perthame (1987), Discontinuous solutions of deterministic optimal stopping time problems, Math. Modeling and Numer. Anal. 21, 557-579.

E.N. Barron (1999), Viscosity solutions and analysis in L^∞, in Nonlinear Analysis, Differential Equations and Control, F.H. Clarke and R.J. Stern (eds.), Kluwer, Dordrecht, pp. 1-60.

E.N. Barron, and R. Jensen (1990), Semicontinuous viscosity solutions of Hamilton-Jacobi equations with convex Hamiltonians, Comm. Partial Diff. Eq. 15, 1713-1742.

E.N. Barron, R. Jensen and W. Liu (1996), Hopf-Lax formula for $u_t + H(u, Du) = 0$, J. Differ. Eq. 126, 48-61.

E.N. Barron, R. Jensen and W. Liu (1997), Hopf-Lax formula for $u_t + H(u, Du) = 0$. II, Comm. Partial Diff. Eq. 22, 1141-1160.

J.M. Borwein and Q.J. Zhu (1996), Viscosity solutions and viscosity subderivatives in smooth Banach spaces with applications to metric regularity, SIAM J. Control Optim. 34, 1568-1591.

J.M. Borwein and Q.J. Zhu (1999), A survey of subdifferential calculus with applications, Nonlinear Anal. Th. Methods Appl. 38, 687-773.

F.H. Clarke, and Yu.S. Ledyaev (1994), Mean value inequalities in Hilbert space, Trans. Amer. Math. Soc. 344, 307-324.

F.H. Clarke, Yu.S. Ledyaev, R.J. Stern and P.R. Wolenski (1998), Nonsmooth analysis and control theory, Springer, New York.

M.G. Crandall, L.C. Evans and P.-L. Lions (1984), Some properties of viscosity solutions of Hamilton-Jacobi equations, Trans. Amer. Math. Soc. 282, 487-502.

M.G. Crandall, H. Ishii and P.-L. Lions (1992), User's guide to viscosity solutions of of second order partial differential equations, Bull. Amer. Math. Soc. 27, 1-67.

M.G. Crandall and P.-L. Lions (1983), Viscosity solutions to Hamilton-Jacobi equations, Trans. Amer. Math. Soc. 277.

R. Deville (1999), Smooth variational principles and nonsmooth analysis in Banach spaces, in Nonlinear Analysis, Differential Equations and Control, F.H. Clarke and R.J. Stern (eds.), Kluwer, Dordrecht, 369-405.

E. El Haddad and R. Deville (1996), The viscosity subdifferential of the sum of two functions in Banach spaces. I First order case, J. Convex Anal. 3, 295-308.

L.C. Evans (1998), Partial differential equations, Amer. Math. Soc., Providence.

H. Frankowska (1987), Equations d'Hamilton-Jacobi contingentes, C.R. Acad. Sci. Paris Serie I 304, 295-298.

H. Frankowska (1993), Lower semicontinuous solutions of Hamilton-Jacobi-Bellman equations, SIAM J. Control Optim. 31 (1), 257-272.

G.N. Galbraith (2000), Extended Hamilton-Jacobi characterization of value functions in optimal control, SIAM J. Control Optim. 39 (1), 281-305.

C. Imbert (1999), Convex analysis techniques for Hopf-Lax' formulae in Hamilton-Jacobi equations with lower semicontinuous initial data, preprint, Univ. P. Sabatier, Toulouse, May.

C. Imbert and M. Volle (1999), First order Hamilton-Jacobi equations with completely convex data, preprint, October.

A.D. Ioffe (1998), Fuzzy principles and characterization of trustworthiness, Set-Val. Anal. 6, 265-276.

P.-L. Lions (1982), Generalized Solutions of Hamilton-Jacobi Equations, Pitman, London.

J.-E. Martinez-Legaz (1988), On lower subdifferentiable functions, in "Trends in Mathematical Optimization", K.H. Hoffmann et al. eds, Birkhauser, Basel, 197-232.

J.-E. Martinez-Legaz (1988), Quasiconvex duality theory by generalized conjugation methods, Optimization, 19, 603-652.

J.-P. Penot (1997), Mean-value theorem with small subdifferentials, J. Opt. Th. Appl. 94 (1), 209-221.

J.-P. Penot (2000), What is quasiconvex analysis? Optimization 47, 35-110.

J.-P. Penot and M. Volle (1987), Dualité de Fenchel et quasi-convexité, C.R. Acad. Sc. Paris série I, 304 (13), 269-272.

J.-P. Penot and M. Volle (1988), Another duality scheme for quasiconvex problems, in "Trends in Mathematical Optimization", K.H. Hoffmann et al. eds, Birkhauser, Basel, 259-275.

J.-P. Penot and M. Volle (1990), On quasi-convex duality, Math. Operat. Research 15 (4), 597-625.

J.-P. Penot and M. Volle (2000), Hamilton-Jacobi equations under mild continuity and convexity assumptions, J. Nonlinear and Convex Anal. 1, 177-199.

J.-P. Penot and M. Volle (1999), Convexity and generalized convexity methods for the study of Hamilton-Jacobi equations, Proc. Sixth Conference on Generalized Convexity and Generalized Monotonicity, Samos, Sept. 1999, N.

Hadjisavvas, J.-E. Martinez-Legaz, J.-P. Penot, eds., Lecture Notes in Econ. and Math. Systems, Springer, Berlin, to appear.

J.-P. Penot and C. Zalinescu, Continuity of usual operations and variational convergences, preprint.

J.-P. Penot and C. Zalinescu, Persistence and stability of solutions of Hamilton-Jacobi equations, preprint.

S. Plaskacz and M. Quincampoix (2000), Value function for differential games and control systems with discontinuous terminal cost, SIAM J. Control Optim. 39, no. 5, 1455-1498.

S. Plaskacz and M. Quincampoix (2000), Discontinuous Mayer control problems under state constraints, Topological Methods in Nonlinear Anal. 15, 91-100.

P. Plazanet (1990), Contributions à l'analyse des fonctions convexes et des différences de fonctions convexes. Application à l'optimisation et à la théorie des E.D.P., thesis, Univ. P. Sabatier, Toulouse.

R.T. Rockafellar and R. J.-B. Wets (1997), Variational Analysis, Springer-Verlag, Berlin.

R.T. Rockafellar and P.R. Wolenski (2000), Convexity in Hamilton-Jacobi theory. I. Dynamics and duality. SIAM J. Control Optim. 39, no. 5, 1323–1350.

R.T. Rockafellar and P.R Wolenski (2000), Convexity in Hamilton-Jacobi theory. II. Envelope representations. SIAM J. Control Optim. 39, no. 5, 1351–1372.

A.I. Subbotin (1995), Generalized solutions of first-order PDE's, Birkhäuser, Basel.

R. Vinter (2000), Optimal Control, Birkhäuser, Boston.

M. Volle (1998), Duality for the level sum of quasiconvex functions and applications, ESAIM: Control, Optimisation and Calculus of Variations, 3, 329-343, http://www.emath.fr/cocv/

M. Volle (1997), Conditions initiales quasiconvexes dans les équations de Hamilton-Jacobi, C.R. Acad. Sci. Paris série I, 325, 167-170.

6   SOME CLASSES OF ABSTRACT CONVEX FUNCTIONS

A.M. Rubinov

School of Information Technology and Mathematical Sciences
University of Ballarat, Victoria 3353 Australia

and A.P. Shveidel

Department of Mathematics, Karaganda State University
Karaganda, 470 074, Kazakhstan

Abstract: We describe classes of abstract convex functions with respect to some sets of functions with the peaking property. We also describe conditions, which guarantee that the corresponding abstract subdifferentials are nonempty.

Key words: Abstract convexity, supremal generator, peaking property, abstract subdifferential.

1   INTRODUCTION

We begin with the following definitions. Let H be a set of functions defined on a set X. A function $f : X \to \mathbb{R}_{+\infty} := \mathbb{R} \cup \{+\infty\}$ is called *abstract convex* with respect to H (H- convex) if

$$f(x) = \sup\{h(x) : x \in \operatorname{supp}(f, H)\} \text{ for all } x \in X, \tag{1.1}$$

where $\operatorname{supp}(f, H) = \{h \in H : h(x) \le f(x) \ (\forall x \in X)\}$ is the *support set* of the function f with respect to H.

A set H is called a *supremal generator* of a set F of functions f defined on X if each $f \in F$ is abstract convex with respect to H. A supremal generator H is a base (in a certain sense) of F, so some properties of H can be extended to the entire set F. (See Rubinov (2000), Chapter 6 and references therein for details.) If H is a "small" set then some of its properties can be verified by the direct calculation. Thus small supremal generators are very helpful in the examination of some problems. This observation explains, why a description of small supremal generators for the given broad class of functions is one of the main problems of abstract convexity. The reverse problem: to describe abstract convex functions with respect to a given set H, is also very interesting.

If H consists of continuous functions then the set of H-convex functions is contained in the set LSC_H of all lower semicontinuous functions f such that $f \ge h$ for some function $h \in H$. (Here $f \ge h$ stands for $f(x) \ge h(x)$ for all $x \in X$.) The set LSC_H is very large. In particular, if constants belong to H, then LSC_H contains all bounded from below lower semicontinuous functions. As it turned out there are very small supremal generators of the very large set LSC_H. These supremal generators can be described by means of the so-called peaking property (see Pallaschke and Rolewicz (1997) and references therein) or the technique based on functions, support to Urysohn peaks (see Rubinov (2000) and references therein).

We present two known examples of such generators (see Rubinov (2000) and references therein).

1) let X be a Hilbert space and H be the set of all quadratic functions h of the form

$$h(x) = -a\|x - x_0\|^2 - c, \qquad x \in X,$$

where $a \ge 0$, $x_0 \in X$ and $c \in \mathbb{R}$. Then H is a supremal generator of LSC_H. The set H can be described by only three parameters: a point $x \in X$, a number

$c \in \mathbb{R}$ and a number $a \in \mathbb{R}_+$. If X coincides with n-dimensional space then the dimension of H is $n + 2$.

2) Let X be a Banach space and H be the set of all functions h of the form

$$h(x) = -a\|x - x_0\| - c, \qquad x \in X,$$

where $a \geq 0$, $x_0 \in X$ and $c \in \mathbb{R}$. The set H is a supremal generator of LSC_H. If $X = \mathbb{R}^n$ then the dimension of H again equal to $n + 2$.

It is interesting to give a direct description of the set LSC_H without references to the class H. We give such a description for a broad class of sets H, which possess the peaking property. Let X be a normed space and p be a continuous sublinear function defined on X and such that

$$\gamma_p := \inf_{\|x\|=1} p(x) > 0. \tag{1.2}$$

Let k be a positive number. We shall study the set H^k of all functions h of the form

$$h(x) = -ap^k(x - x_0) - c \quad (x \in X) \tag{1.3}$$

with $x_0 \in X$, $c \in \mathbb{R}$ and $a > 0$ and show that this set is a supremal generator of the class of functions $\mathcal{P}_k$, which depends only on k and does not depend on p. The class $\mathcal{P}_k$ is very broad. It consists of all lower semicontinuous functions $f : X \to \mathbb{R}_{+\infty}$ such that $\liminf_{\|x\| \to +\infty} f(x)/\|x\| > -\infty$.

Consider the space $X = \mathbb{R}^n$ and a sublinear function p defined on X such that (1.2) holds. It is well-known that there exists a set of linear function U_p such that $p(x) = \max_{l \in U_p}[l, x]$, where $[l, x]$ stands for the inner product of vectors l and x. Let H^1 be the set of functions defined by (1.3) for the given function p and $k = 1$. Since H^1 is a supremal generator of $\mathcal{P}_1$, it follows that each function $f \in \mathcal{P}_1$ can be represented in the following form:

$$f(x) = \sup_{(a,c,x_0) \in V(f)} \left(-a \max_{l \in U_p}([l, x - x_0] - c) \right),$$

where $V(f) = \{(a, c, x_0) : -ap(x - x_0) - c \leq f(x) \ \forall \ x \in X\}$ and U_p does not depend on f. Thus we have the following sup-min presentation of an arbitrary function $f \in \mathcal{P}_1$ through affine functions:

$$f(x) = \sup_{(a,c,x_0) \in V(f)} \min_{l \in U_p}(a[l, x_0 - x] - c). \tag{1.4}$$

Since the class $\mathcal{P}_1$ does not depend on the choice of a sublinear function p with $\gamma_p > 0$ it is interesting to consider such functions p that the corresponding set U_p has the least possible cardinality. Clearly, this cardinality is greater than or equal to $n + 1$, since for each function $p(x) = \max_{i=1,\ldots,j}[l_i, x]$ with $j \leq n$ and nonzero l_i we have $\gamma_p \equiv \min_{\|x\|=1} p(x) < 0$. We discuss this question in details (see Example 3.3, Example 3.4 and Remark 3.1).

We shall also describe conditions, which guarantee that the so-called abstract convex subdifferentials are not empty. Recall the corresponding definitions (Rubinov (2000)). Let X be an arbitrary set. A set L of functions defined on X, is called the set of abstract linear functions if for every $l \in L$ the functions $h_{l,c}(x) = l(x) - c$ do not belong to L for each $c \neq 0$. The set $H_L = \{h_{l,c} : l \in L, c \in \mathbb{R}\}$ is called the set of L-affine functions. Let f be an H_L-convex function. The set

$$\partial_L f(x_0) = \{l \in L : l(x) - l(x_0) \leq f(x) - f(x_0) \text{ for all } x \in X\}$$

is called the L-subdifferential of the function f at a point x_0. Clearly $l \in \partial_L f(x_0)$ if and only if $f(x_0) = h(x_0)$ where $h(x) := l(x) - c$ with $c := f(x_0) - l(x_0)$. Due to the definition of L-subdifferential $h(x) \leq f(x)$ for all x. Thus the L-subdifferential is not empty if and only if the supremum in the equality $f(x_0) = \sup\{h(x) : h \in \text{supp}\,(f, H_L)\}$ is attained.

The set H^k defined by (1.3) can be considered as the set H_{L^k} of L^k-affine functions where

$$L^k = \{l_{k,a,x_0} : a > 0, x_0 \in X, c \in \mathbb{R}\}, \tag{1.5}$$

and $l_{k,a,x_0}(x) = -ap^k(x - x_0)$, $(x \in X)$. We express conditions, which guarantee that L^k-subdifferential is not empty in terms of the calmness of degree k (see Theorem 4.1). Note that if L^1-subdifferential $\partial_{L^1} f(x)$ is not empty for all x then sup-min representation (1.4) become max-min representation.

Note that the classical convex subdifferential $\partial f(x)$ of convex function f at the point x plays two different roles. First, a subgradient (that is an element of $\partial f(x)$) accomplishes a *local approximation* of f in a neighborhood of x. Second, a subgradient permits the construction of a *global affine support*, that is an affine function, which does not exceed f over the entire space and coincides with f at x. Generalizations of the notion of the convex subdifferential based on local approximation (global affine support, respectively) lead to nonsmooth analysis (abstract convexity, respectively). It is very important to unite such different theories as nonsmooth analysis and abstract convexity in the study of some concrete non-convex objects. We provide an example of such a unifications (see Proposition 4.1).

2   SETS $\mathcal{P}_K$

Let X be a normed space and k be a positive number. Denote by $\mathcal{P}_k$ the set of lower semicontinuous functions $f : X \to \mathbb{R}_{+\infty}$ such that f is bounded from below on each ball and

$$\liminf_{\|x\|\to\infty} \frac{f(x)}{\|x\|^k} > -\infty. \tag{2.1}$$

If X is a finite -dimensional space, then $f \in \mathcal{P}_k$ if and only if f is lower semicontinuous and (2.1) holds. We describe some simple properties of the set $\mathcal{P}_k$.

1) $\mathcal{P}_k$ is closed under the pointwise addition;

2) if $f \in \mathcal{P}_k$ and $g \geq f$ is a lower semicontinuous function, then $g \in \mathcal{P}_k$ as well. In particular, if T is an arbitrary index set, $f_t \in \mathcal{P}_k$ for each $t \in T$ and $(\sup_{t\in T} f_t)(x) = \sup_{t\in T} f_t(x)$, then $\sup_{t\in T} f_t \in \mathcal{P}_k$.

3) If $f_1, f_2 \in \mathcal{P}_k$, then the function $x \mapsto \min(f_1(x), f_2(x))$ also belongs to $\mathcal{P}_k$.

4) Let $f \in \mathcal{P}_k$ and $g(x) = af(x - x_0) - c$, where $a > 0$, $x_0 \in X$ and $c \in \mathbb{R}$. Then $g \in \mathcal{P}_k$.

5) Constants belong to $\mathcal{P}_k$ for all $k > 0$. Convex lower semicontinuous functions (in particular, linear continuous functions) belong to $\mathcal{P}_k$ for all $k \geq 1$.

Note also that $\mathcal{P}_l \subset \mathcal{P}_k$ if $l < k$.

Let $f : X \to \mathbb{R}_{+\infty}$ and $k > 0$. Consider the function $\tilde{f}^k$ defined by

$$\tilde{f}^k(x) = \operatorname{sign} f(x)|f(x)|^k, \qquad x \in X.$$

It follows from the definition of the class $\mathcal{P}_k$ that $\mathcal{P}_k = \{\tilde{f}^k : f \in \mathcal{P}_1\}$. We now describe some subsets of the set $\mathcal{P}_1$.

1. Each lower semicontinuous positively homogeneous of degree one function, which is bounded from below on the unit ball, belongs to $\mathcal{P}_1$;

2. Each Lipschitz function belongs to $\mathcal{P}_1$.

3. A function $f : X \to \mathbb{R}_{+\infty}$ is called radiant if $f(\alpha x) \leq \alpha f(x)$ for all $x \in X$ and $\alpha \in (0, 1]$. Equivalent definition: a function f is radiant if $f(\beta y) \geq \beta f(y)$ for all $y \in X$ and $\beta \geq 1$. Recall that a function $f : X \to \mathbb{R}_{+\infty}$ is called proper, if the set $\operatorname{dom} f := \{x : f(x) < +\infty\}$ is nonempty. The inequality $f(0) \leq 0$ holds for each proper lower semicontinuous and radiant function f. Indeed, let $y \in \operatorname{dom} f$. Since $f(\alpha y) \leq \alpha f(y)$ for all $\alpha \in (0, 1)$, it follows that $f(0) \leq \lim_{\alpha \to 0} f(\alpha y) = 0$. It is easy to see that f is radiant if and only if its epigraph $\operatorname{epi} f = \{(x, \lambda) \in X \times \mathbb{R} : x \in X : \lambda \geq f(x)\}$ is a radiant set. (A subset Ω of a vector space is called radiant if this set is star-shaped with respect to zero, that is $(x \in \Omega, \lambda \in (0, 1)) \implies (\lambda x \in \Omega.))$

We now show that every lower semicontinuous radiant function f, such that $d := \inf_{\|x\| \leq 1} f(x) > -\infty$, belongs to $\mathcal{P}_1$. Assume without loss of generality that f is proper. Since $f(0) \leq 0$ it follows that $d \leq 0$. Let $x \in X$ and $\|x\| = \beta > 1$. Then $x = \beta x'$, where $\|x'\| = 1$, so

$$f(x) = f(\beta x') \geq \beta f(x') \geq \beta d = \|x\| d. \tag{2.2}$$

Applying (2.2) we conclude that

$$\liminf_{\|x\| \to \infty} \frac{f(x)}{\|x\|} \geq d > -\infty.$$

Let us check that f is bounded from below on each ball $B_r = \{x : \|x\| \leq r\}$. We can consider only balls with $r > 1$. Let $x \in B_r$ and $\|x\| > 1$. Then due to (2.2) we have $f(x) \geq \|x\| d \geq rd$. If $\|x\| \leq 1$, then $f(x) \geq d$. Thus $\inf_{|x\| \leq r} f(x) \geq dr > -\infty$.

4. Consider a function f with the star-shaped epigraph $\operatorname{epi} f$. This means that there exists a point $(x, \lambda) \in X \times \mathbb{R}$ such that $(y, \mu) \in \operatorname{epi} f \implies \alpha(x, \lambda) + (1 - \alpha)(y, \mu) \in \operatorname{epi} f$ for all $\alpha \in (0, 1)$. An equivalent definition: there is a point x and a number λ such that $\alpha \lambda + (1 - \alpha) f(y) \geq f(\alpha x + (1 - \alpha) y)$ for all $y \in X$ and $\alpha \in (0, 1)$. Since each star-shaped set is a shift of a radiant set, it easily follows from **3** that a lower semicontinuous function with the star-shaped epigraph belongs to $\mathcal{P}_1$.

3   SUPREMAL GENERATORS OF THE SETS $\mathcal{P}_K$

In this section we describe small supremal generators of the set $\mathcal{P}_k$. Let X be a normed space and p be a continuous sublinear function defined on X with the following property: there exists a number $\gamma > 0$ such that

$$p(x) \geq \gamma \|x\| \quad \text{for all } x \in X. \tag{3.1}$$

Let $a > 0, x_0 \in X$ and l_{k,a,x_0} be a function defined on X by $l_{k,a,x_0}(x) = -ap^k(x - x_0)$. Consider the following sets:

$$L^k = \{l : l_{k,a,x_0} : a \in \mathbb{R}_+, \ x_0 \in X\}, \quad H^k = \{h : h = l - c\mathbf{1} : l \in L^k, c \in \mathbb{R}\},$$

where $\mathbf{1}(x) = 1$ for all $x \in X$. First of all we describe abstract convex with respect to H^k functions. Let LSC_{H^k} be the set of all lower semicontinuous functions f such that the set $\operatorname{supp}(f, H^k) = \{h \in H^k : h \leq f\}$ is not empty.

Lemma 3.1 $LSC_{H^k} \subset \mathcal{P}_k$.

Proof: Let $f \in LSC_{H^k}$. Then f is lower semicontinuous and $\operatorname{supp}(f, H^k) \neq \emptyset$. Let $h \in \operatorname{supp}(f, H^k)$, $h(x) = -ap^k(x - x_0) - c$. Since h is bounded from below on each ball, it follows that f enjoys the same property. We have

$$
\liminf_{\|x\| \to +\infty} \frac{f(x)}{\|x\|^k} \geq \liminf_{\|x\| \to +\infty} \frac{h(x)}{\|x\|^k} = -a \limsup_{\|x\| \to +\infty} \frac{p^k(x - x_0) + c/a}{\|x\|^k}
$$

$$
= -a \limsup_{\|x\| \to +\infty} \frac{p^k(x - x_0)}{\|x - x_0\|^k} \frac{\|x - x_0\|^k}{\|x\|^k}.
$$

Note that $p(x - x_0) \leq \|p\| \|x - x_0\|$ where $\|p\| = \sup_{\|x\|=1} |p(x)| < +\infty$. Therefore

$$
\liminf_{\|x\| \to +\infty} \frac{f(x)}{\|x\|^k} \geq -a\|p\|^k > -\infty. \tag{3.2}
$$

Hence $f \in \mathcal{P}_k$. $\qquad\qquad\square$

Corollary 3.1 *Each H^k-convex function belongs to $\mathcal{P}_k$.*

Lemma 3.2 *Each function $f \in \mathcal{P}_k$ is H^k-convex.*

Proof: It is sufficient to consider only finite functions. Indeed, for each $f \in \mathcal{P}_k$ we have $f(x) = \sup_n f_n(x)$, where $f_n(x) = \min(f(x), n)$, $n = 1, 2, \ldots$. So f is H^k-convex if each f_n is H^k-convex.

Let $f \in \mathcal{P}_k$ be a finite function. For each $a > 0$ consider the hypograph $K_a = \{(x, \lambda) : \lambda \leq -ap^k(x)\}$ of the function $l_{k,a,0}(x) = -ap^k(x)$. Assume that for every $x_0 \in X$ and every $\varepsilon > 0$ there exists $a > 0$ such that

$$(x_0, f(x_0) - \varepsilon) + K_a) \cap \operatorname{epi} f = \emptyset. \tag{3.3}$$

Then f is H-convex. To prove it, take a point $x_0 \in X$ and arbitrary $\varepsilon > 0$. Let a be a number such that (3.3) holds. Consider the function $h(x) = -ap^k(x - x_0) + (f(x_0) - \varepsilon)$. Let $(x_0, f(x_0) - \varepsilon) + K_a = \tilde{K}$. We have

$$
\begin{aligned}
\tilde{K} &= \{(x, \nu) : x = x_0 + y, \nu = f(x_0) - \varepsilon + \mu, \ \mu \leq -ap^k(y)\} \\
&= \{(x, \nu) : x = x_0 + y : \nu \leq f(x_0) - \varepsilon - ap^k(y)\} \\
&= \{(x, \nu) : \nu \leq f(x_0) - \varepsilon - ap^k(x - x_0)\} \\
&= \{(x, \nu) : \nu \leq h(x)\}.
\end{aligned}
$$

Let $x \in X$ and $y = x - x_0$. Since $(x, f(x)) \in \operatorname{epi} f$ it follows from (3.3) that $(x, f(x)) \notin (x_0, f(x_0) - \varepsilon) + K_a$, so $f(x) > h(x)$. Since $h(x_0) = f(x_0) - \varepsilon$ and ε is an arbitrary positive number, we conclude that $f(x_0) = \sup\{h(x_0) : h \in \operatorname{supp}(f, H)\}$.

We now prove that (3.3) is valid. Assume in contrary that there exists $x_0 \in X$ and $\varepsilon > 0$ such that for each positive integer n a pair (x_n, λ_n) can be found such that $(x_n, \lambda_n) \in \operatorname{epi} f$ and $(x_n, \lambda_n) \in (x_0, f(x_0) - \varepsilon) + K_n$. Let $(y_n, \mu_n) = (x_n, \lambda_n) - (x_0, f(x_0) - \varepsilon)$. It follows from above that

$$x_n = x_0 + y_n, \ \lambda_n \geq f(x_n); \tag{3.4}$$

$$\lambda_n = f(x_0) - \varepsilon + \mu_n; \quad \mu_n \leq -np^k(y_n). \tag{3.5}$$

We consider separately two cases.

1) The sequence μ_n is bounded. Since $|\mu_n| \geq np^k(y_n)$ it follows that $p^k(y_n) \to 0$. Since $p(y) \geq \gamma\|y\|$ for all y, we conclude that $y_n \to 0$, so $x_n \to x_0$. Due to lower semicontinuity of f and the inequality in (3.4) we have

$$\liminf_n \lambda_n \geq \liminf_n f(x_n) \geq f(x_0).$$

On the other hand

$$\lambda_n = f(x_0) - \varepsilon + \mu_n \leq f(x_0) - \varepsilon,$$

so $\liminf_n \lambda_n < f(x_0)$. We arrive at a contradiction, which shows that the sequence μ_n cannot be bounded.

2) The sequence μ_n is unbounded. Without lost of generality assume that $\lim_n \mu_n = -\infty$. Then $\lim_n \lambda_n = \lim_n(f(x_0) - \varepsilon + \mu_n) = -\infty$, so

$$\lim_n f(x_n) = -\infty. \tag{3.6}$$

The function f is bounded from below on each ball therefore (3.6) implies unboundedness of the sequence x_n. Since $\liminf_{\|x\|\to+\infty} f(x)/\|x\|^k > -\infty$, we conclude that there exists $c > 0$ such that $f(x_n) \geq -c\|x_n\|^k$ for all sufficiently large n. Hence

$$\lambda_n \geq -c\|x_n\|^k \tag{3.7}$$

for these n. On the other hand, applying (3.5) we deduce that

$$\lambda_n \leq f(x_0) - \varepsilon - np^k(y_n) \tag{3.8}$$

for all n. Due to (3.7) and (3.8), we have for all large enough n:

$$\frac{f(x_0) - \varepsilon}{\|x_n\|^k} - n\left(\frac{p(y_n)}{\|x_n\|}\right)^k \geq -c \tag{3.9}$$

It follows from (3.1) that

$$\frac{p(y_n)}{\|x_n\|} = \frac{p(y_n)}{\|y_n\|}\frac{\|y_n\|}{\|x_n\|} \geq \gamma\frac{\|y_n\|}{\|x_n\|}.$$

Since $\|x_n\|/\|y_n\| \to 1$ as $n \to +\infty$ and $\|x_n\|$ is unbounded it follows that the sequence on the left hand of (3.9) is unbounded from below, which contradicts (3.9). $\square$

Theorem 3.1 *Let $P(H^k)$ be the set of all H^k-convex functions. Then*

$$P(H^k) = \mathcal{P}_k = LSC_{H^k}.$$

Proof: The result follows directly from Lemma 3.1, Lemma 3.2 and the obvious inclusion $P(H^k) \subset LSC_{H^k}$. $\square$

We now present some examples.

Example 3.1 Let X be a normed space. The set H^1 of all functions h defined on X by $h(x) = -a\|x - x_0\| - c$ with $x_0 \in X, a \leq 0, c \in \mathbb{R}$ is a supremal generator of $\mathcal{P}_1$. If $\mathcal{P}_1 \supset H \supset H^1$, then H is also a supremal generator of $\mathcal{P}_1$. In particular the set of all concave Lipschitz functions is a supremal generator of $\mathcal{P}_1$.

Example 3.2 Let X be a Hilbert space and H^2 be the set of all functions h defined on X by $h(x) = -a\|x\|^2 + [l, x] - c$ with $a > 0, l \in X, c \in \mathbb{R}$. (Here

$[l, x]$ is the inner product of vectors l and x). It is well known (see, for example, Rubinov (2000)) that H^2 is a supremal generator of the set LSC_{H^2}. Clearly $h \in H^2$ if and only if there exists a point $x_0 \in X$ and a number $c' \in \mathbb{R}$ such that $h(x) = -a\|x - x_0\|^2 - c'$. So Theorem 3.1 shows that $\mathcal{P}_2$ coincides with the set of all H^2-convex functions.

Example 3.3 Let U be a closed convex bounded subset of a normed space X such that $0 \in \operatorname{int} U$. Then the Minkowski gauge μ_U of the set U is a continuous sublinear function, such that $\inf_{\|x\|=1} \mu_U(x) = \gamma > 0$. (Recall that $\mu_U(x) = \inf\{\lambda > 0 : x \in \lambda U\}$ for $x \in X$.) Hence the set H^k of all functions h defined on X by $h(x) = -a\mu_U^k(x - x_0) - c$ with $a > 0, x_0 \in X, c \in \mathbb{R}$ is a supremal generator of $\mathcal{P}_k$.

Example 3.4 The following particular case of Example 3.3 is of special interest. Let X be a n-dimensional space $\mathbb{R}^n$, $I = \{1, \ldots, m\}$ and $l_i \, (i \in I)$ are vectors such that their conic hull $\operatorname{cone}(l_1, \ldots, l_m) := \{\sum_{i \in I} \lambda_i l_i : \lambda \geq 0 \, (i \in I)\}$ coincides with the entire space $\mathbb{R}^n$. Then the set

$$S = \{x \in \mathbb{R}^n : [l_i, x] \leq 1, i \in I\} \tag{3.10}$$

is bounded and contains a ball cB where $B = \{x : \|x\| \leq 1\}$ and $c > 0$. Let μ_S be the Minkowski gauge of S. We have

$$\mu_S(x) \leq \mu_{cB}(x) = \frac{1}{c}\mu_B(x) = \frac{1}{c}\|x\| \text{ for all } x \in \mathbb{R}^n.$$

Thus the function μ_S is finite. The set S is bounded, so there exists $\gamma > 0$ such that $S \subset (1/\gamma)B$. We have

$$\mu_S(x) \geq \mu_{(1/\gamma)B}(x) = \gamma\mu_B(x) = \gamma\|x\| \text{ for all } x \in \mathbb{R}^n. \tag{3.11}$$

Let us calculate the Minkowski gauge μ_S. Since $S = \bigcap_{i \in I} M_i$, where $M_i = \{x : [l_i, x] \leq 1\}$ and $\mu_{M_i}(x) = \max([l_i, x], 0) \, (i \in I)$, it follows that $\mu_S(x) = \max([l_1, x], \ldots, [l_m, x], 0)$. Due to (3.11) we conclude that $\mu_S(x) > 0$ for all nonzero $x \in \mathbb{R}^n$, so

$$\mu_S(x) = \max([l_1, x], \ldots, [l_m, x]) \text{ for all } x \in \mathbb{R}^n. \tag{3.12}$$

It follows from (3.12) and (3.11) that $p = \mu_S$ is a continuous sublinear function, which possesses property (3.1), so the set H^1 of all functions h of the form

$$h(x) = -a\max_{i \in I}([l_i, x - x_0] - c \,\, (x \in \mathbb{R}^n) \tag{3.13}$$

with $a \geq 0, x_0 \in \mathbb{R}^n, c \in \mathbb{R}$ is a supremal generator of the set $\mathcal{P}_1$.

Remark 3.1 Consider a number m with the following property: there exist m vectors $l_1, \ldots, l_m$ such that the sublinear function $p(x) = \max_{i=1,\ldots,m}[l_i, x]$, $(x \in X)$ is strictly positive for all $x \neq 0$. If $m \leq n$ then this property does not hold for the function p. Indeed the system $[l_i, x] = -1$, $i = 1, \ldots, m$ has a solution for arbitrary nonzero vectors $l_1, \ldots, l_m$. It follows from the Example 3.4 that we can find corresponding vectors if $m = n + 1$. Thus the least number m, which possesses mentioned property, is equal to $n + 1$.

4 L^K-SUBDIFFERENTIALS

In this section we describe sufficient conditions, which guarantee that the L_k-subdifferential is not empty. These conditions becomes also necessary for $k = 1$.

Recall the following well known definition (see, for example, Burke (1991)). A function $f : X \to \mathbb{R}_{+\infty}$ is called calm at a point $x_0 \in \mathrm{dom}\, f$ if

$$\liminf_{x \to x_0} \frac{f(x) - f(x_0)}{\|x - x_0\|} > -\infty. \tag{4.1}$$

We say that a function f is calm of degree $k > 0$ at $x_0 \in \mathrm{dom}\, f$ if

$$\liminf_{x \to x_0} \frac{f(x) - f(x_0)}{\|x - x_0\|^k} > -\infty. \tag{4.2}$$

(This definition can be found, for example in Rubinov (2000).)

Assume, we have a continuous sublinear function p defined on a normed space X, which enjoys the property (3.1). Let L^k be the set of abstract linear functions defined by (1.5).

Theorem 4.1 *1) Let $f \in \mathcal{P}_k$ and $x_0 \in X$. If the function f is calm of degree k at the point x_0 then the subdifferential $\partial_{L^k} f(x_0)$ is nonempty and contains the function $l(x) = -cp^k(x - x_0)$ with some $c > 0$.*

2) If $k = 1$ and the subdifferential $\partial_{L^1} f(x_0)$ is nonempty then f is calm (of degree $k = 1$) at a point x_0.

Proof: 1) Assume that f is calm of degree k, that is (4.2) holds. Then there exist numbers c_1 and $d_1 > 0$ such that $f(x) - f(x_0) \geq c_1 \|x - x_0\|^k$ if $\|x - x_0\| < d_1$. Since $\liminf_{x \to \infty} f(x)/\|x\|^k > -\infty$ it follows that there exist numbers c_2' and $d_2' > 0$ such that $f(x) \geq c_2' \|x\|^k$ if $\|x\| > d_2'$. Since

$$\frac{f(x) - f(x_0)}{\|x - x_0\|^k} = \frac{f(x)}{\|x\|^k} \frac{\|x\|^k}{\|x - x_0\|^k} - \frac{f(x_0)}{\|x - x_0\|^k},$$

we can find numbers c_2 and $d_2 > 0$ such that $f(x) - f(x_0) \geq c_2\|x - x_0\|^k$ if $\|x\| > d_2$. Consider the set $D = \{x : \|x - x_0\| \geq d_1, \|x\| \leq d_2\}$. Since the function f is bounded from below on the ball $\{x : \|x\| \leq d_2\}$, it follows that there exists a number c_3 such that $f(x) - f(x_0) \geq c_3\|x - x_0\|^k$ for all $x \in D$. Due to the inequality $-\|x - x_0\|^k \geq -(1/\gamma)^k p^k(x - x_0)$ we can find a number c such that $f(x) - f(x_0) \geq cp^k(x - x_0)$ for all $x \in X$. This means that the function $l(x) = -cp^k(x - x_0)$ belongs to the subdifferential $\partial_{L^k} f(x_0)$.

2) Let $k = 1$ and $\partial_{L^1} f(x_0) \neq \emptyset$. Let $l \in \partial_{L^1} f(x_0)$, where $l(x) = -ap(x - x_1) - c$. We have,

$$
\begin{aligned}
f(x) - f(x_0) &\geq l(x) - l(x_0) = -a(p(x - x_1) - p(x_0 - x_1)) \\
&\geq -a\left(\sup_{\|y\|=1} p(y)\right)\|x - x_0\| \geq -ap(x - x_0),
\end{aligned}
$$

so the function f is calm at the point x_0. $\qquad\qquad\square$

Recall (see, for example, Demyanov and Rubinov (1995)) that a function f defined on a normed space X is called *subdifferentiable* at a point $x \in X$ if there exists the directional derivative

$$
f'_x(u) = \lim_{\alpha \to +0} (1/\alpha)(f(x + \alpha u) - f(x))
$$

for all $u \in X$ and f'_x is a continuous sublinear function. Each convex function f is subdifferentiable at a point $x \in \operatorname{int} \operatorname{dom} f$. A function f is called *quasidifferentiable* (see, for example, Demyanov and Rubinov (1995)) at a point x if the directional derivative f'_x exists and can be represented as the difference of two continuous sublinear functions. If f is the difference of two convex functions, then f is quasidifferentiable.

The support set $\operatorname{supp}(r, X^*)$ of a continuous sublinear function $r : X \to \mathbb{R}$ with respect to the conjugate space X^* will be denoted by ∂r. Note that ∂r coincides with the subdifferential (in the sense of convex analysis) of the sublinear function r at the point 0.

Proposition 4.1 *Let p be a continuous sublinear function, such that (3.1) holds and $L^1 = \{l_{1,a,y} : a > 0, y \in X\}$, where $l_{1,a,y}(x) = -ap(x - y)$. Let $f \in \mathcal{P}_1$ be a locally Lipschitz at a point $x_0 \in X$ function. Then*

1) The L^1-subdifferential $\partial_{L^1} f(x_0)$ is not empty and contains a function l_{1,c,x_0} with some $c > 0$.

2) Let f be quasidifferentiable at the point x_0 and $f'_{x_0}(u) = r_1(u) - r_2(u)$, where r_1, r_2 are continuous sublinear functions and let $l_{1,c,x_0} \in \partial_{L^1} f(x_0)$. Then $\partial r_2 \subset \partial r_1 + c\partial p$.

Proof: 1) The function f is calm at the point x_0, so the result follows from Theorem 4.1.

2) Let $l(x) = -cp(x - x_0)$ be a L^1-subgradient of f at the point x_0. Let $u \in X$ and $\alpha \geq 0$. Then

$$-c\alpha p(u) = -cp((x_0 + \alpha u) - x_0) = l(x_0 + \alpha u) - l(x_0) \leq f(x_0 + \alpha u) - f(x_0).$$

Thus

$$-cp(u) \leq f'(x, u) = r_1(u) - r_2(u) \ \text{ for all } \ u \in X,$$

which leads to the inclusion $\partial r_2 \subset \partial r_1 + c\partial p$. $\square$

Corollary 4.1 *Let f be a locally Lipschitz subdifferentiable at x_0 function and $l_{1,c,x_0} \in \partial_{L^1} f(x_0)$. Then*

$$c\partial p \cap \partial f'_{x_0} \neq \emptyset \tag{4.3}$$

Indeed, it follows from Proposition 4.1 that $0 \in \partial f'_{x_0} + c\partial p$, which is equivalent to (4.3).

Corollary 4.2 *Let f be differentiable at a point x_0 and $l_{1,c,x_0} \in \partial_{L^1} f(x_0)$. Then $\nabla f(x_0) \in c\partial p$.*

Acknowledgments

This research has been supported by Australian Research Council Grant A69701407.

References

Balder E.J.(1977), An extension of duality-stability relations to nonconvex optimization problems, *SIAM J. Control and Optimization*, **15**, 329-343.

Burke, J.V. (1991), Calmness and exact penalization, *SIAM J. Control and Optimization* **29**, 493 -497.

Demyanov V.F. and Rubinov A.M.,(1995) *Constructive Nonsmooth Analysis*, Verlag Peter Lang, Frankfurt on Main.

Dolecki S. and Kursyusz, S. (1978). On Φ-convexity in extremal problems, SIAM J. Control and Optimization, **16**, 277-300.

Kutateladze S.S. and Rubinov A.M. (1972). Minkowski duality and its applications, *Russian Mathem. Surveys* **27**, 137-191.

Pallaschke D. and Rolewicz S. (1997). *Foundations of Mathematical Optimization (Convex analysis without linearity)* Kluwer Academic Publishers, Dordrecht.

Rubinov A. M. (2000) *Abstract convexity and global optimization*, Kluwer Academic Publishers, Dordrecht.

Singer I. (1997) *Abstract Convex Analysis*. Wiley-Interscience Publication, New York.

II OPTIMIZATION ALGORITHMS

7   AN IMPLEMENTATION OF TRAINING DUAL-NU SUPPORT VECTOR MACHINES

Hong-Gunn Chew, Cheng-Chew Lim and Robert E. Bogner

Department of Electrical and Electronic Engineering
The University of Adelaide
and
The Cooperative Research Centre for
Sensor Signal and Information Processing
Australia

Abstract: Dual-ν Support Vector Machine (2ν-SVM) is a SVM extension that reduces the complexity of selecting the right value of the error parameter selection. However, the techniques used for solving the training problem of the original SVM cannot be directly applied to 2ν-SVM. An iterative decomposition method for training this class of SVM is described in this chapter. The training is divided into the initialisation process and the optimisation process, with both processes using similar iterative techniques. Implementation issues, such as caching, which reduces the memory usage and redundant kernel calculations are discussed.

Key words: Training Dual-nu Support Vector Machine, nu-SVM, decision variable initialisation, decomposition.

1   INTRODUCTION

The Support Vector Machine (SVM) is a classification paradigm based on statistical learning that has shown promise in real world applications. Successful use of SVMs in large scale applications include face detection (Osuna et al. (1997b)), target detection (Chew et al. (2000)), and text decoding using Support Vector Regression (Chang and Lin (2001b)).

The setting of the error penalty in the original SVM formulation (Burges (1998)) is essentially based on trial-and-error, which requires additional time consuming training. This shortcoming is partially overcome with the formulation of ν-SVM by Schölkopf et al. (2000). A more general formulation, termed Dual-ν-SVM (Chew et al. (2001a)) or 2ν-SVM, provides better performance when the training class sizes are not the same, or when different class error rates are required. It is introduced in Section 2.

The training of a SVM involves solving a large convex quadratic programming (QP) problem, and can only be solved numerically. There are issues with the computation and memory complexities that need to be addressed, before the training problem can be solved. A decomposition method well suited to handle such implementation concerns is described in Sections 3 and 5.

As in any numerical optimisation, there is a starting point for the decision variables where the algorithm starts searching the path to the optimal point. In the original SVM formulation, the decision variables are set to zero. In 2ν-SVM, setting the decision variables to zero will result in an optimisation constraint being violated. It is therefore necessary to initialise the decision variables properly before the optimisation process can proceed. In Section 4, a systematic process of initialising the decision variables is discussed. Numerous implementation aspects on the initialisation and the decomposition are dealt with in Section 5.

2   DUAL-ν SUPPORT VECTOR MACHINES

Consider a set of l data vectors $\{\mathbf{x}_i, y_i\}$, with $\mathbf{x}_i \in \mathrm{R}^d$, $y_i \in +1, -1$, $i = 1, \ldots, l$, where $\mathbf{x}_i$ is the i-th data vector that belongs to a binary class y_i. We seek the hyperplane that best separates the two classes with the widest margin. More specifically, the objective of training the SVM is to find the hyperplane (Burges (1998))

$$\mathbf{w} \cdot \Phi(\mathbf{x}) + b = 0, \qquad (2.1)$$

subject to

$$y_i(\mathbf{w} \cdot \Phi(\mathbf{x}_i) + b) \geq \rho - \xi_i, \tag{2.2}$$

$$\xi_i \geq 0, \tag{2.3}$$

to minimise

$$\frac{1}{2}\|\mathbf{w}\|^2 - \sum_i C_i(\nu\rho - \xi_i), \tag{2.4}$$

where ρ is the position of the margins, and ν is the error parameter to be defined later in the section. The function Φ is a mapping function from the data space to the feature space to provide generalisation for the decision function that may not be a linear function of the training data. The problem is equivalent to maximising the margin $2/\|\mathbf{w}\|$, while minimising the cost of the errors $C_i(\nu\rho - \xi_i)$, where $\mathbf{w}$ is the normal vector and b is the bias, describing the hyperplane, and ξ_i is the slack variable for classification errors. The margins are defined by $\mathbf{w} \cdot \mathbf{x} + b = \pm\rho$.

In the Dual-ν formulation, we introduce ν_+ and ν_- as the error parameters of training for the positive and negative classes respectively, where

$$\begin{aligned}
0 &< \nu_+ < 1, \\
0 &< \nu_- < 1.
\end{aligned} \tag{2.5}$$

Denoting the error penalty as

$$C_i = \begin{cases} C_+, & y_i = +1 \\ C_-, & y_i = -1 \end{cases}, \tag{2.6}$$

with

$$C_+ = \left[l_+ \left(1 + \frac{\nu_+}{\nu_-} \right) \right]^{-1}, \tag{2.7}$$

$$C_- = \left[l_- \left(1 + \frac{\nu_-}{\nu_+} \right) \right]^{-1}, \tag{2.8}$$

$$\nu = \frac{2\nu_+\nu_-}{\nu_+ + \nu_-}, \tag{2.9}$$

where l_+ and l_- are the numbers of training points for the positive and negative classes respectively. Note that the original ν-SVM formulation by Schölkopf et al. (2000) can be derived from 2ν-SVM by letting $\nu_+ = \frac{\nu l}{2l_+}$ and $\nu_- = \frac{\nu l}{2l_-}$.

The 2ν-SVM training problem can be formulated as a Wolfe dual Lagrangian (Chew et al. (2001a)) problem. The Wolfe dual Lagrangian is explained by

Fletcher (1987). The present problem involves:

$$\max_{\{\alpha_i\}} \left\{ L_d \equiv -\frac{1}{2} \sum_{i,j} \alpha_i \alpha_j y_i y_j K_{ij} \right\} \tag{2.10}$$

subject to

$$0 \leq \alpha_i \leq C_i, \tag{2.11}$$

$$\sum_i \alpha_i y_i = 0, \tag{2.12}$$

$$\sum_i \alpha_i \geq \nu, \tag{2.13}$$

where $i, j \in 1, \ldots, l$, and the kernel function is

$$K_{ij} = K(\mathbf{x}_i, \mathbf{x}_j) = \Phi(\mathbf{x}_i) \cdot \Phi(\mathbf{x}_j). \tag{2.14}$$

In solving the 2ν-SVM problem, constraint (2.13) can be written as an equality (Schölkopf et al. (2000)) during the optimisation process,

$$\sum_i \alpha_i = \nu. \tag{2.15}$$

This property is required in the training process and will be shown in Section 3.3.

3 OPTIMISATION METHOD

Training a SVM involves the maximisation of a quadratic function. As the size of the data matrix Q, where $Q_{ij} = y_i y_j K_{ij}$, scales with l^2, with most problems, it will not be possible to train a SVM using standard QP techniques without some sort of decomposition on the optimisation problem.

Many methods had been suggested to solve the QP problem of the original formulation of SVMs: chunking (Vapnik (1995)), decomposition (Osuna et al. (1997a), and Hsu and Lin (2002)), and Sequential Minimal Optimization (SMO) (Platt (1999)). However, these methods cannot be directly applied to 2ν-SVM. The addition of the variable margin, ρ, increases the difficulty in determining the state of the Karush-Kuhn-Tucker (KKT) conditions that some methods require during the process of optimisation.

In this section, we will describe a decomposition method which has been approached from a pairwise decomposition viewpoint, and results in a method similar to that proposed by Chang and Lin (2001a).

A constrained convex QP problem can be broken down into a smaller QP problem at each iterative step, with a smaller set of decision variables. The solution is found when there is no set of decision variables that can be changed to maximise the objective function. We have chosen to break the problem down to a two decision variable problem, as the sub-problem can be solved analytically. The constraints imposed by the problem are still satisfied and are discussed in Section 3.3.

3.1 Iterative decomposition training

In the iterative optimisation process, the k-th iteration of the Lagrangian of 2ν-SVM (2.10) is

$$
\begin{aligned}
F^{(k)} &= F(\alpha_1^{(k)}, \alpha_2^{(k)}, \ldots, \alpha_l^{(k)}) \\
&= -\frac{1}{2}\sum_{i,j}\alpha_i^{(k)}\alpha_j^{(k)} y_i y_j K_{ij}.
\end{aligned}
\tag{3.1}
$$

The solution to the problem is obtained when no update to $\alpha_i^{(k)}$ can be found that increases $F^{(k)}$ while satisfying the constraints of the problem.

In the decomposition process (Joachims (1999)), the problem is divided into the working set B, and the non-working set N, where B and N are exclusive and $B \cup N = \{1 \ldots l\}$. That is for a chosen working set B,

$$
\begin{aligned}
F &= F_{BB} + 2F_{BN} + F_{NN} \\
F &= -\frac{1}{2}\sum_{i,j\in B}\alpha_i\alpha_j y_i y_j K_{ij} \\
&\quad - \sum_{i\in B, j\in N}\alpha_i\alpha_j y_i y_j K_{ij} - \frac{1}{2}\sum_{i,j\in N}\alpha_i\alpha_j y_i y_j K_{ij},
\end{aligned}
\tag{3.2}
$$

and we desire the solution to the sub-problem F_{BB}. The solution to the problem is obtained when no working set B can be found and updated that increases F_{BB} while satisfying the contraints of the problem.

We can combine the two processes to get a simple and intuitive method of solving the 2ν-SVM training problem. At each interative step, the objective function is maximised with respect to only two variables. That is, we decompose the problem into a sub-problem with a working set of two points p and q. This decomposition simplifies the iterative step while still converges to the solution, due to the convex surface of the problem.

With $p, q \in 1, \ldots, l$, (3.1) can be rewritten to extract the $\alpha_p^{(k)}$ and $\alpha_q^{(k)}$ components of the function, to give

$$
\begin{aligned}
F_{pq}^{(k)} = \; & -\frac{1}{2} \sum_{m,n \neq p,q} \alpha_m^{(k)} \alpha_n^{(k)} y_m y_n K_{mn} \\
& - \alpha_p^{(k)} y_p \sum_{m \neq p,q} \alpha_m^{(k)} y_m K_{mp} \\
& - \alpha_q^{(k)} y_q \sum_{m \neq p,q} \alpha_m^{(k)} y_m K_{mq} \\
& - \frac{1}{2} \left(\alpha_p^{(k)} \right)^2 K_{pp} - \alpha_p^{(k)} \alpha_q^{(k)} y_p y_q K_{pq} - \frac{1}{2} \left(\alpha_q^{(k)} \right)^2 K_{qq}.
\end{aligned} \tag{3.3}
$$

With some change δ_{pq} made on the decision variables, the change in $F_{pq}^{(k+1)}$ can be obtained using the substitutions

$$
\begin{aligned}
\alpha_p^{(k+1)} &= \alpha_p^{(k)} - y_p \delta_{pq}, \\
\alpha_q^{(k+1)} &= \alpha_q^{(k)} + y_q \delta_{pq}, \\
\alpha_i^{(k+1)} &= \alpha_i^{(k)}, \qquad \forall i = 1, \ldots, l, i \neq p, q.
\end{aligned} \tag{3.4}
$$

As only the p-th and q-th decision variables are updated, the change in the objective function at iteration $(k+1)$ is shown in the Appendix to be

$$
\begin{aligned}
\Delta F_{pq}^{(k+1)} &= F_{pq}^{(k+1)} - F^{(k)} \\
&= \delta_{pq} \left(G_p^{(k)} - G_q^{(k)} \right) \\
&\quad - \frac{1}{2} \left(\delta_{pq} \right)^2 \left(K_{pp} - 2K_{pq} + K_{qq} \right),
\end{aligned} \tag{3.5}
$$

where

$$
G_i^{(k)} = \sum_j \alpha_j^{(k)} y_j K_{ji}. \tag{3.6}
$$

We use (3.5) to train 2ν-SVM by selecting the optimal decision variables to update at each iterative step until the objective function is at its maximum.

3.2   Choosing optimising points

In each iterative step, the objective function is increased by the largest amount possible in changing $\alpha_p^{(k)}$ and $\alpha_q^{(k)}$. We seek the pair of decision variables to update in each step by searching for the maximum change in the objective function in updating the pair. The optimal δ_{pq} for each (p, q) pair, denoted by

δ_{pq}^*, is obtained from

$$\frac{\mathrm{d}F_{pq}^{(k+1)}}{\mathrm{d}\delta_{pq}} = \left(G_p^{(k)} - G_q^{(k)}\right) - \delta_{pq}\left(K_{pp} - 2K_{pq} + K_{qq}\right) = 0,$$

resulting in

$$\delta_{pq}^* = \frac{G_p^{(k)} - G_q^{(k)}}{K_{pp} - 2K_{pq} + K_{qq}}. \tag{3.7}$$

This in turn gives the maximum possible change in the objective function for the updated pair as

$$\Delta F_{pg}^{*(k+1)} = \frac{1}{2}\frac{\left[G_p^{(k)} - G_q^{(k)}\right]^2}{K_{pp} - 2K_{pq} + K_{qq}}. \tag{3.8}$$

The process for finding (p, q) is therefore to find

$$\max_{p,q}\left\{\Delta F_{pg}^{*(k+1)}\right\} \tag{3.9}$$

for $p, q \in 1, \ldots, l$.

Using (3.9) requires $O\left(l^2\right)$ operations to search for the (p, q) pair. We can reduce the complexity by simplifying the search to

$$\max_{p,q}\left\{S_{pq}^{(k+1)} = G_p^{(k)} - G_q^{(k)}\right\}. \tag{3.10}$$

Although the new search criterion is a simplification of (3.9), due to the convexity of the objective function, the optimisation process still converges to the same maximum, albeit in a less optimal path, but has a search complexity of only $O\left(l\right)$, as it only searches for the maximum and minimum of $G_i^{(k)}$.

All kernel functions need to satisfy Mercer's condition (Vapnik (1995)). Mercer's condition states that the kernel is actually a dot product in some space and therefore, the denominator of (3.7) is

$$\begin{aligned} K_{pp} - 2K_{pq} + K_{qq} &= \Phi\left(\mathbf{x}_p\right) \cdot \Phi\left(\mathbf{x}_p\right) - 2\Phi\left(\mathbf{x}_p\right) \cdot \Phi\left(\mathbf{x}_q\right) + \Phi\left(\mathbf{x}_q\right) \cdot \Phi\left(\mathbf{x}_q\right) \\ &= \left\|\Phi\left(\mathbf{x}_p\right) - \Phi\left(\mathbf{x}_q\right)\right\|^2 \end{aligned}$$

which is positive for all valid kernel functions. Thus, it is clear that for the (p, q) pair found using either (3.9) or (3.10), we obtain $\delta_{pq}^* \geq 0$. If $\delta_{pq}^* = 0$, the objective function is at its maximum and we have found the trained SVM. The iteration process continues as long as

$$\delta_{pq}^* > 0. \tag{3.11}$$

3.3 Constraints of 2ν-SVM

In maximising the objective function (2.10), constraints (2.11), (2.12) and (2.13) have to be satisfied. We now derive the conditions for meeting each constraint.

Proposition 3.1 *If $\{\alpha_i^{(0)}\}$ satisfies (2.12), and $\{\alpha_i^{(k)}\}$ is updated using (3.4), then $\{\alpha_i^{(k)}\}$ satisfies (2.12) for any k.*

Proof: From (2.12) and using (3.4),

$$
\begin{aligned}
\sum_i \alpha_i^{(k+1)} y_i &= \sum_{m\neq p,q} \alpha_m^{(k+1)} y_m + \alpha_p^{(k+1)} y_p + \alpha_q^{(k+1)} y_q \\
&= \sum_{m\neq p,q} \alpha_m^{(k)} y_m \\
&\quad + \left(\alpha_p^{(k)} - y_p\delta_{pq}\right) y_p + \left(\alpha_q^{(k)} + y_q\delta_{pq}\right) y_q \\
\sum_i \alpha_i^{(k+1)} y_i &= \sum_i \alpha_i^{(k)} y_i.
\end{aligned}
$$

Since $\{\alpha_i^{(0)}\}$ satisfies (2.12), by induction, $\{\alpha_i^{(k)}\}$ satisfies (2.12) for any k. $\square$

Proposition 3.2 *If $\{\alpha_i^{(0)}\}$ satisfies (2.13), and $\{\alpha_i^{(k)}\}$ is updated using (3.4) with*

$$
y_p = y_q, \tag{3.12}
$$

then $\{\alpha_i^{(k)}\}$ satisfies (2.13) for any k.

Proof: From (2.13), using (3.4) and (3.12),

$$
\begin{aligned}
\sum_i \alpha_i^{(k+1)} &= \sum_{m\neq p,q} \alpha_m^{(k+1)} + \alpha_p^{(k+1)} + \alpha_q^{(k+1)} \\
&= \sum_{m\neq p,q} \alpha_m^{(k)} \\
&\quad + \left(\alpha_p^{(k)} - y_p\delta_{pq}\right) + \left(\alpha_q^{(k)} + y_q\delta_{pq}\right) \\
\sum_i \alpha_i^{(k+1)} &= \sum_i \alpha_i^{(k)}.
\end{aligned}
$$

Since $\{\alpha_i^{(0)}\}$ satisfies (2.13), by induction, $\{\alpha_i^{(k)}\}$ satisfies (2.13) for any k. $\square$

Remark 3.1 *Due to (3.12), the search for the update pair (p, q) is divided into two parts, one for each class. The class which returns the higher increase in the objective function is selected for the update process.*

Remark 3.2 *Proposition 3.2 shows that the update process of the optimisation results in*

$$\sum_i \alpha_i^{(k)} = \sum_i \alpha_i^{(0)}.$$

Since the solution of the training problem has the property of (2.15), we need to initialise $\{\alpha_i^{(0)}\}$, such that

$$\sum_i \alpha_i^{(0)} = \nu, \tag{3.13}$$

to enable the optimisation process to reach the solution.

Proposition 3.3 *If $\{\alpha_i^{(0)}\}$ satisfies (2.11), and $\{\alpha_i^{(k)}\}$ is updated using (3.4) with*

$$\delta_{pq} = \begin{cases} \min\left[\delta_{pq}^*, \alpha_p^{(k)}, C_p - \alpha_q^{(k)}\right], & \text{for } y_p = y_q = +1 \\ \min\left[\delta_{pq}^*, C_p - \alpha_p^{(k)}, \alpha_q^{(k)}\right], & \text{for } y_p = y_q = -1 \end{cases} \tag{3.14}$$

then $\{\alpha_i^{(k+1)}\}$ satisfies (2.11) for any k.

Proof: It is clear that since δ_{pq}^* is always positive (3.11), the limiting constraints on δ_{pq} are

$$\alpha_p^{(k)} - \delta_{pq} \geq 0, \qquad \alpha_q^{(k)} + \delta_{pq} \leq C_q, \qquad \text{for } y_p = y_q = +1, \tag{3.15}$$

$$\alpha_p^{(k)} + \delta_{pq} \geq C_p, \qquad \alpha_q^{(k)} - \delta_{pq} \leq 0, \qquad \text{for } y_p = y_q = -1. \tag{3.16}$$

Using δ_{pq} as stated in (3.14) will meet the constraints of (3.15) and (3.16), and therefore (2.11). Thus, if $\{\alpha_i^{(0)}\}$ satisfies (2.11), then by induction, $\{\alpha_i^{(k)}\}$ satisfies (2.11) for any k. $\square$

Remark 3.3 *The selection process requires $\delta_{pq} > 0$ for each iteration. From (3.15) and (3.16), it is clear that we need*

$$\alpha_p^{(k)} > 0, \qquad \alpha_q^{(k)} < C_q, \qquad \text{for } y_p = y_q = +1, \tag{3.17}$$

$$\alpha_p^{(k)} < C_p, \qquad \alpha_q^{(k)} > 0, \qquad \text{for } y_p = y_q = -1. \tag{3.18}$$

From Propositions 3.2 and 3.3, the selection of the (p, q) pair for each iteration therefore requires the search for (3.10) for each class, while satisfying (3.12), (3.17) and (3.18), to find $\delta_{pq} > 0$ with (3.14).

3.4   Algorithm

The optimisation process with iterative decomposition is stated in the following realisable algorithm.

- Given

$$
\begin{aligned}
\mathbf{x}_i &\in \mathbf{R}^d \\
y_i &= \{+1, -1\} \\
i &= 1, \ldots, l \\
C_i &= \begin{cases} C_+, & y_i = +1 \\ C_-, & y_i = -1 \end{cases} \\
K_{ij} &= K\left(\mathbf{x}_i, \mathbf{x}_j\right).
\end{aligned}
$$

- Find the set of decision variables, $\{\alpha_i\}$, that maximises the objective function

$$
L_d \equiv -\frac{1}{2} \sum_{i,j} \alpha_i \alpha_j y_i y_j K_{ij}
$$

subject to

$$
\begin{aligned}
0 &\leq \alpha_i \leq C_i, \\
\sum_i \alpha_i y_i &= 0, \\
\sum_i \alpha_i &\geq \nu.
\end{aligned}
$$

- Define

  - the equations

$$
\begin{aligned}
G_i^{(k)} &= \sum_j \alpha_j^{(k)} y_j K_{ji}, \\
S_{pq}^{(k+1)} &= G_p^{(k)} - G_q^{(k)},
\end{aligned}
$$

  - and the sets

$$
W_+^{(k)} = \left\{ (i,j) \, \middle| \, \begin{array}{l} i, j = 1, \ldots, l, \\ y_i = y_j = +1, \\ \alpha_i^{(k)} > 0, \alpha_j^{(k)} < C_i \end{array} \right\},
$$

$$W_-^{(k)} = \left\{ (i,j) \,\middle|\, \begin{array}{l} i,j = 1,\ldots,l, \\ y_i = y_j = -1, \\ \alpha_i^{(k)} < C_i, \alpha_j^{(k)} > 0 \end{array} \right\}.$$

- Start at $k = 0$, for some set of $\left\{\alpha_i^{(0)}\right\}$ where

$$\sum_i \alpha_i^{(0)} y_i = 0,$$

$$\sum_i \alpha_i^{(0)} = \nu.$$

- While $\exists\, (p,q) \in W_+^{(k)} \cap W_-^{(k)}$ such that $S_{pq}^{(k+1)} > 0$,

  - find $(p_+, q_+) \in W_+^{(k)}$, and $(p_-, q_-) \in W_-^{(k)}$, with

$$\Theta_+ = \max_{(p_+,q_+)\in W_+^{(k)}} \left\{ S_{p_+q_+}^{(k+1)} \right\},$$

$$\Theta_- = \max_{(p_-,q_-)\in W_-^{(k)}} \left\{ S_{p_-q_-}^{(k+1)} \right\},$$

  - if $(\Theta_+ \geq \Theta_-)$

    * then

$$(p,q) = (p_+, q_+),$$

$$\delta_+ = \frac{\Theta_+}{K_{pp} - 2K_{pq} + K_{qq}},$$

$$\delta = \min\left[\delta_+, \alpha_p^{(k)}, C_q - \alpha_q^{(k)} \right],$$

    * else

$$(p,q) = (p_-, q_-),$$

$$\delta_- = \frac{\Theta_-}{K_{pp} - 2K_{pq} + K_{qq}},$$

$$\delta = \min\left[\delta_-, C_p - \alpha_p^{(k)}, \alpha_q^{(k)} \right],$$

  - and updating the corresponding decision variables

$$\alpha_p^{(k+1)} = \alpha_p^{(k)} - y_p\delta,$$

$$\alpha_q^{(k+1)} = \alpha_q^{(k)} + y_q\delta,$$

  - update $k = k + 1$.

- Terminate.

4 INITIALISATION TECHNIQUE

In Section 3.4, the optimisation procedure needs the decision variables to be initialised, such that constraints (2.11), (2.12) and (2.15) are satisfied. One approach is to choose a number of decision variables to change, such that $\alpha_i = C_i$ until the constraints are met. The selection of training points to update usually involves selecting the first n points in the training set, and updating the corresponding decision variables, until the constraints are satisfied. This method of initialising the decision variables is used in most of the available ν-SVM software packages, but it is not systematic.

It is possible to initialise the decision variable using a similar iterative approach to the optimisation process. The method in Sections 4.1, 4.2 and 4.3 will demonstrate that the selection of variables to update is systematic. The initialisation method also reduces the training time of the optimisation, while not taking more computational time to perform (Chew et al. (2001b)).

4.1 Iterative decision variable initialisation

Consider the Lagrangian of 2ν-SVM (2.10). The t-th iteration of the Lagrangian (3.1) is rewritten to extract the $\alpha_r^{(t)}$ component of the function, for some $r \in 1, \dots, l$, as

$$
\begin{aligned}
F_r^{(t)} &= -\frac{1}{2} \sum_{m,n \neq r} \alpha_m^{(t)} \alpha_n^{(t)} y_m y_n K_{mn} \\
&\quad - \alpha_r^{(t)} y_r \sum_{m \neq r} \alpha_m^{(t)} y_m K_{mp} - \frac{1}{2} \left(\alpha_r^{(t)} \right)^2 K_{rr}.
\end{aligned}
\tag{4.1}
$$

As only the r-th decision variable is updated, we can derive the change in $F_r^{(t+1)}$ using the substitutions

$$
\begin{aligned}
\alpha_r^{(t+1)} &= \alpha_r^{(t)} + \delta_r, \\
\alpha_i^{(t+1)} &= \alpha_i^{(t)}, \qquad \forall i = 1, \dots, l, i \neq r,
\end{aligned}
\tag{4.2}
$$

for some change, δ_r. The objective function at iteration $(t+1)$ is shown in the Appendix to be

$$
\begin{aligned}
\Delta F_r^{(t+1)} &= F_r^{(t+1)} - F^{(t)} \\
&= -\delta_r y_r G_r^{(t)} - \frac{1}{2} (\delta_r)^2 K_{rr},
\end{aligned}
\tag{4.3}
$$

where

$$G_i^{(t)} = \sum_j \alpha_j^{(t)} y_j K_{ji}. \qquad (4.4)$$

We use (4.3) to initialise the set of decision variables such that the variables satisfy constraints (2.11), (2.12) and (2.15), for the training of 2ν-SVMs, by selecting the optimal decision variable to update at each iterative step.

4.2 Choosing initialisation points

The training process of 2ν-SVMs requires the maximisation of the objective function. Therefore, during the initialisation process, we seek the minimal decrease in the objective function when changing $\alpha_r^{(t)} = C_r$. We find the variable to update in each step by searching for the maximum increase (or minimum decrease) in the objective function in updating the pair, with $\delta_r = C_r$, which is

$$
\begin{aligned}
\Delta F_r^{(t+1)} &= F_r^{(t+1)} - F^{(t)} \\
&= -C_r y_r G_r^{(t)} - \frac{1}{2} \left(C_r \right)^2 K_{rr}.
\end{aligned}
\qquad (4.5)
$$

The process for finding r is therefore to find

$$\min_r \left\{ I_r^{(t+1)} = -2 \left[\Delta F_r^{(t+1)} \right] \right\} \qquad (4.6)$$

for all $r \in 1, \ldots, l$, and $\alpha_r^{(t)} = 0$.

4.3 Constraints of initialisation

The iterative process for initialising α_i is to vary α_i in order to meet (2.15). However, the two other constraints (2.11) and (2.12) have to be satisfied in the process as well. While (2.11) is clearly never violated, an additional restriction is required to meet (2.12) when the process terminates.

By combining (2.12) and (2.13), the constraints reduce to

$$\sum_{\{i|y_i=+1\}} \alpha_i \geq \frac{\nu}{2}, \qquad \sum_{\{i|y_i=-1\}} \alpha_i \geq \frac{\nu}{2}. \qquad (4.7)$$

Similarly, by combining (2.12) and (2.15), the constraints reduce to

$$\sum_{\{i|y_i=+1\}} \alpha_i = \sum_{\{i|y_i=-1\}} \alpha_i = \frac{\nu}{2}. \qquad (4.8)$$

In the search for r at each iteration, (4.7) allows the exclusion of the class that has met the constraint. The selection process therefore searches for r in a reduced set of training points,

$$N^{(t)} = \left\{ i \;\middle|\; \begin{array}{l} i = 1, \ldots, l, \\ \alpha_i^{(t)} = 0, \\ \sum\limits_{\{j|y_j=y_i\}} \alpha_j^{(t)} < \frac{\nu}{2} \end{array} \right\}. \tag{4.9}$$

Note that (2.15) is more stringent than (2.13), implying that meeting (2.15) is sufficient for (2.13). Consequently, we refine the update equation (4.2) to (4.10) to satisfy (4.8).

Proposition 4.1 *Let $N^{(t)}$ be defined by (4.9). If the search for $r \in N^{(t)}$ uses (4.6) in the $\{\alpha_i\}$ initialisation, and the update of $\alpha_i^{(t)}$ at each iteration t uses*

$$\begin{aligned} \alpha_r^{(t+1)} &= \min\left\{ C_r, \frac{\nu}{2} - \sum_{\{i|y_i=y_r\}} \alpha_i^{(t)} \right\}, \\ \alpha_i^{(t+1)} &= \alpha_i^{(t)}, \qquad \forall i = 1, \ldots, l, i \neq r, \end{aligned} \tag{4.10}$$

the resulting set of α_i will satisfy (4.8) at the end of the initialisation process.

Proof: From (4.9), we see that $\alpha_r^{(t)} = 0$. Therefore the updated $\sum \alpha_i^{(t+1)}$ for class y_r is

$$\begin{aligned} \sum_{\{i|y_i=y_r\}} \alpha_i^{(t+1)} &= \sum_{\{i|y_i=y_r,i\neq r\}} \alpha_i^{(t+1)} + \alpha_r^{(t+1)} \\ &= \sum_{\{i|y_i=y_r\}} \alpha_i^{(t)} + \min\left\{ C_r, \frac{\nu}{2} - \sum_{\{i|y_i=y_r\}} \alpha_i^{(t)} \right\} \\ &= \min\left\{ \sum_{\{i|y_i=y_r\}} \alpha_i^{(t)} + C_r, \frac{\nu}{2} \right\}, \end{aligned}$$

$$\sum_{\{i|y_i=y_r\}} \alpha_i^{(t+1)} \leq \frac{\nu}{2}. \tag{4.11}$$

Since the initialisation process terminates when (4.7) is satisfied, it follows from (4.11) that (4.8) is satisfied. $\square$

Proposition 4.2 *The number of iterations required in the initialisation is only*

$$\lceil \nu_+ l_+ \rceil + \lceil \nu_- l_- \rceil. \tag{4.12}$$

Proof: The derivation of (4.12) is given in the Appendix. $\square$

The iterative process of initialising the decision variables is therefore to find $r \in N^{(t)}$ using (4.6), and updating $\alpha_r^{(t)}$ with (4.10). The result of the initialisation process is a set of decision variables, $\{\alpha_i\}$, that satisfies the training constraints imposed on 2ν-SVMs, as well as reduces the training time required.

4.4 Algorithm

The initialisation process is stated in the following algorithm.

- Given

$$
\begin{aligned}
\mathbf{x}_i &\in \mathbf{R}^d \\
y_i &= \{+1, -1\} \\
i &= 1, \ldots, l \\
C_i &= \begin{cases} C_+, & y_i = +1 \\ C_-, & y_i = -1 \end{cases} \\
K_{ij} &= K(\mathbf{x}_i, \mathbf{x}_j)
\end{aligned}
$$

- Find the initial state of the set of decision variables, $\{\alpha_i\}$, for training such that

$$
\begin{aligned}
0 &\leq \alpha_i \leq C_i, \\
\sum_i \alpha_i y_i &= 0, \\
\sum_i \alpha_i &= \nu.
\end{aligned}
$$

- Define

    - the equations

$$
\begin{aligned}
G_i^{(t)} &= \sum_j \alpha_j^{(t)} y_j K_{ji}, \\
V_i^{(t)} &= \sum_{\{j \mid y_j = y_i\}} \alpha_j^{(t)},
\end{aligned}
$$

- and the set

$$
N^{(t)} \;=\; \left\{\, i \;\middle|\; \begin{array}{l} i = 1, \ldots, l, \\ \alpha_i^{(t)} = 0, \\ V_i^{(t)} < \frac{\nu}{2} \end{array} \right\}.
$$

- Start at $t = 0$, with $\alpha_i^{(0)}$.

- While $N^{(t)} \neq \emptyset$,

  - find $r \in N^{(t)}$ with

$$
\min_{r \in N^{(t)}} \left\{ I_r^{(t+1)} = 2C_r y_r G_r^{(t)} + (C_r)^2 K_{rr} \right\},
$$

  - and updating the corresponding decision variable

$$
\alpha_r^{(t+1)} = \min \left(C_r, \frac{\nu}{2} - V_r^{(t)} \right),
$$

  - update $t = t + 1$.

- Terminate.

5 IMPLEMENTATION ISSUES

Training 2ν-SVMs is a computationally expensive process. The iterative process described in Section 3 provides a working optimiser that allows large problems to be solved without the need for large amounts of memory to store the Hessian, Q. The process can be further improved to reduce the computational complexity.

Three issues are discussed on the possible improvements to the computational performance of the optimisation: the kernel calculations, the optimisation process, and the initialisation process. We shall see in Section 5.3 that the initialisation process, besides initialisation of the decision variables, also initialise the caches for the optimisation, as part of the process.

5.1 Kernel calculations

The kernel calculations form a large proportion of the computational load in the training process, as each stage of both iterative processes requires K_{ij}. K_{ij} usually involves expensive calculations, such as the exponential for radial basis function kernels, and the numeric powers for polynomial kernels. The calculation of K_{ij} each time can be avoided by caching it as a $l \times l$ matrix K, but it is impossible to store the whole matrix for all but the smallest training problems. For example, we would require 800 Megabytes of cache for $l = 10240$ using double precision floating point numbers.

It is, however, possible to cache only K_{ij} for $\alpha_j > 0$. The two algorithms stated here mainly requires only G_i, and can safely ignore the terms where $\alpha_j = 0$. Thus we only need to cache the columns of K where $\alpha_j > 0$. This caching method will improve the performance of the optimisation by reducing the number of kernel calculations required. As the optimisation process is iterative, the set of $\{j|\alpha_j > 0\}$ changes as the iteration progresses. The strategy to manage the caching is to add columns to the cache when $\alpha_j = 0$ becomes non-zero, and to only remove columns when memory is full. The column where $\alpha_j = 0$, and that was unused for the longest time, is removed first.

It is also useful to cache the diagonal of the kernel matrix, K_{ii}, as it only requires l elements and is used in each step of both algorithms.

5.2   Optimisation process

The training process involves the iterative maximisation of the objective function, by way of finding the maximum $S_{pq}^{(k+1)}$ for the positive class and the negative class. It is easy to see that finding the maximum $S_{pq}^{(k+1)}$ is basically finding the maximum and minimum of $G_i^{(k)}$. This is why the process has a complexity of only $O(l)$ instead of $O(l^2)$.

The calculation of $G_i^{(k)}$ can be reduced significantly by updating it at each iteration rather than recalculating it again. Since only the p-th and q-th decision variables are changed, from (3.6) and using (3.4),

$$
\begin{aligned}
G_i^{(k+1)} &= \sum_{m \neq p,q} \alpha_m^{(k+1)} y_m K_{mi} + \alpha_p^{(k+1)} y_p K_{pi} + \alpha_q^{(k+1)} y_q K_{qi} \\
&= \sum_{m \neq p,q} \alpha_m^{(k)} y_m K_{mi} \\
&\quad + \left(\alpha_p^{(k)} - y_p \delta_{pq} \right) y_p K_{pi} + \left(\alpha_q^{(k)} + y_q \delta_{pq} \right) y_q K_{qi} \\
&= \sum_j \alpha_j^{(k)} y_j K_{ji} - \delta_{pq}(K_{pi} - K_{qi}) \\
G_i^{(k+1)} &= G_i^{(k)} - \delta_{pq}(K_{pi} - K_{qi}),
\end{aligned}
$$

(5.1)

which has a complexity of $O(l)$ rather than $O(l^2)$, when updating $G_i^{(k+1)}$, for all $i \in 1, \ldots, l$. Note that the kernel functions required by the update are for columns p and q of the kernel matrix. As $\alpha_p > 0$ and $\alpha_q > 0$ at iteration (k) or $(k+1)$, the columns will be cached by the caching strategy of Section 5.1.

We also have to consider the cost involved in initialising the G_i cache, that is $G_i^{(k=0)}$. Assuming that the number of non-zero decision variables is m, calculating $G_i^{(k=0)}$ would require $O(ml)$ loops and kernel calculations. However, we will see in the next section that the initialisation method provides $G_i^{(k=0)}$ in its process as well.

5.3   Initialisation calculations

The initialisation process seeks the minimum of $I_i^{(t+1)}$, which has one dynamic term $2C_i y_i G_i^{(t)}$, and one static term $(C_i)^2 K_{ii}$. The static term can be calculated at the start of the process, cached, and added to the dynamic term at each iteration.

As in the optimisation process, $G_i^{(t)}$ in the dynamic term, can be updated at each iteration to reduce the computation requirements. From (4.4), and using (4.2) with $\delta_r = C_r$,

$$
\begin{aligned}
G_i^{(t+1)} &= \sum_{m \neq r} \alpha_m^{(t+1)} y_m K_{mi} + \alpha_r^{(t+1)} y_r K_{ri} \\
&= \sum_{m \neq r} \alpha_m^{(t)} y_m K_{mi} + \left(\alpha_r^{(t)} + \delta_r \right) y_r K_{ri} \\
&= \sum_{j} \alpha_j^{(t)} y_j K_{ji} + \delta_r y_r K_{ri} \\
G_i^{(t+1)} &= G_i^{(t)} + \delta_r y_r K_{ri}.
\end{aligned}
\tag{5.2}
$$

Again, updating $G_i^{(t+1)}$ has a complexity of $O(l)$.

Here, we see that at the beginning of the initialisation process, all decision variables are zero, $\alpha_i^{(t=0)} = 0$. This means $G_i^{(t=0)} = 0$ and no columns of the kernel matrix cache are required. At each iteration, a zero decision variable r is updated, resulting in adding column r of the kernel matrix to the cache, and updating $G_i^{(t+1)}$. The process terminates after $n = \lceil \nu_+ l_+ \rceil + \lceil \nu_- l_- \rceil$ iterations (4.12), with $G_i^{(t=n)}$ and kernel cache of $\left\{ j \,\middle|\, \alpha_j^{(t=n)} > 0 \right\}$ columns. Noting that

$$
\begin{aligned}
G_i^{(t=n)} &= G_i^{(k=0)}, \\
\left\{ j \,\middle|\, \alpha_j^{(t=n)} > 0 \right\} &\equiv \left\{ j \,\middle|\, \alpha_j^{(k=0)} > 0 \right\},
\end{aligned}
$$

the initialisation process has provided the $G_i^{(k)}$ cache and kernel column cache needed in the optimisation process.

Note that n is also the minimum number of non-zero decision variables in $\{\alpha_i\}$ that will satisfy the constraints of 2ν-SVM training, that is $n \leq m$. Additionally, the calculation of $G_i^{(t)}$ has a complexity of $O(nl)$. We assume that the calculation of $G_i^{(t)}$ is the major computational proportion of the initialisation process. Thus, the computational complexity of initialising the decision variables is lower than that needed to calculate $G_i^{(k=0)}$ for the optimisation process proper, as $O(nl) \leq O(ml)$. This means that the cost of initialising the decision variable is minimal and comparable to that of the ad hoc initialisation process that many current implementations use. This new initialisation process has the advantage of reducing the number of optimisation iterations to train 2ν-SVM.

6 PERFORMANCE RESULTS

We tested the iterative method against the exisiting ad hoc method using three datasets. The ad hoc method initialises the first few decision variables until the constraints are met before the optimisation step. Each dataset is trained with the error parameters set to $\nu_+ = \nu_- = 0.2$, and using three different kernels: linear, polynomial, and radial basis function (RBF).

The first dataset is a face detection problem with 1,000 training points, with each point being an image. The dataset is trained with the linear, the polynomial (degree 5), and the RBF ($\sigma = 4,000$) kernels. The second dataset is a radar image vehicle detection problem with 10,000 training points. The dataset is trained with the linear, the polynomial (degree 5), and the RBF ($\sigma = 256$) kernels. The third dataset is a handwritten digit recognition problem (digit 7 against other digits) with 100,000 training points, and is trained with the linear, the polynomial (degree 4), and the RBF ($\sigma = 15$) kernels. This dataset is also trained with $\nu_+ = \nu_- = 0.01$, using the polynomial and RBF kernels.

The 2ν-SVMs are trained on an Intel Pentium 4 (2GHz) Linux machine, and the results are given in Table 6.1. The table shows the iterative initisation method provides improvements for most of the tests. The iterative initialisation method may increase the training time for small problems, but this is mainly due to the slow convergence of the optimisation process, as seen in the face detection problem using the polynomial kernel.

Figure 6.1 shows more clearly that while the computational time required for initialisation is not reduced, there is a significant reduction in the optimisation

Table 6.1 Training times for the ad hoc and the iterative methods of initialising the decision variables

Dataset	Initialisation/Optimisation times (seconds)		
Kernel	Ad hoc	Iterative	Change
Face detection			
linear	7 / 28	7 / 21	-1% / -24%
polynomial	7 / 43	7 / 51	-1% / 18%
radial basis function	7 / 104	7 / 95	0% / -9%
Target detection			
linear	34 / 3552	29 / 2699	-14% / -24%
polynomial	40 / 12670	37 / 8866	-7% / -30%
radial basis function	39 / 135	36 / 115	-7% / -15%
Digit recognition ($\nu = 0.2$)			
linear	2988 / 3899	2889 / 399	-3% / -90%
polynomial	3165 / 4133	3064 / 659	-3% / -84%
radial basis function	3223 / 4326	3130 / 469	-3% / -89%
Digit recognition ($\nu = 0.01$)			
polynomial	178 / 11900	154 / 11190	-14% / -6%
radial basis function	180 / 12167	157 / 12061	-13% / -1%

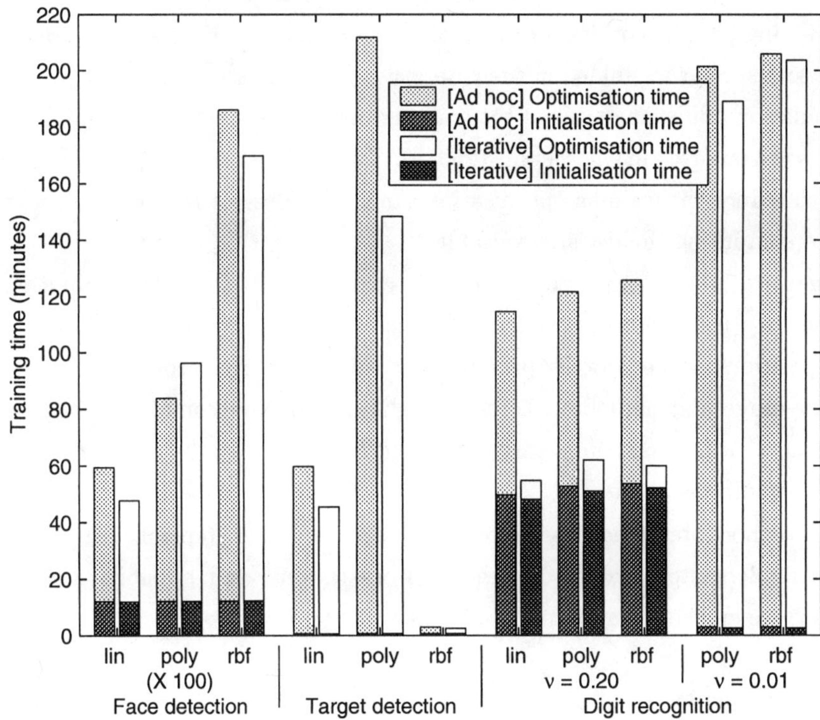

Figure 6.1   Comparison of initialisation and optimisation times for the ad hoc and the iterative methods of initialising the decision variables

time in some cases. These results clearly show that even with the more complex iterative initialisation process, there are time improvements for most of the tested cases.

7 CONCLUSIONS

2ν-SVM is a natural extension to SVM that allows different bounds for each of the binary classes, and compensation for the uneven training class size effects. We have described the process for training 2ν-SVMs using an iterative process. The training process consists of the initialisation and the optimisation proper. Both use a similar technique in their iterative procedures.

Simulation and evaluations of the training process are continuing, and some results were reported in Chew et al. (2001b). The initialisation process has been found to reduce the training optimisation time, and does not require significant costs in computing the decision variable.

In general, the optimisation process is expensive, in terms of computing and memory utilisation, as well as the time it takes. By using caches and decomposition in the process presented in this work, the problems of high memory usage and redundant kernel calculations are overcome. Specifically, the memory utilisation complexity is reduced from $O(l^2)$ to $O(l)$, to cache the kernel calculations.

The method presented has led to an efficient classifier implementation. It can be implemented readily on desktop workstations, and possibly on high performance embedded systems.

Acknowledgments

Hong-Gunn Chew is supported by an University of Adelaide Scholarship, and by a supplementary scholarship from the Cooperative Research Centre for Sensor Signal and Information Processing.

Appendix

Derivation of (3.5). From (3.3) and (3.4),

$$
\begin{aligned}
F_{pq}^{(k+1)} \; = \; & -\frac{1}{2} \sum_{m,n \neq p,q} \alpha_m^{(k)} \alpha_n^{(k)} y_m y_n K_{mn} \\
& - \left(\alpha_p^{(k)} - y_p \delta_{pq} \right) y_p \sum_{m \neq p,q} \alpha_m^{(k)} y_m K_{mp} \\
& - \left(\alpha_q^{(k)} + y_q \delta_{pq} \right) y_q \sum_{m \neq p,q} \alpha_m^{(k)} y_m K_{mq} \\
& - \frac{1}{2} \left(\alpha_p^{(k)} - y_p \delta_{pq} \right)^2 K_{pp} \\
& - \left(\alpha_p^{(k)} - y_p \delta_{pq} \right) \left(\alpha_q^{(k)} + y_q \delta_{pq} \right) y_p y_q K_{pq} \\
& - \frac{1}{2} \left(\alpha_q^{(k)} + y_q \delta_{pq} \right)^2 K_{qq} \\
= \; & -\frac{1}{2} \sum_{m,n \neq p,q} \alpha_m^{(k+1)} \alpha_n^{(k+1)} y_m y_n K_{mn} \\
& - \alpha_p^{(k)} y_p \sum_{m \neq p,q} \alpha_m^{(k)} y_m K_{mp} + \delta_{pq} \sum_{m \neq p,q} \alpha_m^{(k)} y_m K_{mp} \\
& - \alpha_q^{(k)} y_q \sum_{m \neq p,q} \alpha_m^{(k)} y_m K_{mq} - \delta_{pq} \sum_{m \neq p,q} \alpha_m^{(k)} y_m K_{mq} \\
& - \frac{1}{2} \left(\alpha_p^{(k)} \right)^2 K_{pp} + \delta_{pq} \alpha_p^{(k)} y_p K_{pp} - \frac{1}{2} \left(\delta_{pq} \right)^2 K_{pp} \\
& - \alpha_p^{(k)} \alpha_q^{(k)} y_p y_q K_{pq} - \delta_{pq} \alpha_p^{(k)} y_p K_{pq} \\
& + \delta_{pq} \alpha_q^{(k)} y_q K_{pq} - \left(\delta_{pq} \right)^2 K_{pq} \\
& - \frac{1}{2} \left(\alpha_q^{(k)} \right)^2 K_{qq} - \delta_{pq} \alpha_q^{(k)} y_q K_{qq} - \frac{1}{2} \left(\delta_{pq} \right)^2 K_{qq} \\
= \; & -\frac{1}{2} \sum_{i,j} \alpha_i^{(k)} \alpha_j^{(k)} y_i y_j K_{ij} \\
& + \delta_{pq} \sum_i \alpha_i^{(k)} y_i K_{ip} - \delta_{pq} \sum_i \alpha_i^{(k)} y_i K_{iq} \\
& - \frac{1}{2} \left(\delta_{pq} \right)^2 \left[K_{pp} - 2 K_{pq} + K_{qq} \right]
\end{aligned}
$$

Using (3.1) and (3.6),

$$
\begin{aligned}
F_{pq}^{(k+1)} \; = \; & F^{(k)} + \delta_{pq} \left(G_p^{(k)} - G_q^{(k)} \right) \\
& - \frac{1}{2} \left(\delta_{pq} \right)^2 \left[K_{pp} - 2 K_{pq} + K_{qq} \right]
\end{aligned}
$$

Derivation of (4.3). From (4.1) and (4.2),

$$
\begin{aligned}
F_r^{(t+1)} &= -\frac{1}{2} \sum_{m,n \neq r} \alpha_m^{(t)} \alpha_n^{(t)} y_m y_n K_{mn} \\
&\quad - \left(\alpha_r^{(t)} + \delta_r \right) y_r \sum_{m \neq r} \alpha_m^{(t)} y_m K_{mr} - \frac{1}{2} \left(\alpha_r^{(t)} + \delta_r \right)^2 K_{rr} \\
&= -\frac{1}{2} \sum_{m,n \neq r} \alpha_m^{(t)} \alpha_n^{(t)} y_m y_n K_{mn} \\
&\quad - \alpha_r^{(t)} y_r \sum_{m \neq r} \alpha_m^{(t)} y_m K_{mr} - \delta_r \sum_{m \neq r} \alpha_m^{(t)} y_m K_{mr} \\
&\quad - \frac{1}{2} \left(\alpha_r^{(t)} \right)^2 K_{rr} - \delta_r \alpha_r^{(t)} y_r K_{rr} - \frac{1}{2} \left(\delta_r \right)^2 K_{rr} \\
&= -\frac{1}{2} \sum_{i,j} \alpha_i^{(t)} \alpha_j^{(t)} y_i y_j K_{ij} \\
&\quad - \delta_r \sum_{i} \alpha_i^{(t)} y_i K_{ir} - \frac{1}{2} \left(\delta_r \right)^2 K_{rr}
\end{aligned}
$$

Using (3.1) and (4.4),

$$
F_r^{(t+1)} = F^{(t)} - \delta_r y_r G_r^{(t)} - \frac{1}{2} \left(\delta_r \right)^2 K_{rr}.
$$

Derivation of (4.12). The update of α_r at each iteration is C_r, except for the last update of each class. Consider the positive training class. The initialisation process will stop updating the decision variables for the positive class when (4.7) is met. Let n_+ be the number of iterations for updating decision variables for the positive class. From (4.7) and (2.6),

$$
\begin{aligned}
\sum_{\{i|y_i=+1\}} \alpha_i &\geq \frac{\nu}{2} \\
n_+ C_+ &\geq \frac{\nu}{2} \\
n_+ &\geq \frac{\nu}{2} \frac{1}{C_+}.
\end{aligned}
$$

Since n_+ is the smallest integer satisfying the inequality, using (2.7) and (2.9),

$$
\begin{aligned}
n_+ &= \left\lceil \frac{\nu_+ \nu_-}{\nu_+ + \nu_-} \left\{ l_+ \left(1 + \frac{\nu_+}{\nu_-} \right) \right\} \right\rceil \\
&= \left\lceil \frac{\nu_+ \nu_-}{\nu_+ + \nu_-} \left\{ l_+ \frac{\nu_- + \nu_+}{\nu_-} \right\} \right\rceil \\
n_+ &= \left\lceil \nu_+ l_+ \right\rceil.
\end{aligned}
$$

The derivation is similar for the negative class, and therefore the total number of iterations required for initialising the decision variables is

$$n = n_+ + n_- = \lceil \nu_+ l_+ \rceil + \lceil \nu_- l_- \rceil.$$

References

Burges, C.J.C. (1998), A Tutorial on Support Vector Machines for Pattern Recognition, *Data Mining and Knowledge Discovery*, Vol. 2, no. 2.

Chang, C.C. and Lin, C.J. (2001a), Training nu-Support Vector Classifiers: Theory and Algorithms, *Neural Computation*, Vol. 13, no. 9, pp. 2119–2147.

Chang, C.C. and Lin, C.J. (2001b), IJCNN 2001 Challenge: Generalization Ability and Text Decoding, *Proceedings of INNS-IEEE International Joint Conference on Neural Networks, IJCNN2001*, Washington, DC, USA.

Chew, H.G., Crisp, D.J., Bogner, R.E. and Lim, C.C. (2000), Target Detection in Radar Imagery using Support Vector Machines with Training Size Biasing, *Proceedings of the Sixth International Conference on Control, Automation, Robotics and Vision, ICARCV2000*, Singapore.

Chew, H.G., Bogner, R.E. and Lim, C.C. (2001a), Dual-nu Support Vector Machine with Error Rate and Training Size Biasing, *Proceedings of the 26th International Conference on Acoustics, Speech and Signal Processing, ICASSP2001*, Salt Lake City, Utah, USA.

Chew, H.G., Lim, C.C. and Bogner, R.E. (2001b), On Initialising nu-Support Vector Machine Training, *Proceedings of the 5th International Conference on Optimisation: Techniques and Applications, ICOTA2001*, pp. 1740–1747, Hong Kong.

Fletcher, R. (1987), *Practical Methods of Optimization*, John Wiley and Sons, Inc., 2nd edition.

Hsu, C.W. and Lin, C.J (2002), A Simple Decomposition Method for Support Vector Machines, *Machine Learning*, Vol. 46, pp. 291–314.

Joachims, T. (1999), Making Large-scale SVM Learning Practical, in Schölkopf, B., Burges, C.J.C and Smola, A.J. editors, *Advances in Kernel Methods — Support Vector Learning*, MIT Press, Cambridge, MA, USA.

Osuna, E., Freund, R. and Girosi, F. (1997a), An Improved Training Algorithm for Support Vector Machines, in Principe, J, Gile, L, Morgan, N. and Wilson,

E. editors, *Neural Networks for Signal Processing VII — Proceedings of the 1997 IEEE Workshop*, pp. 276–285, New York, USA.

Osuna, E., Freund, R. and Girosi, F. (1997b), Training support vector machines: An application to face detection, *Proceedings of CVPR'97*, Puerto Rico.

Platt, J.C. (1999) Fast Training of Support Vector Machines Using Sequential Minimal Optimization, in Schölkopf, B., Burges, C.J.C and Smola, A.J. editors, *Advances in Kernel Methods - Support Vector Learning*, MIT Press, Cambridge, MA, USA.

Schölkopf, B., Smola, A.J., Williamson, R.C. and Bartlett, P.L. (2000), New Support Vector Algorithms, *Neural Computation*, Vol. 12, pp. 1207–1245.

Vapnik, V. (1995), *The Nature of Statistical Learning Theory*, Springer-Verlag, New York, USA.

8 AN ANALYSIS OF THE BARZILAI AND BORWEIN GRADIENT METHOD FOR UNSYMMETRIC LINEAR EQUATIONS

Yu-Hong Dai,

Institute of Computational Mathematics and Scientific/Engineering Computing
Chinese Academy of Sciences, P.R. China (dyh@lsec.cc.ac.cn)

Li-Zhi Liao

Department of Mathematics, Hong Kong Baptist University,
Kowloon Tong, Kowloon, Hong Kong (liliao@hkbu.edu.hk)

and Duan Li

Department of Systems Engineering and Engineering Management
Chinese University of Hong Kong, Hong Kong (dli@se.cuhk.edu.hk)

Abstract: The Barzilai and Borwein gradient method does not ensure descent in the objective function at each iteration, but performs better than the classical steepest descent method in practical computations. Combined with the technique of nonmonotone line search etc., such a method has found successful applications in unconstrained optimization, convex constrained optimization and stochastic optimization. In this paper, we give an analysis of the Barzilai and Borwein gradient method for two unsymmetric linear equations with only two variables. Under mild conditions, we prove that the convergence rate of the Barzilai and Borwein gradient method is Q-superlinear if the coefficient matrix A has the same eigenvalue; if the eigenvalues of A are different, then the convergence rate is R-superlinear.

Key words: Unsymmetric linear equations, gradient method, convergence, Q-superlinear, R-superlinear.

1   INTRODUCTION

Consider the problem of minimizing a strongly convex quadratic,

$$\min q(x) = \frac{1}{2}x^T A x - b^T x, \tag{1.1}$$

where $A \in R^{n \times n}$ is a real symmetric positive definite matrix and $b \in R^n$. The Barzilai and Borwein gradient method Barzilai & Borwein (1988) for solving (1.1) has the form

$$x_{k+1} = x_k - \alpha_k g_k, \tag{1.2}$$

where g_k is the gradient of f at x_k and α_k is determined by information achieved at the points x_{k-1} and x_k. Denote $s_{k-1} = x_k - x_{k-1}$ and $y_{k-1} = g_k - g_{k-1}$. One choice for the stepsize α_k is such that the matrix $D_k = \alpha_k I$ satisfies a certain quasi-Newton relation:

$$\min \|D_k^{-1} s_{k-1} - y_{k-1}\|_2, \tag{1.3}$$

yielding

$$\alpha_k = \frac{s_{k-1}^T s_{k-1}}{s_{k-1}^T y_{k-1}}. \tag{1.4}$$

Compared with the classical steepest descent method, which can be dated to Cauchy (1847), the Barzilai and Borwein gradient method often requires less computational work and speeds up the convergence greatly (see Akaike (1959); Fletcher (1990)). Theoretically, Raydan (1993) proved that the Barzilai and Borwein gradient method can always converge to the unique solution $x^* = A^{-1}b$ of problem (1.1). If there are only two variables, Barzilai & Borwein (1988) established the R-superlinear convergence of the method; for any dimensional convex quadratics, Dai & Liao (2002) strengthened the analysis of Raydan (1993) and proved the R-linear convergence of the Barzilai and Borwein gradient method. A direct application of the Barzilai and Borwein method in chemistry can be found in Glunt (1993).

To extend the Barzilai and Borwein gradient method to minimize a general smooth function

$$\min f(x), \quad x \in R^n, \tag{1.5}$$

Raydan (1997) considered the use of the nonmonotone line search technique by Grippo et al (1986) for the Barzilai and Borwein gradient method which cannot ensure descent in the objective function at each iteration. The resulting algorithm, called the global Barzilai and Borwein algorithm, is proved to be globally convergent for general functions and is competitive to some standard conjugate gradient codes (see Raydan (1997)). A successful application of the global Barzilai and Borwein algorithm can be found in Birgin et al (1999). The idea of Raydan (1997) was further extended in Birgin et al (2000) for minimizing differentiable functions on closed convex sets, resulting in a more efficient class of projected gradient methods. Liu & Dai (2001) recently provided a powerful scheme for unconstrained optimization problems with strong noises by combining the Barzilai and Borwein gradient method and the stochastic approximation method. Other work related to the Barzilai and Borwein gradient method can be found in Birgin et al (2000). Because of its simplicity and numerical efficiency, the Barzilai and Borwein gradient method has been studied by many researchers.

It is trivial to see that problem (1.1) is equivalent to solving the symmetric positive definite linear system

$$Ax = b. \tag{1.6}$$

Denoting g_k to be the residual at x_k, namely

$$g_k = Ax_k - b, \tag{1.7}$$

one can similarly define the steepest descent method and the Barzilai and Borwein gradient method for problem (1.6). Friedlander et al (1999) presented a generalization of the two methods for problem (1.6), and compared several different strategies of choosing the stepsize α_k. To develop an efficient algorithm based on the Barzilai and Borwein gradient method for solving nonlinear equations, we present in this paper an analysis of the Barzilai and Borwein gradient method for the unsymmetric linear system (1.6), where $A \in R^{n \times n}$ is nonsingular but not necessarily symmetric positive definite.

As shown in the coming sections, the analysis of the Barzilai and Borwein gradient method is difficult for the unsymmetric linear equation (1.6). In this paper, we assume that there are only two variables and A has two real eigenvalues. In addition, we assume that

$$s_k^T y_k \neq 0, \quad \text{for all } k \geq 1. \tag{1.8}$$

The condition (1.8) does not imply the positive definiteness of A, which is required in the analysis of Barzilai & Borwein (1988), Dai & Liao (2002), and

Raydan (1993). Under the above assumptions, we prove that if the eigenvalues of A are the same, the Barzilai and Borwein gradient method is Q-superlinearly convergent (see Section 2). If A has different eigenvalues, the method converges for almost all initial points and the convergence rate is R-superlinear (see Section 4). The two results strongly depend on the analyses of two recurrence relations, namely (2.8) and (4.11) (see Sections 3 and 5, respectively). Some concluding remarks are drawn in Section 6.

2 CASE OF IDENTICAL EIGENVALUES

Assume that λ_1 and λ_2 are the two nonzero real eigenvalues of A. Since the Barzilai and Borwein gradient method is invariant under orthogonal transformations, we assume without loss of generality that the matrix A in (1.6) has the following form

$$A = \begin{pmatrix} \lambda_1 & \delta \\ 0 & \lambda_2 \end{pmatrix}, \tag{2.1}$$

where $\delta \in R^1$. In this section, we will consider the case that A has two equal eigenvalues, namely

$$\lambda_1 = \lambda_2 = \lambda. \tag{2.2}$$

By relations (1.2) and (1.7), we can write

$$g_{k+1} = (I - \alpha_k A)g_k. \tag{2.3}$$

Denoting $g_k = (g_k^{(1)}, g_k^{(2)})^T$, it follows from (2.3), (1.2), (1.4), and (2.2) that

$$g_{k+1}^{(1)} = \frac{\delta g_{k-1}^{(1)} g_{k-1}^{(2)}}{g_{k-1}^T A g_{k-1}} g_k^{(1)} - \frac{\delta \|g_{k-1}\|_2^2}{g_{k-1}^T A g_{k-1}} g_k^{(2)} \tag{2.4}$$

$$g_{k+1}^{(2)} = \frac{\delta g_{k-1}^{(1)} g_{k-1}^{(2)}}{g_{k-1}^T A g_{k-1}} g_k^{(2)}. \tag{2.5}$$

Let us define

$$t_k = g_k^{(1)}/g_k^{(2)}. \tag{2.6}$$

Noting that the condition (1.8) is equivalent to

$$g_k^T A g_k \neq 0, \quad \text{for all } k \geq 0, \tag{2.7}$$

we have by (2.4)-(2.5) that $g_k^{(1)} g_k^{(2)} \neq 0$ for all k. Thus t_k is well defined and $t_k \neq 0$. The relation (2.7) also implies that $\delta \neq 0$, for otherwise the algorithm gives the solution in at most two steps. Furthermore, the division of (2.4) by (2.5) yields the recurrence relation

$$t_{k+1} = t_k - t_{k-1} - t_{k-1}^{-1}. \tag{2.8}$$

By Theorem 3.1, we know that there exists at most a zero measure set $\mathcal{S}$ such that

$$\lim_{k \to \infty} |t_k| = +\infty, \quad \text{for all } (t_1, t_2) \in R^2 \backslash \mathcal{S}. \tag{2.9}$$

Then by (2.4), we can show that for most initial points, the Barzilai and Borwein gradient method converges globally and the convergence rate is Q-superlinear.

Theorem 2.1 *Consider the linear equations (1.6), where $A \in R^{2 \times 2}$ is non-singular and has two identical real eigenvalues. Suppose that the Barzilai and Borwein gradient method (1.2) and (1.4) is used, satisfying (1.8). Then for all $(t_1, t_2) \in R^2 \backslash \mathcal{S}$, where $\mathcal{S} \in R^2$ is some zero measure set in R^2, we have that*

$$\lim_{k \to \infty} g_k = 0. \tag{2.10}$$

Further, the convergence rate is Q-superlinear, namely

$$\lim_{k \to \infty} \frac{\|g_{k+1}\|_2}{\|g_k\|_2} = 0. \tag{2.11}$$

Proof: By Theorem 3.1, we know that relation (2.9) holds for some zero measure set in R^2. It follows from (2.6), (2.9), and (2.2) that

$$\lim_{k \to \infty} \frac{\delta g_{k-1}^{(1)} g_{k-1}^{(2)}}{g_{k-1}^T A g_{k-1}} = \lim_{k \to \infty} \frac{\delta t_{k-1}}{\lambda t_{k-1}^2 + \delta t_{k-1} + \lambda} = 0, \tag{2.12}$$

$$\lim_{k \to \infty} \frac{\delta \|g_{k-1}\|_2^2}{g_{k-1}^T A g_{k-1}} = \lim_{k \to \infty} \frac{\delta(1 + t_{k-1}^2)}{\lambda t_{k-1}^2 + \delta t_{k-1} + \lambda} = \frac{\delta}{\lambda}. \tag{2.13}$$

Then it follows from (2.4) and the above relations that

$$\lim_{k \to \infty} \frac{g_{k+1}^{(1)}}{g_k^{(1)}} = \lim_{k \to \infty} \frac{\delta g_{k-1}^{(1)} g_{k-1}^{(2)}}{g_{k-1}^T A g_{k-1}} - \lim_{k \to \infty} \frac{\delta \|g_{k-1}\|_2^2}{g_{k-1}^T A g_{k-1}} t_k^{-1} = 0. \tag{2.14}$$

Noting that $\|g_k\|_2^2 = (g_k^{(1)})^2(1+t_k^{-2})$, we then get by this, (2.14), and (2.6) that

$$\lim_{k\to\infty} \frac{\|g_{k+1}\|_2}{\|g_k\|_2} = \lim_{k\to\infty} \frac{|g_{k+1}^{(1)}|}{|g_k^{(1)}|} \frac{\sqrt{1+t_{k+1}^{-2}}}{\sqrt{1+t_k^{-2}}} = 0. \qquad (2.15)$$

Therefore this theorem is true.

In practical computations, it was found that the relation $\lim_{k\to\infty} |t_k| = +\infty$ always holds for any t_1 and t_2, and that the sequence $\{\|g_k\|\}$ generated by the Barzilai and Borwein gradient method is Q-superlinearly convergence if $\lambda_1 = \lambda_2$.

In the case of identical eigenvalues, the number $|\delta|$ provides an indication of the extent to which A is close to a symmetric matrix. From relation (2.13), we can see that the smaller $|\delta|$ is, the faster the Barzilai and Borwein gradient method converges to the solution.

3   PROPERTIES OF THE RECURRENCE RELATION (2.8)

In this section, we analyze the recurrence relation (2.8) and establish the property (2.9) for the sequence $\{t_k\}$ after giving some lemmas.

Lemma 3.1 (i) $t_{k+3} = -(t_k + t_k^{-1} + t_{k+1}^{-1})$.

(ii) *Define* $\bar{t}_k = -t_k$. *Then we also have that* $\bar{t}_{k+1} = \bar{t}_k - \bar{t}_{k-1} - \bar{t}_{k-1}^{-1}$.

Proof: The statements follow directly from (2.8) and the definition of $\{\bar{t}_k\}$.

Lemma 3.2 *Suppose that* $\{a_k\}$ *is a positive sequence that satisfies*

$$a_k \geq a_{k-1} + a_{k-1}^{-1}, \quad \text{for all } k \geq 2. \qquad (3.1)$$

Then $\{a_k\}$ *is disconvergent, namely* $\lim_{k\to\infty} a_k = +\infty$.

Proof: Relation (3.1) and $a_k > 0$ indicates that $\{a_k\}$ is monotonically increasing. Assume that

$$\lim_{k\to\infty} a_k = M < +\infty. \qquad (3.2)$$

Then by this and (3.1), we get that

$$\lim_{k\to\infty} a_k \geq M + M^{-1}, \qquad (3.3)$$

which contradicts (3.2). So this Lemma is true.

Lemma 3.3 *Consider the sequence $\{t_k\}$ that satisfies (2.8) and $t_k \neq 0$ for all k. Then there exists at most a zero measure set S in R^2 such that for all $(t_1, t_2) \in R^2 \backslash S$, one of the following relations holds for some integer $\bar{k}$:*

$$t_{\bar{k}} > 0, \quad t_{\bar{k}+1} > 0, \quad t_{\bar{k}+2} > 0 \tag{3.4}$$

or

$$t_{\bar{k}} < 0, \quad t_{\bar{k}+1} < 0, \quad t_{\bar{k}+2} < 0. \tag{3.5}$$

Proof: Assume that neither (3.4) nor (3.5) holds for all $\bar{k} \geq 1$. Then there must exist an integer $\hat{k}$ such that either $t_{\hat{k}} > 0$, $t_{\hat{k}+1} < 0$ or

$$t_{\hat{k}} < 0, \quad t_{\hat{k}+1} > 0. \tag{3.6}$$

Since t_k and $\bar{t}_k = -t_k$ satisfy the same recurrence relation by part (ii) of Lemma 3.2, we assume without loss of generality that (3.6) holds. Then by (3.6) and (2.8), we have that $t_{\hat{k}+2} = t_{\hat{k}+1} - t_{\hat{k}} - t_{\hat{k}}^{-1} > 0$. It follows that $t_{\hat{k}+3} < 0$, for otherwise (3.4) holds with $\bar{k} = \hat{k} + 1$. Similarly, we can prove that

$$t_{\hat{k}+4} < 0, \quad t_{\hat{k}+5} > 0. \tag{3.7}$$

The repetition of the above procedure yields

$$t_{\hat{k}+4i+j} \begin{cases} > 0, & \text{for all } i \geq 0 \text{ and } j = 1, 2; \\ < 0, & \text{for all } i \geq 0 \text{ and } j = 3, 4. \end{cases} \tag{3.8}$$

Noting that

$$t_{\hat{k}+4i+3} = t_{\hat{k}+4i+2} - (t_{\hat{k}+4i+1} + t_{\hat{k}+4i+1}^{-1}) < 0, \tag{3.9}$$

$$t_{\hat{k}+4i+5} = -(t_{\hat{k}+4i+2} + t_{\hat{k}+4i+2}^{-1} + t_{\hat{k}+4i+3}^{-1}) > 0, \tag{3.10}$$

we get that $t_{\hat{k}+4i+2} + t_{\hat{k}+4i+2}^{-1} + [t_{\hat{k}+4i+2} - (t_{\hat{k}+4i+1} + t_{\hat{k}+4i+1}^{-1})]^{-1} < 0$, yielding

$$t_{\hat{k}+4i+1} + t_{\hat{k}+4i+1}^{-1} < t_{\hat{k}+4i+2} + [t_{\hat{k}+4i+2} + t_{\hat{k}+4i+2}^{-1}]^{-1}. \tag{3.11}$$

In addition, it follows by $t_{\hat{k}+4i} < 0$ and (2.8) with k replaced by $\hat{k} + 4i + 1$ that

$$t_{\hat{k}+4i+2} \geq t_{\hat{k}+4i+1} + 2, \tag{3.12}$$

which, with (3.9), implies that $t_{\hat{k}+4i+1} < 1 < t_{\hat{k}+4i+2}$. Since $t + t^{-1}$ is monotonically decreasing for $t \in (0, 1)$, we can conclude from (3.11) that

$$t_{\hat{k}+4i+2} > t_{\hat{k}+4i+1}^{-1}, \tag{3.13}$$

for otherwise we have $t_{\hat{k}+4i+2}^{-1} \geq t_{\hat{k}+4i+1}$ and hence

$$t_{\hat{k}+4i+1} + t_{\hat{k}+4i+1}^{-1} \geq t_{\hat{k}+4i+2} + t_{\hat{k}+4i+2}^{-1} > t_{\hat{k}+4i+2} + (t_{\hat{k}+4i+2} + t_{\hat{k}+4i+2}^{-1})^{-1}, \quad (3.14)$$

contradicting (3.11). By (3.9) and (3.13), we get that

$$t_{\hat{k}+4i+1}^{-1} < t_{\hat{k}+4i+2} < t_{\hat{k}+4i+1} + t_{\hat{k}+4i+1}^{-1}. \quad (3.15)$$

It follows from (3.9) and (3.15) that

$$|t_{\hat{k}+4i+3}| = |t_{\hat{k}+4i+1}| - (t_{\hat{k}+4i+2} - t_{\hat{k}+4i+1}^{-1}) < |t_{\hat{k}+4i+1}|. \quad (3.16)$$

In addition, we can see from the definition of $t_{\hat{k}+4i+4}$ and (3.8) that

$$|t_{\hat{k}+4i+4}| > |t_{\hat{k}+4i+2}| + |t_{\hat{k}+4i+2}|^{-1}. \quad (3.17)$$

Similar to (3.15)-(3.17), we can prove the following three relations

$$|t_{\hat{k}+4i+3}^{-1}| < |t_{\hat{k}+4i+4}| < |t_{\hat{k}+4i+3}| + |t_{\hat{k}+4i+3}^{-1}|, \quad (3.18)$$

$$|t_{\hat{k}+4i+5}| < |t_{\hat{k}+4i+3}|, \quad (3.19)$$

$$|t_{\hat{k}+4i+6}| > |t_{\hat{k}+4i+4}| + |t_{\hat{k}+4i+4}|^{-1}. \quad (3.20)$$

By (3.17), (3.20), and Lemma 3.2, we get that

$$\lim_{i \to \infty} |t_{\hat{k}+2i}| = +\infty. \quad (3.21)$$

This together with (3.15), (3.16), (3.18), and (3.19) indicates that

$$\lim_{i \to \infty} |t_{\hat{k}+2i+1}| = 0. \quad (3.22)$$

By (3.15), we can assume that

$$t_{\hat{k}+4i+2} = t_{\hat{k}+4i+1}^{-1} + (1 - \psi_{\hat{k}+4i+1}) t_{\hat{k}+4i+1}, \quad (3.23)$$

where $\psi(t_{\hat{k}+4i+1}) \in (0,1)$. Then it follows from (3.23) and (2.8) that

$$t_{\hat{k}+4i+3} = -\psi_{\hat{k}+4i+1} t_{\hat{k}+4i+1} \quad (3.24)$$

and hence by part (i) of Lemma 3.1,

$$t_{\hat{k}+4i+5} = -(1 - \psi_{\hat{k}+4i+1}^{-1}) t_{\hat{k}+4i+1}^{-1} - \Gamma_i, \quad (3.25)$$

where

$$\Gamma_i = \frac{(2 - \psi_{\hat{k}+4i+1})t_{\hat{k}+4i+1} + (1 - \psi_{\hat{k}+4i+1})^2 t_{\hat{k}+4i+1}^3}{1 + (1 - \psi_{\hat{k}+4i+1})t_{\hat{k}+4i+1}^2}. \tag{3.26}$$

From (3.19), (3.22), and (3.25), we see that

$$\lim_{i\to\infty} \psi_{\hat{k}+4i+1} = 1 \quad \text{and} \quad \lim_{i\to\infty} (\psi_{\hat{k}+4i+1} - 1)t_{\hat{k}+4i+1}^{-1} = 0. \tag{3.27}$$

From the above relations, we can further deduce that

$$\psi_{\hat{k}+4i+1} = 1 + h_{\hat{k}+4i+1}, \quad \text{where} \quad h_{\hat{k}+4i+1} = o(t_{\hat{k}+4i+1}). \tag{3.28}$$

Then it follows from (3.25), (3.26), and (3.28) that

$$t_{\hat{k}+4i+5} = -t_{\hat{k}+4i+1} - t_{\hat{k}+4i+1}^{-1} h_{\hat{k}+4i+1} + o(t_{\hat{k}+4i+1}). \tag{3.29}$$

On the other hand, by (3.24) and the first part of (3.27), we see that

$$\lim_{i\to\infty} t_{\hat{k}+4i+3}/t_{\hat{k}+4i+1} = -1. \tag{3.30}$$

Noting that $\{-t_k\}$ and $\{t_k\}$ satisfy the same recurrence relation and replacing all $t_{\hat{k}+4i+j}$ with $-t_{\hat{k}+4i+j+2}$ in the previous discussions, we can similarly establish that

$$\lim_{i\to\infty} (-t_{\hat{k}+4i+5})/(-t_{\hat{k}+4i+3}) = -1. \tag{3.31}$$

Relations (3.30) and (3.31) imply that

$$\lim_{i\to\infty} t_{\hat{k}+4i+5}/t_{\hat{k}+4i+1} = 1. \tag{3.32}$$

Then we get by (3.29) and (3.32) that

$$h_{\hat{k}+4i+1} = -2t_{\hat{k}+4i+1}^2 + o(t_{\hat{k}+4i+1}^2), \tag{3.33}$$

which, together with (3.28) and (3.24), gives

$$t_{\hat{k}+4i+3} = -t_{\hat{k}+4i+1} + 2t_{\hat{k}+4i+1}^3 + o(t_{\hat{k}+4i+1}^3). \tag{3.34}$$

Similarly, we also have that

$$t_{\hat{k}+4i+5} = -t_{\hat{k}+4i+3} + 2t_{\hat{k}+4i+3}^3 + o(t_{\hat{k}+4i+3}^3). \tag{3.35}$$

Substituting (3.34) into (3.35) and comparing the resulting expression with (3.29) yield

$$h_{\hat{k}+4i+1} = -2t_{\hat{k}+4i+1}^2 + 6t_{\hat{k}+4i+1}^4 + o(t_{\hat{k}+4i+1}^4). \tag{3.36}$$

Consequently, we have that

$$t_{\hat{k}+4i+3} = -t_{\hat{k}+4i+1} + 2t_{\hat{k}+4i+1}^3 - 6t_{\hat{k}+4i+1}^5 + o(t_{\hat{k}+4i+1}^5), \tag{3.37}$$

$$t_{\hat{k}+4i+5} = -t_{\hat{k}+4i+3} + 2t_{\hat{k}+4i+3}^3 - 6t_{\hat{k}+4i+3}^5 + o(t_{\hat{k}+4i+3}^5). \tag{3.38}$$

Substituting (3.37) into (3.38) and comparing it with (3.29), we then obtain

$$h_{\hat{k}+4i+1} = -2t_{\hat{k}+4i+1}^2 + 6t_{\hat{k}+4i+1}^4 - 38t_{\hat{k}+4i+1}^6 + o(t_{\hat{k}+4i+3}^6). \tag{3.39}$$

Recursively, we know that for any j the coefficient of $t_{\hat{k}+4i+1}^j$ in $h_{\hat{k}+4i+1}$ is uniquely determined. It follows from this and (3.23) that $t_{\hat{k}+4i+2} = t_{\hat{k}+4i+1}^{-1} - h_{\hat{k}+4i+1}t_{\hat{k}+4i+1}$ is uniquely determined. In other words, for any $t_{\hat{k}+4i+1} > 0$, there exists at most one value of $t_{\hat{k}+4i+2}$ such that the cycle (3.8) occurs. Therefore there exists at most a zero measure set $\mathcal{S}$ in R^2 such that the cycle (3.8) occurs for $(t_1, t_2) \in \mathcal{S}$. This completes our proof.

Theorem 3.1 *Consider the sequence $\{t_k\}$ that satisfies (2.8). Then there exists at most a zero measure set $\mathcal{S}$ in R^2 such that*

$$\lim_{k\to\infty} |t_k| = +\infty, \quad \text{for all } (t_1, t_2) \in R^2\backslash\mathcal{S}. \tag{3.40}$$

Proof: By Lemma 3.3, we know that there exists at most a zero measure set $\mathcal{S}$ in R^2 such that (3.4) or (3.5) holds for some $\bar{k}$ if $(t_1, t_2) \in R^2\backslash\mathcal{S}$. Assume without loss of generality that (3.4) holds, for otherwise we may consider the sequence $\bar{t}_k = -t_k$. Then by part (i) of Lemma 3.1 and relation (3.4), we have that

$$t_{\bar{k}+3} = -(t_{\bar{k}} + t_{\bar{k}}^{-1} + t_{\bar{k}+1}^{-1}) < 0, \tag{3.41}$$

$$t_{\bar{k}+4} = -(t_{\bar{k}+1} + t_{\bar{k}+1}^{-1} + t_{\bar{k}+2}^{-1}) < 0. \tag{3.42}$$

Since $t + t^{-1} \geq 2$ for $t > 0$, it follows from (3.41) and (3.4) that $t_{\bar{k}+3} < -2$. Consequently,

$$t_{\bar{k}+5} = -(t_{\bar{k}+2} + t_{\bar{k}+2}^{-1}) - t_{\bar{k}+3}^{-1} \leq -2 - (-2)^{-1} < 0. \tag{3.43}$$

By relations (3.41)–(3.43), we can similarly show that $t_{\bar{k}+6} > 0$, $t_{\bar{k}+7} > 0$ and $t_{\bar{k}+8} > 0$. The repetition of this procedure yields

$$t_{\bar{k}+6i+j} \begin{cases} > 0, & \text{for all } i \geq 0 \text{ and } j = 0, 1, 2; \\ < 0, & \text{for all } i \geq 0 \text{ and } j = 3, 4, 5. \end{cases} \tag{3.44}$$

The above relation, (2.8), (3.41), and (3.42) indicate that

$$|t_{\bar{k}+3(i+1)+j}| \geq |t_{\bar{k}+3i+j}| + |t_{\bar{k}+3i+j}|^{-1}, \quad \text{for all } i \geq 0 \text{ and } j = 0, 1. \tag{3.45}$$

Thus by Lemma 3.2, we obtain

$$\lim_{i \to \infty} |t_{\bar{k}+3i+j}| = +\infty, \quad \text{for } j = 0, 1. \tag{3.46}$$

It remains to prove that

$$\lim_{i \to \infty} |t_{\bar{k}+3i+2}| = +\infty. \tag{3.47}$$

For this aim, we first show that

$$\limsup_{i \to \infty} |t_{\bar{k}+3i+2}| = +\infty. \tag{3.48}$$

In fact, if (3.48) is not true, then there exists some constant $M \geq 1$ such that

$$|t_{\bar{k}+3i+2}| \leq M, \quad \text{for all } i \geq 0. \tag{3.49}$$

Relation (3.46) implies that there exists some integer $\bar{i} \geq 0$ such that

$$|t_{\bar{k}+3i+3}| \geq 2M, \quad \text{for all } i \geq \bar{i}. \tag{3.50}$$

It follows from part (i) of Lemma 3.1, (3.44), (3.49), and (3.50) that for all $i \geq \bar{i}$,

$$|t_{\bar{k}+3i+5}| = |t_{\bar{k}+3i+2}| + |t_{\bar{k}+3i+2}|^{-1} - |t_{\bar{k}+3i+3}|^{-1} \geq |t_{\bar{k}+3i+2}| + (2M)^{-1}. \tag{3.51}$$

This relation gives

$$\lim_{i \to \infty} |t_{\bar{k}+3i+2}| = +\infty, \tag{3.52}$$

contradicting (3.49). So (3.48) must hold. For any $M \geq 1$, we continue to denote $\bar{i} \geq 0$ to be such that (3.50) holds. Let $\{i_l : l = 1, 2, \ldots\}$ be the set of all positive integers such that

$$i_l > i_{l-1} > \cdots > i_1 \geq \bar{i} \tag{3.53}$$

and

$$|t_{\bar{k}+3i_l+2}| \geq 2M. \tag{3.54}$$

Relation (3.48) implies that $\{i_l\}$ is an infinite set. By the choice of $\{i_l\}$, we have that

$$|t_{\bar{k}+3j+2}| < 2M, \quad \text{for } j \in [i_l + 1, i_{l+1} - 1]. \tag{3.55}$$

It follows from this, (3.50) and part (i) of Lemma 3.1 that

$$
\begin{aligned}
|t_{\bar{k}+3(j+1)+2}| &\geq |t_{\bar{k}+3j+2}| \geq \cdots \geq |t_{\bar{k}+3(i_l+1)+2}| \\
&\geq |t_{\bar{k}+3i_l+2}| - |t_{\bar{k}+3i_l+3}| \geq 2M - (2M)^{-1} \geq M \quad (3.56)
\end{aligned}
$$

for all $j \in [i_l + 1, i_{l+1} - 1]$. Therefore we have that

$$
|t_{\bar{k}+3i+2}| \geq M, \quad \text{for all } i \geq i_1. \tag{3.57}
$$

Since M can be arbitrarily large, we know that (3.47) must hold. This together with (3.46) complete our proof.

4   CASE OF DIFFERENT EIGENVALUES

In this section, we analyze the Barzilai and Borwein gradient method for unsymmetric linear equations (1.6), assuming that the coefficient matrix A, in the form of (2.1) has two different real eigenvalues, i.e.,

$$
\lambda_1 \neq \lambda_2. \tag{4.1}
$$

Denote $\beta = \dfrac{\delta}{\lambda_2 - \lambda_1}$, $P = \begin{pmatrix} 1 & -\gamma \\ 0 & 1 \end{pmatrix}$, and $D = \begin{pmatrix} \lambda_1 & 0 \\ 0 & \lambda_2 \end{pmatrix}$. Then the matrix A can be written as

$$
A = P^{-1}DP. \tag{4.2}
$$

Further, defining $u_k = Pg_k$ and $v_k = P^{-T}g_k$, we can get by multiplying (2.3) with P that

$$
u_{k+1} = (1 - \frac{u_{k-1}^T v_{k-1}}{u_{k-1}^T D v_{k-1}} D)u_k. \tag{4.3}
$$

Assume that $u_k = (u_k^{(1)}, u_k^{(2)})^T$ and $v_k = (v_k^{(1)}, v_k^{(2)})^T$. The above relation indicates that

$$
\begin{cases}
u_{k+1}^{(1)} &= \frac{(\lambda_2 - \lambda_1)u_{k-1}^{(2)} v_{k-1}^{(2)}}{u_{k-1}^T D v_{k-1}} u_k^{(1)} \\[2mm]
u_{k+1}^{(2)} &= \frac{(\lambda_1 - \lambda_2)u_{k-1}^{(1)} v_{k-1}^{(1)}}{u_{k-1}^T D v_{k-1}} u_k^{(2)}.
\end{cases} \tag{4.4}
$$

Under the condition (1.8), it is easy to show by (4.4) and the defintions of u_k and v_k that $u_k^{(1)} u_k^{(2)} \neq 0$ and $v_k^{(1)} v_k^{(2)} \neq 0$. Let q_k be the ratio $u_k^{(1)}/u_k^{(2)}$. It follows from (4.4) that

$$
q_{k+1} = -q_k q_{k-1}^{-1}(v_{k-1}^{(2)}/v_{k-1}^{(1)}). \tag{4.5}
$$

On the other hand, by the definitions of u_k and v_k, we have that

$$\frac{v_k^{(2)}}{v_k^{(1)}} = \beta + \frac{g_k^{(2)}}{g_k^{(1)}}, \quad \frac{u_k^{(1)}}{u_k^{(2)}} = -\gamma + \frac{g_k^{(1)}}{g_k^{(2)}}, \tag{4.6}$$

which yields

$$\frac{v_k^{(2)}}{v_k^{(1)}} = \gamma + \frac{1}{\gamma + q_k}. \tag{4.7}$$

Substituting (4.7) into (4.5), we then obtain the following recurrence relation

$$q_{k+1} = -q_k q_{k-1}^{-1} \left(\gamma + \frac{1}{\gamma + q_{k-1}} \right). \tag{4.8}$$

If A is symmetric, namely $\delta = \gamma = 0$, the above relation reduces to

$$q_{k+1} = -q_k q_{k-1}^{-2}, \tag{4.9}$$

from which one can establish the R-superlinear convergence result of the Barzilai and Borwein gradient method (see Barzilai & Borwein (1988)). In this paper, we assume that $\gamma \neq 0$.

For simplicity, we denote $\tau = \gamma^{-2}$ and define the sequence

$$p_k = -\gamma^{-1} q_k. \tag{4.10}$$

Then it follows from (4.8) that

$$p_{k+1} = p_k p_{k-1}^{-1} \left(1 + \frac{\tau}{1 - p_{k-1}} \right). \tag{4.11}$$

For the sequence p_k that satisfies (4.11), if $(|p_1|, |p_2|) \neq (\sqrt{1+\tau}, \sqrt{1+\tau})$, and if $p_k \neq 0, 1$ for all k, we know from Theorems 5.1 and 5.2 that, there exists some integer $\tilde{k}$ such that the following relations hold:

$$\lim_{i \to \infty} \frac{p_{\tilde{k}+6(i+1)+j}}{p_{\tilde{k}+6i+j}} = (1+\tau)^2, \quad \text{for } j = 1, 2, \tag{4.12}$$

$$\lim_{i \to \infty} p_{\tilde{k}+6i+3} = +\infty, \tag{4.13}$$

$$\lim_{i \to \infty} \frac{p_{\tilde{k}+6(i+1)+j}}{p_{\tilde{k}+6i+j}} = (1+\tau)^{-2}, \quad \text{for } j = 4, 5, \tag{4.14}$$

$$\lim_{i \to \infty} p_{\tilde{k}+6i+6} = 0. \tag{4.15}$$

From the above relations, we can show that if A has two different eigenvalues, the Barzilai and Borwein gradient method converges globally and its convergence rate is R-superlinear.

Theorem 4.1 *Consider the unsymmetric linear equations (1.6), where $A \in R^{2\times 2}$ is given in (2.1), with $\lambda_1 \neq \lambda_2$, $\lambda_1\lambda_2 \neq 0$ and $\delta \neq 0$. Assume that*

$$(g_1^{(1)}/g_1^{(2)},\ g_2^{(1)}/g_2^{(2)}) \neq (\gamma \pm \sqrt{1+\gamma^2}, \gamma \pm \sqrt{1+\gamma^2}). \qquad (4.16)$$

Then for the Barzilai and Borwein gradient method (1.2) and (1.4), if (1.8) holds, we have that

$$\lim_{k\to\infty} g_k = 0. \qquad (4.17)$$

Further, the convergence rate is R-superlinear.

Proof: It follows from (4.4), (4.7) and (4.10) that

$$\frac{u_{k+1}^{(1)}}{u_k^{(1)}} = \frac{\lambda_1 - \lambda_2}{-\lambda_2 + \lambda_1 p_{k-1}(1 + \frac{\tau}{1-p_{k-1}})^{-1}} \qquad (4.18)$$

$$\frac{u_{k+1}^{(2)}}{u_k^{(2)}} = \frac{\lambda_1 - \lambda_2}{\lambda_1 - \lambda_2 p_{k-1}^{-1}(1 + \frac{\tau}{1-p_{k-1}})}. \qquad (4.19)$$

The condition (4.16) implies that

$$(p_k, p_{k+1}) \neq (-\sqrt{1+\tau},\ \sqrt{1+\tau}), \quad \text{for all } k \geq 1. \qquad (4.20)$$

Thus by Theorems 5.1 and 5.2, we know that there exists some integer $\bar{i}$ such that the relations (4.12)–(4.15) hold. For any sufficient small $\varepsilon > 0$, let $\bar{\varepsilon} \in (0, \varepsilon)$ to be another small number. For this $\bar{\varepsilon}$, we know from (4.12) and (4.14) that there exists an integer $\hat{i}$ such that for all $i \geq \hat{i}$,

$$p_{\bar{k}+6i+j} \begin{cases} \geq c(1+\tau-\bar{\varepsilon})^{2i}, & \text{for } j = 1, 2; \\ \leq c(1+\tau-\bar{\varepsilon})^{-2i}, & \text{for } j = 4, 5, \end{cases} \qquad (4.21)$$

where $c > 0$ is some constant. Further, since $\bar{\varepsilon} < \varepsilon$, it follows from (4.21), (4.13), and (4.15) that there exist an integer $\bar{i} \geq \hat{i}$ such that for all $i \geq \tilde{i}$,

$$\frac{u_{\tilde{k}+6i+j+1}^{(1)}}{u_{\tilde{k}+6i+j}^{(1)}} \begin{cases} \leq |\frac{\lambda_1-\lambda_2}{c\lambda_1}|(1+\tau-\varepsilon)^{-2i}, & \text{for } j = 1, 2; \\ \leq 1, & \text{for } j = 3; \\ \leq |\frac{\lambda_1-\lambda_2}{\lambda_2}| + \varepsilon, & \text{for } j = 4, 5, 6, \end{cases} \qquad (4.22)$$

$$\frac{u_{\tilde{k}+6i+j+1}^{(2)}}{u_{\tilde{k}+6i+j}^{(2)}} \begin{cases} \leq |\frac{\lambda_1-\lambda_2}{\lambda_1}| + \varepsilon, & \text{for } j = 1, 2, 3; \\ \leq |\frac{\lambda_1-\lambda_2}{c^{-1}\lambda_2}|(1+\tau-\varepsilon)^{-2i}, & \text{for } j = 4, 5; \\ \leq 1, & \text{for } j = 6. \end{cases} \qquad (4.23)$$

Thus for any $\varepsilon > 0$, the following relation holds with some positive constants c_1 and c_2:

$$|u_k^{(l)}| \le c_1 c_2^k (1 + \tau - \varepsilon)^{-\frac{k^2}{18}}, \quad \text{for } l = 1, 2. \tag{4.24}$$

The definition of u_k implies that $g_k = P^{-1} u_k$, this and (4.24) give

$$\|g_k\|_2 \le c_3 c_2^k (1 + \tau - \varepsilon)^{-\frac{k^2}{18}}, \tag{4.25}$$

where $c_3 = \sqrt{2} \|P^{-1}\|_2 c_1$. Relation (4.25) indicates that the sequence $\{\|g_k\|\}$ is globally convergent and the convergence rate is R-superlinear.

Theorem 2.1 tells us that under certain conditions on A, the Barzilai and Borwein gradient method is R-superlinearly convergent for most initial starting points x_1 and x_2. If (4.16) does not hold, namely

$$(g_1^{(1)}/g_1^{(2)}, \ g_2^{(1)}/g_2^{(2)}) = (\gamma \pm \sqrt{1 + \gamma^2}, \gamma \pm \sqrt{1 + \gamma^2}), \tag{4.26}$$

direct calculations show that

$$(p_{\bar{k}+2i}, p_{\bar{k}+2i+1}) = (-\sqrt{1+\tau}, \sqrt{1+\tau}), \quad \text{for some } \bar{k} > 0 \text{ and all } i \ge 0. \tag{4.27}$$

By this relation, (4.18) and (4.19), we see that if $\lambda_1 \lambda_2 > 0$, then the Barzilai and Borwein gradient method is linearly convergent; otherwise, if $\lambda_1 \lambda_2 < 0$, the method needs not converge. Here it is worthwhile pointing out that the case of (4.16) can be avoided if the first stepsize α_1 is computed by an exact line search.

Since

$$\tau = \gamma^{-2} = (\lambda_2 - \lambda_1)^2 / \delta^2, \tag{4.28}$$

the value of τ can be regarded as a quantity that shows the degree to which the matrix A is close to a symmetric matrix. Relation (4.25) indicates that, the bigger τ is, namely if A is closer to a symmetric matrix, the faster the Barzilai and Borwein gradient method converges to the solution.

5 PROPERTIES OF THE RECURRENCE RELATION (4.11)

In this section, we consider the sequence $\{p_k\}$ that satisfies (4.11). Assume that $p_k \ne 0, 1$ for all $k \ge 1$. To expedite our analyses, we introduce functions

$$h(p) = 1 + \tau(1-p)^{-1}, \quad \phi(p) = p^{-1} h(p). \tag{5.1}$$

Lemma 5.1 (i) $p_{k+3} = p_k^{-1}h(p_k)h(p_{k+1}) = \phi(p_k)h(p_{k+1})$;

(ii) $p_{k+6}/p_k = [h(p_{k+3})/h(p_k)]\,[h(p_{k+4})/h(p_{k+1})]$;

(iii) *Define* $r_k = (1+\tau)p_k^{-1}$. *Then we also have that* $r_{k+1} = r_k r_{k-1}^{-1}(1+\frac{\tau}{1-r_{k-1}})$.

Proof: (i) follows from the definitions of $\{p_k\}$, $h(p)$ and $\phi(p)$. (ii) follows from (i). For (iii), noting that $p_k = (1+\tau)r_k^{-1}$, we have that

$$r_{k+1} = (1+\tau)p_{k+1}^{-1} = (1+\tau)p_k^{-1}p_{k-1}\left(1 + \frac{\tau}{1-p_{k-1}}\right)^{-1}$$

$$= (1+\tau)r_k r_{k-1}^{-1}\left(1 + \frac{\tau}{1-(1+\tau)r_{k-1}}\right)^{-1} = r_k r_{k-1}^{-1}\left(1 + \frac{\tau}{1-r_{k-1}}\right).$$

So this lemma is true.

Lemma 5.1 shows that the sequences p_k and r_k satisfy the same recurrence relation. This observation will greatly simplify our coming analyses.

Lemma 5.2 *Consider the sequence* p_k *that satisfies (4.11) and* $p_k \neq 0,1$ *for all* k. *Then for any integer* $\hat{k}$, *the following cycle cannot occur*

$$\lim_{i \to \infty} p_{\hat{k}+4i+1} = 1, \quad \lim_{i \to \infty} p_{\hat{k}+4i+2} = 0, \tag{5.2}$$

$$\lim_{i \to \infty} p_{\hat{k}+4i+3} = 1+\tau, \quad \lim_{i \to \infty} p_{\hat{k}+4i+4} = \infty. \tag{5.3}$$

Proof: We proceed by contradiction and assume that (5.2)-(5.3) hold for some $\hat{k}$. For simplicity, we let $\hat{k} = 0$. By the definition of $\{p_k\}$ and relation (5.2), it is easy to see that

$$\lim_{i \to \infty} p_{4i+2}(1 - p_{4i+1})^{-1} = a, \quad \text{where } a = 1+\tau^{-1}. \tag{5.4}$$

Let us introduce the following infinitesimals

$$\delta_i = p_{4i+2}, \tag{5.5}$$

$$\gamma_i = p_{4i+2}(1 - p_{4i+1})^{-1} - a. \tag{5.6}$$

It follows from (5.6) that

$$p_{4i+1} = 1 - \delta_i(a + \gamma_i)^{-1}. \tag{5.7}$$

By the definition of $\{p_k\}$ and the above relations, we get by direct calculation that

$$p_{4i+3} = \frac{(a+\gamma_i)(\tau a + \tau\gamma_i + \delta_i)}{a + \gamma_i - \delta_i}, \tag{5.8}$$

$$p_{4i+4} = \frac{a+\gamma_i}{a+\gamma_i-\delta_i} \frac{(1+\tau-\delta_i)(\tau a + \tau\gamma_i + \delta_i)}{\delta_i(1-\delta_i)}. \tag{5.9}$$

Let r_k be given in part (iii) of Lemma 5.1 and denote

$$\bar{i} = r_{4i+4}, \tag{5.10}$$

$$\theta_i = r_{4i+4}(1 - r_{4i+3})^{-1} - a. \tag{5.11}$$

Direct calculations yield

$$\bar{i} = \frac{1+o(1)}{1+\tau+o(1)}, \tag{5.12}$$

$$\theta_i = -\frac{(1+\tau)a\gamma_i + (a+\tau)a\delta_i}{\tau a\gamma_i + (\tau+a+1)\delta_i + o(|\gamma_i|+|\delta_i|)} + \Delta_i, \tag{5.13}$$

where

$$\Delta_i = \frac{-(1+\tau)^2\gamma_i{}^2 + (1+\tau)\gamma_i\delta_i + (a^2 - (1+\tau))\delta_i{}^2 + o(\gamma_i{}^2 + |\gamma_i\delta_i| + \delta_i{}^2)}{(1+\tau)(\tau a\gamma_i + (\tau+a+1)\delta_i) + o(|\gamma_i|+|\delta_i|)}.$$

By the definition of $\{r_k\}$ and relations (5.2)–(5.3), we must have

$$\lim_{i\to\infty} \theta_i = 0. \tag{5.14}$$

Thus by (5.13), the following relation holds

$$\lim_{i\to\infty} \frac{\gamma_i}{\delta_i} = -\frac{a+\tau}{1+\tau} = -\frac{1+\tau+\tau^2}{\tau(1+\tau)}. \tag{5.15}$$

Substituting (5.15) into (5.13), we can then obtain

$$\theta_i = -\frac{((\tau+2)^2 + \tau^{-1})\delta_i + o(\delta_i)}{1+\tau+o(1)}. \tag{5.16}$$

Relations (5.12) and (5.16) show that

$$\lim_{i\to\infty} \frac{\bar{i}}{\theta_i} = -\frac{1}{(2+\tau)^2 + \tau^{-1}}. \tag{5.17}$$

However, due to the relation between $\{r_k\}$ and $\{p_k\}$, similarly to (5.15) we have

$$\lim_{i\to\infty} \frac{\bar{i}}{\theta_i} = -\frac{1+\tau+\tau^2}{\tau(1+\tau)}. \tag{5.18}$$

Since the value on the right hand side of (5.17) is not equal to the one of (5.18) for any $\tau > 0$, we see that the two relations contradict each other. Therefore this lemma is true.

Lemma 5.3 *Consider the sequence $\{p_k\}$ that satisfies (4.11) and $p_k \neq 0, 1$ for all k. Assume that*

$$p_k > 0, \quad \text{for all large } k. \tag{5.19}$$

Then there exists some index $\bar{k}$ such that one of the following relations holds:

$$0 < p_{\bar{k}} \leq p_{\bar{k}+1} < 1, \tag{5.20}$$

$$1 + \tau < p_{\bar{k}+1} \leq p_{\bar{k}}. \tag{5.21}$$

Proof: For the function $\phi(p)$ defined in (5.1), it is easy to check that $p_1^* = 1+\tau-\sqrt{\tau(1+\tau)}$ minimizes $\phi(p)$ in $(0,1)$ and $p_2^* = 1+\tau+\sqrt{\tau(1+\tau)}$ maximizes $\phi(p)$ in $(1+\tau, +\infty)$, and

$$\min_{p \in (0,1)} \phi(p) = \phi(p_1^*) > 1, \quad \max_{p > 1+\tau} \phi(p) = \phi(p_2^*) < 1. \tag{5.22}$$

It follows from (4.11) that

$$p_{k+l} = p_{k+l-1}\phi(p_{k+l-2}) = \cdots\cdots = p_k \prod_{i=k-1}^{k+l-2} \phi(p_i), \quad \text{for all } l \geq 1. \tag{5.23}$$

By (5.23), (5.22), and (5.19), we can see that there exists some $\hat{k}$ such that

$$p_{\hat{k}} \in (0,1), \quad p_{\hat{k}+1} > 1+\tau \tag{5.24}$$

or

$$p_{\hat{k}} > 1+\tau, \quad p_{\hat{k}+1} \in (0,1). \tag{5.25}$$

By part (iii) of Lemma 5.1, we assume without loss of generality that (5.24) holds, for otherwise consider the sequence $\{r_k\}$. Now we proceed by contradiction and assume that neither (5.20) nor (5.21) holds. Then by the definition of $\{p_k\}$ and the relation (5.24), it is easy to show that for all $i \geq 1$:

$$p_{\hat{k}+4i-1} \in (0,1), \ p_{\hat{k}+4i} \in (0,1), \ p_{\hat{k}+4i+1} > 1+\tau, \ p_{\hat{k}+4i+2} > 1+\tau. \tag{5.26}$$

In the following, we will prove that the cycle (5.26) cannot occur infinitely. In fact, since $p_{\hat{k}+4i} \in (0,1)$, we have by part (i) of Lemma 5.1 that

$$p_{\hat{k}+4i+2} = \phi(p_{\hat{k}+4i-1})h(p_{\hat{k}+4i}) > (1+\tau)\phi(p_{\hat{k}+4i-1}), \qquad (5.27)$$

this together with $p_{\hat{k}+4i-1} \in (0,1)$ and (5.22) implies that

$$p_{\hat{k}+4i+2} \geq (1+\tau)\phi(p_1^*) \geq p_2^*. \qquad (5.28)$$

It follows from the definition of $p_{\hat{k}+4i+4}$ and $p_{\hat{k}+4i+3} \in (0,1)$ that

$$p_{\hat{k}+4i+4} = p_{\hat{k}+4i+3}\phi(p_{\hat{k}+4i+2}) < \phi(p_{\hat{k}+4i+2}). \qquad (5.29)$$

In addition, by the definition of $p_{\hat{k}+4i+5}$ and $p_{\hat{k}+4i+5} > 1+\tau$, it follows that

$$\phi(p_{\hat{k}+4i+3}) > (1+\tau)p_{\hat{k}+4i+4}^{-1}. \qquad (5.30)$$

Writing $p_{\hat{k}+4i+5} = \phi(p_{\hat{k}+4i+2})h(p_{\hat{k}+4i+3})$, we can prove from $p_{\hat{k}+4i+5} > 1+\tau$ and (5.28) that $p_{\hat{k}+4i+3} \in (p_1^*, 1)$. Noting that $\phi(p)$ is monotonically decreasing for $p \geq p_2^*$ and combining relations (5.27)-(5.30), we can obtain

$$\phi(p_{\hat{k}+4i+3}) > (1+\tau)[\phi((1+\tau)\phi(p_{\hat{k}+4i-1}))]^{-1}. \qquad (5.31)$$

Since, by the definition of $\phi(p)$,

$$\phi((1+\tau)p^{-1}) = \phi(p)^{-1}, \qquad (5.32)$$

we have by this and (5.31) that

$$(1+\tau)\phi(p_{\hat{k}+4i+3})^{-1}\phi(\phi(p_{\hat{k}+4i-1})^{-1}) < 1. \qquad (5.33)$$

Using the fact that $\phi(p)$ is monotonically decreasing for $p \in (p_1^*, 1)$, we get from (5.33) that

$$p_{\hat{k}+4i+3} > p_{\hat{k}+4i-1}, \qquad (5.34)$$

for otherwise we have from the monotonicity of $\phi(p)$ and (5.26) that

$$\begin{aligned}
(1+\tau)&\phi(p_{\hat{k}+4i+3})^{-1}\phi(\phi(p_{\hat{k}+4i-1})^{-1}) \\
&\geq (1+\tau)\phi(p_{\hat{k}+4i-1})^{-1}\phi(\phi(p_{\hat{k}+4i-1})^{-1}) \qquad (5.35) \\
&= (1+\tau)h(\phi(p_{\hat{k}+4i-1})^{-1}) > 1.
\end{aligned}$$

Relations (5.34) and (5.26) indicate that $\lim\limits_{i\to\infty} p_{\hat{k}+4i-1} = c_4 \in (0,1]$. If $c_4 < 1$, then we have that

$$\lim_{i\to\infty}(1+\tau)\phi(p_{\hat{k}+4i+3})^{-1}\phi(\phi(p_{\hat{k}+4i-1})^{-1}) = (1+\tau)\phi(c_4^{-1})\phi(\phi(c_4)^{-1}) > 1,$$

contradicting (5.33). Thus we must have

$$\lim_{i \to \infty} p_{\hat{k}+4i-1} = 1. \tag{5.36}$$

Further, by the definition of $\{p_k\}$, (5.36), and part (iii) of Lemma 5.1, we obtain

$$\lim_{i \to \infty} p_{\hat{k}+4i+2} = +\infty, \quad \lim_{i \to \infty} p_{\hat{k}+4i} = 0, \quad \lim_{i \to \infty} p_{\hat{k}+4i+1} = 1 + \tau. \tag{5.37}$$

However, Lemma 5.2 shows that the cycle (5.36) and (5.37) cannot occur. Thus (5.26) cannot hold for all k and this lemma is true.

Theorem 5.1 *Consider the sequence $\{p_k\}$ that satisfies (4.11) and $p_k \neq 1$ for all k. Assume that relation (5.19) holds. Then there exists some index $\tilde{k}$ such that (4.12)–(4.15) hold.*

Proof: By Lemma 5.3, there exists an integer $\bar{k}$ such that (5.20) or (5.21) holds. By part (iii) of Lemma (5.1), we assume without loss of generality that (5.20) holds, for otherwise we consider the sequence $\{(1 + \tau)p_k^{-1}\}$. By (5.20) and part (i) of Lemma 5.1, it is easy to see that

$$p_{\bar{k}+2} = p_{\bar{k}+1}p_{\bar{k}}^{-1}h(p_{\bar{k}}) \geq h(p_{\bar{k}}) > 1 + \tau, \tag{5.38}$$

$$p_{\bar{k}+3} = p_{\bar{k}+2}\phi(p_{\bar{k}+1}) > p_{\bar{k}+2}. \tag{5.39}$$

Notice that

$$p_{\bar{k}+4} = \phi(p_{\bar{k}+1})h(p_{\bar{k}+2}) = \phi(p_{\bar{k}+1})h(p_{\bar{k}+1}\phi(p_{\bar{k}})). \tag{5.40}$$

By viewing $p_{\bar{k}+4}$ as a function of $p_{\bar{k}}$ and $p_{\bar{k}+1}$ with $0 < p_{\bar{k}} \leq p_{\bar{k}+1} < 1$, we can check that $p_{\bar{k}+4}$ reaches its minimum when $p_{\bar{k}} = p_{\bar{k}+1} = p_1^* = 1+\tau-\sqrt{\tau(1 + \tau)}$, and that its minimum value is

$$\phi(p_1^*)h(p_1^*\phi(p_1^*)) = h(p_1^*) > 1 + \tau. \tag{5.41}$$

It follows from the definition of $p_{\bar{k}+4}$ and (5.38) that

$$p_{\bar{k}+4} = p_{\bar{k}+3}\phi(p_{\bar{k}+2}) < p_{\bar{k}+3}, \tag{5.42}$$

from which and (5.41) we get that

$$p_{\bar{k}+3} > p_{\bar{k}+4} > 1 + \tau. \tag{5.43}$$

By (5.43) and part (iii) of Lemma 5.1, we can similarly prove that

$$p_{\bar{k}+5} \in (0,1), \quad 0 < p_{\bar{k}+6} < p_{\bar{k}+7} < 1. \tag{5.44}$$

Recursively, we have for all $i \geq 1$,

$$\begin{cases} p_{\bar{k}+6i-1} \in (0,1), \quad 0 < p_{\bar{k}+6i} < p_{\bar{k}+6i+1} < 1, \\ p_{\bar{k}+6i+2} > 1+\tau, \quad p_{\bar{k}+6i+3} > p_{\bar{k}+6i+4} > 1+\tau. \end{cases} \tag{5.45}$$

By (5.45) and part (ii) of Lemma 5.1, we then obtain

$$\frac{p_{\bar{k}+6(i+1)+j}}{p_{\bar{k}+6i+j}} \begin{cases} \geq (1+\tau)^2, & \text{for } i \geq 1 \text{ and } j = 2,3; \\ \leq (1+\tau)^{-2}, & \text{for } i \geq 1 \text{ and } j = 5,6. \end{cases} \tag{5.46}$$

In the following, we will show that

$$\lim_{i \to \infty} p_{\bar{k}+6i+1} = 0. \tag{5.47}$$

To do so, we first show by contradiction that

$$\liminf_{i \to \infty} p_{\bar{k}+6i+1} = 0. \tag{5.48}$$

Assume that (5.48) is false. Then there exists some constant $c_5 > 0$ such that

$$p_{\bar{k}+6i+1} \geq c_5, \quad \text{for all } i \geq 1. \tag{5.49}$$

Noting that

$$h(p) \begin{cases} > 1+\tau, & \text{for } p \in (0,1), \\ \in (0,1), & \text{for } p > 1+\tau, \end{cases} \tag{5.50}$$

we have from this, part (ii) of Lemma 5.1, and (5.46) that

$$\limsup_{i \to \infty} \frac{p_{\hat{k}+6i+7}}{p_{\hat{k}+6i+1}} \leq \frac{1+\tau}{h(c_5)} < 1. \tag{5.51}$$

The relation (5.51) implies the truth of (5.47), which contradicts (5.49). Thus (5.48) is true. For any $\varepsilon \in (0,0.5]$, we know from (5.46) and (5.48) that there exists some integer $\hat{i}$ such that

$$p_{\bar{k}+6\hat{i}+1} \leq \varepsilon \tag{5.52}$$

and

$$\frac{h(p_{\bar{k}+6i+5})}{h(p_{\bar{k}+6i+2})} \leq h(\varepsilon), \quad \text{for all } i \geq \hat{i}. \tag{5.53}$$

By part (ii) of Lemma 5.1, (5.50), (5.53), (5.52), and $\varepsilon \leq 0.5$, it is clear that

$$p_{\bar{k}+6\hat{i}+7} \leq p_{\bar{k}+6\hat{i}+1}(h(\varepsilon)/(1+\tau)) \leq 2\varepsilon. \tag{5.54}$$

Let $\tilde{i}$ be the least integer such that $\tilde{i} \geq \hat{i}$ and

$$p_{\bar{k}+6\tilde{i}+1} \leq \varepsilon. \tag{5.55}$$

Then for any $i \in [\hat{i}+1, \tilde{i}-1]$, we have that $p_{\bar{k}+6i+1} > \varepsilon$, and this together with part (ii) of Lemma 5.1, (5.50), (5.53), and (5.54) indicates that

$$p_{\bar{k}+6i+1} \leq p_{\bar{k}+6(i-1)+1} \leq \cdots \leq p_{\bar{k}+6(\hat{i}+1)+1} \leq 2\varepsilon, \quad \text{for all } i \in [\hat{i}+1, \tilde{i}-1].$$

The above relations, (5.52), and (5.55) imply that

$$p_{\bar{k}+6i+1} \leq 2\varepsilon, \quad \text{for all } i \in [\hat{i}, \tilde{i}]. \tag{5.56}$$

Since the $\hat{i}$ in (5.52) can be arbitrarily large, we can then obtain

$$p_{\bar{k}+6i+1} \leq 2\varepsilon, \quad \text{for all } i \geq \hat{i}. \tag{5.57}$$

Thus by the definition of limit and $p_{\bar{k}+6i+1} \geq 0$, relation (5.47) holds. In a similar way, we can show that $\lim_{i\to\infty} p_{\hat{i}+6i+4} = +\infty$. By this, (5.47), and (5.46), we know that (4.12)–(4.15) hold with $\tilde{k} = \bar{k} + 1$.

Lemma 5.4 *Consider the sequence $\{p_k\}$ that satisfies (4.11). Define the function $D(z_1, z_2) = d_1(z_1)d_2(z_2)$, where*

$$d_1(z) = \begin{cases} -\dfrac{z}{\sqrt{1+\tau}}, & \text{if } z \in [-\sqrt{1+\tau}, 0), \\[2mm] -\dfrac{\sqrt{1+\tau}}{z}, & \text{if } z \in (-\infty, -\sqrt{1+\tau}) \end{cases} \tag{5.58}$$

and

$$d_2(z) = \begin{cases} \dfrac{z}{\sqrt{1+\tau}}, & \text{if } z \in (1, \sqrt{1+\tau}], \\[2mm] \dfrac{\sqrt{1+\tau}}{z}, & \text{if } z \in (\sqrt{1+\tau}, 1+\tau). \end{cases} \tag{5.59}$$

If the following relations hold for some $\hat{k}$,

$$p_{\hat{k}} < 0, \ p_{\hat{k}+1} \in (1, 1+\tau), \ p_{\hat{k}+2} < 0, \ p_{\hat{k}+3} \in (1, 1+\tau), \tag{5.60}$$

then we have

$$D(p_{\hat{k}+2}, p_{\hat{k}+3}) \geq D(p_{\hat{k}}, p_{\hat{k}+1}). \tag{5.61}$$

Proof: Noting that the following relation holds for $l = 1, 2$

$$d_l((1+\tau)p_k^{-1}) = d_l(p_k), \tag{5.62}$$

we can assume without loss of generality that

$$p_{\hat{k}} \in [-\sqrt{1+\tau}, 0). \tag{5.63}$$

Relation (5.63), the definition of $\{p_k\}$ and (i) of Lemma 5.1 indicate that

$$\begin{cases} p_{\hat{k}+3} \geq \sqrt{1+\tau}, & \text{if } p_{\hat{k}+1} \in (1, 1+\tau], \\ p_{\hat{k}+2} \leq -\sqrt{1+\tau}, & \text{if } p_{\hat{k}+1} \in [\sqrt{1+\tau}, 1+\tau). \end{cases} \tag{5.64}$$

By (5.60), (5.63), and (5.64), we can divide our proof into four cases:

(i) $p_{\hat{k}+1} \leq \sqrt{1+\tau}, \quad p_{\hat{k}+2} \leq -\sqrt{1+\tau}, \quad p_{\hat{k}+3} \geq \sqrt{1+\tau};$

(ii) $p_{\hat{k}+1} \leq \sqrt{1+\tau}, \quad p_{\hat{k}+2} \geq -\sqrt{1+\tau}, \quad p_{\hat{k}+3} \geq \sqrt{1+\tau};$

(iii) $p_{\hat{k}+1} \geq \sqrt{1+\tau}, \quad p_{\hat{k}+2} \leq -\sqrt{1+\tau}, \quad p_{\hat{k}+3} \leq \sqrt{1+\tau};$

(iv) $p_{\hat{k}+1} \geq \sqrt{1+\tau}, \quad p_{\hat{k}+2} \leq -\sqrt{1+\tau}, \quad p_{\hat{k}+3} \geq \sqrt{1+\tau}.$

For the case (i), noting that

$$p(1 + \frac{\tau}{1-p}) \leq -(1+\tau), \quad \text{if } p \in (1, \sqrt{1+\tau}), \tag{5.65}$$

we have from the definitions of D, d_1, d_2 and part (i) of Lemma 5.1 that

$$\frac{D(p_{\hat{k}+2}, p_{\hat{k}+3})}{D(p_{\hat{k}}, p_{\hat{k}+1})} = \frac{-\frac{\sqrt{1+\tau}}{p_{\hat{k}+2}} \cdot \frac{\sqrt{1+\tau}}{p_{\hat{k}+3}}}{-\frac{p_{\hat{k}}}{\sqrt{1+\tau}} \cdot \frac{p_{\hat{k}+1}}{\sqrt{1+\tau}}} = \frac{(1+\tau)^2}{p_{\hat{k}} p_{\hat{k}+1} p_{\hat{k}+2} p_{\hat{k}+3}}$$

$$= \frac{(1+\tau)^2}{p_{\hat{k}+2}(1 + \frac{\tau}{1-p_{\hat{k}}})[p_{\hat{k}+1}(1 + \frac{\tau}{1-p_{\hat{k}+1}})]}$$

$$\leq \frac{(1+\tau)^2}{\sqrt{1+\tau} \cdot \sqrt{1+\tau} \cdot 1 + \tau} = 1. \tag{5.66}$$

For the case (ii), we also have that

$$\frac{D(p_{\hat{k}+2}, p_{\hat{k}+3})}{D(p_{\hat{k}}, p_{\hat{k}+1})} = \frac{-\frac{p_{\hat{k}+2}}{\sqrt{1+\tau}} \cdot \frac{\sqrt{1+\tau}}{p_{\hat{k}+3}}}{-\frac{p_{\hat{k}}}{\sqrt{1+\tau}} \cdot \frac{p_{\hat{k}+1}}{\sqrt{1+\tau}}} = \frac{(1+\tau)p_{\hat{k}+2}}{p_{\hat{k}} p_{\hat{k}+1} p_{\hat{k}+3}}$$

$$= \frac{(1+\tau)(-p_{\hat{k}+2})}{(1 + \frac{\tau}{1-p_{\hat{k}}})[-p_{\hat{k}+1}(1 + \frac{\tau}{1-p_{\hat{k}+1}})]}$$

$$\leq \frac{(1+\tau) \cdot \sqrt{1+\tau}}{\sqrt{1+\tau} \cdot (1+\tau)} = 1. \tag{5.67}$$

The cases (iii)-(iv) can be similarly shown. So (5.61) is true.

Lemma 5.5 *Consider the sequence $\{p_k\}$ that satisfies (4.11) and $p_k \neq 0, 1$. Assume that there exists an infinitely subsequence $\{k_i\}$ such that $p_{k_i} < 0$. If*

$$(p_k, p_{k+1}) \neq (-\sqrt{1+\tau}, \sqrt{1+\tau}), \quad \text{for all } k \geq 1, \tag{5.68}$$

then there exists some index $\bar{k}$ such that at least one of the following relations holds

$$p_{\bar{k}} < 0, \quad p_{\bar{k}+1} \in (0,1), \tag{5.69}$$

$$p_{\bar{k}} < 0, \quad p_{\bar{k}+1} > 1 + \tau. \tag{5.70}$$

We proceed by contradiction and assume that neither (5.69) nor (5.70) holds for any $\bar{k}$. By the definition of $\{p_k\}$, we see that if $p_k < 0$ and $p_{k+1} < 0$, then $p_{k+2} > 0$. So there must exist some integer $\hat{k}$ such that

$$p_{\hat{k}} < 0, \quad p_{\hat{k}+1} \in (1, 1+\tau). \tag{5.71}$$

It follows that $p_{\hat{k}+2} < 0$ and $p_{\hat{k}+3} \in (1, 1+\tau)$. Recursively, we obtain

$$p_{\hat{k}+2i} < 0, \quad p_{\hat{k}+2i+1} \in (1, 1+\tau), \quad \text{for all } i \geq 0. \tag{5.72}$$

Define d_1, d_2, and D as in Lemma 5.4. Then by Lemma 5.4 and (5.72), we know that $D(p_{\hat{k}+2i}, p_{\hat{k}+2i+1})$ is monotonically decreasing with i. If (5.68) holds, one can strengthen the analyses in Lemma 5.4 and obtain a constant $c_6 \in (0,1)$ which depends only on $p_{\hat{k}}$ and $p_{\hat{k}+1}$ such that

$$D(p_{\hat{k}+2i+2}, p_{\hat{k}+2i+3}) \geq c_6 D(p_{\hat{k}+2i}, p_{\hat{k}+2i+1}), \quad \text{for all } i \geq 0. \tag{5.73}$$

Since $D(p_{\hat{k}+2i}, p_{\hat{k}+2i+1}) < 0$ for all i, we then have that

$$\lim_{i \to \infty} D(p_{\hat{k}+2i}, p_{\hat{k}+2i+1}) = 0. \tag{5.74}$$

It follows by (5.58) and (5.72) that

$$d_2(p_{\hat{k}+2i+1}) \geq (1+\tau)^{-\frac{1}{2}}, \tag{5.75}$$

and from this, (5.74), and the definition of D, we get

$$\lim_{i \to \infty} d_1(p_{\hat{k}+2i}) = 0. \tag{5.76}$$

On the other hand, the definition of $p_{\hat{k}+2i+2}$ and (5.72) imply that

$$p_{\hat{k}+2i} p_{\hat{k}+2i+2} = p_{\hat{k}+2i+1} h(p_{\hat{k}+2i}) \in (1, (1+\tau)^2). \tag{5.77}$$

By (5.76), (5.77), and (5.58), we know that there exists some integer, which we continue to denote by $\hat{k}$, such that

$$\lim_{i\to\infty} p_{\hat{k}+4i} = -\infty, \quad \lim_{i\to\infty} p_{\hat{k}+4i+2} = 0. \tag{5.78}$$

Further, from (5.78), (5.72), and the definition of $\{p_k\}$, we can obtain

$$\lim_{i\to\infty} p_{\hat{k}+4i+1} = 1, \quad \lim_{i\to\infty} p_{\hat{k}+4i+3} = 1 + \tau. \tag{5.79}$$

However, Lemma 5.2 shows that the cycle (5.78) and (5.79) cannot occur. The contradiction shows the truth of this Lemma.

Theorem 5.2 *Consider the sequence $\{p_k\}$ that satisfies (4.11) and $p_k \neq 0, 1$ for all k. Assume that there exists an infinite subsequence $\{k_i\}$ such that $p_{k_i} < 0$ and that (5.68) holds. Then there exists some integer $\tilde{k}$ such that relations (4.12)–(4.15) hold.*

Proof: By Lemma 5.5, there exists an index $\bar{k}$ such that (5.69) or (5.70) holds. By part (iii) of Lemma 5.1, we assume without loss of generality that (5.69) holds. Then it follows by the definition of $\{p_k\}$ that

$$p_{\bar{k}+2} < 0, \quad p_{\bar{k}+3} < -(1+\tau), \quad p_{\bar{k}+4} > \phi(p_1^*), \quad p_{\bar{k}+5} < 0, \tag{5.80}$$

where $p_1^* = 1 + \tau - \sqrt{\tau(1+\tau)}$ is same as the one in (5.41). Note that

$$h(p) \in (1, 1+\tau), \quad \text{for } p < 0. \tag{5.81}$$

By this relation, (5.80), and part (ii) of Lemma 5.1, we can get

$$\frac{p_{\bar{k}+6}}{p_{\bar{k}}} = \frac{h(p_{\bar{k}+3})}{h(p_{\bar{k}})} \frac{h(p_{\bar{k}+4})}{h(p_{\bar{k}+1})} < \frac{1+\tau}{1} \cdot \frac{h(\phi(p_1^*))}{1+\tau} = h(\phi(p_1^*)), \tag{5.82}$$

$$\frac{p_{\bar{k}+7}}{p_{\bar{k}+1}} = \frac{h(p_{\bar{k}+4})}{h(p_{\bar{k}+1})} \frac{h(p_{\bar{k}+5})}{h(p_{\bar{k}+2})} < \frac{1}{1+\tau} \cdot \frac{1+\tau}{1} = 1. \tag{5.83}$$

Thus we also have

$$p_{\bar{k}+6} < 0, \quad p_{\bar{k}+7} \in (0, 1). \tag{5.84}$$

The recursion of the above procedure yields for all $i \geq 1$,

$$p_{\bar{k}+6i} < 0, \quad p_{\bar{k}+6i+1} \in (0, 1) \tag{5.85}$$

and

$$\frac{p_{\bar{k}+6(i+1)}}{p_{\bar{k}+6i}} < h(\phi(p_1^*)), \quad \frac{p_{\bar{k}+6(i+1)+1}}{p_{\bar{k}+6i+1}} < 1. \tag{5.86}$$

Relation (5.86) and the fact that $h(\phi(p_1^*)) < 1$ indicate that

$$\lim_{i \to \infty} p_{\bar{k}+6i} = 0. \tag{5.87}$$

Similar to (5.86), we can establish the relation

$$\frac{p_{\bar{k}+6(i+1)+3}}{p_{\bar{k}+6i+3}} > (h(\phi(p_1^*)))^{-1}, \quad \frac{p_{\bar{k}+6(i+1)+4}}{p_{\bar{k}+6i+4}} > 1. \tag{5.88}$$

This together with $p_{\bar{k}+6i+3} < 0$ and $h(\phi(p_1^*)) < 1$ implies that

$$\lim_{i \to \infty} p_{\bar{k}+6i+3} = -\infty. \tag{5.89}$$

We shall now prove that

$$\lim_{i \to \infty} p_{\bar{k}+6i+2} = -\infty. \tag{5.90}$$

In fact, by part (ii) of Lemma 5.1, (5.87), (5.89) and (5.81), we can get

$$\liminf_{i \to \infty} \frac{p_{\bar{k}+6i+8}}{p_{\bar{k}+6i+2}} \geq 1. \tag{5.91}$$

In a similar way, we can show

$$\limsup_{i \to \infty} \frac{p_{\bar{k}+6i+11}}{p_{\bar{k}+6i+5}} \leq 1. \tag{5.92}$$

Relations (5.86), (5.88), (5.91) and (5.92) imply that there exists some integer $\hat{i}$ such that the following relations hold for all $i \geq \hat{i}$,

$$p_{\bar{k}+6i+3} < p_{\bar{k}+6i+5} < 0, \quad p_{\bar{k}+6i+2} < p_{\bar{k}+6i+6} < 0. \tag{5.93}$$

By part (ii) of Lemma 5.1 and (5.93), we can obtain that

$$p_{\bar{k}+6i+8} < p_{\bar{k}+6i+2}, \quad \text{for all } i \geq \hat{i}. \tag{5.94}$$

Similar to (5.94), we can verify that

$$p_{\bar{k}+6i+11} > p_{\bar{k}+6\hat{i}+5}, \quad \text{for all } i \geq \hat{i}. \tag{5.95}$$

Thus by part (ii) of Lemma 5.1, (5.87), (5.89), (5.94) and (5.95), it follows that

$$\liminf_{i \to \infty} \frac{p_{\bar{k}+6i+8}}{p_{\bar{k}+6i+2}} \geq \frac{h(p_{\bar{k}+6\hat{i}+5})}{h(p_{\bar{k}+6\hat{i}+2})} \cdot \frac{1+\tau}{1} > 1, \tag{5.96}$$

and this together with $p_{\bar{k}+6i+2} < 0$ implies that the truth of (5.90). We now prove that

$$\lim_{i \to \infty} p_{\bar{k}+6i+1} = 0. \tag{5.97}$$

In fact, by (5.86), we know that $p_{\bar{k}+6i+1}$ is monotonically decreasing and hence its limit exists. Assume that

$$\lim_{i\to\infty} p_{\bar{k}+6i+1} = c_7 > 0. \tag{5.98}$$

Then from this and part (ii) of Lemma (5.1) we can get

$$\frac{p_{\bar{k}+6i+7}}{p_{\bar{k}+6i+1}} \leq \frac{h(p_{\bar{k}+6i+4})}{h(p_{\bar{k}+6i+1})} \frac{h(p_{\bar{k}+6i+5})}{h(p_{\bar{k}+6i+2})} \leq \frac{1+\tau}{h(c_7)} < 1, \tag{5.99}$$

which implies the truth of (5.97). However this contradicts (5.98). Thus (5.97) is true. Similar to (5.90) and (5.97), we can show

$$\lim_{i\to\infty} p_{\bar{k}+6i+5} = 0, \quad \lim_{i\to\infty} p_{\bar{i}+6i+4} = +\infty. \tag{5.100}$$

By part (ii) of Lemma 5.1 and relations (5.87), (5.89), (5.90), (5.97), and (5.100), we know that (4.12)–(4.15) hold with $\tilde{k} = \bar{k}$.

6 CONCLUDING REMARKS

In this paper, we have given an analysis of the Barzilai and Borwein gradient method for unsymmetric linear equations, assuming that the dimension $n = 2$. Under mild assumptions, we have proved that the convergence rate of the Barzilai and Borwein gradient method is Q-superlinear if the coefficient matrix A has two same eigenvalues; if the eigenvalues of A are different, then the method converges for almost all starting points x_1 and x_2 and the convergence rate is R-superlinear. These results strongly depend on the study of the two nonlinear recurrence relations, namely (2.8) and (4.11), making the analyses difficult.

From the relations (2.13) and (4.25), we can see that the convergence of the Barzilai and Borwein gradient is related to the symmetric degree of the coefficient matrix A. If A is close to a symmetric matrix, then the method converges rapidly; conversely, if the matrix A is markedly unsymmetric, then the method will converge slowly. The convergence of the Barzilai and Borwein gradient method for unsymmetric linear equations is slower than that for symmetric linear equations. In the symmetric case, the Barzilai and Borwein gradient method can give the solution in at most two steps if A has two same eigenvalues; if the eigenvalues of A are different, we know from Barzilai & Borwein (1988) that the R-superliner convergence order of the method is $\sqrt{2}-\varepsilon$, where $\varepsilon > 0$ is any small number. In the unsymmetric case, however, relation

(4.25) indicates that the R-superlinear order of the method for unsymmetric linear equations is only 1 if A has two different eigenvalues. Thus to accelerate the Barzilai and Borwein gradient method for unsymmetric linear equations, it may be worthwhile studying how to make a transformation to an unsymmetric matrix that improves its symmetric degree.

This paper have made some efforts on directly extending the convergence result in Barzilai & Borwein (1988) of the Barzilai and Borwein gradient method to the unsymmetric linear equations. As pointed out by the referee, another possibility is to apply the method to the symmetric (least square) system

$$A^T A x = A^T b, \tag{6.1}$$

that is equivalent to (1.6) in theory if A is nonsingular. In this case, since $A^T A$ is symmetric and positive definite, we know by Barzilai & Borwein (1988) and Dai & Liao (2002) that the Barzilai and Borwein gradient method for (6.1) is R-linearly convergent and if $n = 2$, the convergence rate is R-superlinear. In practical computations, transforming (1.6) into (6.1) has two main disadvantages; namely the condition number of the problem will be squared and the multiplications of A transpose with vectors will be unavoidable. It is not known yet which is the better approach in extending the Barzilai and Borwein gradient method for solving unsymmetric linear equations.

Acknowledgments

This research was supported by the Chinese NSF grant (no. 19801033 and 10171104) and Hong Kong Research Grants Council (no. CUHK 4392/99E). The authors would like to thank the anonymous referee for his many valuable comments on this paper.

References

H. Akaike (1959), On a successive transformation of probability distribution and its application to the analysis of the optimum gradient method, *Ann. Inst. Statist. Math. Tokyo*, Vol. 11, pp. 1-17.

J. Barzilai and J. M. Borwein (1988), Two-point step size gradient methods, *IMA J. Numer. Anal.*, Vol. 8, pp. 141-148.

E. G. Birgin, I. Chambouleyron, and J. M. Martínez (1999), Estimation of the optical constants and the thickness of thin films using unconstrained optimization, *J. Comput. Phys.*, Vol. 151, pp. 862-880.

E. G. Birgin and Y. G. Evtushenko (1998), Automatic differentiation and spectral projected gradient methods for optimal control problems, *Optim. Methods Softw.*, Vol. 10, pp. 125-146.

E. G. Birgin, J. M. Martínez, and M. Raydan (2000), Nonmonotone spectral projected gradient methods on convex sets, *SIAM J. Optim.*, Vol. 10, pp. 1196-1211.

A. Cauchy (1847), Méthode générale pour la résolution des systèmes d'équations simultanées, *Comp. Rend. Acad. Sci. Paris*, Vol. 25, pp. 46-89.

Y. H. Dai and L.-Z. Liao (2002), R-linear convergence of the Barzilai and Borwein gradient method, *IMA J. Numer. Anal.*, Vol. 22, No. 1, pp. 1-10.

R. Fletcher (1990), Low storage methods for unconstrained optimization, *Lectures in Applied Mathematics (AMS)*, Vol. 26, pp. 165-179.

A. Friedlander, J. M. Martínez, B. Molina, and M. Raydan (1999), Gradient method with retards and generalizations, *SIAM J. Numer. Anal.*, Vol. 36, 275-289.

W. Glunt, T. L. Hayden, and M. Raydan (1993), Molecular conformations from distance matrices, *J. Comput. Chem.*, Vol. 14, pp. 114-120.

L. Grippo, F. Lampariello, and S. Lucidi (1986), A nonmonotone line search technique for Newton's method, *SIAM J. Numer. Anal.*, Vol. 23, pp. 707-716.

W. B. Liu and Y. H. Dai (2001), Minimization Algorithms based on Supervisor and Searcher Cooperation, *Journal of Optimization Theory and Applications*, Vol. 111, No. 2, pp. 359-379.

M. Raydan (1993), On the Barzilai and Borwein choice of steplength for the gradient method, *IMA J. Numer. Anal.*, Vol. 13, pp. 321-326.

M. Raydan (1997), The Barzilai and Borwein gradient method for the large scale unconstrained minimization problem, *SIAM J. Optim.*, Vol. 7, pp. 26-33.

9   AN EXCHANGE ALGORITHM FOR MINIMIZING SUM-MIN FUNCTIONS

Alexei V. Demyanov

Faculty of Mathematics and Mechanics
St.Petersburg State University
Staryi Peterhof, St.Petersburg
198904 RUSSIA

Abstract: The problem of minimizing maxmin-type functions or the sum of minimum-type functions appears to be increasingly interesting from both theoretical and practical considerations. Such functions are essentially nonsmooth and, in general, they can successfully be treated by existing tools of Nonsmooth Analysis. In some cases the problem of finding a minimizer of such a function can be reduced to solving some mixed combinatorial–continuous problem.

In the present paper the problem of minimizing the sum of minima of a finite number of functions is discussed. It is shown that this problem is equivalent to solving a finite (though may be quite large) number of simpler (and sometimes quite trivial) optimization problems. Necessary conditions for global a minimum and sufficient conditions for a local minimum are stated. An algorithm for finding a local minimizer (the so-called exchange algorithm) is proposed. It converges to a local minimizer in a finite number of steps. A more general algorithm (called ε–exchange algorithm) is described which allows one to escape from a local minimum point. Numerical examples demonstrate the algorithm.

Key words: Sum-min function, necessary conditions for a minimum, sufficient conditions, exchange algorithm, ε–exchange algorithm.

1 INTRODUCTION

Let functions $\varphi_{ij}(x) : R^n \to R$ be given, where $i \in I := 1 : m, j \in J_i := 1 : N_i$. Construct the functions

$$\varphi_i(x) = \min_{j \in J_i} \varphi_{ij}(x) \quad \forall i \in 1 : m \tag{1.1}$$

and the function

$$F(x) = \sum_{i \in I} \varphi_i(x). \tag{1.2}$$

The function F defined by (1.2) is called a sum-min function. Many practical problems arising, e.g., in Mathematical Diagnostics, Data Mining, Clustering, Network Allocation etc. (see Mangasarian (1997), Rao (1971), Bagirov, Rubinov (1999), Bagirov, Rubinov, Yearwood (2001), Rubinov (2000)) can be described by mathematical models where it is required to find a minimizer of F. The problem of minimizing the function F is, first of all, generally speaking, nonsmooth (even if all the functions φ_{ij}'s are quite good) and "extremely" multiextremal. The nonsmoothness can be treated by the existing tools of Nonsmooth Analysis and Nondifferentiable Optimization (for example, local properties of F can be studied, under some conditions, imposed on φ_{ij}, by considering the directional derivatives of F). But the multiextremality remains the main unavoidable obstacle.

In the paper we consider the problem of minimizing one class of sum-min functions (the so-called separable-like sum-min functions). It is known that this problem is equivalent to solving a finite number of "simpler" problems. Namely, it is shown that

$$\inf_{x \in R^n} F(x) = \min_{j \in J} \inf_{x \in R^n} F_j(x) \tag{1.3}$$

where

$$J = \{j = (j_1, ..., j_m) \mid j_i \in J_i \quad \forall i \in J\},$$

$$F_j(x) = \sum_{i \in I} \varphi_{ij_i}(x).$$

A similar result for the best piece-wise polynomial approximation problem was proved in Vershik, Malozemov, Pevnyi (1975).

For every $j \in J$ the function $F_j(x)$ is called an elementary function, and it is assumed that one is able to find a minimizer of $F_j(x)$ (exactly or approximately). The problem of minimizing $F_j(x)$ will also be referred to as an elementary problem. In some cases the problem of minimizing F_j is quite simple (see Examples in Section 7).

Theoretically, it is possible to find a global minimizer of the function F by solving all elementary problems, however, the number of such elementary problems is too large, therefore it is important to be able to find a proper local minimizer at a reasonable computational price.

The paper is organized as follows. In Section 2 a special case of the above problem is stated (namely, the case $m = 2$ and $\varphi_{ij}(x) = \varphi_{ij}(x_i)$, where $x = (x_1, ..., x_m)$, $x_i \in R^{n_i}$, $\sum_{i\in I} n_i = n$). A result analogous to (1.3) is formulated in Section 3. Necessary conditions for a point to be a global minimizer and sufficient conditions for a point to be a local minimizer is proved in Section 4. An algorithm for finding a local minimizer is described in Section 5. This algorithm (called an exchange algorithm) is based on the necessary minimality conditions. In a finite number of steps a local minimizer is constructed. In Section 6 a modification of the exchange algorithm (ε–exchange algorithm) is proposed to "escape" from a local minimum point. Illustrative numerical examples are described in Section 7.

2   STATEMENT OF THE PROBLEM

Let a set of points $\Omega = \{t_1, \ldots, t_N\} \subset Y$ (where Y is some space) and functions $\varphi_i : \Omega \times R^{n_i} \to R$, $i \in 1 : m$, be given. Put $x = (x_1, \ldots, x_m) \in S := R^{n_1} \times \ldots \times R^{n_m}$.

Let

$$\varphi(t, x) = \min_{i\in 1:m} \varphi_i(t, x_i)$$

and

$$F(x) = \sum_{t\in\Omega} \varphi(t, x) = \sum_{t\in\Omega} \min_{i\in 1:m} \varphi_i(t, x_i).$$

Problem P: Find a point $x^* = (x_1^*, \ldots, x_m^*) \in S$, such that

$$F(x^*) = \min_{x\in S} F(x).$$

In the paper the case $m = 2$ is described in detail. Note that no specific requirements on φ_i are imposed (the functions φ_i's are not even assumed to be continuous).

Thus, we consider the case $m = 2$, $S = R^{n_1} \times R^{n_2}$. Then

$$F(x) = \sum_{t\in\Omega} \min\left\{\varphi_1(t, x_1), \varphi_2(t, x_2)\right\}.$$

For every $\gamma_1 \subset \Omega$ and $\gamma_2 \subset \Omega$ put

$$c_1(x_1, \gamma_1) = \sum_{t \in \gamma_1} \varphi_1(t, x_1), \quad c_2(x_2, \gamma_2) = \sum_{t \in \gamma_2} \varphi_2(t, x_2). \qquad (2.1)$$

If $\gamma = \emptyset$, then by definition $c_1(x_1, \emptyset) = c_2(x_2, \emptyset) = 0$.

Let us introduce the set

$$T(\Omega) = \left\{ \Gamma = (\gamma_1, \gamma_2) \middle| \gamma_1 \subset \Omega, \gamma_2 \subset \Omega, \ \gamma_1 \cap \gamma_2 = \emptyset, \ \gamma_1 \cup \gamma_2 = \Omega \right\} \subset 2^\Omega \times 2^\Omega.$$

For any $\Gamma = (\gamma_1, \gamma_2) \in T(\Omega)$ let us consider the function

$$F_\Gamma(x) = c_1(x_1, \gamma_1) + c_2(x_2, \gamma_2).$$

It is clear, that for every $\Gamma \in T(\Omega)$

$$F(x) \le F_\Gamma(x) \quad \forall x \in S. \qquad (2.2)$$

For each fixed point $x \in S$ let us introduce the sets

$$\widehat{\sigma}_1(x) = \left\{ t \in \Omega \middle| \varphi_1(t, x_1) < \varphi_2(t, x_2) \right\},$$

$$\widehat{\sigma}_2(x) = \left\{ t \in \Omega \middle| \varphi_1(t, x_1) > \varphi_2(t, x_2) \right\},$$

$$\Sigma(x) = \left\{ t \in \Omega \middle| \varphi_1(t, x_1) = \varphi_2(t, x_2) \right\}.$$

The set $\Sigma(x)$ is called the set of "common points".

Let sets $\widetilde{\sigma}_1, \widetilde{\sigma}_2$ be such that

$$\widetilde{\sigma}_1 \subset \Sigma(x), \ \widetilde{\sigma}_2 \subset \Sigma(x), \ \widetilde{\sigma}_1 \cap \widetilde{\sigma}_2 = \emptyset, \ \widetilde{\sigma}_1 \cup \widetilde{\sigma}_2 = \Sigma(x).$$

Put

$$\sigma_1 = \widetilde{\sigma}_1 \cup \widehat{\sigma}_1(x), \quad \sigma_2 = \widetilde{\sigma}_2 \cup \widehat{\sigma}_2(x). \qquad (2.3)$$

Clearly, $\sigma_1 \cup \sigma_2 = \Omega$, $\sigma_1 \cap \sigma_2 = \emptyset$. Hence, any distribution $(\widetilde{\sigma}_1, \widetilde{\sigma}_2)$ of the common points generates the corresponding disjoint partition (σ_1, σ_2) of the set Ω.

Definition 2.1 *Every such a partition (it depends upon x and, evidently, is not unique) is called an x-proper partition of the set Ω.*

Let $T(\Omega, x)$ denote the family of all x-proper partitions of the set Ω. Clearly,

$$T(\Omega, x) \subset T(\Omega). \tag{2.4}$$

Besides, it is easy to see that for any $\Gamma = (\gamma_1, \gamma_2) \in T(\Omega, x)$ the relation

$$F_\Gamma(x) = F(x) \tag{2.5}$$

holds.

For $\Gamma = (\gamma_1, \gamma_2) \in T(\Omega)$ let us introduce the function

$$\Phi(\Gamma) = \inf_{x \in S} F_\Gamma(x) = \inf_{x_1 \in R^{n_1}} c_1(x_1, \gamma_1) + \inf_{x_2 \in R^{n_2}} c_2(x_2, \gamma_2).$$

Now it is possible to formulate the following

Problem P1: Find $\Gamma^* \in T(\Omega)$ such that

$$\Phi(\Gamma^*) = \min_{\Gamma \in T(\Omega)} \Phi(\Gamma).$$

3   EQUIVALENCE OF THE TWO PROBLEMS

It will now be shown that the problem P is equivalent to the problem P1, that is the problem of minimizing the function $F(x)$ on S is equivalent to the problem of minimizing the function $\Phi(\Gamma)$ on the set $T(\Omega)$.

Let us suppose that for any $\sigma_1 \subset \Omega$ and $\sigma_2 \subset \Omega$ the functions $c_1(x_1, \sigma_1)$ and $c_2(x_2, \sigma_2)$ attain their minimal values on R^{n_1} and R^{n_2}, respectively.

Theorem 3.1 *The following equality holds:*

$$\inf_{x \in S} F(x) = \min_{\Gamma \in T(\Omega)} \Phi(\Gamma). \tag{3.1}$$

Proof: It is clear from (2.2) that for any $\Gamma = (\gamma_1, \gamma_2) \in T(\Omega)$

$$\inf_{x \in S} F(x) \leq \inf_{x \in S} F_\Gamma(x) = \inf_{x_1 \in R^{n_1}} c_1(x_1, \gamma_1) + \inf_{x_2 \in R^{n_2}} c_2(x_2, \gamma_2) = \Phi(\Gamma).$$

Hence,

$$\inf_{x \in S} F(x) \leq \min_{\Gamma \in T(\Omega)} \Phi(\Gamma). \tag{3.2}$$

Let us take an arbitrary $\overline{x} = (\overline{x}_1, \overline{x}_2) \in S$. For every $\Gamma_0 = (\sigma_1, \sigma_2) \in T(\Omega, \overline{x})$, the conditions (2.4) and (2.5) yield the relations

$$F(\overline{x}) = c_1(\overline{x}_1, \sigma_1) + c_2(\overline{x}_2, \sigma_2) \geq$$

$$\geq \inf_{x_1 \in R^{n_1}} c_1(x_1, \sigma_1) + \inf_{x_2 \in R^{n_2}} c_2(x_2, \sigma_2) = \Phi(\Gamma_0) \geq \min_{\Gamma \in T(\Omega)} \Phi(\Gamma).$$

Since an arbitrary $\overline{x} \in S$ was chosen,

$$\inf_{x \in S} F(x) \geq \min_{\Gamma \in T(\Omega)} \Phi(\Gamma). \tag{3.3}$$

The inequalities (3.2) and (3.3) imply (3.1). $\triangle$

Take any $\Gamma = (\sigma_1, \sigma_2) \in T(\Omega)$ and find $x_1(\sigma_1) \in R^{n_1}$, $x_2(\sigma_2) \in R^{n_2}$ such that

$$\min_{x_1 \in R^{n_1}} c_1(x_1, \sigma_1) = c_1(x_1(\sigma_1), \sigma_1), \tag{3.4}$$

$$\min_{x_2 \in R^{n_2}} c_2(x_2, \sigma_2) = c_2(x_2(\sigma_2), \sigma_2). \tag{3.5}$$

The point $x(\Gamma) = (x_1(\sigma_1), x_2(\sigma_2))$ is not unique, if the minima in (3.4) or (3.5) are attained at more than one point.

Remark 3.1 *Theorem 3.1 implies that the problem of minimizing the function F on S is reduced to the problem of solving a finite number (precisely, $|T(\Omega)|$) of problems of minimizing the functions of the form*

$$F_\Gamma(x) = c_1(x_1, \sigma_1) + c_2(x_2, \sigma_2),$$

where $\Gamma = (\sigma_1, \sigma_2) \in T(\Omega)$. However, since $|T(\Omega)| = 2^{|\Omega|}$, then a quite large number of points in the set Ω can annihilate the practical value of this fact.

Here, as usual, $|A|$ stands for the number of points in a set A.

Thus, the problem of minimizing F on Ω becomes a combinatorial one (if the problem of minimizing $F_\Gamma(x)$ is assumed to be an elementary one).

4   MINIMALITY CONDITIONS

Theorem 4.1 *For a point $x^* \in S$ to be a global minimizer of the function F it is necessary that for any x^*-proper partition $\Gamma = (\sigma_1, \sigma_2) \in T(\Omega, x^*)$ the following conditions hold:*

$$i) \; c_1(x_1^*, \sigma_1) = \min_{x_1 \in R^{n_1}} c_1(x_1, \sigma_1), \tag{4.1}$$

$$ii) \; c_2(x_2^*, \sigma_2) = \min_{x_2 \in R^{n_2}} c_2(x_2, \sigma_2). \tag{4.2}$$

If, in addition, the functions φ_i's, $i \in 1 : 2$, are continuous (respectively, in x_i), then the above conditions are sufficient conditions for a point x^* to be a local minimizer.

Proof: Necessity Let x^* be a global minimizer of the function F. Assume that the theorem is not valid. Let, for example, the condition (4.1) don't hold. Then there exists a point $\overline{x}_1 \in R^{n_1}$ such that

$$c_1(x_1^*, \sigma_1) > c_1(\overline{x}_1, \sigma_1). \tag{4.3}$$

Then for the point $\overline{x} = (\overline{x}_1, x_2^*) \in S$ by (4.3) and (2.2) one gets

$$F(x^*) = c_1(x_1^*, \sigma_1) + c_2(x_2^*, \sigma_2) > c_1(\overline{x}_1, \sigma_1) + c_2(x_2^*, \sigma_2) \geq F(\overline{x}),$$

which contradicts the fact, that the point x^* is a global minimizer of F.
Sufficiency Let the functions φ_i's be continuous in x_i and the conditions (4.1), (4.2) hold. Then the function

$$\varphi(t, x) = \min\left\{\varphi_1(t, x_1), \varphi_2(t, x_2)\right\}$$

is also continuous. The continuity of the function φ with respect to $x \in S$ implies that

- for every $t_i \in \widehat{\sigma}_1(x^*)$ there exists $\delta_i > 0$ such that $t_i \in \widehat{\sigma}_1(x)$ for any $x \in B(x^*, \delta_i)$,

- for every $t_j \in \widehat{\sigma}_2(x^*)$ there exists $\varepsilon_j > 0$ such that $t_j \in \widehat{\sigma}_2(x)$ for any $x \in B(x^*, \varepsilon_j)$.

Here $B(x^*, \delta) = \{x \in S \mid \|x - x^*\| \leq \delta\}$.

Put

$$\delta = \min\{\delta_i, \varepsilon_j | t_i \in \widehat{\sigma}_1(x^*), \; t_j \in \widehat{\sigma}_2(x^*)\}.$$

Thus, for all $\overline{x} \in B(x^*, \delta)$ we have

$$\widehat{\sigma}_1(x^*) \subset \widehat{\sigma}_1(\overline{x}), \quad \widehat{\sigma}_2(x^*) \subset \widehat{\sigma}_2(\overline{x}). \tag{4.4}$$

Let us consider the sets

$$A_1 = \widehat{\sigma}_1(\overline{x}) \setminus \widehat{\sigma}_1(x^*), \quad A_2 = \widehat{\sigma}_2(\overline{x}) \setminus \widehat{\sigma}_2(x^*).$$

Now we will prove the inclusions $A_1 \subset \Sigma(x^*), \quad A_2 \subset \Sigma(x^*)$.

Let $t \in A_1$, then, since

$$\Omega = \widehat{\sigma}_1(x^*) \cup \widehat{\sigma}_2(x^*) \cup \Sigma(x^*),$$

either $t \in \widehat{\sigma}_2(x^*)$, or $t \in \Sigma(x^*)$. However, if $t \in \widehat{\sigma}_2(x^*) \subset \widehat{\sigma}_2(\overline{x})$, then we have

$$t \in \widehat{\sigma}_1(\overline{x}) \cap \widehat{\sigma}_2(\overline{x}),$$

which contradicts the fact that $\widehat{\sigma}_1(\overline{x}) \cap \widehat{\sigma}_2(\overline{x}) = \emptyset$. So, $t \in \Sigma(x^*)$, and, hence, $A_1 \subset \Sigma(x^*)$. Similarly we get $A_2 \subset \Sigma(x^*)$.

Now let us take an arbitrary disjoint partition $(\widetilde{\sigma}_1, \widetilde{\sigma}_2)$ of the set $\Sigma(\overline{x})$. Since for the corresponding $\overline{x}$-proper partition $(\sigma_1, \sigma_2) \in T(\Omega, \overline{x})$ we obtain

$$\sigma_1 = \widehat{\sigma}_1(\overline{x}) \cup \widetilde{\sigma}_1 = \widehat{\sigma}_1(x^*) \cup A_1 \cup \widetilde{\sigma}_1, \tag{4.5}$$

$$\sigma_2 = \widehat{\sigma}_2(\overline{x}) \cup \widetilde{\sigma}_2 = \widehat{\sigma}_2(x^*) \cup A_2 \cup \widetilde{\sigma}_2. \tag{4.6}$$

It is easy to see, that from (4.4) it follows, that $\widetilde{\sigma}_1 \subset \Sigma(\overline{x}) \subset \Sigma(x^*)$, and $\widetilde{\sigma}_2 \subset \Sigma(\overline{x}) \subset \Sigma(x^*)$, therefore

$$A_1 \cup \widetilde{\sigma}_1 \subset \Sigma(x^*), \quad A_2 \cup \widetilde{\sigma}_2 \subset \Sigma(x^*),$$

and $(A_1 \cup \widetilde{\sigma}_1, A_2 \cup \widetilde{\sigma}_2)$ is a disjoint partition of the set $\Sigma(x^*)$.

So, the equalities (4.5) and (4.6) ensure us, that (σ_1, σ_2) is an x^*-proper partition of the set Ω, and the hypotheses of the theorem are satisfied. Hence, the condition (4.1) implies that

$$c_1(x_1^*, \sigma_1) \le c_1(\overline{x}_1, \sigma_1), \tag{4.7}$$

and the condition (4.2) implies that

$$c_2(x_2^*, \sigma_2) \le c_2(\overline{x}_2, \sigma_2). \tag{4.8}$$

It follows from (4.7) and (4.8) that x^* is a local minimizer of the function F.

$$\triangle$$

Definition 4.1 *A point $x^* \in S$ satisfying the conditions (4.1) and (4.2) is called a stationary point.*

If φ_i are continuous, a stationary point is a local minimizer. The opposite is not true: not every local minimizer is a stationary point.

Remark 4.1 *The notion of stationary point is closely related to the necessary condition used. Since the conditions (4.1) and (4.2) are necessary conditions for a global minimum then it is natural to expect that not every local minimizer is a stationary point. Note, that conditions (4.1) and (4.2) are of nonlocal nature.*

5 AN EXCHANGE ALGORITHM

Let us suppose that for every $\Gamma = (\sigma_1, \sigma_2) \in T(\Omega)$ infima

$$\inf_{x_i \in R^{n_i}} c_i(x_i, \sigma_i), \ i = 1, 2,$$

are attained, that is there exists a point $x(\Gamma) = (x_1(\sigma_1), x_2(\sigma_2)) \in S$, such that

$$c_i(x_i(\sigma_i), \sigma_i) = \min_{x_i \in R^{n_i}} c_i(x_i, \sigma_i), \ i = 1, 2. \tag{5.1}$$

The following algorithm allows one to find a stationary point of the function F (and if φ_i, $i = 1, 2$, are continuous, then the resulting point will be a local minimizer of F).

1. Take an arbitrary $x^0 = (x_1^0, x_2^0) \in S$.

2. Let $x^k = (x_1^k, x_2^k) \in S$ have already been found. Construct the sets $\widehat{\sigma}_1(x^k)$, $\widehat{\sigma}_2(x^k)$ and $\Sigma(x^k)$.

3. Check the conditions (4.1) and (4.2) for all $\Gamma = (\sigma_1, \sigma_2) \in T(\Omega, x^k)$.

4. If the conditions (4.1) and (4.2) are satisfied for all $\Gamma \in T(\Omega, x^k)$ then the point x^k is stationary, and the process terminates.

5. Otherwise, find any $\Gamma_k = (\sigma_1^k, \sigma_2^k) \in T(\Omega, x^k)$ for which one of the conditions (4.1), (4.2) is violated.

6. ■ If the condition (4.1) holds, then put $\overline{x}_1^k = x_1^k$.

   ■ If not, then find $\overline{x}_1^k \in R^{n_1}$ such that

$$\min_{x_1 \in R^{n_1}} c_1(x_1, \sigma_1^k) = c_1(\overline{x}_1^k, \sigma_1^k) < c_1(x_1^k, \sigma_1^k).$$

- ■ If the condition (4.2) holds, then put $\overline{x}_2^k = x_2^k$.

- ■ If not, then find $\overline{x}_2^k \in R^{n_2}$ such that

$$\min_{x_2 \in R^{n_2}} c_2(x_2, \sigma_2^k) = c_2(\overline{x}_2^k, \sigma_2^k) < c_2(x_2^k, \sigma_2^k).$$

7. Put $x^{k+1} = (\overline{x}_1^k, \overline{x}_2^k)$. Go to step 2.

Clearly,

$$F(x^{k+1}) < F(x^k). \tag{5.2}$$

As a result, a finite sequence $\{x^k\}$ is constructed such that condition (5.2) holds. Since every x-proper partition (σ_1^k, σ_2^k) may occur only once (due to (5.2)), then, taking into account the fact that $|T(\Omega)|$ is finite, one concludes that the algorithm converges to a stationary point in a finite number of steps.

Remark 5.1 *The algorithm described above may require at some steps complete enumeration of the set $|T(\Omega, x^k)| = 2^{|\Sigma(x^k)|}$. So, in practice the algorithm is effective if $|\Sigma(x^k)|$ is not very large. Theoretically, the case of complete enumeration of $T(\Omega)$ is possible (as for every algorithm of discrete mathematics).*

6 AN ε-EXCHANGE ALGORITHM

In this section it is assumed that φ_i's are continuous. For every fixed point $x \in S$ and $\varepsilon > 0$ let us introduce the sets

$$\widehat{\sigma}_{\varepsilon 1}(x) = \left\{ t \in \Omega \,\middle|\, \varphi_1(t, x_1) < \varphi_2(t, x_2) - \varepsilon \right\},$$

$$\widehat{\sigma}_{\varepsilon 2}(x) = \left\{ t \in \Omega \,\middle|\, \varphi_1(t, x_1) - \varepsilon > \varphi_2(t, x_2) \right\},$$

$$\Sigma_\varepsilon(x) = \left\{ t \in \Omega \,\middle|\, |\varphi_1(t, x_1) - \varphi_2(t, x_2)| \leq \varepsilon \right\}.$$

The set $\Sigma_\varepsilon(x)$ is called the set of "ε-common points".
Let sets $\widetilde{\sigma}_1, \widetilde{\sigma}_2$ be such that

$$\widetilde{\sigma}_1 \subset \Sigma_\varepsilon(x), \ \widetilde{\sigma}_2 \subset \Sigma_\varepsilon(x), \ \widetilde{\sigma}_1 \cap \widetilde{\sigma}_2 = \emptyset, \ \widetilde{\sigma}_1 \cup \widetilde{\sigma}_2 = \Sigma_\varepsilon(x).$$

Put

$$\sigma_1 = \widetilde{\sigma}_1 \cup \widehat{\sigma}_{\varepsilon 1}(x), \ \sigma_2 = \widetilde{\sigma}_2 \cup \widehat{\sigma}_{\varepsilon 2}(x).$$

It is easy to see, that $\sigma_1 \cup \sigma_2 = \Omega$, $\sigma_1 \cap \sigma_2 = \emptyset$. Hence, any distribution $(\tilde{\sigma}_1, \tilde{\sigma}_2)$ of the set of ε-common points generates the corresponding disjoint partition (σ_1, σ_2) of the set Ω.

Definition 6.1 *Every such a partition (it depends upon x, ε and, evidently, it is not unique) is called an (x, ε)-proper partition of the set Ω.*

Let us denote by $T_\varepsilon(\Omega, x)$ the family of all (x, ε)-proper partitions of the set Ω. Clearly,

$$T(\Omega, x^*) \subset T_\varepsilon(\Omega, x) \subset T(\Omega).$$

Take $\Gamma = (\gamma_1, \gamma_2) \in T(\Omega)$ and construct the point $x(\Gamma)$ (see (5.1)). By applying the exchange algorithm (described in Section 5), and taking $x(\Gamma)$ as the initial point, in a finite number of steps we get a point $x^*(\Gamma)$ which is a stationary point of the function F.

Definition 6.2 *Let all φ_i be continuous and $\varepsilon > 0$. A point $x^* \in S$ is called an ε-local minimizer of the function F, if*

- *x^* is a stationary point of the function F,*

- *for every $(x^*, \varepsilon)-$proper partition $\Gamma = (\sigma_1, \sigma_2) \in T_\varepsilon(\Omega, x^*)$ of the set Ω for the point $x^*(\Gamma)$ (that is a local minimizer of F, delivered by the exchange algorithm with the initial point $x(\Gamma)$) the inequality*

$$F(x^*(\Gamma)) \geq F(x^*) \qquad (6.1)$$

holds.

Remark 6.1 *Since φ_i's are continuous, every stationary point is a local minimizer, the converse is not true: a local minimizer is not necessarily a stationary point.*

Remark 6.2 *The stationarity of a point is based on the necessary condition (see Rem. 4.1) while the property of being an ε-local minimizer depends also on the exchange algorithm. We will use this notion for the sake of convenience.*

Definition 6.3 *A point $x^* \in S$ is called a strict ε-local minimizer of the function F, if it is an ε-local minimizer of the function F, and, in addition, the inequality (6.1) is strict, i.e.*

$$F(x^*(\Gamma)) > F(x^*) \quad \forall \Gamma \in T_\varepsilon(\Omega, x^*) \setminus T(\Omega, x^*).$$

Note that for $\Gamma \in T(\Omega, x^*)$ the inequality (6.1) becomes the equality.

Let us describe an algorithm for finding ε-local minimizers. Let $\varepsilon > 0$ be fixed.

1. Choose an arbitrary stationary point $x^0 \in S$ (constructed, for example, by the exchange algorithm).

2. Let $x^k = (x_1^k, x_2^k) \in S$ have already been found.

3. Construct the sets $\widehat{\sigma}_{\varepsilon 1}(x^k), \widehat{\sigma}_{\varepsilon 2}(x^k)$ and $\Sigma_\varepsilon(x^k)$.

4. Check the ε-local minimality of x^k (since x^k is a stationary point, it remains to verify the condition (6.1) for all $\Gamma \in T_\varepsilon(\Omega, x^k)$).

   If the point x^k is an ε-local minimizer, then the process terminates. Otherwise, go to step 5.

5. Find any $\Gamma_k \in T_\varepsilon(\Omega, x^k)$ such that

$$F(x^*(\Gamma_k)) < F(x^k).$$

   Such a Γ_k exists, since x^k is not an ε-local minimizer.

6. Put $x^{k+1} = x^*(\Gamma_k)$ and go to step 3.

As a result, in a finite number of steps an ε-local minimizer is constructed.

Remark 6.3 *If ε is sufficiently large then an ε-local minimizer is a global minimizer of the function F. In this case $T_\varepsilon(\Omega, x) = T(\Omega)$, and hence if ε is large, from the computational point of view, the ε-exchange algorithm is not effective (it is equivalent to the complete enumeration of the elements of the set $T(\Omega))$.*

If ε is fairly small, the above described algorithm allows one to escape from a local minimum point (if the point itself is not yet an ε-local minimizer).

7   AN APPLICATION TO ONE CLUSTERING PROBLEM

Let a set of points $\Omega = \{t_1, \ldots, t_N\} \subset R^n$ be given. Introduce the functions

$$\varphi_1(t, x_1) = \|t - x_1\|^2, \quad \varphi_2(t, x_2) = \|t - x_2\|^2, \quad x_i \in R^n, \ i = 1, 2,$$

where $\|x\|^2 = \langle x, x \rangle$. Put $x = (x_1, x_2)$,

$$\varphi(t, x) = \min\{\|t - x_1\|^2, \|t - x_2\|^2\}.$$

Note that φ_i's are continuously differentiable and convex. Consider the following clustering problem.

Problem CP: Find $x^* = (x_1^*, x_2^*) \in R^n \times R^n$ such that

$$F(x^*) = \min_{x \in R^n \times R^n} F(x),$$

where

$$F(x) = \sum_{t \in \Omega} \varphi(t, x).$$

Take any $\Gamma = (\sigma_1, \sigma_2) \in T(\Omega, x)$, the functions $c_i(x_i, \sigma_i)$ defined in (2.1) take the form

$$c_i(x_i, \sigma_i) = \sum_{t \in \sigma_i} \|t - x_i\|^2, \ i = 1, 2. \tag{7.1}$$

Clearly,

$$\min_{x_i \in R^n} c_i(x_i, \sigma_i) = c_i(x_i(\sigma_i), \sigma_i), \tag{7.2}$$

where

$$x_i(\sigma_i) = \frac{1}{|\sigma_i|} \sum_{t \in \sigma_i} t. \tag{7.3}$$

Many allocation problems can be described by the above model, maybe with slightly different performance functionals (for example, in **?** the functions $\varphi_i's$ have the form $\varphi_i(t, x) = \|t - x_i\|$). We have chosen the quadratic functions φ_i (see (7.1)) for the simplicity reasons (since then the auxiliary problems (see (7.2), (7.3)) have the explicit solutions), because our main intention here is to demonstrate the Algorithm.

Remark 7.1 *Let*

$$x_1^* = \frac{1}{N} \sum_{t \in \Omega} t, \ M = \max_{t \in \Omega} \|t - x_1^*\|.$$

Then for every $x_2 \in R^n$ such that $\|x_2 - x_1^*\| > 2M$, the point (x_1^*, x_2) is stationary (that is, a local minimizer as well). Such local minimizers will be called trivial stationary point, and we shall ignore them, looking for better ones.

7.1   Example 1

Let $n = 2$, and Ω be as shown in Table 7.1. The set Ω contains 32 points in R^2, and we have to find two clusters $x_1^* \in R^2$ and $x_2^* \in R^2$, which minimize the functional

$$F(x) = F(x_1, x_2) = \sum_{i \in I} \min\{\|t_i - x_1\|^2, \|t_i - x_2\|^2\},$$

where $I = 1 : 32$. First, let us solve the 1-cluster problem:

Find

$$\min_{x_1 \in R^2} F_1(x_1) = F_1(x_1^*),$$

where

$$F_1(x_1) = \sum_{i \in I} \|t_i - x_1\|^2.$$

It follows from (7.3) that $x_1^* = (-0.5, 2.0656)$ and $F_1(x_1^*) = 782.2722$.

The first and quite natural idea is to take the point $x^0 = (x_1^*, x_1^*)$ as the initial point for our 2-cluster problem. In this case $\Sigma(x^0) = \Omega$ (and $F(x^0) = F_1(x_1^*) = 782.2722$), and at the first step of the exchange algorithm (only to check the conditions (4.1) and (4.2)!!!) it is required to solve 2^{32} elementary problems (which is absolutely unacceptable from practical considerations). However, it is sufficient to slightly perturb x^0 to overcome this difficulty: e.g., take $\overline{x}^0 = (x_1^* + z_1, x_1^*)$, where $z_1 = (0, 0.01)$.

Applying the exchange algorithm with $\overline{x}^0$ as the initial point, in two steps we get the local minimizer $\overline{x} = (\overline{x}_1, \overline{x}_2)$, where

$$\overline{x}_1 = (-0.53846, 5.5), \quad \overline{x}_2 = (-0.47368, 0.28421)$$

and $F(\overline{x}) = 523.9929$.

Now take $\overline{\overline{x}}^0 = (x_1^* + z_2, x_1^*)$ where x_1^* is as above, $z_2 = (0.01, 0)$.

Applying the exchange algorithm with $\overline{\overline{x}}^0$ as the initial point, in four steps we get another local minimizer $\overline{\overline{x}} = (\overline{\overline{x}}_1, \overline{\overline{x}}_2)$, where

$$\overline{\overline{x}}_1 = (1.95, 2.98), \quad \overline{\overline{x}}_2 = (-4.5833, 0.54167)$$

and $F(\overline{\overline{x}}) = 417.5478$.

The point $\overline{x}^0$ and the corresponding partition (σ_1, σ_2) of the set Ω at $\overline{x}^0$ are shown in Figure 7.2. The notations used in the Figures are explained in Figure 7.1. The points in $\hat{\sigma}_1(\overline{x}^0)$ are referred to as the points in the first cluster, and the points in $\hat{\sigma}_2(\overline{x}^0)$ are referred to as the points in the second cluster.

For the initial point $x^0 = (x_1^0, x_2^0) \in R^2 \times R^2$, where $x_1^0 = (-5, 10)$, $x_2^0 = (5, -5)$ (the function value $F(x^0) = 1707.81$), the exchange algorithm terminated in three steps, resulting in the local minimizer $x^* = (x_1^*, x_2^*)$, where

$$x_1^* = (-1.8421, 4.1316), \quad x_2^* = (1.4615, -0.9538).$$

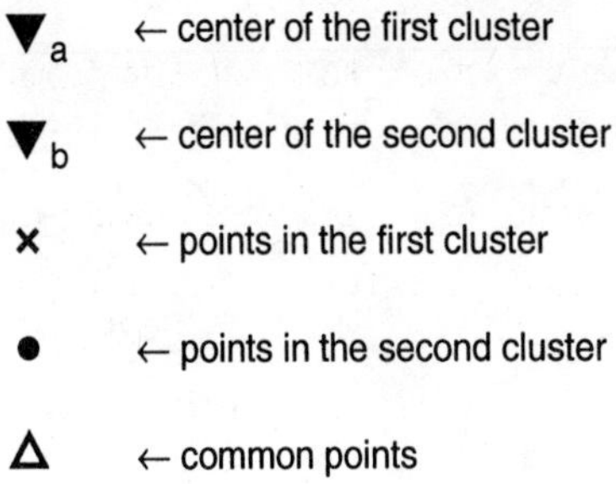

$$\blacktriangledown_a \quad \leftarrow \text{center of the first cluster}$$

$$\blacktriangledown_b \quad \leftarrow \text{center of the second cluster}$$

$$\times \quad \leftarrow \text{points in the first cluster}$$

$$\bullet \quad \leftarrow \text{points in the second cluster}$$

$$\triangle \quad \leftarrow \text{common points}$$

Figure 7.1 Legend.

The function value is $F(x^*) = 498.4104$. The point x^* and the partition (σ_1, σ_2) of the set Ω at x^* are shown in Figure 7.3.

For the initial point $x^0 = (x_1^0, x_2^0)$ with $x_1^0 = (10, 10)$, $x_2^0 = (-10, -10)$ in three steps the local minimizer $x^* = (x_1^*, x_2^*)$, where $x_1^* = (1.9500, 2.9800)$, $x_2^* = (-4.5833, 0.5417)$, is found with $F(x^*) = 417.5478$.

Table 7.1 The set of points $\Omega = \{t_1, \ldots, t_{32}\} \subset R^2$.

i	t_i	i	t_i	i	t_i	i	t_i
1	(3, 1)	9	(-5, -2)	17	(-1, 6)	25	(2, 5)
2	(3, -1)	10	(-6, 1)	18	(1, 6)	26	(2, 4)
3	(5, 1)	11	(-6, -1)	19	(0, 7)	27	(2, -3.4)
4	(-2, 1)	12	(2, 0)	20	(0, 5)	28	(-2, 7)
5	(-2, -1)	13	(6, 0)	21	(-1, 7)	29	(-5, 4)
6	(-3, 2)	14	(4, 2)	22	(1, 7)	30	(-5, 4)
7	(-3, -2)	15	(4, -2)	23	(2, 6)	31	(-6, 2)
8	(-5, 2)	16	(5, -1)	24	(1, 3)	32	(-7, 4.5)

7.2 Example 2

Consider again the problem discussed in Example 1, and apply the ε-exchange algorithm. As the initial point for the ε-exchange algorithm let us choose one of the local minimizers, obtained in Example 1.

The results of numerical experiments for the ε-exchange algorithm with different ε are presented in Table 7.2. The local minimizer $x^0 = (x_1^0, x_2^0)$, where

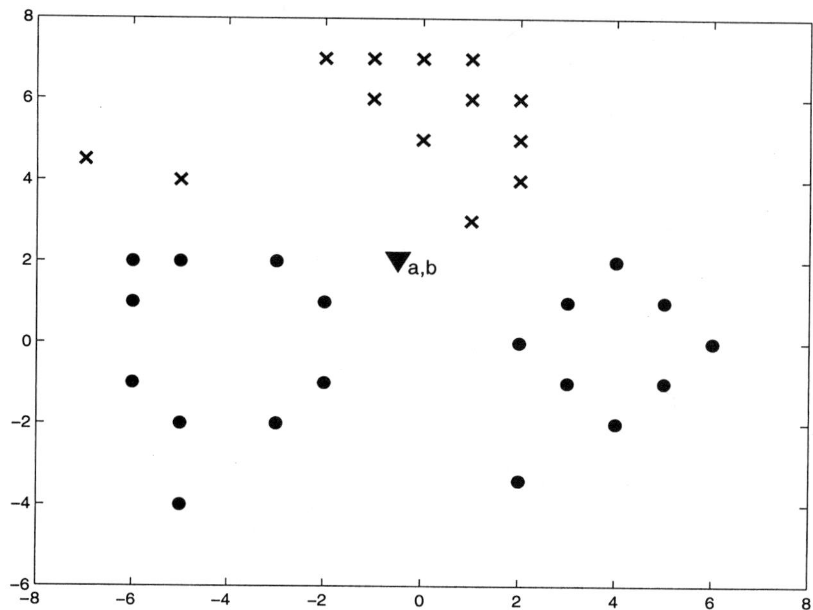

Figure 7.2   The partition of the set Ω at $\overline{x}^0 = (x_1^* + z_1, x_1^*)$.

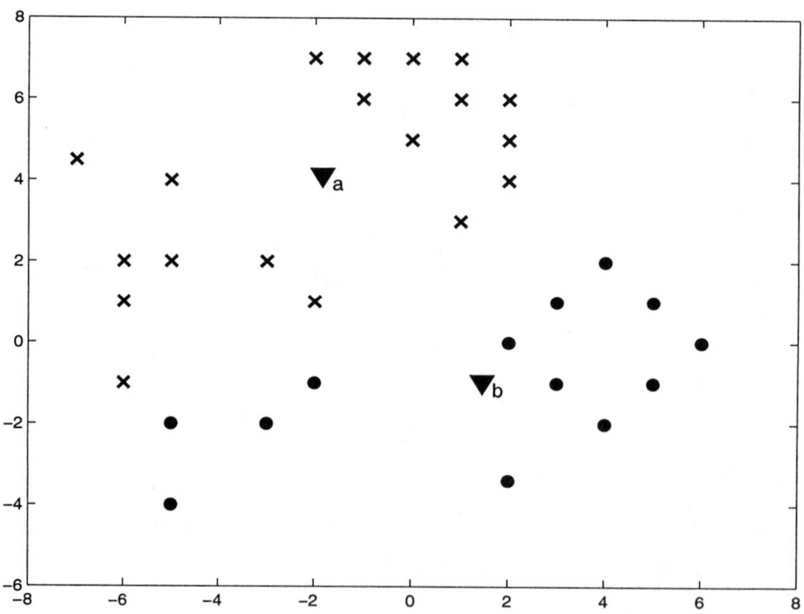

Figure 7.3   Results of the exchange algorithm.

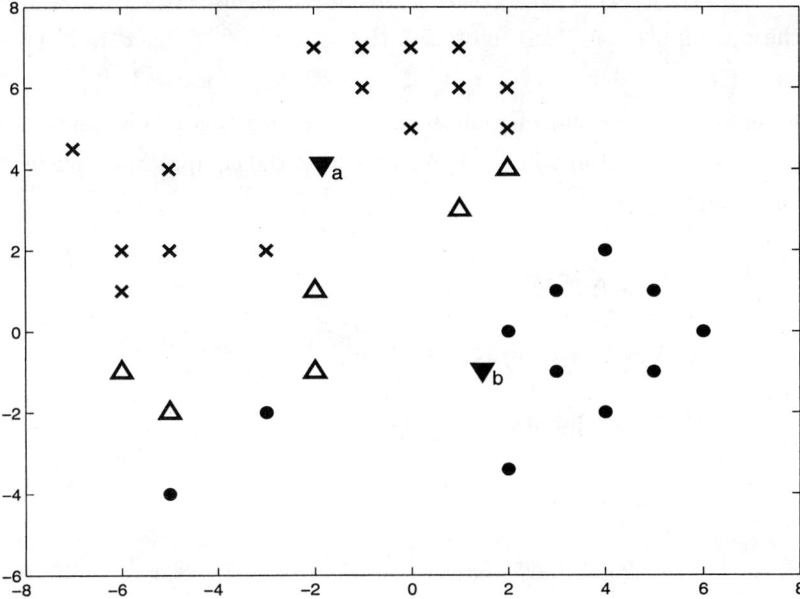

Figure 7.4   The first step of the ε-exchange algorithm with $\varepsilon = 15$ (six ε-common points).

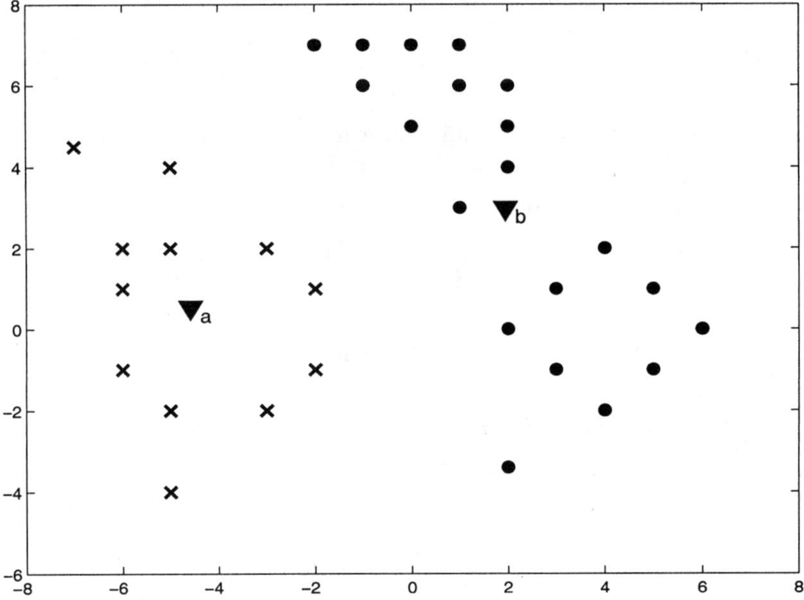

Figure 7.5   An ε-local minimizer with $\varepsilon = 15$ for the initial point $x^0 = (x_1^0, x_2^0)$, $x_1^0 = (-1.8421, 4.1316)$, $x_2^0 = (1.4615, -0.95385)$, $F(x^*) = 417.5478$.

$x_1^0 = (-1.8421, 4.1316)$, $x_2^0 = (1.4615, -0.95385)$, was taken as the initial point for these computations. In Figure 7.4 the sets $\widehat{\sigma}_{\varepsilon 1}(x^0)$, $\widehat{\sigma}_{\varepsilon 2}(x^0)$, $\Sigma_\varepsilon(x^0)$ are depicted (they are denoted by $\times$, $\bullet$, $\triangle$, respectively) for $\varepsilon = 15$.

In Figure 7.5 the results of application of the ε-exchange algorithm are shown (for $\varepsilon=15$ and the initial point x^0). As a result of the computations we received four local minimizers:

- $x^{*1} = (x_1^{*1}, x_2^{*1})$, where

$$x_1^{*1} = (-1.8421, 4.1316), \quad x_2^{*1} = (1.4615, -0.95385), \quad F(x^{*1}) = 498.4104.$$

  It is an ε-local minimizer for ε up to 4.

- $x^{*2} = (x_1^{*2}, x_2^{*2})$, where

$$x_1^{*2} = (-2.0000, 3.825), \quad x_2^{*2} = (2.0000, -0.8667), \quad F(x^{*2}) = 497.1842.$$

  This point is an ε-local minimizer for ε up to 8.

- $x^{*3} = (x_1^{*3}, x_2^{*3})$, where

$$x_1^{*3} = (-2.1739, 3.0217), \quad x_2^{*3} = (3.7778, -0.3778), \quad F(x^{*3}) = 478.3746.$$

  The point x^{*3} is an ε-local minimizer for ε up to 10.

- $x^{*4} = (x_1^{*4}, x_2^{*4})$, where

$$x_1^{*4} = (-4.5833, 0.5417), \quad x_2^{*4} = (1.9500, 2.9800), \quad F(x^{*4}) = 417.5478.$$

  The point x^{*4} is an ε-local minimizer for ε up to 30.

Remark 7.2 *We have computed only integer values of ε up to 30. It seems that the point x^{*4} is a global minimizer of F. It is interesting to note that for $\varepsilon = 9$ at the first step the set of common points was the same as for $\varepsilon = 8$. However, at further steps the increase of ε affected the result, since due to deeper "ε-diving" we were able to pick up a better ε-minimizer.*

Remark 7.3 *Numerical experiments with several real databases demonstrated the effectiveness of the exchange algorithm.*

8   CONCLUSIONS

Thus, we have described two algorithms: the exchange algorithm for constructing a stationary point and the ε–exchange algorithm for finding, may be, a better minimizer. The ε–exchange algorithm allows one to "escape" from a local minimum. These algorithms are conceptual (in the terminology of E.Polak) (see Polak (1971)), though in some cases (as is demonstrated in section 7) they are directly applicable.

It may happen, that the number of common (or ε-common points) is large. In such a case it is useful to perform some preliminary aggregation of these points reducing their number to a reasonable quantity (as well as to reduce or increase the value of ε). The aggregation idea was proposed by Prof. M. Gaudioso.

Computationally implementable modifications of the above algorithms for specific classes of functions will be reported elsewhere.

At each step of the both algorithms an elementary problem of minimizing a function of the form $F_\Gamma(x)$ is to be solved. The algorithms converge in a finite number of steps to, at least, a local minimizer. If ε is sufficiently large, the ε–exchange algorithm will produce a global minimizer, however, theoretically it may require the complete enumeration (and solution) of all elementary problems. The hope and expectation are that if we take a reasonable ε, then (at least statistically) the price asked for a fairly good local minimizer will not be too high.

The case $m > 2$ can be studied in a similar way. Analogous results and algorithms can be formulated. The number of "elementary" problems becomes $m^{|\Omega|}$ (cf. Rem. 3.1) and, of course, all calculations are more complicated. However, the exchange and ε-exchange algorithms can be constructed.

Acknowledgments

The author is thankful to an anonymous referee for his careful reading of the manuscript, useful advice, remarks and suggestions.

References

Bagirov, A.M., Rubinov, A.M. (1999), Global Optimization of IPH functions over the unit simplex with applications to cluster analysis, In *Proceedings of the Third Australia-Japan Joint Workshop on Intelligent and Evolutionary Systems*, Canberra, Australia, pp. 191-192.

Bagirov, A.M., Rubinov, A.M. and Yearwood J. (2001), Using global optimization to improve classification for medical diagnosis, *Topics in Health Information Management*, Vol. 22, pp. 65-74.

Mangasarian, O.L. (1997), Mathematical Programming in Data Mining, *Data Mining and Knowledge Discovery*, Vol. 1, pp. 183-201.

Polak E. (1971), *Computational Methods in Optimization; A unified approach*, Academic Press, New York, NY.

Rao, M.R. (1971), Cluster analysis and mathematical programming, *Journal of the American Statistical Association*, Vol. 66, pp. 622-626.

Rubinov, A.M. (2000), *Abstract Convexity and Global Optimization*, Kluwer Academic Publishers, Dordrecht.

Vershik, A.M., Malozemov, V.N., Pevnyi, A.B. (1975), The best piecewise polynomial approximation, *Siberian Mathematical Journal*, Vol. XVI, Nr. 5, pp. 925-938.

Table 7.2   The behavior of the ε-exchange Algorithm, depending on ε.

ε	$|\Sigma_\varepsilon(x^0)|$	$F(x_\varepsilon^*)^a$	$x_{1\varepsilon}^*$	$x_{2\varepsilon}^*$
0	0	498.4104	(-1.8421, 4.1316)	(1.4615, -0.95385)
1	0	498.4104	(-1.8421, 4.1316)	(1.4615, -0.95385)
2	0	498.4104	(-1.8421, 4.1316)	(1.4615, -0.95385)
3	0	498.4104	(-1.8421, 4.1316)	(1.4615, -0.95385)
4	0	498.4104	(-1.8421, 4.1316)	(1.4615, -0.95385)
5	1	497.1842	(-2.0000, 3.825)	(2.0000, -0.8667)
6	2	497.1842	(-2.0000, 3.825)	(2.0000, -0.8667)
7	3	497.1842	(-2.0000, 3.825)	(2.0000, -0.8667)
8	3	497.1842	(-2.0000, 3.825)	(2.0000, -0.8667)
9^b	3	478.3746	(-2.1739, 3.0217)	(3.7778, -0.3778)
10	3	478.3746	(-2.1739, 3.0217)	(3.7778, -0.3778)
11	4	417.5478	(-4.5833, 0.5417)	(1.9500, 2.9800)
12	4	417.5478	(-4.5833, 0.5417)	(1.9500, 2.9800)
13	5	417.5478	(-4.5833, 0.5417)	(1.9500, 2.9800)
14	5	417.5478	(-4.5833, 0.5417)	(1.9500, 2.9800)
15	6	417.5478	(-4.5833, 0.5417)	(1.9500, 2.9800)
16	6	417.5478	(-4.5833, 0.5417)	(1.9500, 2.9800)
17	6	417.5478	(-4.5833, 0.5417)	(1.9500, 2.9800)
18	7	417.5478	(-4.5833, 0.5417)	(1.9500, 2.9800)
19	7	417.5478	(-4.5833, 0.5417)	(1.9500, 2.9800)
20	7	417.5478	(-4.5833, 0.5417)	(1.9500, 2.9800)
21	8	417.5478	(-4.5833, 0.5417)	(1.9500, 2.9800)
22	8	417.5478	(-4.5833, 0.5417)	(1.9500, 2.9800)
23	9	417.5478	(-4.5833, 0.5417)	(1.9500, 2.9800)
24	10	417.5478	(-4.5833, 0.5417)	(1.9500, 2.9800)
25	10	417.5478	(-4.5833, 0.5417)	(1.9500, 2.9800)
26	11	417.5478	(-4.5833, 0.5417)	(1.9500, 2.9800)
27	11	417.5478	(-4.5833, 0.5417)	(1.9500, 2.9800)
28	12	417.5478	(-4.5833, 0.5417)	(1.9500, 2.9800)
29	12	417.5478	(-4.5833, 0.5417)	(1.9500, 2.9800)
30^c	14	417.5478	(-4.5833, 0.5417)	(1.9500, 2.9800)

[a] $x_\varepsilon^* = (x_{1\varepsilon}^*, x_{2\varepsilon}^*)$ denotes the ε-local minimizer constructed by the ε-exchange Algorithm.

[b] Although at the first step the set of common points was the same as for $\varepsilon = 8$, at further steps the increase of ε affected the result, allowing to find a better ε-local minimizer.

[c] At this point the computations were terminated.

10 ON THE BARZILAI-BORWEIN METHOD

Roger Fletcher

Department of Mathematics
University of Dundee
Dundee DD1 4HN, Scotland, UK
Email: fletcher@maths.dundee.ac.uk

Abstract: A review is given of the underlying theory and recent developments in regard to the Barzilai-Borwein steepest descent method for large scale unconstrained optimization. One aim is to assess why the method seems to be comparable in practical efficiency to conjugate gradient methods. The importance of using a non-monotone line search is stressed, although some suggestions are made as to why the modification proposed by Raydan (1997) often does not usually perform well for an ill-conditioned problem. Extensions for box constraints are discussed. A number of interesting open questions are put forward.

Key words: Barzilai-Borwein method, steepest descent, elliptic systems, unconstrained optimization.

1 INTRODUCTION

The context of this paper is the solution of the unconstrained minimization problem

$$\text{minimize} \quad f(\mathbf{x}) \qquad \mathbf{x} \in \mathbb{R}^n \tag{1.1}$$

where the number of variables n is very large, typically 10^6 or so. The case of minimization subject to simple bounds is also considered later in the paper. A related problem is that of the solution of a nonlinear self-adjoint elliptic system of equations

$$\gamma(\mathbf{x}) = \mathbf{0}, \tag{1.2}$$

in which $\gamma = \nabla f$ is the gradient of some variational function. The case in which $f(\mathbf{x})$ is a strictly convex quadratic function

$$f(\mathbf{x}) = \tfrac{1}{2}\mathbf{x}^T A \mathbf{x} - \beta^T \mathbf{x} \tag{1.3}$$

is also studied, which is equivalent to the solution of the linear system of equations

$$A\mathbf{x} = \beta \tag{1.4}$$

in which A is a positive definite symmetric matrix. This is referred to as the *quadratic case* and is important, not only as a model problem to analyse properties of methods, but also in its own right as a problem that becomes difficult to solve for large n, when (sparse) Choleski factorisation is impractical due to lack of time or storage capacity. It is the large scale situation (both for quadratic or non-quadratic problems) that we study in this paper.

The methods that we study are all iterative methods, since in the quadratic case we cannot expect to be able to carry out enough iterations to obtain an exact solution, even if the theory allows this possibility, due to the size of n or the build up of round-off error. Early methods such as relaxation methods or SOR have mostly been supplanted by use of the conjugate gradient method or variations thereof. For the quadratic case the conjugate gradient (CG) method itself (Hestenes and Stiefel (1952)), or some preconditioned conjugate gradient (PCG) method (see for example Golub and Van Loan (1966)), is usually the method of choice, although there are other variants such as the minimum residual (MR) algorithm that are also applicable to the case that A is symmetric and indefinite. A particular feature of these methods is that they terminate in at most n iterations. This is not particularly exciting when n is large, but Reid (1971) gives reasons why the methods are effective as iterative methods in that they are able to deliver a reasonably accurate estimate of the solution in substantially fewer than n iterations, particularly if A has a favourable eigenvalue structure. The CG method is particularly attractive because it only requires $4n$ storage locations for its implementation. PCG methods need to store and solve linear systems with some matrix that approximates A and makes tolerable demands on time and storage.

For non-quadratic systems there are various methods of line search type that are based on using the search direction formula of the CG iteration. The simplest methods are those of Fletcher and Reeves (1964) ($3n$ locations) and Polak (1971) ($4n$ locations), the latter being more usually preferred in practice. These can also be preconditioned in a manner similar to the quadratic case. Then there are also methods that use rather more storage, such as CONMIN (Shanno and Phua (1980)) ($7n$ locations), the Limited Memory BFGS method (Nocedal (1980)), ($9n+$ locations), the Truncated Newton method (Dembo, Eisenstat and Steihaug (1982)), and many others.

Amongst all of these, steepest descent methods hardly rate a mention in a modern text-book on optimization, even though the storage requirements are minimal ($3n$ locations). Indeed, the 'classical' steepest descent method with exact line seach (Cauchy (1847)) is known to behave increasingly badly in the quadratic case as the condition number of A deteriorates. Early attempts to modify the method led to the introduction of CG methods, with much superior performance.

In 1988, a paper by Barzilai and Borwein (1988) proposed a steepest descent method (the BB method) that uses a different strategy for choosing the step length. The main result of the paper is to show the surprising result that for $n = 2$, the method converges R-superlinearly. Barzilai and Borwein also show that their method is considerably superior to the classical steepest descent method for one instance of a quadratic function with $n = 4$, but no other numerical results are given. Fletcher (1990) investigates some connections with the spectrum of A in the quadratic case, and an ingenious proof by Raydan (1993) demonstrates convergence in the quadratic case. However, neither of these papers gives any numerical results and the method attracted little attention until the seminal paper of Raydan (1997). This paper introduces a globalization strategy based on the non-monotone line search technique of Grippo, Lampariello and Lucidi (1986), which enables global convergence of the BB method to be established for non-quadratic functions. Of equal importance, a wide range of numerical experience is reported on problems of up to 10^4 variables, showing that the method compares reasonably well against the Polak-Ribière and CONMIN techniques. Earlier papers by Glunt, Hayden and Raydan (1993) and Glunt, Hayden and Raydan (1994) also report promising numerical results on a distance matrix problem. The paper Glunt, Hayden and Raydan (1994) reports on the possibilities for preconditioning the BB method, and this theme is also taken up by Molina and Raydan (1996). Of particular interest is the possibility of applying the BB method to box-constrained optimization problems, and this is considered by Friedlander, Martínez and Raydan (1995) (for quadratic functions) and by Birgin, Martínez and Raydan (2000). The latter paper considers the BB method in the context of projection on to a convex set. Another recent theoretical development has been the result that the unmodified BB method is R-linearly convergent in the quadratic case (Dai and Liao (1999)).

Despite all these advances, there is still much to be learned about the BB method and its modifications. This paper reviews what is known about the method, and advances some reasons that partially explain why the method is competitive with CG based methods. The importance of maintaining the non-monotonicity property of the basic method is stressed. It is argued that the use of the line search technique of Grippo, Lampariello and Lucidi (1986) in the manner proposed by Raydan (1997) may not be the best way of globalizing the BB method, and some tentative alternatives are suggested. Some other interesting observations about the distribution of the BB steplengths are also made. Many open questions still remain about the BB method and its potential, and these are discussed towards the end of the paper.

2   THE BB METHOD FOR QUADRATIC FUNCTIONS

The theory and practice of line search methods for minimizing $f(\mathbf{x})$ has been well explored. In such a method, a search direction $\mathbf{s}^{(k)}$ is chosen at the start of iteration k, and a step length θ_k is chosen to (approximately) minimize $f(\mathbf{x}^{(k)} + \theta\mathbf{s}^{(k)})$ with respect to θ. Then $\mathbf{x}^{(k+1} = \mathbf{x}^{(k)} + \theta_k\mathbf{s}^{(k)}$ is set. Usually

$\mathbf{s}^{(k)T}\gamma^{(k)} < 0$ for all k (a descent method) and it is possible to guarantee that the method is *monotonic* in the sense that the sequence $\{f^{(k)}\}$ is strictly monotonically decreasing unless a stationary point is exactly located. The classical steepest descent method belongs to this class, with $\mathbf{s}^{(k)} = -\gamma^{(k)}$. CG methods have $\mathbf{s}^{(1)} = -\gamma^{(1)}$ and $\mathbf{s}^{(k)} = -\gamma^{(k)} + \beta_k \mathbf{s}^{(k-1)}$ for $k > 1$, where $\beta_k = \gamma^{(k)T}\gamma^{(k)}/\gamma^{(k-1)T}\gamma^{(k-1)}$ in the Fletcher-Reeves (FR) method, and $\beta_k = \gamma^{(k)T}(\gamma^{(k)} - \gamma^{(k-1)})/\gamma^{(k-1)T}\gamma^{(k-1)}$ in the (Polak-Ribière) (PR) method. When $f(\mathbf{x})$ is the quadratic function (1.3), the step θ_k is readily calculated from the expression $\theta_k = -\mathbf{s}^{(k)T}\gamma^{(k)}/\mathbf{s}^{(k)T}A\mathbf{s}^{(k)}$. For non-quadratic functions it is in general only possible to find an approximate solution of the line search problem, and for CG methods it seems better if the solution is reasonably accurate.

In contrast, the BB method is a fixed step gradient method, which we choose to write in the form

$$\mathbf{x}^{(k+1)} = \mathbf{x}^{(k)} + \mathbf{d}^{(k)} \qquad \text{where} \qquad \mathbf{d}^{(k)} = -\gamma^{(k)}/\alpha_k. \tag{2.1}$$

Initially, $\alpha_1 > 0$ is arbitrary, and Barzilai and Borwein give two alternative formulae,

$$\alpha_k = \mathbf{d}^{(k-1)T}\mathbf{y}^{(k-1)}/\mathbf{d}^{(k-1)T}\mathbf{d}^{(k-1)} \tag{2.2}$$

and

$$\alpha_k = \mathbf{y}^{(k-1)T}\mathbf{y}^{(k-1)}/\mathbf{y}^{(k-1)T}\mathbf{d}^{(k-1)}, \tag{2.3}$$

for $k > 1$, where we denote $\mathbf{y}^{(k-1)} = \gamma^{(k)} - \gamma^{(k-1)}$. In fact, attention has largely been focussed on (2.2) and it is this formula that is discussed here, although there seems to be some evidence that the properties of (2.3) are not all that dissimilar.

In the rest of this section, we explore the properties of the BB method and other gradient methods for minimizing a strictly convex quadratic function. For the BB method, (2.2) can be expressed in the form

$$\alpha_k = \gamma^{(k-1)T}A\gamma^{(k-1)}/\gamma^{(k-1)T}\gamma^{(k-1)} \tag{2.4}$$

and can be regarded as a Rayleigh quotient, calculated from the previous gradient vector. This is in contrast to the classical steepest descent method which is equivalent to using a similar formula, but with $\gamma^{(k-1)}$ replaced by $\gamma^{(k)}$. Another relevant property, possessed by all gradient methods, and also the conjugate gradient method, is that

$$\mathbf{x}^{(k+1)} - \mathbf{x}^{(1)} \in \mathrm{span}\{\gamma^{(1)},\, A\gamma^{(1)},\, A^2\gamma^{(1)},\, \ldots A^{k-1}\gamma^{(1)}\}. \tag{2.5}$$

That is to say, the total step lies in the span of the so-called *Krylov sequence*. Also for quadratic functions, the BB method has been shown to converge (Raydan (1993)), and convergence is R-linear (Dai and Liao (1999)). However the sequences $\{f(\mathbf{x}^{(k)})\}$ and $\{\|\gamma(\mathbf{x}^{(k)})\|_2\}$ are non-monotonic, an explanation of which is given below, and no realistic estimate of the R-linear rate is known. However the case $n = 2$ is special, and it is shown in Barzilai and Borwein (1988) that the rate of convergence is R-superlinear.

To analyse the convergence of any gradient method for a quadratic function, we can assume without loss of generality that an orthogonal transformation is made that transforms A to a diagonal matrix of eigenvalues $\mathrm{diag}(\lambda_i)$. Moreover, if there are any eigenvalues of multiplicity $m > 1$, then we can choose the corresponding eigenvectors so that $g_i^{(1)} = 0$ for at least $m - 1$ corresponding indices of $\gamma^{(1)}$. It follows from (2.1) and the properties of a quadratic function that $\gamma^{(k+1)} = \gamma^{(k)} - A\gamma^{(k)}/\alpha_k$ and hence using $A = \mathrm{diag}(\lambda_i)$ that

$$g_i^{(k+1)} = \left(1 - \frac{\lambda_i}{\alpha_k}\right) g_i^{(k)} \qquad i = 1, 2, \ldots, n. \tag{2.6}$$

It is clear from this recurrence that if $g_i^{(k)} = 0$ for any i and $k = k'$, then this property will persist for all $k > k'$. Thus, without any loss of generality, we can assume that A has distinct eigenvalues

$$0 < \lambda_1 < \lambda_2 < \ldots < \lambda_n, \tag{2.7}$$

and that $g_i^{(1)} \neq 0$ for all $i = 1, 2, \ldots, n$.

Many things can be deduced from these conditions and (2.6). First, if α_k is equal to any eigenvalue λ_i, then $g_i^{(k+1)} = 0$ and this property persists subsequently. If both

$$g_1^{(k-1)} \neq 0 \qquad \text{and} \qquad g_n^{(k-1)} \neq 0 \tag{2.8}$$

then it follows from (2.4) and the extremal properties of the Rayleigh quotient that

$$\lambda_1 < \alpha_k < \lambda_n. \tag{2.9}$$

Thus, for the BB method, and assuming that α_1 is not equal to λ_1 or λ_n, then a simple inductive argument shows that (2.8) and (2.9) hold for all $k > 1$. It follows, for example, that the BB method does not have the property of finite termination.

From (2.6), it follows for any eigenvalue λ_i close to α_k that $|g_i^{(k+1)}| \ll |g_i^{(k)}|$. It also follows that the values $|g_1^{(k)}|$ are monotonically decreasing. However, if on any iteration $\alpha_k < \frac{1}{2}\lambda_n$, then $|g_n^{(k+1)}| > |g_n^{(k)}|$ and if α_k is close to λ_1 then the ratio of $|g_n^{(k+1)}/g_n^{(k)}|$ can approach $\lambda_n/\lambda_1 - 1$. Thus we see the potential for non-monotonic behaviour in the sequences $\{f(\mathbf{x}^{(k)})\}$ and $\{\|\gamma(\mathbf{x}^{(k)})\|_2\}$, and the extent of the non-monotonicity depends in some way on the size of the condition number of A. On the other hand, if α_k is close to λ_n then all the coefficients g_i decrease in modulus, but the change in g_1 is negligible if the condition number is large. Moreover, small values of α_k tend to diminish the components $|g_i|$ for small i and hence enhance the relative contribution of components for large i. This in turn leads to large values of α_k on a subsequent iteration, if the step is calculated from (2.4). Thus, in the BB method, we see values of α_k being selected from all parts of the interior of the spectrum, with no apparent pattern, with jumps in the values of $f(\mathbf{x}^{(k)})$ and $\|\gamma(\mathbf{x}^{(k)})\|_2$ occurring when α_k is small.

There are a number of reasons that might lead one to doubt whether the BB method could be effective in practice. Although a nice convergence proof is given by Raydan (1993), we have to recognise the fact that although both the CG and BB methods select iterates that satisfy the Krylov sequence property (2.5), it is the CG method that gives the minimum possible value of $f(\mathbf{x}^{(k+1)})$. Likewise the Minimum Residual (MR) method gives the minimum possible value of $\|\gamma^{(k+1)}\|_2$. Thus we must accept that the BB method is necessarily inferior in regard to these measures in exact arithmetic, and there is limited scope for the BB method to improve as regards elapsed time, for example. Also the possibility of non-monotonic behaviour of the BB method might seem to give further reason to prefer the CG method.

To see just how inferior the BB method is, a large scale test problem is devised, based on the solution of an elliptic system of linear equations arising from a 3D Laplacian on a box, discretized using a standard 7-point finite difference stencil. Thus we define the matrices

$$
T = \begin{bmatrix} 6 & -1 & & & \\ -1 & 6 & -1 & & \\ & -1 & 6 & \ddots & \\ & & \ddots & \ddots & -1 \\ & & & -1 & 6 \end{bmatrix}, \quad W = \begin{bmatrix} T & -I & & & \\ -I & T & -I & & \\ & -I & T & \ddots & \\ & & \ddots & \ddots & -I \\ & & & -I & T \end{bmatrix}
$$

and

$$
A = \begin{bmatrix} W & -I & & & \\ -I & W & -I & & \\ & -I & W & \ddots & \\ & & \ddots & \ddots & -I \\ & & & -I & W \end{bmatrix} \tag{2.10}
$$

where T is $l \times l$, W is block $m \times m$ and A is block $n \times n$ where l, m, n are the number of interior nodes in each coordinate direction, and are specified by the user. The interval length is taken to be $h = 1/(l+1)$ and is the same in each direction. Hence the dimensions of the box are $1 \times Y \times Z$ where $Y = (m+1)h$ and $Z = (n+1)h$. We fix the solution of the problem to be function

$$
u(x, y, z) =
$$
$$
x(x-1)y(y-Y)z(z-Z)\exp(-\tfrac{1}{2}\sigma^2((x-\alpha)^2 + (y-\beta)^2 + (z-\gamma)^2)), \tag{2.11}
$$

evaluated at the nodal points. This function is a Gaussian centered on the point (α, β, γ), multiplied by quadratics $x(x-1)$ etc., that give $u = 0$ on the boundary. The parameter σ controls the rate of decay of the Gaussian. The problem has lmn variables, and we denote $\mathbf{u}^*$ to be the solution derived from (2.11) and calculate the right hand side $\beta = A\mathbf{u}^*$. In our tests we choose $l = m = n = 100$ giving a problem with 10^6 variables, in which case the condition number of A is $\lambda_n/\lambda_1 = 4133.6 = 10^{3.61}$. We also choose $\mathbf{u}^{(1)} = \mathbf{0}$. We denote the resulting test problem to be **Laplace1** and choose the parameters in two different ways, that is

(a) $\sigma = 20$, $\alpha = \beta = \gamma = 0.5$ (b) $\sigma = 50$, $\alpha = 0.4$, $\beta = 0.7$, $\gamma = 0.5$.

The problem **Laplace1(a)** has the centre of the Gaussian in the centre of the box, giving the problem a high degree of symmetry. Also the smaller value of σ gives a smoother solution. Hence this problem is more easy to solve than **Laplace1(b)**.

The results for this problem are given in Table 2.1 below. The CG method is coded as recommended by Reid (1971). Times are given in seconds and double precision Fortran is used on a SUN Ultra 10 at 440 MHz. The iteration is terminated when $\|\gamma^{(k)}\|_2$ is less than 10^{-6} of its initial value.

Table 2.1   Double length comparison (quadratic case)

Problem	BB		CG		MR	
	Time	Iterations	Time	Iterations	Time	Iterations
Laplace1(a)	543	859	162	178	179	171
Laplace1(b)	640	1009	285	306	302	290

We see from the table that there is little to choose between the CG and MR methods, and the elapsed time improves on the BB method by a factor of over 3 for Laplace1(a) and a factor of over 2 for Laplace1(b). For comparsion purposes, the classical steepest descent method was manually terminated after 2000 iterations (1355 seconds), by which time it had only reduced the initial gradient norm by a factor of 0.18, so that not even one significant figure improvement had been obtained. Thus we see that, although the performance of the BB does not quite match that of the CG method, it is able to solve the problem in reasonable time, and significantly improves on the classical steepest descent method.

Nonetheless, in view of the above, we might ask if there are any circumstances under which the BB method might be worth considering as an alternative to the CG method. The answer lies in the fact that the success of the CG iteration depends very much on the search direction calculation $\mathbf{s}^{(k)} = -\gamma^{(k)} + \beta_k \mathbf{s}^{(k-1)}$ being consistent with data arising from a quadratic model. Any deviation from the quadratic model can seriously degrade performance. To illustrate that relatively small perturbations can cause this to happen, we repeat the calculations of Table 2.1 using single precision arithmetic. The results are displayed in Table 2.2.

We see that the CG and MR methods now take more than twice as many iterations for Laplace1(a), with a similar, but not quite as bad, outcome for Laplace1(b). The comparison in time is less marked, presumably because of

Table 2.2   Single length comparison (quadratic case)

Problem	BB		CG		MR	
	Time	Iterations	Time	Iterations	Time	Iterations
Laplace1(a)	462	964	254	387	340	448
Laplace1(b)	310	645	290	443	397	523

the cost savings associated with using single rather than double precision. For the BB method a different picture emerges. For Laplace1(a), somewhat more iterations are required, whereas for Laplace1(b), considerably fewer iterations are required. Again the time comparison is improved by using single precision, to such an extent that there is now little difference between the performance of the BB and CG methods on the Laplace1(b) problem. My interpretation of this is that the BB method is affected in a much more random way by round off errors, and small departures of $\gamma^{(k)}$ and α_k from the values arising from a quadratic problem are not necessarily detrimental.

This has implications for the likely success of the BB method in other contexts. For example, if $f(\mathbf{x})$ is made up of a quadratic function plus a small non-quadratic term, we might expect the BB method to still converge, and show improved performance relative to the CG method. Another situation is in the minimization of a quadratic function subject to simple bounds by an active set or projection type of method. If the number of active constraints changes, as is often the case, then it is usually not possible to continue to use the standard CG formula for the search direction and yet preserve the termination and optimality properties. To do this it is necessary to restart using the steepest descent direction when a new active set is obtained. Thus it is more attractive to use the BB method in some way in this situation.

3   THE BB METHOD FOR NON-QUADRATIC FUNCTIONS

If the deviation of $f(\mathbf{x})$ from a quadratic function is small then it may still be possible to use the unmodified BB method successfully. However, in general it is possible for the method to diverge. This is illustrated by using the test problem referred to as **Strictly Convex 2** by Raydan (1997), in which $f(\mathbf{x})$ is defined by

$$f(\mathbf{x}) = \sum_{i=1}^{n} \tfrac{1}{10}i\left(e^{x_i} - x_i\right). \tag{3.1}$$

The initial point is $\mathbf{x}^{(1)} = (1, 1, \ldots, 1)^T$ and the solution is $\mathbf{x}^* = \mathbf{0}$. The Hessian matrix at $\mathbf{x}^*$ is $\tfrac{1}{10}\operatorname{diag}((1, 2, \ldots, n)$ so that the condition number is n. It is readily verified that the unmodified BB method converges for $n = 20$ but diverges for $n = 30$, even though (3.1) is a strictly convex function and has a positive definite Hessian matrix for all $\mathbf{x}$.

It is therefore necessary to modify the BB method if it is to be used as a general purpose solver for non-quadratic problems. An important contribution is that of Raydan (1997) who suggests using the non-monotonic line search of Grippo, Lampariello and Lucidi (1986). In Raydan's method (the BB-Raydan method, say) the initial estimate of the line search step is $\theta = \alpha_k^{-1}$, with some adjustment if α_k^{-1} turns out to be unreasonably large or small (and θ is required to be positive). An Armijo-type line search on θ is then carried out until the acceptance condition

$$f(\mathbf{x}^{(k)} + \mathbf{d}) \le \max_{\max(k-M,1)\le j\le k} f^{(j)} - \gamma\gamma^{(k)T}\mathbf{d} \tag{3.2}$$

is met, where $\mathbf{d} = -\theta\gamma^{(k)}$ is the displacement along the steepest descent direction. This allows any point to be accepted if it improves sufficiently on the largest of the $M + 1$ (or k if $k \le M$) most recent function values. As usual $\gamma > 0$ is a small preset constant, and the integer M controls the amount of monotonicity that is allowed. Raydan recommends the value $M = 10$ and presents a lot of encouraging numerical evidence on test problems with up to $n = 10^4$ variables. His results are competitive with CG methods but he observes some poorer results on ill-conditioned problems.

To obtain more insight, a non-quadratic test problem of 10^6 variables is derived, based on a 3D Laplacian, in which the objective function is

$$\tfrac{1}{2}\mathbf{u}^T A\mathbf{u} - \beta^T\mathbf{u} + \tfrac{1}{4}h^2 \sum_{ijk} u_{ijk}^4, \tag{3.3}$$

which is not untypical of what might arise from a nonlinear partial differential equation. This problem is referred to as **Laplace2**. The matrix A is that defined in (2.10), and the vector β is chosen so that the minimizer $\mathbf{u}^*$ of (3.3) is the function $u(x, y, z)$ in (2.11), evaluated at the discretization points. The non-quadratic term in (3.3) includes a factor h^2, and $0 \leq u_{ijk} < 1$ which also makes the u_{ijk}^4 term small. Thus the relative contribution of the non-quadratic term is small, and as we shall see, the unmodified BB method is able to solve instances of this problem.

The progress of various methods for solving Laplace2(b) is given in Table 3.1. The columns headed **#ls**, **#f** and **#g** give the numbers of line searches,

Table 3.1   Time (minutes) and numbers of evaluations to solve Laplace2(b)

Problem	5 figures				6 figures			
	Time	#ls	#f	#g	Time	#ls	#f	#g
Polak-Ribière CG	20.6	445	697	684	∞			
Limited mem. BFGS	35.4	315	711	669	∞			
BB-Raydan M=10	29.0	274	1140	866	40.4	394	1595	1201
Unmodified BB	14.4	-	487	487	16.7	-	572	572
BB method (γ only)	8.8	-	-	487	10.3	-	-	572

function evaluations and gradient evaluations required to solve the problem. The calculations are carried out in double precision Fortran and the non-BB methods both use the same line search based on standard strong Wolfe conditions with a relative slope tolerance of 0.1. For the BB-Raydan method the column **#ls** gives the number of occasions on which the Armijo line search is used. In the limited memory method, 3 back pairs of vectors are stored. Initially an accuracy tolerance of better than 10^{-6} of the initial gradient norm was required (the column headed '6 figures' in the table) but only the BB methods were able to find the solution to this accuracy. Therefore the comparison is carried out on the basis of 5 figure accuracy (10^{-5} of the initial gradient norm required).

It can be seen that here the unmodified BB method actually improves on the PR-CG method, but that this improvement is not maintained for the BB-Raydan method. The limited memory BFGS method shows up worst in the tests. The line search for the non-BB methods is seen to be reasonably efficient with only about two function and gradients calls per line search. The BB method (γ only), to be described below, gives the best performance. One reason for the improvement of the unmodified BB method over the PR-CG method might be the effect of non-quadratic terms degrading the performance of the CG method. Another possibility is that the CG line search now requires additional evaluations of the function and gradient to attain the required accuracy in the line search. One would not like draw any firm conclusions on the basis of just one set of results, but these results do reinforce Raydan's conclusion that the BB method, suitably modified, can match or even improve on the PR-CG method.

Probably the most interesting outcome to emerge is the difference in performance of the unmodified BB method and the BB-Raydan method. The reasons for this are readily seen by examining the performance of the unmodified method shown in Figure 3.1. Here the difference $f^{(k)} - f^*$ is plotted on a log scale against the number of iterations. A noticeable feature is the four occasions

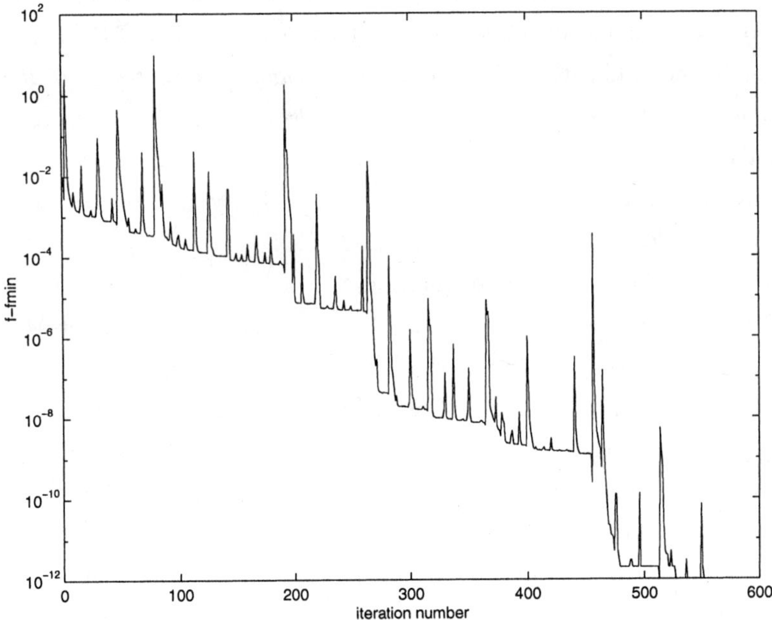

Figure 3.1   Performance of unmodified BB on Laplace2(b)

on which a huge jump is seen in $f^{(k)} - f^*$ above the slowly varying part of the graph. In particular the jump around iteration 460 is over 10^5 in magnitude.

Although somewhat disconcerting, these jumps are actually very beneficial in that they are soon followed by a considerable overall improvement in $f^{(k)} - f^*$. For example, before the spike at around iterations 260-270, $f^{(k)} - f^*$ is slowly varying at around $10^{-5.5}$, whereas afterwards it is around $10^{-7.5}$, an improvement of two orders of magnitude. Similar improvements can be seen either side of the other large spikes. I think the explanation for this is as follows. Consider the quadratic case of the previous section and the effect of 'small' components of the gradient (by which I mean components $g_i^{(k)}$ for small i) in the case that the condition number λ_n/λ_1 is large. In this case, large values of α_k have very little effect on the small components, but they diminish significantly the size of the large components, by virtue of (2.6). Subsequently therefore, an iteration occurs on which the small components dominate $\gamma^{(k-1)}$, and this gives rise to a small Rayleigh quotient for α_k (see (2.4)), which in turn causes a large increase in the large components of the gradient. The effect over this, possibly over two or three iterations, is to cause the huge spike in the graph of $f^{(k)} - f^*$. (A similar effect is observed if $\|\gamma^{(k)}\|_2$ is graphed.) However the effect of these iterations is to significantly reduce the relative contribution of the small components in the gradient, *and it is only by allowing large increases in the large components that these small components can be efficiently removed.* Gradient methods which do not permit non-monotonic steps, or which limit their effect, are only able to remove the small components slowly, and hence suffer from slow convergence.

Looking at the spike around iteration 460 in Figure 3.1, this value of $f^{(k)}$ could only be accepted by Raydan's modification for a value of about $M = 200$, corresponding to the position of the previous higher spike. Therefore we see that the value of $M = 10$ used by Raydan severely restricts the amount of non-monotonicity that can occur. Moreover, the test (3.2) does not allow values of $f^{(k)}$ that are larger than $f^{(1)}$ to be accepted. For Laplace2(b), the value of $f^{(1)} - f^*$ is about 0.94×10^{-2}. Thus we see from Figure 3.1 that many of the early spike values would not be acceptable, and it is only after iteration 270 or thereabouts that there are no spike values for which $f^{(k)}$ is greater than $f^{(1)}$. Therefore this is another feature of Raydan's modification that restricts the amount of non-monotonicity. The above interpretation also explains why the numerical results obtained by Raydan are poor for very ill-conditioned problems. This is because, from (2.6), the non-monotonicity effects are most noticeable if the condition number is very large. We have seen that the value of $M = 10$ fails to allow the very large spikes to be accepted, which, as we argue above, is important for avoiding slow convergence in a gradient method.

Obvious suggestions to improve the performance of Raydan's modification are first to choose much larger values of M, especially if the problem is likely to be ill-conditioned. Another suggestion is to allow increases in $f^{(k)}$ up to a user supplied value $\bar{f} > f^{(1)}$ on early iterations. This is readily implemented by defining $f^{(k)} = \bar{f}$ for $k < 1$ and changing (3.2) so that the range of indices j is $k - M \le j \le k$. These changes make it more likely that the non-monotonic steps observed in Figure 3.1 are able to be accepted. On the other hand, although the convergence proof presented by Raydan would still hold, this extra freedom to accept 'bad' points might cause difficulties for very non-quadratic problems, and further research on how best to devise a non-monotone line search is needed.

Another idea to speed up Raydan's method is based on the observation that the unmodified BB method does not need to refer to values of the objective function. These are only needed when the non-monotone line search based on (3.2) is used. Therefore it is suggested that a non-monotone line search based on $\|\gamma\|$ is used. As in (3.2), an Armijo-type search is used along the line $\mathbf{x}^{(k)} + \mathbf{d}$ where $\mathbf{d} = -\theta \gamma^{(k)}$, using a sequence of values such as $\theta = 1, \frac{1}{10}, \frac{1}{100}, \dots$. An acceptance test such as

$$\|\gamma\|_2 \le \max_{k - M \le j \le k} \|\gamma^{(j)}\|_2 (1 - \gamma \theta \alpha) \tag{3.4}$$

would be used, where $\gamma \in (0, 1)$ is a small constant and we denote $\gamma = \gamma(\mathbf{x}^{(k)} + \mathbf{d})$ and $\alpha = \mathbf{d}^T(\gamma - \gamma^{(k)})/\mathbf{d}^T\mathbf{d}$, which is the prospective value of α_{k+1} (see (2.2)) if the step is accepted. Also we denote $\|\gamma^{(k)}\|_2 = \bar{g}$ for $k < 1$, where $\bar{g}$ is a user supplied upper limit on $\|\gamma^{(k)}\|_2$. The calculation shown in the last line of Table 3.1 is obtained by choosing M and $\bar{g}$ sufficiently large so that $\theta_k = 1$ for all k, and shows the benefit to be gained by not evaluating $f(\mathbf{x})$. To prove convergence it is necessary to show that $\mathbf{x}^{(k)} + \mathbf{d}$ would always be accepted for sufficiently small θ in the Armijo sequence. This is readily proved if $f(\mathbf{x})$ is a strictly convex function, as follows. Using the identity

$$\gamma^T \gamma = \gamma^{(k)T} \gamma^{(k)} + 2\gamma^{(k)T}(\gamma - \gamma^{(k)}) + (\gamma - \gamma^{(k)})^T(\gamma - \gamma^{(k)})$$

and a Taylor series for $\gamma(\mathbf{x}^{(k)} + \mathbf{d})$ about $\mathbf{x}^{(k)}$ we may obtain

$$\gamma^T \gamma = \gamma^{(k)T} \gamma^{(k)} + 2\gamma^{(k)T}(\gamma - \gamma^{(k)}) + o(\theta)$$

and hence using the binomial theorem that

$$\|\gamma\|_2 = \|\gamma^{(k)}\|_2(1 - \theta\alpha) + o(\theta),$$

where α is defined above. It follows from the Taylor series and the strict convexity of $f(\mathbf{x})$ that there exists a constant $\lambda > 0$ such that $\alpha \geq \lambda$, and consequently that

$$\|\gamma\|_2 \leq \|\gamma^{(k)}\|_2(1 - \theta\gamma\alpha)$$

if θ is sufficiently small. Thus we can improve on the most recent value $\|\gamma^{(k)}\|_2$ in this case, and hence (3.4) holds *a fortiori*.

4 DISCUSSION

One thing that I think emerges from this review is just how little we understand about the BB method. In the non-quadratic case, all the proofs of convergence use standard ideas for convergence of the steepest descent method with a line search, so do not tell us much about the BB method itself, so we shall restrict this discussion to the quadratic case. Here we have Raydan's ingenious proof of convergence Raydan (1993), but this is a proof by contradiction and does not explain for example why the method is significantly better than the classical steepest descent method. For the latter method we have the much more telling result of Akaike (1959) that the asymptotic rate of convergence is linear and the rate constant in the worst case (under the assumptions of Section 2) is $(\lambda_n - \lambda_1)/(\lambda_n + \lambda_1)$. This exactly matches what is observed in practice. This result is obtained by defining the vector $\mathbf{p}^{(k)}$ by $p_i^{(k)} = (g_i^{(k)})^2/\gamma^{(k)T}\gamma^{(k)}$ in the notation of Section 2. This vector acts like a probability distribution for $\gamma^{(k)}$ over the eigenvectors of A, insofar as it satisfies the conditions $\mathbf{p}^{(k)} \geq \mathbf{0}$ and $\mathbf{e}^T\mathbf{p}^{(k)} = 1$. Akaike shows that the components of $\mathbf{p}^{(k)}$ satisfy the recurrence relation

$$p_i^{(k+1)} = \frac{p_i^{(k)}(\lambda_i - \boldsymbol{\lambda}^T\mathbf{p}^{(k)})^2}{\sum_i p_i^{(k)}(\lambda_i - \boldsymbol{\lambda}^T\mathbf{p}^{(k)})^2} \tag{4.1}$$

and that the sequence $\{\mathbf{p}^{(k)}\}$ in the worst case oscillates between two accumulation points $\mathbf{e}_1$ and $\mathbf{e}_n$. Here the scalar product $\boldsymbol{\lambda}^T\mathbf{p}^{(k)}$ is just the Rayleigh quotient calculated from $\gamma^{(k)}$ (like (2.4) but using $\gamma^{(k)}$ on the right hand side). A similar analysis is possible for the BB method, in which a superficially similar

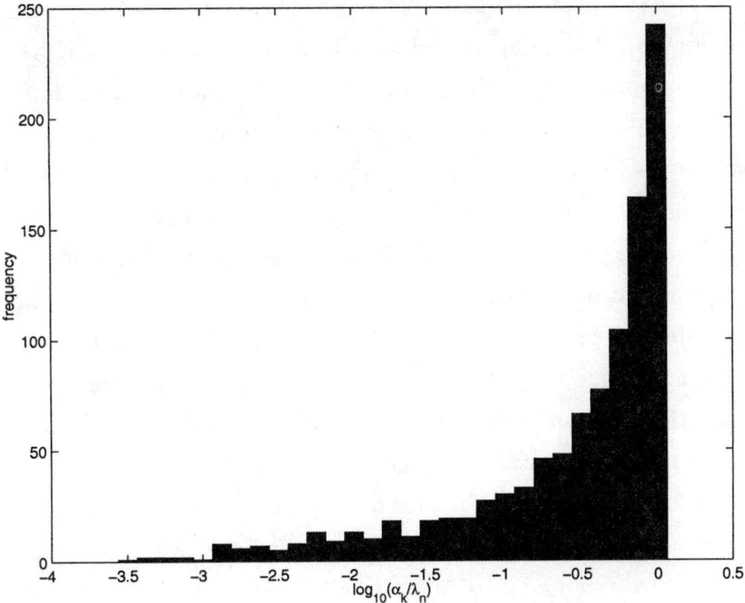

Figure 4.1   Distribution of α_k values for Laplace1(b)

two term recurrence

$$p_i^{(k+1)} = \frac{p_i^{(k)}(\lambda_i - \boldsymbol{\lambda}^T\mathbf{p}^{(k-1)})^2}{\sum_i p_i^{(k)}(\lambda_i - \boldsymbol{\lambda}^T\mathbf{p}^{(k-1)})^2} \tag{4.2}$$

is obtained. However the resulting sequence $\{\mathbf{p}^{(k)}\}$ shows no obvious pattern, and although it must have accumulation points, it is not obvious what they are (probably they include $\mathbf{e}_1$ and $\mathbf{e}_n$, but there may well be others), and the oscillatory behaviour of classical steepest descent is certainly not seen.

In an attempt to obtain further insight into the behaviour of the BB method, the distribution of the 1009 values of α_k obtained in Table 2.1 is graphed in Figure 3.2. It can be seen that a very characteristic pattern

is obtained, with most of the α_k values being generated in the vicinity of λ_n. The range of values is consistent with a condition number of $\lambda_n/\lambda_1 = 10^{3.61}$. It is also seen that there are very few values close to λ_1, and it is values at this end of the spectrum that give rise to the large non-monotonic spikes such as are seen in Figure 3.1. It is easily shown from (2.6) that any $\alpha_k \in (\frac{1}{2}\lambda_n, \lambda_n)$ guarantees to reduce *all* components $|g_i|$, so we see that the great majority of iterations cause an improvement in f, and only relatively few iterations give rise to an increase in f. I have observed the pattern of behaviour shown in Figure 3.2 on a number of ill-conditioned problems, although N. Gould (private communication) indicates that he has generated problems for which the distribution of the α_k does not show this pattern.

What we would like, and what we do not have, is a comprehensive theory that explains these phenomena, and gives a realistic estimate of the rate of convergence averaged over a large number of steps. A useful piece of information at any $\mathbf{x}^{(k)}$ would be a realistic bound on the number of iterations needed to obtain a sufficient improvement on the best value of $f(\mathbf{x})$ that has currently been obtained. This for example could be used in a watchdog-type test for non-quadratic functions, returning to the best previous iterate if the required improvement were not obtained in the said number of iterations. It would also be useful to have a theory that relates to how the eigenvalues λ_i are distributed within the spectrum. For example, if the eigenvalues are distributed in two clusters close to λ_1 and λ_n, then we would expect to be able to show rapid linear convergence, by virtue of the R-superlinear result for $n = 2$.

Then there is the possibility of alternative choices of the step in a steepest descent method. A range of possibilities have been suggested by Friedlander, Martínez, Molina and Raydan (1999), amongst which is the repeated use of a group of iterations in which a classical steepest descent step with $\alpha_k = \gamma^{(k)T} A \gamma^{(k)} / \gamma^{(k)T} \gamma^{(k)}$ is followed by the use of $\alpha_j = \alpha_k$ for the subsequent m iterations, $j = k + 1, \ldots, k + m$. For $m > 1$, this method can considerably increase the non-monotonic behaviour observed in the sequences $\{f(\mathbf{x}^{(k)})\}$ and $\{\|\gamma(\mathbf{x}^{(k)})\|_2\}$. This is shown from (2.6) because term $(1 - \lambda_i/\alpha_k)$ is repeated m times, so that the effect of a value of α_k close to λ_1 is to increase large components $|g_i|$ by a factor close to $(\lambda_n/\lambda_1)^m$ over the m iterations. Nonetheless, it seems overall that the effect of this modification is beneficial, and values up to say $m = 4$ can work well (Y-H. Dai, private communication). Clearly further study of these possibilities is called for.

The success of the BB and related methods for unconstrained optimization leads us to consider how it might be used for constrained optimization. This has already been considered for optimization subject to box constraints, and we review current progress in the next section. An important advance would be to find an effective BB-like method for large-scale linear systems involving the KKT matrix

$$\begin{bmatrix} A & B \\ B^T & O \end{bmatrix}.$$

Such an advance would be an important step in developing methods suitable for large scale quadratic programming, and this could lead to the development of methods for large scale nonlinear programming.

5 OPTIMIZATION WITH BOX CONSTRAINTS

Many methods have been suggested for solving optimization problems in which the constraints are just the simple bounds

$$\mathbf{l} \leq \mathbf{x} \leq \mathbf{u} \tag{5.1}$$

(see the references in Friedlander, Martínez and Raydan (1995) and Birgin, Martínez and Raydan (2000) for a comprehensive review). Use of the BB methodology is considered in two recent papers. That of Friedlander, Martínez and Raydan (1995) is applicable only to quadratic functions and uses an active set type strategy in which the iterates only leave the current face if the norm of reduced gradient is sufficiently small. No numerical results are given, and to me it seems preferable to be able to leave the current face at any time if the components of the gradient vector have the appropriate sign. Such an approach is allowed in the BB-like projected gradient method of Birgin, Martínez and Raydan (2000). This method is applicable to the minimization of a non-quadratic function on any closed convex set, although here we just consider the case of box constraints for which the required projections are readily calculated.

Birgin, Martinéz and Raydan give two methods, both of which use an Armijo-type search on a parameter θ. Both methods use an acceptance test similar to (3.2) which only require sufficient improvement on the largest of the $M+1$ most recent function values in the iteration. In Method 1, the projection

$$\mathbf{x}^+ = P(\mathbf{x}^{(k)} - \theta \gamma^{(k)} / \alpha_k)$$

is calculated, where α_k is the BB quotient given in (2.2). Then an Armijo search on θ is carried out until an acceptable point is obtained. In Method 2 the point

$$\mathbf{y} = P(\mathbf{x}^{(k)} - \gamma^{(k)} / \alpha_k)$$

is calculated, and an Armijo search is carried out along the line $\mathbf{x} = \mathbf{x}^{(k)} + \theta(\mathbf{y} - \mathbf{x}^{(k)})$. Both methods are proved to be globally convergent, by using a sufficient reduction property related to the projected gradient. Numerical results on a wide variety of CUTE test problems of dimension up to about 10^4 are described. These suggest that there is little to choose in practice between Methods 1 and 2, and that the performance is comparable with that of the LANCELOT method of Conn, Gould and Toint (1988), (1989).

There are a number of aspects in which improvements to Methods 1 and 2 might be sought. For a quadratic function we no longer have the assurance that the unmodified BB method converges (in contrast to Raydan's proof in Raydan (1993)), so that the methods rely on the Armijo search, and so are open to the criticisms described in Section 3. It would be nice therefore if a convergence theory for some BB-type projected gradient algorithm for the box constrained QP problem could be developed that does not rely on an Armijo search. Similar remarks hold in the non-quadratic case, and any developments for box constrained QP problems can be expected to have implications for the non-quadratic case. However, it will certainly be necessary to have modifications to allow for non-quadratic effects. Any developments for unconstrained optimization, of the sort referred to in Section 3, may well be relevant here. This could include for example the use of a watchdog-type algorithm that requires sufficient improvement over a fixed number of steps. Thus there are many challenging research topics in regard to BB-like methods that suggest themselves, and we can look forward to interesting developments in the future.

References

Akaike, H. (1959), On a successive transformation of probability distribution and its application to the analysis of the optimum gradient method, *Ann. Inst. Statist. Math. Tokyo*, Vol. 11, pp. 1-17.

Barzilai, J. and Borwein, J.M. (1988), Two-point step size gradient methods, *IMA J. Numerical Analysis*, Vol. 8, pp. 141-148.

Birgin, E.G., Martínez, J.M. and Raydan, M., Nonmonotone spectral projected gradient methods on convex sets, *SIAM J. Optimization*, Vol. 10, pp. 1196-1211.

Cauchy, A., (1847), Méthode générale pour la résolution des systèms d'équations simultanées, *Comp. Rend. Sci. Paris*, Vol. 25, pp. 536-538.

Conn, A.R., Gould, N.I.M. and Toint, Ph.L., (1988, 1989), Global convergence of a class of trust region algorithms for optimization with simple bounds, *SIAM J. Numerical Analysis*, Vol. 25, pp. 433-460, and Vol. 26, pp. 764-767.

Dai, Y.H. and Liao, L.-Z., (1999), R-linear convergence of the Barzilai and Borwein gradient method, Research report, (accepted by *IMA J. Numerical Analysis*).

Dembo, R.S., Eisenstat, S.C. and Steihaug, T., (1982), Inexact Newton Methods, *SIAM J. Numerical Analysis*, Vol. 19, pp. 400-408.

Fletcher, R., (1990), Low storage methods for unconstrained optimization, *Lectures in Applied Mathematics (AMS)* Vol. 26, pp. 165-179.

Fletcher, R. and Reeves, C.M., (1964), Function minimization by conjugate gradients, *Computer J.* Vol. 7, pp. 149-154.

Friedlander, A., Martínez, J.M. and Raydan, M., (1995), A new method for large-scale box constrained convex quadratic minimization problems, *Optimization Methods and Software*, Vol. 5, pp. 57-74.

Friedlander, A., Martínez, J.M., Molina, B., and Raydan, M., (1999), Gradient method with retards and generalizations, *SIAM J. Numerical Analysis*, Vol. 36, pp. 275-289.

Glunt, W., Hayden, T.L. and Raydan, M., (1993), Molecular conformations from distance matrices, *J. Comput. Chem.*, Vol. 14, pp. 114-120.

Glunt, W., Hayden, T.L. and Raydan, M., (1994), Preconditioners for Distance Matrix Algorithms, *J. Comput. Chem.*, Vol. 15, pp. 227-232.

G. H. Golub and C. F. Van Loan, (1996), *Matrix Computations*, 3rd Edition, The Johns Hopkins Press, Baltimore.

Grippo, L., Lampariello, F. and Lucidi, S., (1986), A nonmonotone line search technique for Newton's method, *SIAM J. Numerical Analysis*, Vol. 23, pp. 707-716.

Hestenes, M.R. and Stiefel, E.L., (1952), Methods of conjugate gradients for solving linear systems, *J. Res. Nat. Bur. Standards*, Sect. 5:49, pp. 409-436.

Molina, B. and Raydan, M., (1996), Preconditioned Barzilai-Borwein method for the numerical solution of partial differential equations, *Numerical Algorithms*, Vol. 13, pp. 45-60.

Nocedal, J., (1980), Updating quasi-Newton matrices with limited storage, *Math. of Comp.*, Vol. 35, pp. 773-782.

Polak, E., (1971), *Computational Methods in Optimization: A Unified Approach*, Academic Press, New York.

Shanno, D.F., and Phua, K.H., (1980), Remark on Algorithm 500: Minimization of unconstrained multivariate functions, *ACM Trans. Math. Software*, Vol. 6, pp. 618-622.

Raydan, M., (1993), On the Barzilai and Borwein choice of steplength for the gradient method, *IMA J. Numerical Analysis*, Vol. 13, pp. 321-326.

Raydan, M., (1997), The Barzilai and Borwein gradient method for the large scale unconstrained minimization problem, *SIAM J. Optimization*, Vol. 7, pp. 26-33.

Reid, J.K., (1971), *Large Sparse Sets of Linear Equations*, Academic Press, London, Chapter 11, pp. 231-254.

11 THE MODIFIED SUBGRAIDENT METHOD FOR EQUALITY CONSTRAINED NONCONVEX OPTIMIZATION PROBLEMS

Rafail N. Gasimov and Nergiz A. Ismayilova

Osmangazi University, Department of Industrial Engineering,
Bademlik 26030, Eskişehir, Turkey

Abstract: In this paper we use a sharp Lagrangian function to construct a dual problem to the nonconvex minimization problem with equality constraints. By using the strong duality results we modify the subgradient method for solving a dual problem constructed. The algorithm proposed in this paper has some advantages. In contrast with the penalty or multiplier methods, for improving the value of the dual function, one need not to take the "penalty like parameter" to infinity in the new method. The value of the dual function strongly increases at each iteration. The subgradient of the dual function along which its value increases is calculated explicitly. In contrast, by using the primal-dual gap, the proposed algorithm possesses a natural stopping criteria. We do not use any convexity and differentiability conditions, and show that the sequence of the values of dual function converges to the optimal value. Finally, we demonstrate the presented method on numerical examples.

Key words: Nonconvex programming, augmented Lagrangian, duality with zero gap, subgradient method.

1 INTRODUCTION

In recent years the duality of nonconvex optimization problems have attracted intensive attention. There are several approaches for constructing dual problems. Two of them are most popular: augmented Larangian functional approach (see, for example, Rockafellar (1993), Rockafellar and Wets (1998)) and nonlinear Lagrangian functional approach (see, for example, Rubinov (2000), Rubinov et al (to appear), Yang and Huang (2001)). The construction of dual problems and what is more the zero duality gap property are important as such the optimal value and sometimes the optimal solution of the original constrained optimization problem can be found by solving unconstrained optimization problem. The main purpose of both augmented and nonlinear Lagrangians is to reduce the problem in hand to a problem which is easier to solve. Therefore the justification of theoretical studies in this field is creating new numerical approaches and algorithms: for example, finding a way to compute the multipliers that achieve the zero duality gap or finding a Lagrangian function that is tractable numerically.

The theory of augmented Lagrangians, represented as the sum of the ordinary Lagrangian and an augmenting function, have been well developed for very general problems. Different augmented Lagrangians, including sharp Lagrangian function, can be obtained by taking special augmenting functions (see, for example, Rockafellar and Wets (1998)). In this paper we consider the nonconvex minimization problem with equality constraints and calculate the sharp Lagrangian function for this problem explicitly. By using the strong duality results and the simplicity of the obtained Lagrangian, we modify the subgradient method for solving the dual problem constructed. Note that subgradient methods were first introduced in the middle 60s; the works of Demyanov (1968), Poljak (1969a), (1969b), (1970) and Shor (1985), and (1995) were particularly influental. The convergence rate of subgradient methods is discussed in Goffin (1977). For further study of this subject see Bertsekas (1995) and Bazaraa et al (1993). This method were used for solving dual problems obtained by using ordinary Lagrangians or problems satisfying convexity conditions. However, our main purpose is to find an optimal value and an optimal solution to a nonconvex primal problem. We show that a dual function constructed in this paper is always concave. Without solving any additional problem, we explicitly calculate the subgradient of the dual function along which its value strongly increases. Therefore, in contrast with the penalty or multiplier methods, for improving the value of the dual function, one need not to take the "penalty like parameter" to infinity in the new method. The paper is outlined as follows. The construction of sharp Lagrangian function and zero duality gap properties are presented in Section 2. In Section 3 we explain the modified subgradient method, give convergence theorems and present the results of numerical experiments obtained by applying the presented method.

2 DUALITY

Consider the following equality constrained optimization problem (P):

$$\inf_{x \in X} \quad f_0(x)$$
$$\text{subject to} \quad f(x) = 0,$$

where $X \subseteq R^n$, $f(x) = (f_1(x), f_2(x), \cdots, f_m(x))$ and $f_i : X \to R$, $i = 0, 1, 2, \cdots, m$, are real-valued functions.

Let $\Phi : R^n \times R^m \to \overline{R}$ be a dualizing parametrization function defined as

$$\Phi(x, y) = \begin{cases} f_0(x) & \text{if } x \in X \text{ and } f(x) = y, \\ +\infty & \text{otherwise.} \end{cases}$$

The function $\beta : R^m \to \overline{R}$ defined by

$$\beta(y) = \inf_{x \in X} \Phi(x, y) \tag{2.1}$$

is called the *perturbation function*, corresponding to dualizing parametrization Φ. The augmented Lagrangian $L : R^n \times R^m \times R_+ \to \overline{R}$ associated with the problem (P) will be defined as

$$L(x, u, c) = \inf_{y \in Y} \left\{ \Phi(x, y) + c \, \|y\| - [y, u] \right\},$$

where $\|\cdot\|$ is any norm in R^m, $[y, u] = \sum_{i=1}^{m} y_i u_i$. By using the definition of Φ, we can calculate the augmented Lagrangian associated with (P) explicitly. For every $x \in X$ we have:

$$L(x, u, c) = \inf_{y = f(x)} \left\{ \Phi(x, y) + c \, \|y\| - [y, u] \right\} = f_0(x) + c \, \|f(x)\| - \sum_{i=1}^{m} u_i f_i(x).$$

Every element $x \in X_0$, where X_0 is a feasible set defined by

$$X_0 = [x \in X \,|\, f(x) = 0],$$

such that $f_0(x) = \inf P$ will be termed a solution of (P).

The dual function H is defined as:

$$H(u,c) = \inf_{x \in X} L(x,u,c), \quad \text{for} \ \ u \in R^m, c \in [0, +\infty).$$

Then, a dual problem of (P) is given by

$$(P^*) \qquad \sup P^* = \sup_{(u,c) \in R^m \times R_+} H(u,c).$$

Any element $(u,c) \in R^m \times R_+$ with $H(u,c) = \sup P^*$ is termed a solution of (P^*). The following lemma allows us to represent the primal problem (P) as an " inf sup" of the augmented Lagrangian L.

Lemma 2.1 *For every* $u, y \in R^m$, $y \neq 0$ *and for every* $r \in R_+$ *there exists* $c \in R_+$ *such that* $c\,\|y\| - [y, u] > r$.

Proof: Let $u, y \in R^m$, $y \neq 0$ and $r \in R_+$. We choose $c \in R_+$ with $c > \|u\| + r/\|y\|$. Then $c\,\|y\| - \|u\| \cdot \|y\| > r$. Since $\|u\| \cdot \|y\| \geq [y, u]$, we have $c\,\|y\| - [y, u] > r$.

It follows from this lemma that

$$\sup_{(u,c) \in R^m \times R_+} L(x,u,c) = \begin{cases} f_0(x), & x \in X_0, \\ +\infty, & x \notin X_0. \end{cases}$$

Hence,

$$\inf_{x \in X} \ \sup_{(u,c) \in R^m \times R_+} L(x,u,c) = \inf\{f_0(x) \,|\, x \in X_0\} = \inf P. \qquad (2.2)$$

This means that the value of a mathematical programming problem with equality constraints can be represented as (2.2), regardless of properties the original problem satisfies.

Proofs of the following four theorems are analogous to the proofs of similar theorems earlier presented for augmented Lagrangian functions with quadratic or general augmenting functions. See, for example, Rockafellar (1993) and Rockafellar and Wets (1998).

Theorem 2.1 $\inf P \geq \sup P^*$.

Theorem 2.2 *Suppose that* $\inf P$ *is finite. Then a pair of elements* $\overline{x} \in X$ *and* $(\overline{u}, \overline{c}) \in R^m \times R_+$ *furnishes a saddle point of the augmented Lagrangian* L *on* $X \times (R^m \times R_+)$ *if and only if* $\overline{x}$ *is a solution to* (P), $(\overline{u}, \overline{c})$ *is a solution to* (P^*) *and* $\inf P = \sup P^*$.

Theorem 2.3 *A pair of vectors* $x \in X$ *and* $(u, c) \in R^m \times R_+$ *furnishes a saddle point of the augmented Lagrangian* L *on* $X \times (R^m \times R_+)$ *if and only if*

$$\left. \begin{array}{c} x \text{ is a solution to } (P), \\ \beta\left(y\right) \geq \beta\left(0\right) + \left[y, u\right] - c\left\|y\right\| \text{ for all } y, \end{array} \right\}$$

where β *is a perturbation function defined by* (??)*. When this holds, any* $a > c$ *will have the property that*

$$\left[x \text{ solves } (P)\right] \leftrightarrow \left[x \text{ minimizes } L\left(z, u, a\right) \text{ over } z \in X\right].$$

Theorem 2.4 *Suppose in* (P) *that* f_0 *and* f *are continuous,* X *is compact, and a feasible solution exists. Then* $\inf P = \sup P^*$ *and there exists a solution to* (P)*. Furthermore, in this case, the dual function* H *in* (P^*) *is concave and finite everywhere on* $R^m \times R_+$*, so this maximization problem is effectively unconstrained.*

The following theorem will also be used as a stopping criteria in solution algorithm for dual problem in the next section.

Theorem 2.5 *Let* $\inf P = \sup P^*$ *and suppose that for some* $(\overline{u}, \overline{c}) \in R^m \times R_+$*, and* $\overline{x} \in X$*,*

$$\min_{x \in X} L\left(x, \overline{u}, \overline{c}\right) = f_0\left(\overline{x}\right) + \overline{c}\left\|f\left(\overline{x}\right)\right\| - \left[f\left(\overline{x}\right), \overline{u}\right]. \tag{2.3}$$

Then $\overline{x}$ *is a solution to* (P) *and* $(\overline{u}, \overline{c})$ *is a solution to* (P^*) *if and only if*

$$f\left(\overline{x}\right) = 0. \tag{2.4}$$

Proof: Necessity. If (2.3) holds and $\overline{x}$ is a solution to (P) then $\overline{x}$ is feasible and therefore $f\left(\overline{x}\right) = 0$.

Sufficiency. Suppose to the contrary that (2.3) and (2.4) are satisfied but $\overline{x}$ and $(\overline{u}, \overline{c})$ are not solutions. Then, there exists $\widetilde{x} \in X_0$ such that $f_0(\widetilde{x}) < f_0(\overline{x})$. Hence

$$
\begin{aligned}
f_0(\widetilde{x}) \ &< \ f_0(\overline{x}) = f_0(\overline{x}) + \overline{c} \, \|f(\overline{x})\| - [f(\overline{x}), \overline{u}] = H(\overline{u}, \overline{c}) \\
&= \ \min_{x \in X} L(x, \overline{u}, \overline{c}) \leq \sup_{(u,c) \in R^m \times R_+} \min_{x \in X} L(x, u, c) = \sup P^* \\
&= \ \inf P \leq f_0(\widetilde{x}),
\end{aligned}
$$

which proves the theorem.

3   SOLVING THE DUAL PROBLEM

We have described several properties of the dual function in the previous section. In this section, we utilize these properties to modify the subgradient method for maximizing the dual function H. Theorem 2.2, Theorem 2.3 and Theorem 2.4 give necessary and sufficient conditions for an equality between $\inf P$ and $\sup P^*$. Therefore, when the hypotheses of these theorems are satisfied, the maximization of the dual function H will give us the optimal value of the primal problem.

We consider the dual problem

$$
maximize \quad H(u, c) = \min_{x \in X} L(x, u, c) = \min_{x \in X} \{ f_0(x) + c\,\|f(x)\| - [u, f(x)] \}
$$

$$
\text{subject to } (u, c) \in F = R^m \times R_+,
$$

It will be convenient to introduce the following set:

$$
X(u, c) = Arg \min_{x \in X} \{ f_0(x) + c\,\|f(x)\| - [u, f(x)] \}.
$$

The assertion of the following theorem can be obtained from the known theorems on the subdifferentials of the continuous maximum and minimum functions. See, for example, Polak (1997).

Theorem 3.1 *Let X be a nonempty compact set in R^n and let f_0 and f be continuous, so that for any $(\overline{u}, \overline{c}) \in R^m \times R_+$, $X(\overline{u}, \overline{c})$ is not empty. If $\overline{x} \in X(\overline{u}, \overline{c})$, then $(-f(\overline{x}), \|f(\overline{x})\|)$ is a subgradient of H at $(\overline{u}, \overline{c})$.*

3.1   Subgradient Method

Initialization Step Choose a vector (u_1, c_1) with $c_1 \geq 0$, let $k = 1$, and go to the main step.

Main Step 1. Given (u_k, c_k), solve the following subproblem:

$$minimize \qquad f_0(x) + c_k \|f(x)\| - [u_k, f(x)]$$
$$\text{subject to} \qquad x \in X.$$

Let x_k be any solution. If $f(x_k) = 0$, then stop; by Theorem 2.5, (u_k, c_k) is a solution to (P^*), x_k is a solution to (P). Otherwise, go to step 2.

2. Let

$$u_{k+1} = u_k - s_k f(x_k), c_{k+1} = c_k + (s_k + \varepsilon_k)\|f(x_k)\|, \qquad (3.1)$$

where s_k and ε_k are positive scalar stepsizes, replace k by $k+1$, and repeat step 1.

The following theorem shows that in contrast with the subgradient methods developed for dual problems formulated by using ordinary Lagrangians, the new iterate strictly improves the cost for all values of the stepsizes s_k and ε_k.

Theorem 3.2 *Suppose that the pair* $(u_k, c_k) \in R^m \times R_+$ *is not a solution to the dual problem and* $x_k \in X(u_k, c_k)$. *Then for a new iterate* (u_{k+1}, c_{k+1}) *calculated from (3.1) for all positive scalar stepsizes* s_k *and* ε_k *we have:*

$$0 < H(u_{k+1}, c_{k+1}) - H(u_k, c_k) \leq (2s_k + \varepsilon_k)\|f(x_k)\|^2.$$

Proof: Let $(u_k, c_k) \in R^m \times R_+$, $x_k \in X(u_k, c_k)$ and (u_{k+1}, c_{k+1}) is a new iterate calculated from (3.1) for arbitrary positive scalar stepsize s_k and ε_k. Then by Theorem 3.1, the vector $(-f(x_k), \|f(x_k)\|) \in R^m \times R_+$ is a subgradient of a concave function H at (u_k, c_k) and by definition of subgradients we have:

$$H(u_{k+1}, c_{k+1}) - H(u_k, c_k)$$
$$\leq \quad [(u_{k+1} - u_k), (-f(x_k))] + (c_{k+1} - c_k)\|f(x_k)\|$$
$$= \quad s_k \|f(x_k)\|^2 + (s_k + \varepsilon_k)\|f(x_k)\|^2 = (2s_k + \varepsilon_k)\|f(x_k)\|^2$$

On the other hand

$$H(u_{k+1}, c_{k+1})$$

$$
\begin{aligned}
&= \min_{x \in X} \left\{ f_0\left(x\right) + c_{k+1} \left\| f\left(x\right) \right\| - \left[u_{k+1}, f\left(x\right) \right] \right\} \\
&\geq \min_{x \in X} \left\{ f_0\left(x\right) + c_k \left\| f\left(x\right) \right\| - \left[u_k, f\left(x\right) \right] + \varepsilon_k \left\| f\left(x_k\right) \right\| \left\| f\left(x\right) \right\| \right\} \\
&= \min_{x \in X} \left\{ f_0\left(x\right) + \left(c_k + \varepsilon_k \left\| f\left(x_k\right) \right\| \right) \left\| f\left(x\right) \right\| - \left[u_k, f\left(x\right) \right] \right\} .
\end{aligned}
$$

Now suppose that the last minimum attains for a some $\widetilde{x} \in X$. If $f\left(\widetilde{x}\right)$ were zero, then by Theorem 2.5, the pair $\left(u_k, c_k + \varepsilon_k \left\| f\left(x_k\right) \right\| \right)$ would be a solution to the dual problem and therefore

$$
\begin{aligned}
&\min_{x \in X} \left\{ f_0\left(x\right) + \left(c_k + \varepsilon_k \left\| f\left(x_k\right) \right\| \right) \left\| f\left(x\right) \right\| - \left[u_k, f\left(x\right) \right] \right\} \\
&> \min_{x \in X} \left\{ f_0\left(x\right) + c_k \left\| f\left(x\right) \right\| - \left[u_k, f\left(x\right) \right] \right\} = H\left(u_k, c_k \right),
\end{aligned}
$$

because of $\left(u_k, c_k \right)$ is not a solution. When $f\left(\widetilde{x}\right) \neq 0$ then

$$
\begin{aligned}
&\min_{x \in X} \left\{ f_0\left(x\right) + \left(c_k + \varepsilon_k \left\| f\left(x_k\right) \right\| \right) \left\| f\left(x\right) \right\| - \left[u_k, f\left(x\right) \right] \right\} \\
&= f_0\left(\widetilde{x}\right) + \left(c_k + \varepsilon_k \left\| f\left(x_k\right) \right\| \right) \left\| f\left(\widetilde{x}\right) \right\| - \left[u_k, f\left(\widetilde{x}\right) \right] \\
&> f_0\left(\widetilde{x}\right) + c_k \left\| f\left(\widetilde{x}\right) \right\| - \left[u_k, f\left(\widetilde{x}\right) \right] \\
&\geq \min_{x \in X} \left\{ f_0\left(x\right) + c_k \left\| f\left(x\right) \right\| - \left[u_k, f\left(x\right) \right] \right\} = H\left(u_k, c_k \right).
\end{aligned}
$$

Thus we have established that $H\left(u_{k+1}, c_{k+1} \right) > H\left(u_k, c_k \right)$. The following theorem demonstrates that for the certain values of stepsizes s_k and ε_k, the distance between the points generated by the algorithm and the solution to the dual problem decreases at each iteration (cf. Bertsekas (1995), Proposition 6.3.1).

Theorem 3.3 *Let $\left(u_k, c_k \right)$ be any iteration, which is not a solution to the dual problem, so $f\left(x_k\right) \neq 0$. Then for any dual solution $\left(\overline{u}, \overline{c} \right)$, we have*

$$
\left\| \left(\overline{u}, \overline{c} \right) - \left(u_{k+1}, c_{k+1} \right) \right\| < \left\| \left(\overline{u}, \overline{c} \right) - \left(u_k, c_k \right) \right\|
$$

for all stepsizes s_k such that

$$
0 < s_k < \frac{2 \left(H\left(\overline{u}, \overline{c} \right) - H\left(u_k, c_k \right) \right)}{5 \left\| f\left(x_k\right) \right\|^2}, \tag{3.2}
$$

and $0 < \varepsilon_k < s_k$.

Proof: We have

$$
\left\| \left(\overline{u}, \overline{c} \right) - \left(u_{k+1}, c_{k+1} \right) \right\|^2 = \left\| \overline{u} - u_{k+1} \right\|^2 + \left| \overline{c} - c_{k+1} \right|^2
$$

$$
\begin{aligned}
&= \ \left\| \overline{u} - (u_k - s_k f(x_k)) \right\|^2 + \left| \overline{c} - (c_k + (s_k + \varepsilon_k) \| f(x_k) \|) \right|^2 \\
&= \ \left\| \overline{u} - u_k \right\|^2 + 2 s_k \left[(\overline{u} - u_k), f(x_k) \right] + (s_k)^2 \| f(x_k) \|^2 \\
&\quad + (\overline{c} - c_k)^2 - 2 (s_k + \varepsilon_k)(\overline{c} - c_k) \| f(x_k) \| + (s_k + \varepsilon_k)^2 \| f(x_k) \|^2 \\
&< \ \left\| \overline{u} - u_k \right\|^2 + 2 s_k \left[(\overline{u} - u_k), f(x_k) \right] + (s_k)^2 \| f(x_k) \|^2 \\
&\quad + (\overline{c} - c_k)^2 - 2 s_k (\overline{c} - c_k) \| f(x_k) \| + (2 s_k)^2 \| f(x_k) \|^2 ,
\end{aligned}
$$

where the last inequality is a result of inequalities $\overline{c} - c_k > 0$, $\| f(x_k) \| > 0$, and $0 < \varepsilon_k < s_k$. Now, by using the subgradient inequality

$$
H(\overline{u}, \overline{c}) - H(u_k, c_k) \leq \left[(\overline{u} - u_k), (-f(x_k)) \right] + (\overline{c} - c_k) \| f(x_k) \| ,
$$

we obtain

$$
\begin{aligned}
& \left\| \overline{u} - u_{k+1} \right\|^2 + \left| \overline{c} - c_{k+1} \right|^2 \\
< \ & \left\| \overline{u} - u_k \right\|^2 + \left| \overline{c} - c_k \right|^2 - 2 s_k \left(H(\overline{u}, \overline{c}) - H(u_k, c_k) \right) \\
+ \ & 5 (s_k)^2 \| f(x_k) \|^2 .
\end{aligned}
\tag{3.3}
$$

It is straightforward to verify that for the range of stepsize of equation (3.2) the sum of the last two terms in the above relation is negative. Thus,

$$
\left\| \overline{u} - u_{k+1} \right\|^2 + \left| \overline{c} - c_{k+1} \right|^2 < \left\| \overline{u} - u_k \right\|^2 + \left| \overline{c} - c_k \right|^2 ,
$$

and the theorem is proved. The inequality (3.3) can also be used to establish the convergence theorem for the subgradient method.

Theorem 3.4 *Assume that all conditions of Theorem 2.4 are satisfied. Let (u_k, c_k) be any iteration of the subgradient method. Suppose that each new iteration (u_{k+1}, c_{k+1}) is calculated from (3.1) for the stepsizes*

$$
s_k = \frac{\overline{H} - H(u_k, c_k)}{5 \| f(x_k) \|^2} \quad \text{and } 0 < \varepsilon_k < s_k,
$$

where $\overline{H} = H(\overline{u}, \overline{c})$ denotes the optimal dual value. Then $H(u_k, c_k) \to \overline{H}$.

Proof: By taking $s_k = \frac{\overline{H} - H(u_k, c_k)}{5 \| f(x_k) \|^2}$ in (3.3) we obtain:

$$
\left\| \overline{u} - u_{k+1} \right\|^2 + \left| \overline{c} - c_{k+1} \right|^2 < \left\| \overline{u} - u_k \right\|^2 + \left| \overline{c} - c_k \right|^2 - \frac{\left(\overline{H} - H(u_k, c_k) \right)^2}{5 \| f(x_k) \|^2},
$$

which can be written in the form

$$\left(\overline{H} - H\left(u_k, c_k\right)\right)^2 / 5$$
$$< \left\|f\left(x_k\right)\right\|^2 \left[\left(\left\|\overline{u} - u_k\right\|^2 + \left|\overline{c} - c_k\right|^2\right) - \left(\left\|\overline{u} - u_{k+1}\right\|^2 + \left|\overline{c} - c_{k+1}\right|^2\right)\right]. \quad (3.4)$$

It is obvious that, the sequence $\left\{\left\|\overline{u} - u_k\right\|^2 + \left|\overline{c} - c_k\right|^2\right\}$ is bounded from below (for example, by zero), and by Theorem 3.3, it is decreasing. Thus, $\left\{\left\|\overline{u} - u_k\right\|^2 + \left|\overline{c} - c_k\right|^2\right\}$ is a convergent sequence. Hence

$$\lim_{k \to \infty} \left\{\left(\left\|\overline{u} - u_k\right\|^2 + \left|\overline{c} - c_k\right|^2\right) - \left(\left\|\overline{u} - u_{k+1}\right\|^2 + \left|\overline{c} - c_{k+1}\right|^2\right)\right\} = 0.$$

On the other hand, since X is a compact set and f is continuous, $\left\{5\left\|f\left(x_k\right)\right\|^2\right\}$ is a bounded sequence. Thus, (3.4) implies

$$H\left(u_k, c_k\right) \to \overline{H}.$$

Unfortunately, however, unless we know the dual optimal value $H\left(\overline{u}, \overline{c}\right)$, which is rare, the range of stepsize is unknown. In practice, one can use the stepsize formula

$$s_k = \frac{\alpha_k \left(H_k - H\left(u_k, c_k\right)\right)}{5\left\|f\left(x_k\right)\right\|^2}, \quad (3.5)$$

where H_k is an approximation to the optimal dual value and $0 < \alpha_k < 2$. By Theorem 3.2, the sequence $\left\{H\left(u_k, c_k\right)\right\}$ is increasing, therefore to estimate the optimal dual value from below, we can use the current dual value $H\left(u_k, c_k\right)$. As an upper bound, we can use any primal value $f_0\left(\overline{x}\right)$ corresponding to a primal feasible solution $\overline{x}$.

Now we demonstrate the proposed algorithm on some examples. In examples 3.1, 3.2 and 3.3, below, for finding an $x^k \in X\left(u_k, c_k\right)$, the MATLAB function m-file fminsearch is utilized. In all examples the stepsize parameter ε^k was taken as $0.95s^k$ and s^k is calculated from (3.5) with $\alpha_k = 1$.

Example 3.1 *(see Himmelblau (1972))*

$$f_0\left(x\right) = 1000 - x_1^2 - 2x_2^2 - x_3^2 - x_1 x_2 - x_1 x_3 \to \min$$

subject to

$$f_1\left(x\right) = x_1^2 + x_2^2 + x_3^2 - 25 = 0,$$
$$f_2\left(x\right) = 8x_1 + 14x_2 + 7x_3 - 56 = 0$$

and

$$x_i \geq 0, i = 1, 2, 3.$$

The result reported in Himmelblau (1972) is

$$x^* = (3.512, 0.217, 3.552), \ f_0^* = 961.715.$$

where the constraint is satisfied with

$$\|f(x^*)\| = 2.9 \times 10^{-3}.$$

Through this implementation of the modified subgradient algorithm given above, the result is obtained in a single iteration, starting with the initial guesses $x_0 = 0$, $u_0 = 0$ and $c_0 = 0$. The positivity constraints $x_i \geq 0$, $i = 1, 2, 3$, are eliminated by defining new variables y_i, such that $y_i^2 = x_i$, and then the problem is solved for these y_i s. The obtained result is $x^* = (3.5120790172, \ 0.2169913063, \ 3.5522127962), f_0^* = 961.7151721335$. The constraint is satisfied with this x^* as $\|f(x^*)\| = 9.2 \times 10^{-10}$.

Example 3.2 *(see Khenkin (1976))*

$$f_0(x) = 0.5(x_1 + x_2)^2 + 50(x_2 - x_1)^2 + \sin^2(x_1 + x_2) \to \min$$

subject to

$$f(x) = (x_1 - 1)^2 + (x_2 - 1)^2 + (\sin(x_1 + x_2) - 1)^2 - 1.5 \leq 0.$$

The result reported in Khenkin (1976) is

$$x^* = (0.229014, 0.229014), f_0^* = 0.3004190265$$

where the constraint is satisfied with

$$\|f(x^*)\| = 1.7 \times 10^{-6}.$$

Through this implementation of the modified subgradient algorithm, the result is obtained in a single iteration, starting with the initial guesses $x_0 = 0$, $u_0 = 0$ and $c_0 = 0$. A slack variable has been added to the constraint to convert it to an equality constraint. The obtained result is $x^* = (0.22901434, 0.22901434), f_0^* = 0.300419026502$. The constraint is satisfied with this x^* as $\|f(x^*)\| = 8.1 \times 10^{-9}$.

Example 3.3 *(see Khenkin (1976))*

$$f_0\left(x\right) = 0.5\left(x_1 + x_2\right)^2 + 50\left(x_2 - x_1\right)^2 + x_3^2 + \left|x_3 - \sin\left(x_1 + x_2\right)\right| \to \min$$

subject to

$$f\left(x\right) = \left(x_1 - 1\right)^2 + \left(x_2 - 1\right)^2 + \left(x_3 - 1\right)^2 - 1.5 \leq 0.$$

The result reported in Khenkin (1976) is

$$x^* = \left(0.229014, 0.229014, 0.4421181\right), f_0^* = 0.3004190265$$

where the constraint is satisfied with

$$\left\|f\left(x^*\right)\right\| = 4.1 \times 10^{-4}\ .$$

Through this implementation of the subgradient algorithm, the result is obtained in two iterations, starting with the initial guesses $x_0 = 0$, $u_0 = 0$ and $c_0 = 0$. A slack variable has been added to the constraint to convert it to an equality constraint. The obtained result is

$$x^* = \left(0.22901434, 0.22901434, 0.44218084\right), f_0^* = 0.300419026502.$$

The constraint is satisfied with the above x^* as

$$\left\|f\left(x^*\right)\right\| = 8.9 \times 10^{-9}\ .$$

Note that only the range 0.7-1.2 for an estimate of H as an upper bound in the formula for s_k seems to be giving the correct answer. Outside this range, the number of iterations gets bigger.

Example 3.4 *(see Floudas (1999))*

$$f_0\left(x, y\right) = \left[a, x\right] - 0.5\left[x, Qx\right] + by \to \min$$

subject to

$$f_1\left(x\right) = 6x_1 + 3x_2 + 3x_3 + 2x_4 + x_5 - 6.5 \leq 0$$

$$f_2\left(x\right) = 10x_1 + 10x_3 + y - 20 \leq 0$$

and $0 \leq x_i \leq 1$, $i = 1, 2, 3, 4, 5$, $y \geq 0$, where $a = \left(-10.5, -7.5, -3.5, -2.5, -1.5\right)$, $b = -10$, $Q = 100I$, and I is the identity matrix.

The result reported in Floudas (1999) is

$$x^* = (0, 1, 0, 1, 1), f_0^* = -361.5.$$

where the constraint is satisfied with

$$\|f(x^*)\| = 0.$$

Through this implementation of the subgradient algorithm, the result is obtained in a single iteration, starting with the initial guesses $x_0 = 0$, $u_0 = 0$ and $c_0 = 0$. A slack variable has been added to the constraint to convert it to an equality constraint. For finding an $x^* = x^1 \in X(u_1, c_1)$, the LINGO 6.0 is utilized. The obtained result is the

$$u_1^1 = -1.7117, u_2^1 = -4.0275, c^1 = 8.5335, x^* = (0, 1, 0, 1, 1), f_0^* = -361.5.$$

The upper bound for H was taken as $H = 20$.

Acknowledgments

The authors wish to thank Drs. Y. Kaya and R.S. Burachik for useful discussions and comments.

References

Andramanov, M.Yu., Rubinov, A.M., and Glover, B.M. (1997), Cutting angle method for minimizing increasing convex-along-rays functions, *Research Report* 97/7, SITMS, University of Ballarat, Australia.

Andramanov, M.Yu., Rubinov, A.M. and Glover, B.M. (1999), Cutting angle methods in global optimization, *Applied Mathematics Letters*, Vol. 12, pp. 95-100.

Bazaraa, M.S., Sherali, H.D. and Shetty, C.M. (1993), *Nonlinear programming. Theory and Algorithms* , JohnWiley& Sons, Inc., New York.

Bertsekas, D.P. (1995), *Nonlinear Programming*, Athena Scientific, Belmont, Massachusetts.

Demyanov, V. F. (1968), Algorithms for some Minimax Problems, J. Computer and System Sciences, 2, 342-380.

Floudas, C.A., et al. (1999), *Handbook of test problems in local and global optimization*, Kluwer Academic Publishers, Dordrecht.

Goffin, J.L. (1977), On convergent rates of subgradient optimization methods, *Mathematical Programming* , Vol. 13, pp. 329-347.

Himmelblau, D.M. (1972), *Applied nonlinear optimization*, McGraw-Hill Book Company.

Khenkin, E.I. (1976), A search algorithm for general problem of mathematical programming, *USSR Journal of Computational Mathematics and Mathematical Physics* , Vol. 16, pp. 61-71, (in Russian).

Polak, E. (1997), *Optimization. Algorithms and consistent approximations*, Springer-Verlag.

Poljak, B.T. (1969a), Minimization of unsmooth functionals, *Z. Vychislitelnoy Matematiki i Matematicheskoy Fiziki* , Vol. 9, pp. 14-29.

Poljak, B.T. (1969b), The conjugate gradient method in extremal problems, *Z. Vychislitelnoy Matematiki i Matematicheskoy Fiziki* , Vol. 9, pp. 94-112.

Poljak, B.T. (1970), Iterative methods using Lagrange multipliers for solving extremal problems with constraints of the equation type,*Z. Vychislitelnoy Matematiki i Matematicheskoy Fiziki*, Vol. 10, pp. 1098-1106.

Rockafellar, R.T. (1993), Lagrange Multipliers and Optimality, *SIAM Review*, Vol. 35, pp. 183-238.

Rockafellar, R.T. and Wets, R.J-B. (1998), *Variational Analysis*, Springer Verlag, Berlin.

Rubinov, A.M. (2000), *Abstract convexity and global optimization* , Kluwer Academic Publishers.

Rubinov, A.M., Glover, B.M. and Yang, X.Q. (1999), Decreasing Functions with Applications to Penalizations, *SIAM J. Optimization* , Vol. 10, No. 1, pp.289-313.

Rubinov, A. M., Yang, X.Q., Bagirov, A.M. and Gasimov, R. N., Lagrange-type functions in constrained optimization, *Journal of Mathematical Sciences* , (to appear).

Shor, N.Z. (1985), *Minimization methods for nondifferentiable functions*, Springer Verlag, Berlin.

Shor, N.Z. (1995), Dual Estimates in Multiextremal Problems, *Journal of Global optimization*, Vol. 7, pp. 75-91.

Yang, X.Q. and Huang X.X. (2001), A nonlinear Lagrangian approach to constrained optimization problems, *SIAM J. Optimization* , Vol. 14, pp. 1119 - 1144.

12   INEXACT RESTORATION METHODS FOR NONLINEAR PROGRAMMING: ADVANCES AND PERSPECTIVES

José Mario Martínez

Departamento de Matemática Aplicada,
IMECC-UNICAMP, CP 6065, 13081-970 Campinas SP, Brazil. (martinez@ime.unicamp.br)

and Elvio A. Pilotta

Facultad de Matemática, Astronomía y Física, FaMAF, Universidad Nacional de
Córdoba, CIEM, Cdad. Universitaria (5000) Córdoba, Argentina. (pilotta@mate.uncor.edu)

Abstract: Inexact Restoration methods have been introduced in the last few years for solving nonlinear programming problems. These methods are related to classical restoration algorithms but also have some remarkable differences. They generate a sequence of generally infeasible iterates with intermediate iterations that consist of inexactly restored points. The convergence theory allows one to use arbitrary algorithms for performing the restoration. This feature is appealing because it allows one to use the structure of the problem in quite opportunistic ways. Different Inexact Restoration algorithms are available. The most recent ones use the trust-region approach. However, unlike the algorithms based on sequential quadratic programming, the trust regions are centered not in the current point but in the inexactly restored intermediate one. Global convergence has been proved, based on merit functions of augmented Lagrangian type. In this survey we point out some applications and we relate recent advances in the theory.

Key words: Nonlinear programming, trust regions, GRG methods, SGRA methods, projected gradients, global convergence.

1 INTRODUCTION

A classical problem in numerical optimization is the minimization of a general function with nonlinear constraints. Except in very particular cases, analytical solutions for this problem are not available. Therefore, iterative methods must be used. One of the most natural ideas is to generate a sequence of feasible points $\{x^k\}$ that satisfy the constraints and such that the objective function f is progressively decreased. In this way, one expects that, in the limit, a solution of the minimization problem will be obtained. However, maintaining feasibility of the iterates when the constraints are nonlinear is difficult. Usually, one takes a direction d^k for moving away from the "current point" x^k but, if the constraints are nonlinear, the points of the form $x^k + \alpha d^k$ might be infeasible, no matter how small the parameter α could be. On the other hand, in the linearly constrained case, one can always take a "feasible direction", that guarantees feasibility of $x^k + \alpha d^k$ for $\alpha > 0$ small enough.

The observations above imply that, when nonlinear constraints are present, one must be prepared to lose feasibility and to restore feasibility from time to time. The process of coming back to a feasible point from a nonfeasible one is called "restoration". Usually, it involves the solution of a nonlinear system of equations, perhaps underdetermined.

The most classical restoration methods are Rosen's gradient-projection method Rosen (1960); Rosen (1961), the sequential gradient-restoration algorithms (SGRA) of Miele and coworkers Miele et al (1969); Miele et al (1971); Miele et al (1983) and the GRG method introduced by Abadie and Carpentier. See Abadie and Carpentier (1968); Drud (1985); Lasdon (1982). Roughly speaking, these methods proceed in the following way: given a feasible point x^k, a trial point w is found satisfying a linear approximation of the constraints and such that the functional (or the Lagrangian) value at w is sufficiently smaller than the corresponding value at x^k. Since, in general, w is not feasible, a restoration process is necessary to obtain a feasible z^k. If $f(z^k)$ is not sufficiently smaller than $f(x^k)$, the allowed distance between x^k and w must be decreased.

The main difference between the different classical methods is the way in which restoration is performed. See Rom and Avriel (1989a); Rom and Avriel (1989b). In the gradient-projection method of Rosen the Jacobian of the constraints is fixed at w and we seek the restored point in the orthogonal subspace,

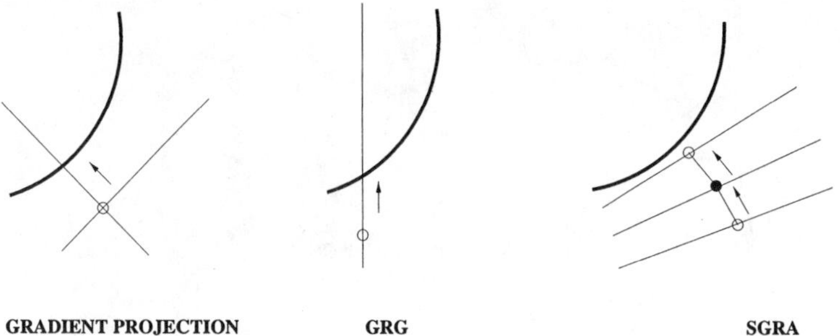

Figure 1.1   Restoration in classical methods.

by means of a modified Newton-like procedure. In the GRG method some (non-basic) variables are fixed and only the remaining "basic" variables are moved in order to get the feasible point. A Newton method with fixed Jacobian is also used for solving the resulting square system of nonlinear equations. The SGRA methods consider the problem of restoration as an underdetermined nonlinear system of equations and solve it using an underdetermined Newton method with damping, for which global convergence can be proved. See Figure 1.1.

The nonlinearity of the constraints impose that restoration procedures must be iterative. Therefore, strictly speaking, a feasible point is rarely obtained and restoration is always inexact. This means that practical implementations include careful decisions with respect to the accuracy at which the restored points must be declared feasible. The tolerances related to feasibility were incorporated in the rigorous definition of an algorithm by Mukai and Polak (1978). Rom and Avriel (1989b) used the Mukai-Polak scheme to prove unified convergence theorems for the classical restoration algorithms using progressive reduction of the feasibility tolerance.

This survey is organized as follows. The main Inexact Restoration ideas are given in Section 2. In Section 3 we give a detailed description of a recently introduced IR algorithm and we state convergence results. In Section 4 we comment the AGP optimality condition, which is useful to analyze limit points of IR methods. In Section 5 we describe the application to Order-Value Optimization. In Section 6 we comment the application to Bilevel programming. In Section 7 we describe an application that identifies feasible regions with homotopy curves for solving nonlinear systems. Conclusions are stated in Section 8.

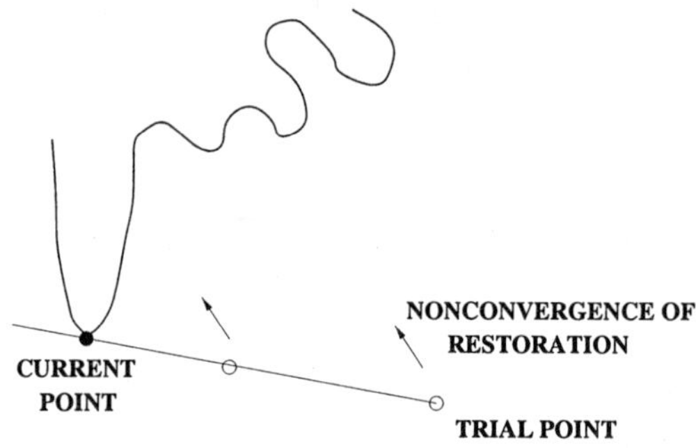

Figure 2.1   Short steps far from solution in classical restoration.

2   MAIN INEXACT RESTORATION IDEAS

The main drawback of feasible methods is that they tend to behave badly in the presence of strong nonlinearities, usually represented by very curved constraints. In these cases, it is not possible to perform large steps far from the solution, because the nonlinearity forces the distance between consecutive feasible iterates to be very short. If a large distance occurs in the tangent space, the newtonian procedure for restoring feasibility might not converge, and the tangent step must be decreased. See Figure 2.1. Short steps far from the solution of an optimization problem are undesirable and, frequently, the practical performance of an optimization method is linked to the ability of leaving quickly the zones where there is no chance of finding a solution. This fact leads one to develop Inexact Restoration algoritms.

The convergence theory of Inexact Restoration methods is inspired in the convergence theory of recent sequential quadratic programming (SQP) algorithms. See Gomes et al (1999). The analogies between IR and the SQP method presented in Gomes et al (1999) are:

1. both are trust-region methods;

2. in both methods the iteration is composed by two phases, the first related to feasibility and the second to optimality;

3. the optimality phase seeks a "more optimal" point in a "tangent approximation" to the constraints;

4. the same type of merit function is used.

However, there exist very important differences, which allow one to relate IR to the classical feasible methods:

1. both in the restoration phase and in the optimality phase, IR deals with the true function and constraints, while SQP deals with a model of both;

2. the trust region in SQP is centered in the current point. In IR the trust region is centered in the restored point.

Because of these differences, we say that IR has a more defined preference for feasibility than SQP.

A very important characteristic of modern IR methods is that we are free to choose the algorithm used for restoration. This allows one to exploit characteristics of the constraints, which would not be taken into account in other nonlinear programming algorithms.

3   DEFINITION OF AN IR ALGORITHM

In this section we give a rigorous definition of the algorithm given in Martínez (2001).

In this algorithm, the trust region is centered in the intermediate point, as in Martínez and Pilotta (2000); Martínez (1997), but, unlike the algorithms introduced in those papers, the Lagrangian function is used in the "tangent set" (which approximates the feasible region), as in Martínez (1998). Accordingly, we define a new merit function that fits well both requirements: one of its terms is the Lagrangian, as in Martínez (1998), but the second term is a nonsmooth measure of infeasibility as in Martínez and Pilotta (2000); Martínez (1997). Inexact Restoration can be applied to problems with quite general constraint structure but, for simplicity, we restrict here to the case "nonlinear equalities and bounds", which is the one considered in Martínez (2001). Every nonlinear programming problem is equivalent to a problem with this structure by means of the addition of slack variables and bounds. So, we consider the problem

$$\min \quad f(x)$$
$$\text{subject to} \quad C(x) = 0, \ x \in \Omega$$

where $\Omega \subset \mathbb{R}^n$ is a closed and convex subset of $\mathbb{R}^n$, $f : \mathbb{R}^n \to \mathbb{R}$ and $C : \mathbb{R}^n \to \mathbb{R}^m$.

We denote $C'(x)$ the Jacobian matrix of C evaluated at x. Throughout the paper we assume that $\nabla f(x)$ and $C'(x)$ exist and are continuous in Ω.

The algorithm is iterative and generates a sequence $\{x^k\} \subset \Omega$. The parameters $\eta > 0, r_{-1} \in [0,1), \beta > 0, M > 0, \theta_{-1} \in (0,1), \delta_{min} > 0, \tau_1 > 0, \tau_2 > 0$ are given, as well as the initial approximation $x^0 \in \Omega$, the initial vector of Lagrange multipliers $\lambda^0 \in \mathbb{R}^m$ and a sequence of positive numbers $\{\omega_k\}$ such that $\sum_{k=0}^{\infty} \omega_k < \infty$. All along this paper $\|\cdot\|$ will denote the Euclidean norm, although in many cases it can be replaced by an arbitrary norm.

Definition 3.1 *The Lagrangian function of the minimization problem is defined as*

$$L(x,\lambda) = f(x) + \langle C(x), \lambda \rangle \tag{3.1}$$

for all $x \in \Omega$, $\lambda \in \mathbb{R}^m$.

Assume that $k \in \{0,1,2,\ldots\}$, $x^k \in \Omega, \lambda^k \in \mathbb{R}^m$ and $r_{k-1}, \theta_{k-1}, \theta_{k-2}, \ldots, \theta_{-1}$ have been computed. The steps for obtaining x^{k+1}, λ^{k+1} and θ_k are given below.

Algorithm 3.1

Step 1. Initialize penalty parameter.

Define

$$\theta_k^{min} = \min\{1, \theta_{k-1}, \ldots, \theta_{-1}\}, \tag{3.2}$$

$$\theta_k^{large} = \min\{1, \theta_k^{min} + \omega_k\} \tag{3.3}$$

and

$$\theta_{k,-1} = \theta_k^{large}.$$

Step 2. Feasibility phase of the iteration.

Set $r_k = r_{k-1}$.

Compute $y^k \in \Omega$ such that

$$\|C(y^k)\| \le r_k\|C(x^k)\| \tag{3.4}$$

and

$$\|y^k - x^k\| \le \beta\|C(x^k)\|. \tag{3.5}$$

If this is not possible, replace r_k by $(r_k+1)/2$ and repeat Step 2. (In Martínez (2001) sufficient conditions are given for ensuring that this loop finishes.)

Step 3. Tangent Cauchy direction.

Compute

$$d_{tan}^k = P_k[y^k - \eta \nabla L(y^k, \lambda_k)] - y^k, \tag{3.6}$$

where $P_k(z)$ is the orthogonal projection of z on π_k and

$$\pi_k = \{z \in \Omega \mid C'(y^k)(z - y^k) = 0\}. \tag{3.7}$$

If $y^k = x^k$ (so $C(x^k) = C(y^k) = 0$) and $d_{tan}^k = 0$, terminate the execution of the algorithm returning x^k as "the solution".

If $d_{tan}^k = 0$ compute $\lambda^{k+1} \in \mathbb{R}^m$ such that $\|\lambda^{k+1}\| \le M$, define

$$x^{k+1} = y^k, \quad \theta_k = \theta_{k-1},$$

$$\text{Ared}_k = (1 - \theta_k)[\|C(x^k)\| - \|C(y^k)\|]$$

and terminate the iteration.

Else, set $i \leftarrow 0$, choose $\delta_{k,0} \ge \delta_{min}$ and continue.

Step 4. Trial point in the tangent set.

Compute, using Algorithm 3.2 below, $z^{k,i} \in \pi_k$ such that

$$\|z^{k,i} - y^k\| \le \delta_{k,i} \quad \text{and} \quad L(z^{k,i}, \lambda^k) < L(y^k, \lambda^k).$$

Step 5. Trial multipliers.

Compute $\lambda_{trial}^{k,i} \in \mathbb{R}^m$ such that $\|\lambda_{trial}^{k,i}\| \le M$.

Step 6. Predicted reduction.

Define, for all $\theta \in [0, 1]$,

$$\text{Pred}_{k,i}(\theta) = \theta[L(x^k, \lambda^k) - L(z^{k,i}, \lambda^k) - \langle C(y^k), \lambda_{trial}^{k,i} - \lambda^k \rangle] \tag{3.8}$$
$$+ (1 - \theta)[\|C(x^k)\| - \|C(y^k)\|]$$

Compute $\theta_{k,i}$, the maximum of the elements $\theta \in [0, \theta_{k,i-1}]$ that verify

$$\text{Pred}_{k,i}(\theta) \ge \frac{1}{2}[\|C(x^k)\| - \|C(y^k)\|]. \tag{3.9}$$

Define

$$\mathrm{Pred}_{k,i} = \mathrm{Pred}_{k,i}(\theta_{k,i}).$$

Step 7. Compare actual and predicted reduction.

Compute

$$\mathrm{Ared}_{k,i} = \theta_{k,i}[L(x^k,\lambda^k) - L(z^{k,i},\lambda^{k,i}_{trial})] + (1 - \theta_{k,i})[\|C(x^k)\| - \|C(z^{k,i})\|]$$

If

$$\mathrm{Ared}_{k,i} \geq 0.1 \ \mathrm{Pred}_{k,i}$$

define

$$x^{k+1} = z^{k,i}, \ \lambda^{k+1} = \lambda^{k,i}_{trial}, \ \theta_k = \theta_{k,i}, \ \delta_k = \delta_{k,i}, \ iacc(k) = i,$$

$$\mathrm{Ared}_k = \mathrm{Ared}_{k,i}, \quad \mathrm{Pred}_k = \mathrm{Pred}_{k,i}$$

and terminate iteration k.

Else, choose $\delta_{k,i+1} \in [0.1\delta_{k,i}, 0.9\delta_{k,i}]$, set $i \leftarrow i+1$ and go to Step 4.

Algorithm 3.2
Step 1.

Compute $t^{k,i}_{break} = \min \ \{1, \delta_{k,i}/\|d^k_{tan}\|\}$.

Step 2.

Set $t \leftarrow t^{k,i}_{break}$.

Step 3.

If

$$L(y^k + td^k_{tan}, \lambda^k) \leq L(y^k, \lambda^k) + 0.1t\langle \nabla L(y^k, \lambda^k), d^k_{tan}\rangle, \qquad (3.10)$$

define $z^{k,i} \in \Omega$ such that $\|z^{k,i} - y^k\| \leq \delta_{k,i}$ and

$$L(z^{k,i}, \lambda^k) \leq \max\{L(y^k + td^k_{tan}, \lambda^k), L(y^k, \lambda^k) - \tau_1\delta_{k,i}, L(y^k, \lambda^k - \tau_2\}. \quad (3.11)$$

and terminate. (Observe that the choice $z^{k,i} = y^k + td^k_{tan}$ is admissible but, very likely, it is not the most efficient choice.)

Step 4.

If (3.10) does not hold, choose $t_{new} \in [0.1t, 0.9t]$, set $t \leftarrow t_{new}$ and go to Step 3.

The following conditions are assumed for proving convergence of the IR algorithm.

A1. Ω is convex and compact.

A2. There exists $L_1 > 0$ such that, for all $x, y \in \Omega$,

$$\|C'(x) - C'(y)\| \leq L_1 \|x - y\| \tag{3.12}$$

A3. There exists $L_2 > 0$ such that, for all $x, y \in \Omega$,

$$\|\nabla f(x) - \nabla f(y)\| \leq L_2 \|x - y\|. \tag{3.13}$$

Under these conditions it can be proved that the algorithm is well defined. We can also prove the convergence theorems stated below.

Theorem 3.1 *Every limit point of a sequence generated by Algorithm IR is a stationary point of*

$$\min \quad \|C(x)\|^2$$
$$\text{subject to} \quad \ell \leq x \leq u.$$

Theorem 3.2 *If $C(x^k) \to 0$, there exists a limit point of the sequence that satisfies the Fritz-John optimality conditions of nonlinear programming.*

The algorithm presented in Martínez (2001) uses a fixed $r \equiv r_k$ for all $k = 0, 1, 2, \ldots$ In Martínez (2001) it was proved that, if Step 2 can always be completed, then $C(x^k) \to 0$, which implies that every limit point is feasible. Clearly, this is not always possible. For example, in some cases the feasible region might be empty. Therefore, we find it useful to show that, even when Step 2 is eventually impossible, one finds a stationary point of the squared norm of infeasibilities. The same result can be proved for most practical nonlinear programming algorithms.

The optimality theorem can be improved. In fact, one can prove that a limit point can be found that satisfies a stronger optimality condition than Fritz-John. This is called the AGP optimality condition and will be the subject of the following section.

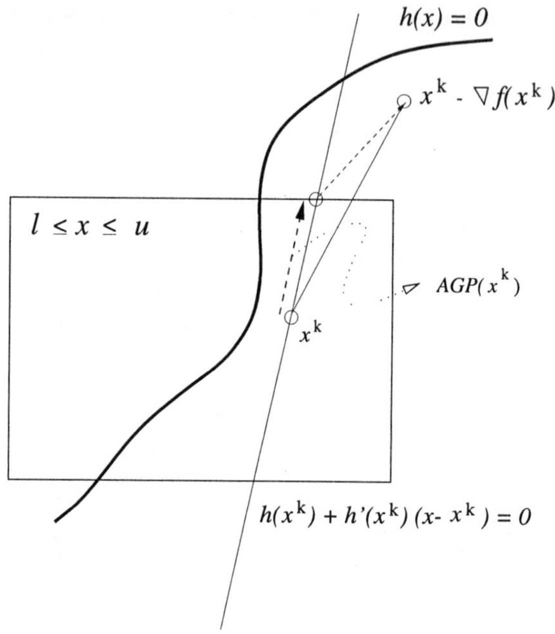

Figure 4.1   AGP vector: equalities and bounds.

4   AGP OPTIMALITY CONDITION

Consider the nonlinear programming problem

$$\min \quad f(x) \tag{4.1}$$

$$\text{subject to} \quad h(x) = 0, \ g(x) \leq 0, \tag{4.2}$$

where, for simplicity, we leave to the reader the definition of the correct dimensions for x, h, g. Given $x \in \mathbb{R}^n$ and the tolerance parameter $\eta > 0$, we divide the inequality constraints in three groups:

1. N: constraints not satisfied at x, defined by $g_i(x) > 0$.

2. A: constraints almost active at x, given by $0 \geq g_i(x) \geq -\eta$.

3. I: constraints strongly satisfied (or inactive) at x, given by $g_i(x) < -\eta$. Clearly, if η is very large, this set is empty.

Definition 4.1 *We define $T(x)$, the "tangent approximation" to the feasible set as the polytope given by*

$$\Omega(x) = \{z \in \mathbb{R}^n \mid \quad h'(x)(z-x) = 0, \ g_i'(x)(z-x) \le 0 \ \textit{if} \ i \in N,$$
$$g_i(x) + g_i'(x)(z-x) \le 0 \ \ \textit{if} \ i \in A\}.$$

Observe that $x \in T(x)$.

Definition 4.2 *The AGP (approximate gradient projection) vector $\bar{g}(x)$ is defined by*

$$\bar{g}(x) = P(x - \nabla f(x)) - x,$$

where $P(z)$ is the Euclidean projection of z on $T(x)$.

(see Martínez and Svaiter (2000)). See Figures 4.1 and 4.2.

We say that a feasible point x^* of (4.1) satisfies the AGP optimality condition if there exists a sequence $\{x^k\}$ that converges to x^* and such that $\bar{g}(x^k) \to 0$. The points x that satisfy the AGP optimality conditions are called "AGP points". It has been proved in Martínez and Svaiter (2000) that the set of local minimizers of a nonlinear programming problem is contained in the set of AGP points and that this set is strictly contained in the set of Fritz-John points. Therefore, the AGP optimality condition is stronger than the Fritz-John optimality conditions, traditionally used in algorithms. When equalities are not present and the problem is convex it can be proved that AGP is sufficient for guaranteeing that a point is a minimizer Martínez and Svaiter (2000).

A careful analysis of the convergence proofs in Martínez and Pilotta (2000); Martínez (2001) shows that Inexact Restoration guaranteeing convergence to points that satisfy the AGP optimality condition. This fact has interesting consequences for applications, as we will see later.

5   ORDER-VALUE OPTIMIZATION

The Order-Value optimization problem (OVO) has been introduced recently in Andreani et al (2001).

Definition 5.1 *Given m functions $f_1, \ldots, f_m$, defined in a domain $\Omega \subset \mathbb{R}^n$ and an integer $p \in \{1, \ldots, m\}$, the (p) order-value function $f(x)$ is defined by*

$$f(x) = f_{i_p(x)}(x),$$

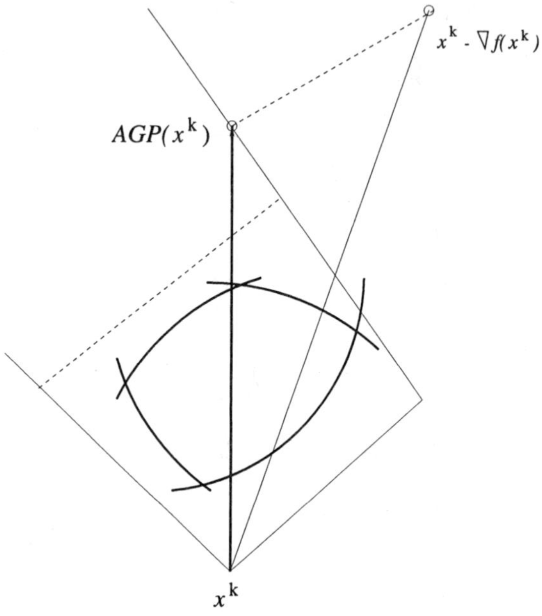

Figure 4.2   AGP vector: inequalities.

where

$$f_{i_1(x)}(x) \le f_{i_2(x)}(x) \le \cdots \le f_{i_p(x)}(x) \le \cdots \le f_{i_m(x)}(x).$$

If $p = 1$, $f(x) = \min\{f_1(x), \ldots, f_m(x)\}$ whereas for $p = m$ we have that $f(x) = \max\{f_1(x), \ldots, f_m(x)\}$.

The OVO problem consists in the maximization of the order-value function:

$$\max \quad f(x) \tag{5.1}$$

$$\text{subject to} \quad x \in \Omega \tag{5.2}$$

The definition of the OVO problem was motivated by applications to decision problems and robust regression with systematic errors. See Figure 5.1.

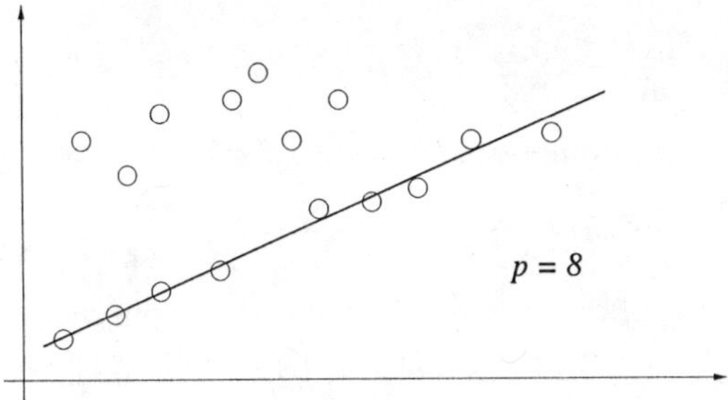

Figure 5.1   OVO linear regression.

It has been proved in Andreani et al (2001) that (5.1) is equivalent to

$$
\begin{aligned}
\max \quad & z \\
\text{subject to} \quad & \sum_{i=1}^{m} r_i(u_i - z + f_i(x)) = 0 \\
& \sum_{i=1}^{m} (1 - r_i)u_i = 0 \\
& \sum_{i=1}^{m} r_i = p - 1 \\
& u_i - z + f_i(x) \geq 0 \quad i = 1,\ldots,m \\
& u \geq 0, \ \ 0 \leq r \leq e, \\
& x \in \Omega.
\end{aligned}
\tag{5.3}
$$

Stationary points of the sum of squares of infeasibilities are feasible points and that, although all the feasible points of this problem are Fritz-John, this is not a serious inconvenient for IR algorithms, due to the AGP condition. Finally, it can be easily verified that an obvious restoration procedure exists, which encourages the use of restoration methods. See Andreani and Martínez (2001).

6   BILEVEL PROGRAMMING

The Bilevel Programming problem is

$$
\begin{aligned}
\min \quad & f(x,y) \\
\text{subject to} \quad & y \text{ solves a nonlinear programming problem} \\
& \quad \text{that depends on } x \\
& \qquad \text{and} \\
& \text{Ordinary constraints.}
\end{aligned}
\tag{6.1}
$$

In other words:

$$\min \quad f(x, y)$$

$$\text{subject to} \quad y \text{ minimizes } P(x, y) \text{ s.t. } t(x, y) = 0, \ s(x, y) \leq 0 \quad (6.2)$$

$$\text{and} \quad h(x, y) = 0, \ g(x, y) \leq 0. \quad (6.3)$$

We omit the dimensions of x, y, t, s, h, g in order to simplify the notation.

Usually, the constraints (6.2) are replaced by the optimality (KKT) conditions of the minimization problems, so that they take the form

$$F(x, y) + \nabla_y t(x, y)\lambda + \nabla_y s(x, y)z = 0, \quad (6.4)$$

$$t(x, y) = 0, \quad (6.5)$$

$$z \geq 0, \ s(x, y) \leq 0, \ z_i \, [s(x, y)]_i = 0 \ \forall i. \quad (6.6)$$

In (6.4) we denote $F(x, y) \equiv \nabla_y f(x, y)$. When F is not a gradient we say that the minimization of $f(x, y)$ subject to (6.3–6.6) is an MPEC (mathematical programming problem with equilibrium constraints). The equilibrium constraints are (6.4–6.6). Completing the inequalities with slack variables and bounds, we obtain the nonlinear programming problem:

$$\min \quad f(x, y) \quad (6.7)$$

$$\text{subject to} \quad h(x, y) = 0, \ g(x, y) + v = 0, \ v \geq 0, \quad (6.8)$$

$$F(x, y) + \nabla_y t(x, y)\lambda + \nabla_y s(x, y)z = 0, \quad (6.9)$$

$$t(x, y) = 0, \quad (6.10)$$

$$s(x, y) + w = 0, \ w_i z_i = 0 \ \forall i, w \geq 0, z \geq 0. \quad (6.11)$$

Nonlinear programming algorithms can be used for solving (6.7-6.11). Since many of these algorithms, including IR, have the property that limit points are stationary points of the squared norm of infeasibilities, it is interesting to find sufficient conditions under which stationary points of the following problem are feasible:

$$\min \quad \|h(x, y)\|^2 + \|g(x, y) + v\|^2 \quad (6.12)$$

$$+\|F(x,y) + \nabla_y t(x,y)\lambda + \nabla_y s(x,y)z\|^2$$

$$+\|t(x,y)\|^2 + \|s(x,y) + w\|^2 + \sum(w_i z_i)^2 \qquad (6.13)$$

$$\text{subject to} \qquad v \geq 0, z \geq 0, w \geq 0. \qquad (6.14)$$

An answer is given by the following theorem (see Andreani and Martínez (2001)).

Theorem 6.1 *Assume that the following conditions hold:*

(i) $(x^*, y^*, v^*, w^*, \lambda^*, z^*)$ *is a KKT point of (6.12),(6.13), (6.14);*

(ii) *The matrix* $\nabla_y F(x^*, y^*) + \sum_{i=1}^{q} [z^*]_i \nabla_{yy}^2 [s(x^*, y^*)]_i$ *is positive definite in the null-space of* $\nabla_y^T t(x^*, y^*)$;

(iii) $h(x^*, y^*) = 0$, $g(x^*, y^*) \leq 0$;

(iv) $t(x^*, y)$ *is an affine function (of y) and the functions* $[s(x^*, y)]_i$ *are convex (as functions of y);*

(v) *There exists* $\tilde{y}$ *such that* $t(x^*, \tilde{y}) = 0$ *and* $s(x^*, \tilde{y}) \leq 0$;

Then, (x^*, y^*) *is a feasible point of (6.8)-(6.11).*

Roughly speaking, the sufficient conditions are related with the convexity of the lower-level problem, when the MPEC under consideration is a Bilevel programming problem.

As in the case of the OVO problem, no feasible point of MPEC satisfies the Mangasarian-Fromowitz constraint qualification and, consequently, all the feasible points are Fritz-John. So, it is mandatory to find stronger optimality conditions in order to explain the practical behavior of algorithms. Fortunately, it can also be proved that the set of AGP points is only a small part of the set of Fritz-John points. The characterization of AGP points is given in Andreani and Martínez (2001). Since IR converges to AGP points, we have an additional reason for using IR in MPEC.

As in Nash-equilibrium problems (see Vicente and Calamai (1994)) the use of the optimization structure instead of optimality conditions in the lower level encourages the use of specific restoration algorithms.

7 HOMOTOPY METHODS

Nonlinear systems of equations often represent mathematical models of practical engineering problems. Homotopic techniques are used for enhancing convergence to solutions, especially when a good initial estimate is not available. In Birgin et al (2001), the homotopy curve is considered as the feasible set of a mathematical programming problem, where the objective is to find the optimal value of the homotopic parameter. Inexact Restoration techniques can then be used to generate approximations in a neighborhood of the homotopy curve, the size of which is theoretically justified.

Assume that $\Omega = \{x \in \mathbb{R}^n \mid \ell \leq x \leq u\}$, the mapping $F : \Omega \to \mathbb{R}^n$ has continuous first partial derivatives, $\ell, u \in \mathbb{R}^n$ and $\ell < u$. Then, the mathematical problem consists in finding $x \in \Omega$ such that

$$F(x) = 0. \tag{7.1}$$

Homotopic methods are used when strictly local Newton-like methods for solving (7.1) fail because a sufficiently good initial guess of the solution is not available. Moreover, in some applications areas, homotopy methods are the rule of choice. See Watson et al (1997). The homotopic idea consists in defining

$$H : \Omega \times \mathbb{R} \to \mathbb{R}^n$$

such that $H(x, 1) \equiv F(x)$ and the system $H(x, 0) = 0$ is easy to solve. The solution of $H(x, t) = 0$ is used as initial guess for solving $H(x, t') = 0$, with $t' > t$. In this way, the solution of the original problem is progressively approximated. For many engineering problems, natural homotopies are suggested by the very essence of the physical situation. In other cases, artificial homotopies can be useful.

In general, the set of points (x, t) that satisfy $H(x, t) = 0$ define a curve in $\mathbb{R}^{n+1}$. Homotopic methods are procedures to "track" this curve in such a way that its "end point" ($t = 1$) can be safely reached. Since the intermediate points of the curve (for which $t < 1$) are of no interest by themselves, it is not necessary to compute them very accurately. So, an interesting theoretical problem with practical relevance is to choose the accuracy to which intermediate points are to be computed. If an intermediate point (x, t) is computed with high accuracy, the tangent line to the curve that passes through this point can be efficiently used to predict the points corresponding to larger values of the parameter t. This prediction can be very poor if (x, t) is far from the true zero-curve of $H(x, t)$. On the other hand, accurate computing of all intermediate points can be unaffordable.

In Birgin et al (2001) a relation between the homotopic framework for solving nonlinear equations and Inexact Restoration methods was established. The idea is to look at the homotopic problem as the nonlinear optimization problem

$$\min \quad (t - 1)^2 \tag{7.2}$$
$$\text{subject to} \quad H(x, t) = 0, \quad x \in \Omega.$$

Therefore, the homotopic curve is the feasible set of (7.2). The nonlinear programming problem (7.2) could be solved by any constrained optimization method, but Inexact Restoration algorithms seem to be closely connected to the classical predictor-corrector procedure used in the homotopic approach. Moreover, they give theoretically justified answers to the accuracy problem.

The identification of the homotopy path with the feasible set of a nonlinear programming problem allows one to use IR criteria to define the closeness of corrected steps to the path. The solutions of (7.1) correspond exactly to those of (7.2), so the identification proposed here is quite natural.

The correspondence between the feasibility phase of IR and the corrector phase of predictor-corrector continuation methods is immediate. The IR technique provides a criterion for declaring convergence of the subalgorithm used in the correction. The optimality phase of IR corresponds to the predictor phase of continuation methods. The IR technique determines how long predictor steps can be taken and establishes a criterion for deciding whether they should be accepted or not.

8   CONCLUSIONS

Some features of the Inexact Restoration framework are reasonably consolidated. Among them, we can cite:

1. freedom for choosing different algorithms in both phases of the method, so that problem characteristics can be exploited and large problems can be solved using appropriate sub-algorithms.

2. loose tolerances for restoration discourage the danger of "over-restoring", a procedure that could demand a large amount of work in feasible methods and that, potentially, leads to short steps far from the solution.

3. trust regions centered in the intermediate point adequately reflect the "preference for feasibility" of IR methods.

4. the use of the Lagrangian in the optimality phase favors practical fast convergence near the solution.

The above characteristics should be preserved in future IR implementations. It is not so clear for us which is the best merit function for IR methods. The one introduced in Martínez (2001) seems to deal well with the feasibility and the optimality requirements but it is certainly necessary to pay attention to other alternatives or, even to the possibility of not using merit functions at all, as proposed, for example, in Bielschowsky (1996); Bielschowsky and Gomes (1998); Fletcher and Leyffer (1997). A remarkable contribution has been recently made by Gonzaga et al(2001).

The philosophy of Inexact Restoration encourages case-oriented applications. Very relevant is the use of IR methods for Bilevel Programming, because we think that, in this important family of problems, the strategy of using opportunistic methods in the lower level (Feasibility Phase) could be useful. A related family of problems for which IR could be interesting is MPEC (Mathematical Programming with Equilibrium Constraints) which, of course, it is closely related to Bilevel Programming. The discretization of control problems also offers an interesting field of IR applications because, in this case, the structure of the state equations suggests ad hoc restoration methods.

Acknowledgments

The authors wish to thank to FAPESP (Grant 90-3724-6), CNPq, FAEP-UNICAMP and SeCYT-UNC (Grant 194-2000) for their support. We also acknowledge an anonymous referee for helpful comments.

References

Abadie, J. and Carpentier, J. (1968), Generalization of the Wolfe reduced-gradient method to the case of nonlinear constraints, in *Optimization*, Edited by R. Fletcher, Academic Press, New York, pp. 37-47.

Andreani, R., Dunder, C. and Martínez, J. M. (2001), Order-Value optimization, *Technical Report* , Institute of Mathematics, University of Campinas, Brazil.

Andreani, R. and Martínez, J. M. (2001), On the solution of mathematical programming problems with equilibrium constraints, *Mathematical Methods of Operations Research* 54, pp. 345-358.

Bielschowsky, R. H. (1996), Nonlinear Programming Algorithms with Dynamic Definition of Near-Feasibility: Theory and Implementations, *Doctoral Dissertation*, Institute of Mathematics, University of Campinas, Campinas, SP, Brazil, 1996.

Bielschowsky, R. H. and Gomes, F. (1998), Dynamical Control of Infeasibility in Constrained Optimization, Contributed Presentation in *Optimization 98*, Coimbra, Portugal.

Birgin, E., Krejić N. and Martínez, J. M. (2001), Solution of bounded nonlinear systems of equations using homotopies with Inexact Restoration, *Technical Report*, Institute of Mathematics, University of Campinas, Brazil.

Drud, A. (1985), CONOPT - A GRG code for large sparse dynamic nonlinear optimization problems, *Mathematical Programming* 31, pp. 153-191.

Fletcher, R. and Leyffer, S. (1997), Nonlinear programming without penalty function, *Numerical Analysis Report NA/171*, University of Dundee.

Gomes, F. M., Maciel, M. C. and Martínez, J. M. (1999), Nonlinear programming algorithms using trust regions and augmented Lagrangians with non-monotone penalty parameters, *Mathematical Programming* 84, pp. 161-200.

Gonzaga, C., Karas, E. and Vanti, M. (2001), A globally convergent filter method for nonlinear programming, *Technical Report*, Federal University of Santa Catarina, Brazil.

Lasdon, L. (1982), Reduced gradient methods, in *Nonlinear Optimization* 1981, edited by M. J. D. Powell, Academic Press, New York, pp. 235-242.

Martínez, J. M. (1997), A Trust-region SLCP model algorithm for nonlinear programming, in *Foundations of Computational Mathematics*. Edited by F. Cucker and M. Shub. Springer-Verlag, pp. 246-255.

Martínez, J. M. (1998), Two-phase model algorithm with global convergence for nonlinear programming, *Journal of Optimization Theory and Applications* 96, pp. 397-436.

Martínez, J. M. (2001), Inexact Restoration method with Lagrangian tangent decrease and new merit function for nonlinear programming, *Journal of Optimization Theory and Applications* 111, pp. 39-58.

Martínez, J. M. and Pilotta, E. A. (2000), Inexact Restoration algorithm for constrained optimization, *Journal of Optimization Theory and Applications* 104, pp. 135-163.

Martínez, J. M. and Svaiter, B. F. (2000), A sequential optimality condition for nonlinear programming, *Technical Report*, Institute of Mathematics, University of Campinas, Brazil.

Miele, A., Yuang, H. Y and Heideman, J. C. (1969), Sequential gradient-restoration algorithm for the minimization of constrained functions, ordinary and conjugate gradient version, *Journal of Optimization Theory and Applications*, Vol. 4, pp. 213-246.

Miele, A., Levy, V. and Cragg, E. (1971), Modifications and extensions of the conjugate-gradient restoration algorithm for mathematical programming problems, *Journal of Optimization Theory and Applications*, Vol. 7, pp. 450-472.

Miele, A., Sims, E. and Basapur, V. (1983), Sequential gradient-restoration algorithm for mathematical programming problems with inequality constraints, Part 1, Theory, *Report No. 168*, Rice University, Aero-Astronautics.

Mukai, H. and Polak, E. (1978), On the use of approximations in algorithm for optimization problems with equality and inequality constraints, *SIAM Journal on Numerical Analysis* 15, pp. 674-693.

Rom, M. and Avriel, M. (1989), Properties of the sequential gradient-restoration algorithm (SGRA), Part 1: introduction and comparison with related methods, *Journal of Optimization Theory and Applications*, Vol. 62, pp. 77-98.

Rom, M. and Avriel, M. (1989), Properties of the sequential gradient-restoration algorithm (SGRA), Part 2: convergence analysis, *Journal of Optimization Theory and Applications*, Vol. 62, pp. 99-126.

Rosen, J. B. (1960), The gradient projection method for nonlinear programming, Part 1, Linear Constraints, *SIAM Journal on Applied Mathematics*, Vol. 8, pp. 181-217.

Rosen, J. B. (1961), The gradient projection method for nonlinear programming, Part 2, Nonlinear constraints, *SIAM Journal on Applied Mathematics*, Vol. 9, pp. 514-532.

Rosen, J. B. (1978), Two-phase algorithm for nonlinear constraint problems, *Nonlinear Programming 3*, Edited by O. L. Mangasarian, R. R. Meyer and S. M. Robinson, Academic Press, London and New York, pp. 97-124.

Vicente, L. and Calamai, P.(1994), Bilevel and multilevel programming: a bibliography review, *Journal of Global Optimization* 5, pp. 291-306.

Watson, L. T., Sosonkina, M., Melville, R. C., Morgan, A. P., Walker, H. F. (1997), Algorithm 777. HOMPACK90: A suite of Fortran 90 codes for globally convergent homotopy algorithms, *ACM Transactions on Mathematical Software* 23, 514-549.

13   QUANTUM ALGORITHM FOR CONTINUOUS GLOBAL OPTIMIZATION

V. Protopopescu and J. Barhen

Center for Engineering Science Advanced Research
Computing and Computational Sciences Directorate
Oak Ridge National Laboratory
Oak Ridge, TN 37831-6355

Abstract: We investigate the entwined roles played by information and quantum algorithms in reducing the complexity of the global optimization problem (GOP). We show that: (i) a modest amount of additional information is sufficient to map the general continuous GOP into the (discrete) Grover problem; (ii) while this additional information is actually available in some classes of GOPs, it cannot be taken advantage of within classical optimization algorithms; (iii) on the contrary, quantum algorithms offer a natural framework for the efficient use of this information, resulting in a speed-up of the solution of the GOP.

Key words: Global optimization, search, Grover's quantum algorithm.

1 GLOBAL OPTIMIZATION PROBLEM

Optimization problems are ubiquitous and extremely consequential. The theoretical and practical interest they have generated has continued to grow from the first recorded instance of Queen Dido's problem (Smith (1974)) to present day forays in complexity theory or large scale logistics applications (see Refs. (Törn (1989)), (Horst (1993)), (Hager (1994)), (Floudas (1996)), and references therein). The formulation of almost any optimization problem is deceptively simple: find the absolute minimum (maximum) of a given function - called the objective function - over the allowed range of its variables. Sometimes, the function to be optimized is not specified in analytic form and must be evaluated point-wise by a computer program, a physical device, or other construct. Such a black-box tool is called an *oracle*. Of course, the brute force approach of evaluating the function on its whole domain is either impossible - if the variables are continuous, or prohibitively expensive - if the variables are discrete, but have large ranges in high dimensional spaces. Since in general each oracle invocation (function evaluation) involves an expensive computational sequence, the number of function evaluations needs to be kept to a minimum. The number of invocations of the oracle measures the query complexity of the problem and gives a fair - although by no means unique - idea of its difficulty or "hardness" (Deng (1996)). Therefore, the number of oracle invocations is one of the paramount criteria in comparing the efficiency of competing optimization algorithms.

The primary difficulty in solving the GOP stems from the fact that the familiar condition for determining extrema (namely, annulment of the gradient of the objective function) is only *necessary* (the function may have a maximim, a minimum, or not have an extremum at all !) and *local* (it does not distinguish between local and global extrema). Indeed, the generic strategy to find the global minimum involves two main operations, namely: (i) descent to a local minimum and (ii) search for the new descent region. Usually, the former operation is deterministic and the latter stochastic. This strategy is marred by additional problems. First, descent assumes a certain degree of smoothness, which is not always warranted. When the dimensionality of the problem is large, the search of the phase space becomes more and more responsible for increasing the query complexity of the problem. Finally, after determining a local minimum, the algorithm is usually trapped in it and special operations have to be designed to restart the search. The "hardness" of the GOP is well illustrated by the following "golf course" example for which the approach described above seems powerless. Define the function $f : [0,1] \to \{0,1\}$ as follows:

$$f(x) = \begin{cases} 1 \text{ for } 0 \leq x \leq a - \epsilon/2 \\ 0 \text{ for } a - \epsilon/2 < x < a + \epsilon/2 \\ 1 \text{ for } a + \epsilon/2 \leq x \leq 1. \end{cases} \tag{1.1}$$

where $a \in (\epsilon/2, 1 - \epsilon/2)$. To obtain the minimum of this function, one should evaluate it within the ϵ interval around the *unknown* number a. If this function is defined like an oracle (i.e., if one does *not* know the position of the point a), the probability of choosing an x within this interval is ϵ. For the n-dimensional version of this oracle, the probability becomes ϵ^n, and the query complexity of the problem grows exponentially with n (the dimensionality curse). Of course, this is an extreme case, for which knowledge about the derivatives (they are all zero whenever defined !) would not help. This and related issues have been deftly discussed by Wolpert and Macready in connection with their "No Free Lunch" (NFL) theorem (Wolpert (1996)).

In the light of the previous example, it seems that without additional knowledge about the structure of the function there is no hope to decide upon an intelligent optimization strategy and one is left with either strategies that have limited albeit efficient applicability or the exhaustive search option.

Thus, new approaches are needed to reduce the complexity of the problem to manageable complexity. Recently, quantum computing has been hailed as the possible solution to some of the computationally hard classical problems (Nielsen (2000)). Indeed, Grover's (Grover (1997)) and Shor's (Shor (1994)) algorithms provide such solutions to the problems of finding a given element in an unsorted set and the prime factorization of very large numbers, respectively. Here we present a solution to the continuous GOP in polynomial time, by developing a generalization of Grover's algorithm to continuous problems. This generalization requires additional information on the objective function. In many optimization problems, some of this additional information *is* available (see below). While other required information may be more difficult to obtain in practical applications, it is important to understand - from a theoretical point of view - the role of the information in connection to the difficulty of the problem, and to be able to assess *a priori what various information is relevant and for what.* For instance, if the objective function were an analytic function, the knowledge of all its derivatives at a given point would allow, in principle, the "knowledge" of the function everywhere else in the domain of analyticity. However, to actually *find* the global minimum, the function would still have to be calculated everywhere! In other words, the (additional) knowledge of all the derivatives at a given point cannot be *efficiently* used to *locate* the global minimum, although in principle it is equivalent to the knowledge of the function at all points. In fact, to locate the global minimum, both methods would require exhaustive calculations.

2 GROVER'S QUANTUM ALGORITHM

A quantum computation is a sequence of unitary transformations on the initial state of the wave function, ψ. As such, quantum computation is purely deterministic and reversible. It requires the initialization or preparation of the initial state, the actual "computation", and the read out of the result effected through a measurement of the final state. If the algorithm is efficient, then, with probability (much) higher than $1/2$, the measurement would collapse the final state onto the desired result. Computer architectures needed to implement classical or quantum algorithm are realized in terms of gates. As opposed to classical gates that operate on bits taking values in the set $\{0, 1\}$, quantum gates operate on normalized vectors in a finite-dimensional complex Euclidian space. In principle, any quantum computer can be viewed as an assembly of elementary quantum gates, such as the NOT and CNOT gates. The NOT gate is the 2×2 matrix $\begin{pmatrix} 0 & 1 \\ 1 & 0 \end{pmatrix}$. It acts on a *qubit*, q, which is the normalized state in a two dimensional Euclidian space, $\mathbb{C}^2$:

$$q = \alpha|0> + \beta|1>, \quad |\alpha|^2 + |\beta|^2 = 1, \tag{2.1}$$

by exchanging the level populations. The CNOT gate acts on four dimensional vectors in $\mathbb{C}^4$. Obviously, some of these vectors can be represented as a tensor product of two two-dimensional vectors; however other vectors in $\mathbb{C}^4$ cannot be written in this form. These latter states are called *entangled states* and play a crucial role in quantum algorithms (Nielsen (2000)). Quantum algorithms are (i) intrinsically parallel and (ii) yield probabilistic results. These properties reflect the facts that: (i) the wave function, ψ, is nonlocal and, in fact, ubiquituous and (ii) the quantity $|\psi|^2$ is interpreted as a probability density.

Grover's original algorithm provides a solution to the following problem. Suppose we have a set of N unsorted objects, $E = \{x_1, x_2, ...x_N\}$, and an oracle function $f : E \to \{0,1\}$, such that $f(x_1) = 1$ and $f(x_i) = 0$, $i = 2, ...N$. Using the oracle, find the element x_1 in the unsorted set E.

On average, the classical solution will involve $\sim N/2 \sim O(N)$ evaluations of the oracle. The quantum algorithm proposed by Grover (Grover (1997)) reduces this number to $O(\sqrt{N})$. In a generalized version of the problem, there may be L "special" elements for which the oracle returns the value one; then the number of evaluations required to find one of them is of the order $O(\sqrt{N/L})$.

We give a brief presentation of Grover's quantum algorithm (Grover (1997)). First, we identify the set E with the complex Euclidean space $\mathbb{C}^N$ and the elements $x_i \in E$ with the unit vectors in $\mathbb{C}^N$, $< x_i | x_j >= \delta_{ij}$, where δ_{ij} is the Kronecker symbol. Then construct the normalized average state of all the elements $|x_i >$:

$$|w >= \frac{1}{\sqrt{N}} \sum_{i=1}^{N} |x_i >= \frac{1}{\sqrt{N}} |x_1 > + \frac{\sqrt{N-1}}{\sqrt{N}} |x_1^{\perp} > . \qquad (2.2)$$

In the second representation, the unit vectors orthogonal to $|x_1 >$ are lumped together in the unit vector $|x_1^{\perp} >$, which formally reduces the problem to a bidimensional space and simplifies the presentation and interpretation of the algorithm.

We note that the scalar product $< x_1 | w >= \frac{1}{\sqrt{N}} := cos\beta = sin\alpha$ where β denotes the angle between the vectors $|w >$ and $|x_1 >$ and α denotes its complement, i.e. the angle between the vectors $|w >$ and $|x_1^{\perp} >$.

The construction of the state $|w >$ can be done in $log_2 N = n$ steps by applying (in parallel) n Hadamard transformations, $H = \frac{1}{\sqrt{2}} \begin{pmatrix} 1 & 1 \\ 1 & -1 \end{pmatrix}$, on the initial zero state, $\underbrace{|0 > \otimes ...... \otimes |0 >}_{n \text{ times}}$. In the $\{|x_1 >, |x_1^{\perp} >\}$ basis, we construct the operators:

$$I_{x_1} := I - 2|x_1 >< x_1| = \begin{pmatrix} -1 & 0 \\ 0 & 1 \end{pmatrix}, \qquad (2.3)$$

which executes a reflection (sign inversion) of the x_1-component of a vector and

$$I_w := I - 2|w><w| = \begin{pmatrix} 1 - 2/N & -2\sqrt{N-1}/N \\ -2\sqrt{N-1}/N & -1 + 2/N \end{pmatrix}, \qquad (2.4)$$

which represents a reflection (sign inversion) of the w-component of a vector. At the level of the oracle, the operator I_{x_1} is implemented by $(-1)^{f(\cdot)}(.)$, which does not depend explicitly on the unknown element x_1, while the application of the operator I_w is obvious, since the average state is known. We define now the *amplitude amplification operator*:

$$Q := -I_w I_{x_1} = \begin{pmatrix} 1 - 2/N & 2\sqrt{N-1}/N \\ -2\sqrt{N-1}/N & 1 - 2/N \end{pmatrix} = \begin{pmatrix} \cos 2\alpha & \sin 2\alpha \\ -\sin 2\alpha & \cos 2\alpha \end{pmatrix} \qquad (2.5)$$

which, in the compressed, two-dimensional representation of the problem, represents a rotation of the state vector with an angle 2α towards $|x_1>$. This means that each application of the operator Q will increase the weight of the unknown vector $|x_1>$ (which explains the name of the operator Q) and after roughly $\frac{\pi/2 - \alpha}{2\alpha} \sim \frac{\pi/2 - 1/\sqrt{N}}{2/\sqrt{N}} \sim \frac{\pi}{4}\sqrt{N}$ applications the state vector will be essentially parallel to $|x_1>$, whereupon a measurement of the state will yield the result $|x_1>$ with a probability very close to unity. We mention that for (and only for) $N = 4$, the result is obtained with certainty, after only one application. In general, if one continues the application of Q, the state vector continues its rotation and the weight of $|x_1>$ decreases; eventually, the evolution of the state is cyclic as prescribed by the unitary evolution. In the original, N-dimensional representation, the operator I_w has the representation:

$$I_w = \begin{pmatrix} 1 - 2/N & 2/N \dots & 2/N \\ 2/N & 1 - 2/N \dots 2/N \\ \vdots & \ddots & \vdots \\ 2/N & \dots & 1 - 2/N \end{pmatrix}. \qquad (2.6)$$

Using this representation, one can show explicitly that the algorithm can be implemented as a sequence of local operations such as rotations, Hadamard transforms, etc. (Grover (1997))

It is easy to check that if the oracle returns the same value for all the elements, i.e. there is no "special" element in the set E, the amplification operator

Q reduces to the identity operator I and, after the required number of applications, the measurement will return any of the states with the same probability, namely $1/N$. In other words, the algorithm behaves consistently.

3 SOLUTION OF THE CONTINUOUS GLOBAL OPTIMIZATION PROBLEM

Grover's search algorithm described in the previous section has been applied to a *discrete* optimization problem, namely finding the minimum among an unsorted set of N different objects. Dürr and Hoyer (Dürr) adapted Grover's original algorithm and solved this problem with probability strictly greater than $1/2$, using $O(\sqrt{N})$ function evaluations (oracle invocations).

In this article, we map the *continuous* GOP to the Grover problem. Once this is achieved, one can apply either Grover's algorithm and obtain an almost certain result with $O(\sqrt{N})$ function evaluations. However, the mapping of the GOP to Grover's problem is not automatic, but requires additional information.

Before spelling out the required information, let us revisit the "pathological" example (1). Without loss of generality, we can take $\epsilon = 1/N$ and divide the segment $[0, 1]$ into N equal intervals. By evaluating the function at the midpoint of the N intervals, we obtain a discrete function that is equal to 1 in $N - 1$ points and equal to 0 in one point, which - up to an unessential transformation - is equivalent to Grover's problem. Direct application of Grover's algorithm yields the corresponding result. Of course, generalization to any dimensionality d is trivial. Thus, a clssically intactable problem becomes much easier within the quantum computing framework. We shall return to this example after discussing the general case.

Consider now a real function of d variables, $f(x_1, x_2, ..., x_d)$. Without restricting generality, we can assume that f is defined on $[0, 1]^d$ and takes values in $[0, 1]$. Assume now that: (i) there is a unique global minimum which is reached at zero; (ii) there are no local minima whose value is infinitesimally close to zero; in other words, the values of the other minima are larger than a constant $\delta > 0$, and (iii) the size of the basin of attraction for the global minimum measured at height δ is known; we shall denote it Δ.

Then our implementation paradigm is the following: (i) instead of $f(.)$, consider the transformation $g(.) := (f(.))^{1/m}$. For sufficiently large m, this function will take values very close to one, except in the vicinity of the global minimum, which will maintain its original value, namely zero. Of course, other transformations can be used to achieve essentially the same result. We calculate m such that $\delta^{1/m} = 1/2$. To avoid technical complications that would not change the tenure and conclusions of the argument, we assume that $\Delta = 1/M$ where M is a natural number, and divide the hypercube $[0,1]^d$ in small d-dimensional hypercubes with sides Δ. At the midpoint of each of these hypercubes, define the function $h(x) := int[g(.) + 1/2]$ (here int denotes the integer part). The function $h(.)$ is defined on a discrete set of N points, $N = M^d$, and takes only values one and zero; by our choice of constants, the region on which it takes value zero is a hypercube with side Δ. Thus we have reduced the problem to the Grover setting. Application of Grover's algorithm to the function $h(.)$, will result in a point that returns the value zero; by construction, this point belongs to the basin of attraction of the global minimum. We return then to the original function $f(.)$ and apply the descent technique of choice that will lead to the global minimum. If the basin of attraction of the global minimum is narrow, the gradients of the function $f(.)$ may reach very large values which may cause overshots. Once that phase of the algorithm is reached, one can proceed to apply a scaling (dilation) transformation that maintains the descent mode but moderates the gradients. On the other hand, as one approaches the global minimum, the gradients become very small and certain acceleration techniques based on non-Lipschitzian dynamics may be required (Barhen (1996)); (Barhen (1997)). If the global minimum is attained at the boundary of the domain, the algorithm above will find it without additional complications.

4   PRACTICAL IMPLEMENTATION CONSIDERATIONS

It is clear that, in general, the conditions imposed on the functions $f(.)$ are rather strong, *sufficient* conditions. However: (a) these conditions *are* both satisfied and *explicitly given* for the academic "golf course" example (1) and (b) while they do not help reduce the complexity of the classical descent/search algorithm, they make a remarkable difference in the quantum framework.

In fact, assumption (i) is satisfied by a large class of important practical problems, namely parameter identification encountered e.g. in remote sensing, pattern recognition, and, in general, inverse problems. In these problems the absolute minimum, namely zero, is attained for the correct values of the parameters, matching of patterns, and fitting of output to input. Assumption (i) can be relaxed in the sense that the function may have multiple global minima, all equal to zero. Functions with multiple global minima will simply result in Grover problems with multiple "special" elements and can be treated accordingly if the number of global minima is known.

Assumption (ii) can be replaced with the much more reasonable assumption that f has a finite number of local minima. This would prevent the value of any local minimum to be infinitesimally close to the value of the global minimum.

Assumption (iii) is the most difficult to fulfill in practical problems. However, this assumption could also be relaxed with no significant performance loss, if more efficient (e.g. exponentially fast) unstructured search quantum algorithms were available. For the time being, the likelihood of exponentially fast search algorithms is uncertain (Protopopescu).

Recently, Chen and Diao (Chen), proposed an algorithm which was supposed to achieve an exponential (as opposed to polynomial) speed-up of the unstructured search. Unfortunately, subtle complexity hidden in one of the proposed steps makes this algorithm unusuitable for exponentially fast search.

Despite their scarcity and still elusive implementation in a practical quantum computer, quantum algorithms could bring very promising solutions to hard computational problems. It seems likely that - like the algorithms proposed so far - future quantum algorithms will be much more "problem tailored" than their classical counterparts. Therefore, specific additional information is crucial. In general, this information may be difficult to obtain, but - as illustrated above - its benefits may significantly outweigh its cost. Indeed, for very high dimensional, computationally intensive problems, even the polynomial reduction of complexity offered by the Grover's algorithm *is* extremely significant.

Acknowledgments

This work was partially supported by the Material Sciences and Engineering Division Program of the DOE Office of Science under contract DE-AC05-00OR22725 with UT-Battelle, LLC. We thank Drs. Robert Price and Iran Thomas from DOE for their support. V. P. thanks Dr. Cassius D'Helon for an enlightening discussion on Ref. 3 and for a careful reading of the manuscript.

References

Barhen, J. and V. Protopopescu, Generalized TRUST Algorithm for Global Optimization, *State of the Art in Global Optimization*, C.A. Floudas and P.M. Pardalos eds., pp. 163-180, Kluwer Academic Press, Dordrecht, Boston 1996.

Barhen, J., V. Protopopescu, and D. Reister, TRUST: A Deterministic Algorithm for Global Optimization, *Science* **276**, 1094-1097 (1997).

Chen, G. and Z. Diao, Exponentially Fast Quantum Search Algorithm, quant-ph/0011109.

Deng, H. Lydia and J. A. Scales, Characterizing Complexity in Generic Optimization Problems, Preprint, Center of Wave Phenomena, Golden, Colorado, CWP-208P, October 1996.

Dürr, C. and P. Hoyer, A Quantum Algorithm for Finding the Minimum, quant-ph/9607014

Floudas, C. A. and P. M. Pardalos, eds. *State of the Art in Global Optimization: Computational Methods and Applications*, Kluwer Academic Publishers, Dordrecht, Boston, 1996.

Grover, L. K., Quantum Mechanics Helps in Searching a Needle in a Haystack, *Phys. Rev. Lett* **78**, 325-328 (1997).

Hager, W. W., D. W. Hearn, and P. M. Pardalos, eds. *Large Scale Optimization: State of the Art*, Kluwer Academic Publishers, Dordrecht, Boston, 1994.

Horst, R. and H. Tuy, *Global Optimization*, 2d ed., Springer-Verlag, Berlin, 1993.

Nielsen, M. and I. Chuang, *Quantm Computation and Quantum Information*, Cambridge University Press, Cambridge, UK, 2000; A. O. Pittenger, *An Intyroduction to Quantum Computing Algorithms*, Birkhäuser, Boston, 2000.

Protopopescu, V. and J. Barhen, to be published.

Shor, P., *Algorithms for Quantum Computation: Discrete Logarithms and Factoring*, Proceedings of the 35th Annual Symposium on Foundations of Computer Science, 1994, p. 124-134.

Smith, D. R., *Variational Methods in Optimization*, Prentice Hall, Inc., Englewood Cliffs, N.J., 1974.

Törn, A. and A. Zilinskas, *Global Optimization*, Springer-Verlag, Berlin, 1989.

Traub, J. F., G. W. Wasilkowski, and H. Wozniakowski, *Information-Based Complexity*, Academic Press, Boston, 1988.

Wolpert, D. H. and W. G. Macready, No Free Lunch Theorems for Optimization, *IEEE Trans. on Evolutionary Computing*, **1** 67–82 (1997); see also W. G. Macready and D. H. Wolpert, What Makes an Optimization Problem Hard ?, *Complexity*, **5**, 40–46 (1996).

14   SQP VERSUS SCP METHODS FOR NONLINEAR PROGRAMMING

Klaus Schittkowski and Christian Zillober

Dept. of Mathematics
University of Bayreuth
D-95440 Bayreuth
Germany

Abstract: We introduce two classes of methods for constrained smooth non-linear programming that are widely used in practice and that are known under the names SQP for sequential quadratic programming and SCP for sequential convex programming. In both cases, convex subproblems are formulated, in the first case a convex quadratic programming problem, in the second case a convex and separable nonlinear program. An augmented Lagrangian merit function can be applied for stabilization and for guaranteeing convergence. The methods are outlined in a uniform way, convergence results are cited, and the results of a comparative performance evaluation are shown based on a set of 306 standard test problems. In addition a few industrial applications and case studies are listed that are obtained for the two computer codes under consideration, i.e., NLPQLP and SCPIP.

Key words: Sequential quadratic programming, SQP, sequential convex programming, SCP, global convergence, augmented Lagrangian merit function, comparative tests.

1 INTRODUCTION

Whenever a mathematical model is available to simulate a *real-life* application, a straightforward idea is to apply mathematical optimization algorithms for minimizing a so-called cost function subject to constraints.

A typical example is the minimization of the weight of a mechanical structure under certain loads and constraints for admissible stresses, displacements, or dynamic responses. Highly complex industrial and academic design problems are solved today by means of nonlinear programming algorithms without any chance to get equally qualified results by traditional empirical approaches.

We consider the smooth, constrained optimization problem to minimize a scalar objective function $f(x)$ under nonlinear inequality constraints,

$$\min_{x \in \mathbb{R}^n} f(x)$$
$$g(x) \leq 0, \tag{1.1}$$

where x is an n-dimensional parameter vector, and the vector-valued function $g(x)$ defines m inequality constraints, $g(x) = (g_1(x), \ldots, g_m(x))^T$. To facilitate the subsequent analysis, upper and lower bounds are not handled separately, i.e., they are considered as general inequality constraints, and there are no equality constraints. They would be linearized in both situations we are investigating and would not lead to any new insight. We assume that the feasible domain of (1.1) is non-empty and bounded.

Sequential quadratic programming (SQP) methods are very well known and are considered as the standard general purpose algorithms for solving smooth nonlinear optimization problems, at least under the following assumptions:

- The problem is not too big.

- The functions and gradients can be evaluated with sufficiently high precision.

- The problem is smooth and well-scaled.

SQP methods have their roots in unconstrained optimization, and can be considered as extensions of quasi-Newton methods by taking constraints into account. The basic idea is to establish a quadratic approximation based on second order information, with the goal to achieve a fast local convergence speed. A quadratic approximation of the Lagrangian is formulated and the constraints are linearized. Second order information about the Hessian of the Lagrangian is updated by a quasi-Newton formula. The convex quadratic program must be solved in each iteration step by an available *black box* solver. For a review, see for example Stoer (1985) and Spellucci (1993) or any textbook about nonlinear programming.

Despite of the success of SQP methods, another class of efficient optimization algorithm was proposed by engineers mainly, where the motivation is found in mechanical structural optimization. The first method is known under the name CONLIN or convex linearization, see Fleury and Braibant (1986) and Fleury (1989), and is based on the observation that in some special cases, typical structural constraints become linear in the inverse variables. Although this special situation is rarely observed in practice, a suitable substitution by inverse variables depending on the sign of the corresponding partial derivatives and subsequent linearization is expected to linearize constraints somehow.

More general convex approximations are introduced by Svanberg (1987) known under the name *moving asymptotes* (MMA). The goal is always to construct convex and separable subproblems, for which efficient solvers are available. Thus, we denote this class of methods by SCP, an abbreviation for *sequential convex programming*. The resulting algorithm is very efficient for mechanical engineering problems, if there are many constraints, if a good starting point is available, and if only a crude approximation of the optimal solution needs to be computed because of certain side conditions, for example calculation time or large round-off errors in objective function and constraints.

In other words, SQP methods are based on local second order approximations, whereas SCP methods are applying global approximations. Some comparative numerical tests of both approaches are available for mechanical structural optimization, see Schittkowski et al. (1994). The underlying finite element formulation uses the software system MBB-LAGRANGE (Kneppe et al. (1987)). However we do not know of any direct comparisons of computer codes of both methods for more general classes of test problems, particularly for standard benchmark examples.

Thus, the purpose of the paper can be summarized as follows. First we outline a general framework for stabilizing the algorithms under consideration by a line search. Merit function is the augmented Lagrangian function, where violation of constraints is penalized in the L_2-norm. SQP and SCP methods are introduced in the two subsequent sections, where we outline some common ideas and some existing convergence results. Section 5 shows the results of comparative numerical experiments based on the 306 test examples of the test problem collections of Hock and Schittkowski (1981) and Schittkowski (1987a). The two computer codes under investigation, are the SQP subroutine NLPQLP of Schittkowski (2001) and the SCP routine SCPIP of Zillober (2001c), Zillober (2002). To give an impression on the convergence of the algorithms in case of structural design optimization, we repeat a few results of the comparative study Schittkowski et al. (1994). A couple of typical industrial applications and case studies are found in Section 6, to show the complexity of modern optimization problems, for which reliable and efficient software is needed.

2   A GENERAL FRAMEWORK

The fundamental tool for deriving optimality conditions and optimization algorithms is the *Lagrange function*

$$L(x, u) := f(x) + u^T g(x) \qquad (2.1)$$

defined for all $x \in \mathbb{R}^n$ and $u = (u_1, \ldots, u_m)^T \in \mathbb{R}^m$. The purpose of $L(x, u)$ is to link objective function $f(x)$ and constraints $g(x)$. The variables in u are called the *Lagrangian multipliers* of the nonlinear programming problem (1.1). Especially, the necessary Karush-Kuhn-Tucker (KKT) optimality conditions are easily formulated by the equations

$$
\begin{aligned}
u^\star &\geq 0 \ , \\
g(x^\star) &\leq 0 \ , \\
\nabla_x L(x^\star, u^\star) &= 0 \ , \\
g(x^\star)^T u^\star &= 0 \ .
\end{aligned}
\qquad (2.2)
$$

Since we assume that (1.1) is nonconvex and nonlinear in general, the basic idea is to replace (1.1) by a sequence of *simpler* problems. Starting from an initial design vector $x_0 \in \mathbb{R}^n$ and an initial multiplier estimate $u_0 \in \mathbb{R}^m$, iterates $x_k \in \mathbb{R}^n$ and $u_k \in \mathbb{R}^m$ are computed successively by solving subproblems of the form

$$
\begin{aligned}
&\min_{y \in \mathbb{R}^n} f^k(y) \\
&g^k(y) \leq 0,
\end{aligned}
\qquad (2.3)
$$

Let y_k be the optimal solution and v_k the corresponding Lagrangian multiplier of (2.3). A new iterate is computed by

$$
\begin{aligned}
x_{k+1} &= x_k + \alpha_k(y_k - x_k) \; , \\
u_{k+1} &= u_k + \alpha_k(v_k - u_k) \; ,
\end{aligned}
\tag{2.4}
$$

where α_k is a steplength parameter discussed subsequently.

Simpler means in this case that the subproblem is solvable by an available *black box* technique, more or less independently from the underlying model structure. In particular, it is assumed that the numerical algorithm for solving (2.3) does not require any additional function or gradient evaluations of the original functions $f(x)$ and $g_j(x)$, $j = 1, \ldots, m$. The approach indicates also that we are looking for a simultaneous approximation of an optimal solution $x^\star$ and of the corresponding multiplier $u^\star$.

Now we summarize the requirements to describe at least the SQP and SCP algorithms in a uniform manner:

1. (2.3) is strictly convex and smooth, i.e. the functions $f^k(x)$ and $g_j^k(x)$ are twice continuously differentiable, $j = 1, \ldots, m$.

2. (2.3) is a first order approximation of (1.1) at x_k, i.e. $f(x_k) = f^k(x_k)$, $\nabla f(x_k) = \nabla f^k(x_k)$, $g(x_k) = g^k(x_k)$, and $\nabla g(x_k) = \nabla g^k(x_k)$.

3. The search direction $(y_k - x_k, v_k - u_k)$ is a descent direction for an augmented Lagrangian merit function introduced below.

4. The feasible domain of (2.3) is bounded.

Strict convexity of (2.3) means that the objective function $f^k(x)$ is strictly convex and that the constraints $g_j^k(x)$ are convex functions for all iterates k and $j = 1, \ldots, m$. If the feasible domain is non-empty, (2.3) has a unique solution $y_k \in \mathbb{R}^n$ with Lagrangian multiplier $v_k \in \mathbb{R}^m$.

A further important consequence is that if $y_k = x_k$ at least conceptually, then x_k and v_k solve the general nonlinear programming problem (1.1) in the sense of a stationary solution. The Karush-Kuhn-Tucker optimality conditions for the subproblem are given by

$$
\begin{aligned}
v_k & \geq 0 \;, \\
g^k(y_k) & \leq 0 \;, \\
\nabla_x L^k(y_k, v_k) & = 0 \;, \\
g^k(y_k)^T v_k & = 0
\end{aligned}
\tag{2.5}
$$

with corresponding Lagrangian

$$
L^k(y, v) := f^k(y) + v^T g^k(y) \;.
\tag{2.6}
$$

If $y_k = x_k$, the current iterate x_k is feasible, and satisfies the complementary slackness condition $g(x_k)^T v_k = g^k(y_k)^T v_k = 0$ as well as the stationary condition $\nabla_x L(x_k, v_k) = \nabla_x L^k(y_k, v_k) = 0$. In other words, the pair x_k and v_k is a stationary point of (1.1).

A line search is introduced to stabilize the solution process, particularly helpful when starting from a bad initial guess. We are looking for an α_k, see (2.4), $0 < \alpha_k \leq 1$, so that a step along a merit function $\Psi_k(\alpha)$ from the current iterate to the new one becomes *acceptable*. The idea is to penalize the Lagrange function in the L_2 norm, as soon as constraints are violated, by defining

$$
\Phi_r(x, u) = f(x) + \sum_{j \in J} \left(u_j g_j(x) + \frac{1}{2} r_j g_j(x)^2 \right) - \frac{1}{2} \sum_{j \in K} u_j^2 / r_j \;,
\tag{2.7}
$$

and we set

$$
\Psi_k(\alpha) = \Phi_{r_k}\left(\begin{pmatrix} x_k \\ u_k \end{pmatrix} + \alpha \begin{pmatrix} y_k - x_k \\ v_k - u_k \end{pmatrix} \right) \;,
\tag{2.8}
$$

where $J = \{j : g_j(x) \geq -u_j / r_j\}$ and $K = \{1, \ldots, m\} \setminus J$ define the constraints considered as active or inactive, respectively.

The steplength parameter α_k is required in (2.4) to enforce global convergence of the optimization method, i.e. the approximation of a point satisfying the necessary Karush-Kuhn-Tucker optimality conditions when starting from arbitrary initial values, e.g. a user-provided $x_0 \in \mathbb{R}^n$ and $u_0 = 0$. The merit function defined by (2.7) is also called augmented Lagrange function, see for example Rockafellar (1974). The corresponding penalty parameter r_k at the k-th iterate that controls the degree of constraint violation, must be chosen carefully to guarantee a descent direction of the merit function, so that the line search is well-defined,

$$\Psi'_k(0) = \nabla\Phi_{r_k}(x_k, u_k)^T \begin{pmatrix} y_k - x_k \\ v_k - u_k \end{pmatrix} < 0 \ . \tag{2.9}$$

The line search consists of a successive reduction of α starting at 1, usually combined with a quadratic interpolation, until a sufficient decrease condition

$$\Psi_k(\alpha_k) < \Psi(0) + \nu\alpha_k\Psi'_k(0) \tag{2.10}$$

is obtained, $0 < \nu < 0.5$ arbitrarily chosen. To prove convergence, however, we need a stronger estimate typically of the form

$$\Psi'_k(0) < -\mu\|y_k - x_k\|^2 \tag{2.11}$$

with $\mu > 0$, which is satisfied for SQP and SCP methods, see Schittkowski (1981a), Schittkowski (1983b) or Zillober (1993), respectively. A more general framework is introduced in Schittkowski (1985a). For a more detailed discussion of line search and different convergence aspects see Ortega and Rheinboldt (1970).

If the constraints of (2.3) become inconsistent, it is possible to introduce an additional variable and to modify objective function and constraints, for example in the simplest form

$$y \in \mathbb{R}^{n+1} : \quad \begin{aligned} \min\ & f^k(y) + \rho_k y_{n+1}^2 \\ & g^k(y) - y_{n+1} \le 0 \quad , \\ & -y_{n+1} \le 0 \ . \end{aligned} \tag{2.12}$$

The penalty term ρ_k is added to the objective function to reduce the influence of the additional variable y_{n+1} as much as possible. The index k implies that this parameter also needs to be updated during the algorithm. It is obvious that (2.12) does always possess a feasible solution.

3 SQP METHODS

Sequential quadratic programming or SQP methods belong to the most powerful nonlinear programming algorithms we know today for solving differentiable nonlinear programming problems of the form (1.1). The theoretical background is described e.g. in Stoer (1985) in form of a review, or in Spellucci (1993) in form of an extensive text book. ¿From the more practical point of view, SQP methods are also introduced in the books of Papalambros and Wilde (2000) or Edgar and Himmelblau (1988). Their excellent numerical performance is tested and compared with other methods in Schittkowski (1980), Schittkowski (1983a), and Hock and Schittkowski (1981), and since many years they belong to the most frequently used algorithms to solve practical optimization problems.

The basic idea is to formulate and solve a quadratic programming subproblem in each iteration which is obtained by linearizing the constraints and approximating the Lagrange function (2.1) quadratically.

To formulate the quadratic programming subproblem, we proceed from given iterates $x_k \in \mathbb{R}^n$, an approximation of the solution, $u_k \in \mathbb{R}^m$ an approximation of the multipliers, and $B_k \in \mathbb{R}^{n \times n}$, an approximation of the Hessian of the Lagrange function in a certain sense. Then we obtain subproblem (2.3) by defining

$$
\begin{aligned}
f^k(y) &= \tfrac{1}{2}\,(y - x_k)^T B_k (y - x_k) + \nabla f(x_k)^T (y - x_k) + f(x_k) \ , \\
g_j^k(y) &= \nabla g_j(x_k)^T (y - x_k) + g_j(x_k) \ , \quad j = 1, \ldots, m \ .
\end{aligned}
\tag{3.1}
$$

It is immediately seen that the requirements of the previous section for (2.3) are satisfied. The key idea is to approximate also second order information to get a fast final convergence speed. The update of the matrix B_k can be performed by standard quasi-Newton techniques known from unconstrained optimization. In most cases, the BFGS-method is applied, see Powell (1978a), Powell (1978b), or Stoer (1985). Starting from the identity or any other positive definite matrix B_0, the difference vectors

$$
\begin{aligned}
q_k &= \nabla_x L(x_{k+1}, v_k) - \nabla_x L(x_k, v_k) \ , \\
p_k &= x_{k+1} - x_k
\end{aligned}
\tag{3.2}
$$

are used to update B_k in the form

$$
B_{k+1} = \Pi(B_k, q_k, w_k) \ ,
\tag{3.3}
$$

where

$$
\Pi(B, q, w) := B + \frac{qq^T}{q^T w} - \frac{Bww^T B}{w^T Bw} \ .
\tag{3.4}
$$

The above formula yields a positive definite matrix B_{k+1} provided that B_k is positive definite and $q_k^T w_k > 0$. A simple modification of Powell (1978a) guarantees positive definite matrices even if the latter condition is violated.

Among the most attractive features of sequential quadratic programming methods is the superlinear convergence speed in the neighborhood of a solution, i.e.

$$\|x_{k+1} - x^\star\| < \gamma_k \|x_k - x^\star\| \tag{3.5}$$

with $\gamma_k \to 0$.

The motivation for the fast convergence speed of SQP methods is based on the following observation: An SQP method is identical to Newton's method to solve the necessary optimality conditions (2.2), if B_k is the Hessian of the Lagrange function at x_k and u_k and if we start sufficiently close to a solution. The statement is easily derived in case of equality constraints only, but holds also for inequality restrictions.

In fact there exist numerous theoretical convergence results in the literature, see for example Spellucci (1993). In the frame of our discussion, we consider the global convergence behaviour, i.e., the question, whether the SQP method converges when starting from an arbitrary initial point. Suppose that the augmented Lagrangian merit function (2.7) is implemented as merit function and that the corresponding penalty parameters that control the degree of constraint violation are determined in the way

$$r_k = \max\left(\frac{2\|v_k - u_k\|^2}{(y_k - x_k)^T B_k (y_k - x_k)}, r_{k-1}\right) \tag{3.6}$$

for the augmented Lagrangian function (2.7), see Schittkowski (1983a). Moreover we need an additional assumption concerning the choice of the matrix B_k, if we neglect the special update mechanism shown in (3.3). We require that the eigenvalues of the matrices B_k remain bounded away from 0, i.e. that $(y_k - x_k)^T B_k (y_k - x_k) \geq \gamma \|y_k - x_k\|^2$ for all k and a $\gamma > 0$. If the iteration data $\{(x_k, u_k, B_k)\}$ are bounded, then it can be shown that there is an accumulation point of $\{(x_k, u_k)\}$ satisfying the Karush-Kuhn-Tucker conditions (2.2) for (1.1), see Schittkowski (1983a).

The statement is quite weak, but without any further information about second derivatives, we cannot guarantee that the approximated point is indeed a local minimizer. ¿From the practical point of view, we need a finite stopping criterion based on the optimality conditions for the subproblem, see (2.5), based on a suitable tolerance $\epsilon > 0$. For example, we could try to test the KKT condition

$$\|\nabla_x L^k(y_k, v_k)\| < \epsilon \tag{3.7}$$

together with sufficient feasibility and complementary slackness satisfaction. The above convergence result ensures that the stopping conditions are satisfied after a finite number of iterations. It should be noted, however, that implemented criteria are much more complex and sophisticated, see for example Schittkowski (1985b).

There remain a few final comments to summarize the most interesting features of SQP methods:

- Linear constraints and bounds of variables remain satisfied.

- In case of n active constraints, the SQP method behaves like Newton's method for solving the corresponding system of equations, i.e., the local convergence speed is even quadratically.

- The algorithm is globally convergent and the local convergence speed is superlinear.

- A simple reformulation allows the efficient solution of constrained nonlinear least squares problems, see Schittkowski (1988), Schittkowski (1994), or Schittkowski (2002).

- A large number of constraints can be treated by an active set strategy, see Schittkowski (1992). In particular, the computation of gradients for inactive restrictions can be omitted.

- There exists a large variety of different extensions to solve also large scale problems, see Gould and Toint (2000) for a review.

4   SCP METHODS

Sequential convex programming methods are developed mainly for mechanical structural optimization. The first approach of Fleury and Braibant (1986) and Fleury (1989) is known under the name convex linearization (CONLIN), and exploits the observation that in some special cases, typical structural constraints become linear in the inverse variables.

To illustrate this motivation, we consider the most simple example, two bars fixed at a wall and connected at the other end. An external load p is applied at this node, see Figure 4.1. The two design variables are the cross sectional areas a_i scaled by elasticity modulus E and length l_i, $i = 1, 2$, i.e., $x_i = Ea_i/l_i$.

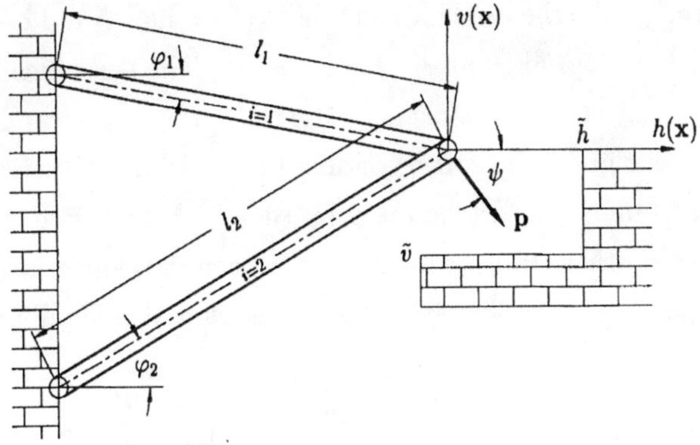

Figure 4.1   2-bar-truss.

If s_i and c_i denote the sinus and co-sinus function values of the corresponding angles of the trusses, $i = 1, 2$, the horizontal and vertical displacements are given in the form

$$h(x) = |p|(\cos\psi(s_1^2/x_2 + s_2^2/x_1) - \sin\psi(s_1 c_1/x_2 + s_2 c_2/x_1))/\sin^2(\phi_1 - \phi_2) \;,$$

$$v(x) = |p|(\sin\psi(c_1^2/x_2 + c_2^2/x_1) - \cos\psi(s_1 c_1/x_2 + s_2 c_2/x_1))/\sin^2(\phi_1 - \phi_2) \;.$$

If we assume now that our optimization problem consists of minimizing the weight of the structure under some given upper bounds for these displacements, we get nonlinear constraints that are linear in the reciprocal design variables.

Although this special situation is always found in case of statically determinate structures, it is rarely observed in practice. However, a suitable substitution by inverse variables depending on the sign of the corresponding partial derivatives and subsequent linearization is expected to linearize constraints somehow.

For the CONLIN method, Nguyen et al. (1987) gave a convergence proof but only for the case that (1.1) consists of a concave objective function and concave constraints which is of minor practical interest. They showed also that a generalization to non-concave constraints is not possible. More general convex approximations are introduced by Svanberg (1987) known under the name *method of moving asymptotes* (MMA). The goal is always to construct nonlinear convex and separable subproblems, for which efficient solvers are available. Using the flexibility of the asymptotes which influence the curvature of the approximations, it is possible to avoid the concavity assumption.

Given an iterate x_k, the model functions of (1.1), i.e., f and g_j, are approximated by functions f^k and g_j^k at x_k, $j = 1, \ldots, m$. The basic idea is to linearize f and g_j with respect to transformed variables $(U_i^k - x_i)^{-1}$ and $(x_i - L_i^k)^{-1}$ depending on the sign of the corresponding first partial derivative. U_i^k and L_i^k are reasonable bounds and are adapted by the algorithm after each successful step. Also several other transformations have been developed in the past.

The corresponding approximating functions that define subproblem (2.3), are

$$
\begin{aligned}
f^k(y) &= \alpha_0^k + \sum_{i \in I_k^+} \frac{\beta_{i,0}^k}{U_i^k - y_i} - \sum_{i \in I_k^-} \frac{\beta_{i,0}^k}{y_i - L_i^k} , \\
g_j^k(y) &= \alpha_j^k + \sum_{i \in I_{jk}^+} \frac{\beta_{i,j}^k}{U_i^k - y_i} - \sum_{i \in I_{jk}^-} \frac{\beta_{i,j}^k}{y_i - L_i^k} ,
\end{aligned}
\tag{4.1}
$$

$j = 1, \ldots, m$, where $y = (y_1, \ldots, y_n)^T$. The index sets are defined by

$$
I_k^+ = \{i : 1 \le i \le n, \frac{\partial}{\partial x_i} f(x_k) \ge 0\} , \quad I_k^- = \{i : 1 \le i \le n, \frac{\partial}{\partial x_i} f(x_k) < 0\} .
$$

In a similar way, I_{jk}^+ and I_{jk}^- are defined. The coefficients α_j^k and $\beta_{i,j}^k$, $j = 0, \ldots,$ m are chosen to satisfy the requirements of Section 2.1, i.e., that (2.3) is convex and that (2.3) is a first order approximation of (1.1) at x_k. By an appropriate regularization of the objective function, strict convexity of $f^k(x)$ is guaranteed, see Zillober (2001a). As shown there, the search direction $(y_k - x_k, v_k - u_k)$ is a descent direction for the augmented Lagrangian merit function. If the adoption rule for the parameters L_i^k and U_i^k fulfills the conditions that the absolute value of their difference to the corresponding component of the current iteration point x_k is uniformly bounded away from 0 and that their absolute value is bounded, global convergence can be shown for the SCP method if a similar update rule for the penalty parameters r_k as (3.6) is applied.

The choice of the asymptotes L_i^k and U_i^k, is crucial for the computational behavior of the method. If additional lower bounds x_l and upper bounds x_u on the variables are given, an efficient update scheme for the i-th coefficient, $i = 1, \ldots, n$, and the k-th iteration step is given as follows:

$$
k = 0, 1 \; : L_i^k = x_{k,i} - \lambda_1(x_{u,i} - x_{l,i}) ,
$$

$$
U_i^k = x_{k,i} + \lambda_1(x_{u,i} - x_{l,i}) .
$$

$$
k = 2, 3, \ldots \; : \text{If } \text{sign}(x_{k,i} - x_{k-1,i}) = \text{sign}(x_{k-1,i} - x_{k-2,i}) :
$$

$$L_i^k = x_{k,i} - \lambda_2(x_{k-1,i} - L_i^{k-1}) \ ,$$

$$U_i^k = x_{k,i} + \lambda_2(U_i^{k-1} - x_{k-1,i}) \ .$$

If $\text{sign}(x_{k,i} - x_{k-1,i}) \neq \text{sign}(x_{k-1,i} - x_{k-1,i})$:

$$L_i^k = x_{k,i} - \lambda_3(x_{k-1,i} - L_i^{k-1}) \ ,$$

$$U_i^k = x_{k,i} + \lambda_3(U_i^{k-1} - x_{k-1,i}) \ .$$

A suitable choice of the constants is $\lambda_1 = 0.5$, $\lambda_2 = 1.15$, $\lambda_3 = 0.7$. If there is no change in the sign of a component of two successive iterations, this situation is interpreted as *smooth* convergence and allows a relaxation of the asymptotes. If there are changes of the sign in two successive iterations, we are afraid of cycling. The asymptotes are chosen closer to the iteration point leading to more conservative approximations.

Additional safeguards ensure the compatibility of this procedure with the overall scheme and guarantee global convergence. A small positive constant is introduced to avoid that the difference between the asymptotes and the current iteration point becomes too small. However, these safeguards are rarely used in practice, see Zillober (2001a) for more details.

For the first SCP codes, the convex and separable subproblems are solved very efficiently by a dual approach, where dense linear systems of equations with m rows and columns are solved, cf. Svanberg (1987) or Fleury (1989). Recently, an interior point method for the solution of the subproblems is proposed by Zillober (2001b). The advantage is to formulate either $n \times n$ or $m \times m$ linear systems of equations leading to a more flexible treatment of large problems. The resulting algorithm is very efficient especially for large scale mechanical engineering problems, and sparsity patterns of the original problem data can be exploited.

To summarize, the most important features of SCP methods are:

- Linear constraints and bounds of variables remain satisfied.

- The algorithm is globally convergent.

- As for SQP methods, a large number of constraints can be treated by an active set strategy, see Zillober (2001c), Zillober (2002). In particular, the computation of gradients for inactive restrictions can be omitted.

- Large scale problems can be handled more flexible by different variations of the solution procedure for the subproblem, see Zillober et al. (2002) for test results with up to 10^6 variables.

- Sparsity in the problem data can be exploited, see again Zillober et al. (2002) for a series of numerical results for elliptic control problems.

5 COMPARATIVE PERFORMANCE EVALUATION

Our numerical tests use all 306 academic and real-life test problems published in Hock and Schittkowski (1983) and in Schittkowski (1987a). Part of them are also available in the CUTE library, see Bongartz et al. (1995). The distribution of the dimension parameter n, the number of variables, is shown in Figure 5.1. We see, for example, that about 270 of 306 test problems have not more than 10 variables. In a similar way, the distribution of the number of constraints is shown in Figure 5.2. The test problems possess also nonlinear equality constraints and additional lower and upper bounds for the variables. The two codes under consideration, NLPQLP and SCPIP, are able to solve more general problems

$$
x \in \mathbb{R}^n : \quad
\begin{aligned}
&\min f(x) \\[4pt]
&h(x) = 0 \quad , \\[4pt]
&g(x) \leq 0 \quad , \\[4pt]
&x_l \leq x \leq x_u \quad ,
\end{aligned}
\tag{5.1}
$$

with smooth functions $h(x) = (h_1, \ldots, h_{m_e})^T$ and $x_l < x_u$.

Since analytical derivatives are not available for all problems, we approximate them numerically. The test examples are provided with exact solutions, either known from analytical solutions or from the best numerical data found so far. The Fortran codes are compiled by the Compaq Visual Fortran Optimizing Compiler, Version 6.5, under Windows 2000, and executed on a Pentium IV processor with 1.0 GHz. Since the calculation times are very short, about 10 sec for solving all 306 test problems with high order derivative approximations, we count only function and gradient evaluations. This is a realistic assumption, since for the practical applications in mind calculation times for evaluating model functions dominate and the numerical efforts within an optimization code are negligible.

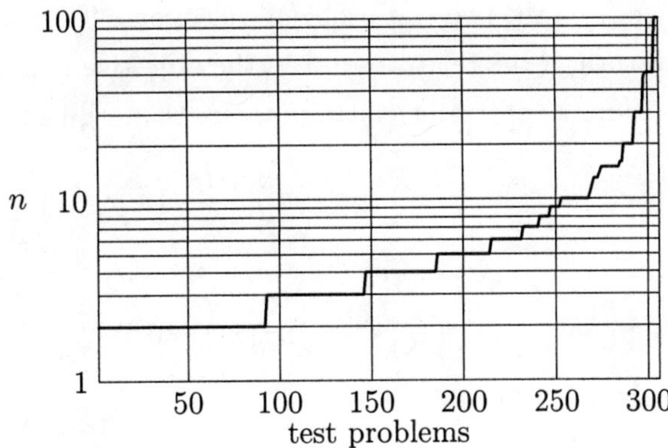

Figure 5.1   Distribution of number of variables.

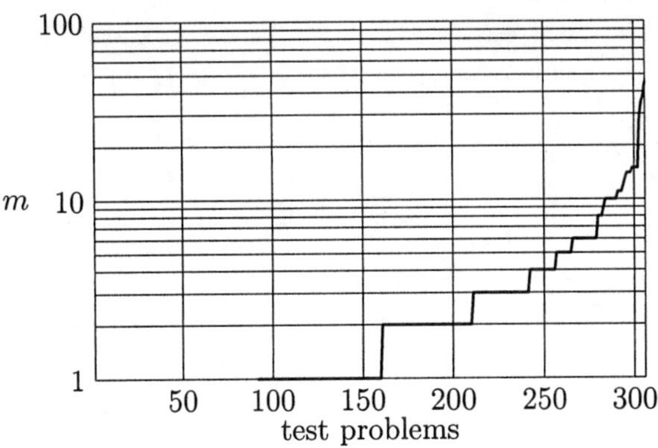

Figure 5.2   Distribution of number of constraints.

First we need a criterion to decide, whether the result of a test run is considered as a successful return or not. Let $\epsilon > 0$ be a tolerance for defining the relative accuracy, x_k the final iterate of a test run, and $x^\star$ the supposed exact solution known from the two test problem collections. Then we call the output a successful return, if the relative error in the objective function is less than ϵ and if the maximum constraint violation is less than ϵ^2, i.e., if

$$f(x_k) - f(x^\star) < \epsilon |f(x^\star)| \, , \ \text{if } f(x^\star) <> 0$$

or

$$f(x_k) < \epsilon \, , \ \text{if } f(x^\star) = 0$$

and

$$r(x_k) = \max(\|h(x_k)\|_\infty, \|g(x_k)^+\|_\infty) < \epsilon^2 \, ,$$

where $\| \ldots \|_\infty$ denotes the maximum norm and $g_j(x_k)^+ = \max(0, g_j(x_k))$.

We take into account that a code returns a solution with a better function value than the known one, subject to the error tolerance of the allowed constraint violation. However, there is still the possibility that an algorithm terminates at a local solution different from the one known in advance. Thus, we call a test run a successful one, if the internal termination conditions are satisfied subject to a reasonably small tolerance ($IFAIL=0$), and if in addition to the above decision,

$$f(x_k) - f(x^\star) \geq \epsilon |f(x^\star)| \, , \ \text{if } f(x^\star) <> 0$$

or

$$f(x_k) \geq \epsilon \, , \ \text{if } f(x^\star) = 0$$

and

$$r(x_k) < \epsilon^2 \, .$$

For our numerical tests, we use $\epsilon = 0.01$, i.e., we require a final accuracy of one per cent. Gradients are approximated by a fourth-order difference formula

$$\frac{\partial}{\partial x_i} f(x) \approx \frac{1}{4! \eta_i} \left(2f(x - 2\eta_i e_i) - 16 f(x - \eta_i e_i) + 16 f(x + \eta_i e_i) - 2f(x + 2\eta_i e_i) \right) \, ,$$

$$(5.2)$$

where $\eta_i = \eta \max(10^{-5}, |x_i|)$, $\eta = 10^{-7}$, e_i the i-th unit vector, and $i = 1, \ldots, n$. In a similar way, derivatives of constraints are computed.

Table 5.1 Performance results for standard test problems.

code	NSUCC	NF	NIT
NLPQLP	100 %	39	25
SCPIP	93 %	74	42

The Fortran implementation of the SQP method introduced in Section 3, is called NLPQLP, see Schittkowski (2001). The code represents the most recent version of NLPQL which is frequently used in academic and commercial institutions, see Schittkowski (1985b). NLPQLP is prepared to run also under distributed systems, but behaves in exactly the same way as the previous version, if the number of simulated processors is set to one. Functions and gradients must be provided by reverse communication and the quadratic programming subproblems are solved by the primal-dual method of Goldfarb and Idnani (1983) based on numerically stable orthogonal decompositions. NLPQLP is executed with termination accuracy $ACC=10^{-8}$ and a maximum number of iterations $MAXIT=500$.

The SCP method has been implemented in Fortran and is called SCPIP, see Zillober (2001c) and Zillober (2002). The convex subproblems are solved by the predictor-corrector interior point method described in Zillober (2001b). All input variables have been chosen such that the calling conventions for SCPIP and NLPQLP are comparable.

Table 5.1 shows

- the percentage of successful test runs $NSUCC$,

- the average number of function calls NF,

- the average number of iterations NIT.

When evaluating NF, we count each single function call. However, function evaluations needed for gradient approximations, are not counted. Their average number is $4 \times NIT$.

Many test problems are unconstrained or possess a highly nonlinear objective function preventing SCP from converging as fast as SQP methods. Moreover, bounds are often set far away from the optimal solution, leading to initial asymptotes too far away from the region of interest. Since SCP methods do not possess fast local convergence properties, SCPIP needs more iterations and function evaluations in the neighborhood of a solution.

The situation is different in mechanical structural optimization, where the SCP methods have been invented. In the numerical study of Schittkowski et al. (1994), 79 finite element formulations of academic and practical problems are collected based on the simulation package MBB-LAGRANGE, see Kneppe et al. (1987). The maximum number of variables is 144 and a maximum number of constraints 1020 without box-constraints. NLPQL, see Schittkowski (1985b), and MMA, former versions of NLPQLP and SCPIP, respectively, are among the 11 optimization algorithms under consideration. To give an impression on the behavior of SQP versus MMA, we repeat some results of Schittkowski et al. (1994), see Table 5.2. We compare the performance with respect to percentage of successful test runs ($NSUCC$), number of function calls (NF), and number of iterations (NIT), respectively.

For the evaluation of performance indices by the priority theory of Saaty, see Schittkowski et al. (1994). The main difficulty is that the optimization algorithms solve only a certain subset of test problems successfully, which differs from code to code. Thus, mean values of a performance criterion are evaluated only pairwise over the set of successfully solved test problems of two algorithms, and then compared in form of a matrix. The decision whether the result of a test run is considered as a successful one or not, depends on a tolerance ϵ which is set to $\epsilon = 0.01$ and $\epsilon = 0.00001$, respectively.

Table 5.2   Performance results for structural optimization test problems

	$\epsilon = 0.01$			$\epsilon = 0.00001$		
code	$NSUCC$	NF	NIT	$NSUCC$	NF	NIT
$NLPQL$	84 %	2.0	1.6	77 %	1.3	1.3
MMA	73 %	1.0	1.0	73 %	1.0	1.0

The figures of Table 5.2 represent the scaled relative performance when comparing the codes among each other. We conclude for example that for $\epsilon = 0.01$, NLPQL requires about twice as many gradient evaluations or iterations, respectively, as MMA. When requiring a higher termination accuracy, however,

NLPQL needs only about 30 % as many gradient calls. On the other hand, NLPQL is a bit more reliable than MMA.

6 SOME ACADEMIC AND COMMERCIAL APPLICATIONS

A few typical academic and commercial applications of the SQP algorithm NLPQL are

- mechanical structural optimization, for example airplane and space structures, Kneppe et al. (1987),

- optimal feed control of tubular chemical reactors, Birk et al. (1999),

- design of surface acoustic wave filters for signal processing, Bünner et al. (2002),

- design of horn radiators for satellite communication, Hartwanger et al. (2000),

- modeling of maltodextrin DE12 drying process in a convection oven, Frias et al. (2001),

- fitting drug dissolution measurements of immediate release solid dosage, Loth et al. (2001),

- diffusion and concurrent metabolism in cutaneous tissue, Boderke et al. (2000).

In some cases, the underlying simulation software is highly complex and developed over many years, in some others NLPQL is used in a special way to solve data fitting problems.

Typical applications of the SCP code SCPIP are

- minimum weight of a cruise ship, Zillober and Vogel (2000a),

- minimum weight of an exhaust pipe in truck design, Zillober (2001b),

- optimal design of the bulk head of an aircraft, Zillober and Vogel (2000b),

- optimal construction of supporting tub in automotive industry, Zillober (2001b),

- topology design of mechanical structures, see below.

To give an impression about the capabilities of an SCP implementation for solving very large scale nonlinear programming problems, we consider topology optimization. Given a predefined domain in the 2D/3D space with boundary conditions and external load, the intention is to distribute a percentage of the initial mass on the given domain such that a global measure takes a minimum, see Bendsøe (1995) for a broad introduction. The so-called power law approach, see also Bendsøe (1989) or Mlejnek (1992), leads to a nonlinear programming problem of the form

$$\begin{aligned}
&\min_{x \in \mathbb{R}^n} \ u^T f \\
&V(x) \leq pV_0 \quad , \\
&K(x)u = f \quad , \\
&0 < x_l \leq x \leq 1 \quad ,
\end{aligned} \qquad (6.1)$$

where x denotes the relative densities, u the displacement vector computed from the linear system of equations $K(x)u = f$ with positive definite stiffness matrix $K(x)$ and external load vector f. The relative densities and the elementary stiffness matrices K_i^0 define $K(x)$ by

$$K(x) = \sum_{i=1}^{n} x_i^3 \, K_i^0 \ .$$

$V(x)$ is the volume of the structure, usually a linear function of the design variables. x_l is a vector of small positive numbers to avoid singularities. The power 3 in the state equation is found heuristically and usually applied in practice. Its role is to penalize intermediate values between the lower bound and 1. Topology optimization problems lead easily to very large scale, highly nonlinear optimization problems. The probably most simple example is a half beam, for our test discretized by 390×260 linear four-nodes square elements leading to $101,400$ optimization variables. SCPIP computes the solution shown in Figure 7.1 after 30 iterations.

7 CONCLUSIONS

In this paper we try to describe SQP and SCP methods in a uniform way. In both cases, convex subproblems are formulated from which a suitable search direction with respect to the design variables and the multipliers are computed. A subsequent line search based on the augmented Lagrangian merit function stabilizes the optimization algorithm and allows to prove global convergence.

Starting from an arbitrary initial design, a stationary point satisfying the necessary Karush-Kuhn-Tucker conditions is approximated.

However, both methods differ significantly in their underlying motivation. SQP algorithms proceed from a local quadratic model with the goal to achieve a fast local convergence speed, superlinearly in case of quasi-Newton updates for the Hessian of the Lagrangian. On the other hand, SCP methods apply a global convex and nonlinear approximation of objective function and constraints based on linearized inverse variables, by using first order information only.

Numerical results based on standard, small scale benchmark problems and on structural design optimization problems are included, also a brief summary of some industrial applications the authors are involved in.

Advantages and disadvantages of SQP versus SCP methods can be summarized as follows.

	Advantages	*Disadvantages*
SQP	highly accurate solutions with fast final convergence speed, able to solve highly nonlinear problems, robust, general purpose tool	fast convergence only in case of accurate gradients, large storage requirements
SCP	tuned for solving structural mechanical optimization problems, no heredity of round-off errors in function and gradient calculations, excellent convergence speed in special situations, able to solve very large problems	slow final convergence possible, not reliable without stabilization, less robust with respect to default tolerances

References

Bendsøe M.P. (1989): *Optimal shape design as a material distribution problems*, Structural Optimization, Vol. 1, 193-202.

Bendsøe M.P. (1995): *Optimization of Structural Topology, Shape and Material*, Springer, Heidelberg.

Figure 7.1   Half beam.

Boderke P., Schittkowski K., Wolf M., Merkle H.P. (2000): *Modeling of diffusion and concurrent metabolism in cutaneous tissue*, Journal on Theoretical Biology, Vol. 204, No. 3, 393-407.

Birk J., Liepelt M., Schittkowski K., Vogel F. (1999): *Computation of optimal feed rates and operation intervals for tubular reactors*, Journal of Process Control, Vol. 9, 325-336.

Blatt M., Schittkowski K. (1998): *Optimal Control of One-Dimensional Partial Differential Equations Applied to Transdermal Diffusion of Substrates*, in: Optimization Techniques and Applications, L. Caccetta, K.L. Teo, P.F. Siew, Y.H. Leung, L.S. Jennings, V. Rehbock eds., School of Mathematics and Statistics, Curtin University of Technology, Perth, Australia, Vol. 1, 81 - 93.

Bongartz I., Conn A.R., Gould N., Toint Ph. (1995): *CUTE: Constrained and unconstrained testing environment*, Transactions on Mathematical Software, Vol. 21, No. 1, 123-160.

Bünner M., Schittkowski K., van de Braak G. (2002): *Optimal design of surface acoustic wave filters for signal processing by mixed integer nonlinear programming*, submitted for publication.

Edgar T.F., Himmelblau D.M. (1988): *Optimization of Chemical Processes*, McGraw Hill.

Fleury C. (1989): *An efficient dual optimizer based on convex approximation concepts*, Structural Optimization, Vol. 1, 81-89.

Fleury C., Braibant V. (1986): *Structural Optimization – a new dual method using mixed variables*, International Journal for Numerical Methods in Engineering, Vol. 23, 409-428.

Frias J.M., Oliveira J.C, Schittkowski K. (2001): *Modelling of maltodextrin DE12 drying process in a convection oven*, Applied Mathematical Modelling, Vol. 24, 449-462.

Goldfarb D., Idnani A. (1983): *A numerically stable method for solving strictly convex quadratic programs*, Mathematical Programming, Vol. 27, 1-33.

Gould N.I.M., Toint P.L. (2000): *SQP methods for large-scale nonlinear programming*, in: *System Modelling and Optimization: Methods, Theory and Applications*, M.J.D. Powell, S. Scholtes eds., Kluwer.

Han S.-P. (1976): *Superlinearly convergent variable metric algorithms for general nonlinear programming problems*, Mathematical Programming, Vol. 11, 263-282.

Han S.-P. (1977): *A globally convergent method for nonlinear programming*, Journal of Optimization Theory and Applications, Vol. 22, 297–309.

Hartwanger C., Schittkowski K., Wolf H. (2000): *Computer aided optimal design of horn radiators for satellite communication*, Engineering Optimization, Vol. 33, 221-244.

Hock W., Schittkowski K. (1981): *Test Examples for Nonlinear Programming Codes*, Lecture Notes in Economics and Mathematical Systems, Vol. 187, Springer.

Hock W., Schittkowski K. (1983): *A comparative performance evaluation of 27 nonlinear programming codes*, Computing, Vol. 30, 335-358.

Kneppe G., Krammer J., Winkler E. (1987): *Structural optimization of large scale problems using MBB-LAGRANGE*, Report MBB-S-PUB-305, Messerschmitt-Bölkow-Blohm, Munich.

Loth H., Schreiner Th., Wolf M., Schittkowski K., Schäfer U. (2001): *Fitting drug dissolution measurements of immediate release solid dosage forms by numerical solution of differential equations*, submitted for publication.

Mlejnek H.P. (1992): *Some aspects of the genesis of structures*, Structural Optimization, Vol. 5, 64-69.

Nguyen V.H., Strodiot J.J., Fleury C. (1987): *A mathematical convergence analysis for the convex linearization method for engineering design optimization*, Engineering Optimization, Vol. 11, 195-216.

Ortega J.M., Rheinboldt W.C. (1970): *Iterative Solution of Nonlinear Equations in Several Variables*, Academic Press, New York-San Francisco-London.

Papalambros P.Y., Wilde D.J. (2000): *Principles of Optimal Design*, Cambridge University Press.

Powell M.J.D. (1978): *A fast algorithm for nonlinearly constraint optimization calculations*, in: Numerical Analysis, G.A. Watson ed., Lecture Notes in Mathematics, Vol. 630, Springer.

Powell M.J.D. (1978): *The convergence of variable metric methods for nonlinearly constrained optimization calculations*, in: Nonlinear Programming 3, O.L. Mangasarian, R.R. Meyer, S.M. Robinson eds., Academic Press.

Powell M.J.D. (1983): *On the quadratic programming algorithm of Goldfarb and Idnani.* Report DAMTP 1983/Na 19, University of Cambridge, Cambridge.

Rockafellar R.T. (1974): *Augmented Lagrange multiplier functions and duality in non-convex programming*, Journal on Control, Vol. 12, 268-285.

Schittkowski K. (1980): *Nonlinear Programming Codes*, Lecture Notes in Economics and Mathematical Systems, Vol. 183 Springer.

Schittkowski K. (1981): *The nonlinear programming method of Wilson, Han and Powell. Part 1: Convergence analysis*, Numerische Mathematik, Vol. 38, 83-114.

Schittkowski K. (1981): *The nonlinear programming method of Wilson, Han and Powell. Part 2: An efficient implementation with linear least squares subproblems*, Numerische Mathematik, Vol. 38, 115-127.

Schittkowski K. (1983): *Theory, implementation and test of a nonlinear programming algorithm*, in: Optimization Methods in Structural Design, H. Eschenauer, N. Olhoff eds., Wissenschaftsverlag.

Schittkowski K. (1983): *On the convergence of a sequential quadratic programming method with an augmented Lagrangian search direction*, Optimization, Vol. 14, 197-216.

Schittkowski K. (1985): *On the global convergence of nonlinear programming algorithms*, ASME Journal of Mechanics, Transmissions, and Automation in Design, Vol. 107, 454-458.

Schittkowski K. (1985/86): *NLPQL: A Fortran subroutine solving constrained nonlinear programming problems*, Annals of Operations Research, Vol. 5, 485-500.

Schittkowski K. (1987a): *More Test Examples for Nonlinear Programming*, Lecture Notes in Economics and Mathematical Systems, Vol. 182, Springer.

Schittkowski K. (1987): *New routines in MATH/LIBRARY for nonlinear programming problems*, IMSL Directions, Vol. 4, No. 3.

Schittkowski K. (1988): *Solving nonlinear least squares problems by a general purpose SQP-method*, in: Trends in Mathematical Optimization, K.-H. Hoffmann, J.-B. Hiriart-Urruty, C. Lemarechal, J. Zowe eds., International Series of Numerical Mathematics, Vol. 84, Birkhäuser, 295-309.

Schittkowski K. (1992): *Solving nonlinear programming problems with very many constraints*, Optimization, Vol. 25, 179-196.

Schittkowski K. (1994): *Parameter estimation in systems of nonlinear equations*, Numerische Mathematik, Vol. 68, 129-142.

Schittkowski K. (2001): *NLPQLP: A New Fortran Implementation of a Sequential Quadratic Programming Algorithm for Parallel Computing*, Report, Department of Mathematics, University of Bayreuth.

Schittkowski K. (2002): *Numerical Data Fitting in Dynamical Systems*, Kluwer.

Schittkowski K., Zillober C., Zotemantel R. (1994): *Numerical comparison of nonlinear programming algorithms for structural optimization*, Structural Optimization, Vol. 7, No. 1, 1-28.

Spellucci P. (1993): *Numerische Verfahren der nichtlinearen Optimierung*, Birkhäuser.

Stoer J. (1985): *Foundations of recursive quadratic programming methods for solving nonlinear programs*, in: Computational Mathematical Programming, K. Schittkowski, ed., NATO ASI Series, Series F: Computer and Systems Sciences, Vol. 15, Springer.

Svanberg K. (1987): *The Method of Moving Asymptotes – a new method for Structural Optimization*, International Journal for Numerical Methods in Engineering, Vol. 24, 359-373.

Zillober Ch. (1993): *A globally convergent version of the Method of Moving Asymptotes*, Structural Optimization, Vol. 6, 166-174.

Zillober Ch. (2001a): *Global convergence of a nonlinear programming method using convex approximations*, Numerical Algorithms, Vol. 27, 256-289.

Zillober Ch. (2001b): *A combined convex approximation – interior point approach for large scale nonlinear programming*, Optimization and Engineering, Vol. 2, 51-73.

Zillober Ch. (2001c): *Software manual for SCPIP 2.3*, Report, Department of Mathematics, University of Bayreuth.

Zillober Ch. (2002): *SCPIP – an efficient software tool for the solution of structural optimization problems*, Structural and Multidisciplinary Optimization, Vol. 24, No. 5, 362-371.

Zillober Ch., Vogel F. (2000a): *Solving large scale structural optimization problems*, in: Proceedings of the 2nd ASMO UK/ISSMO Conference on Engineering Design Optimization, J. Sienz ed., University of Swansea, Wales, 273-280.

Zillober Ch., Vogel F. (2000b): *Adaptive strategies for large scale optimization problems in mechanical engineering*, in: Recent Advances in Applied and Theoretical Mathematics, N. Mastorakis ed., World Scientific and Engineering Society Press, 156-161.

Zillober Ch., Schittkowski K., Moritzen K. (2002): *Very large scale optimization by sequential convex programming*, submitted for publication.

15 AN APPROXIMATION APPROACH FOR LINEAR PROGRAMMING IN MEASURE SPACE

C.F. Wen and S.Y. Wu

Institute of Applied Mathematics
National Cheng Kung University
Tainan 701, TAIWAN

Abstract: The purpose of this paper is to present some results, of a study of linear programming in measure spaces (LPM). We prove that under certain conditions, there exists a solution for LPM. We develop an approximation scheme to solve LPM and prove the convergence properties of the approximation scheme.

Key words: Linear programming in measure spaces, optimal solution, approximation scheme, linear semi-infinite programming problems.

1 INTRODUCTION

Let (E, F) and (Z, W) be two dual pairs of ordered vector spaces. Let E_+ and Z_+ be the positive cones for E and Z respectively, and E_+^* and Z_+^* be the polar cones of E_+ and Z_+ respectively. The general linear programming and its dual problem may be stated as follows: Given $b^* \in F$, $c \in Z$, and a linear map $A : E \to Z$, then linear programming problem and its dual problem can be formulated as:

$$(LP) \quad \text{minimize } \langle x, b^* \rangle$$
$$\text{subject to } Ax - c \in Z_+ \quad \text{and } x \in E_+;$$

$$(DLP) \quad \text{maximize } \langle c, y^* \rangle$$
$$\text{subject to } b^* - A^* y^* \in E_+^* \quad \text{and } y^* \in Z_+^*.$$

We now discuss this kind of linear programming in measure spaces. Glashoff and Gustafson (1982) discuss linear semi-infinite programming (LSIP) in which the constraint inequalities are subject to the relationships of finite linear combinations of functions to functions. The theory and algorithms for linear semi-infinite programming are discussed in Glashoff and Gustafson (1982); Hettich and Kortanek (1993); Lai and Wu (1992a); Reemtsen and Görner (1998); Wu et al (1998). The generalized capacity problem is an infinite dimensional linear programming problem. The generalized capacity problem extends the linear semi-infinite programming from the variable space R^n to variable space of the regular Borel measure space. Lai and Wu (1992) investigate the generalized capacity problem (GCAP) in which the constraint inequalities are subject to the relationships of measures to a function. Kellerer (1988) explores linear programming in measure spaces using a theoretical model. He consideres the linear programming problems for measure spaces of the forms:

$$(P') \quad \text{minimize } \int_X h \, d\mu$$
$$\text{subject to } \mu p_* \geq \nu \quad \text{in } M(Y),$$
$$\text{where } p_* \text{ maps } \mu \in M^+(X) \text{ to } M(Y), \text{ and}$$

$$(D') \quad \text{maximize } \int_Y g d\nu$$
$$\text{subject to all measurable functions } g \geq 0 \text{ on } Y, \text{ and } pg \leq h,$$
$$\text{where } p \text{ maps the set of nonnegative measurable functions}$$
$$\text{on } Y \text{ to measurable functions on } X.$$

Here X and Y are topological spaces endowed with their Borel algebras, and $M^+(X)$ denotes the set of nonnegative measures in $M(X)$. Lai and Wu

(1994) discuss LPM and DLPM (defined in this section) with constraint inequalities on the relationships of measures to measures. They prove that under certain conditions the LPM problem can be reformulated as a general capacity problem as well as a linear semi-infinite programming problem. Therefore LPM is a generalization of the general capacity problem and linear semi-infinite programming problem. Lai and Wu (1994); Wen and Wu (2001) develope algorithms for solving LPM when certain conditions are added to an LPM. In the present paper, we develop an approximation scheme to solve LPM in section 3. This scheme is a discritization method. Bascially, this approach finds a sequence of optimal solutions of corresponding linear semi-infinite programs and shows that the sequence of optimal solutions converges to an optimal solution of LPM. In section 4, we give an algorithm to find the optimal solution of linear semi-infinite programming problem.

We now formulate a linear program for measure spaces (LPM). As in Lai and Wu (1994), X and Y are compact Hausdorff spaces, $C(X)$ and $M(X)$ are, respectively, spaces of continuous real valued function and spaces of regular Borel measures on X. We denote the totality of non-negative Borel mersures on X as $M^+(X)$ and the subset of $C(X)$ consisting of non-negative functions as $C^+(X)$. Given $\nu, \nu^* \in M(Y)$, $\varphi \in C(X \times Y)$ and $h \in C(X)$, we know from Lai and Wu (1994) that LPM can be formulated as follows:

$$
\begin{aligned}
LPM : \quad &\text{minimize } \int_X h \, d\mu(x) \\
&\text{subject to } \mu \in M^+(X), \quad \text{and} \\
&\int_B \int_X \varphi(x,y) d\mu(x) d\nu^*(y) \geq \nu(B) \quad \text{for } B \in B(Y).
\end{aligned}
$$

Here $B(Y)$ stands for the Borel field of Y. We define the bilinear functionals $\langle \cdot, \cdot \rangle_1$, and $\langle \cdot, \cdot \rangle_2$ as follows:

$$
\begin{aligned}
\langle \mu, f \rangle_1 &= \int_X f(x) d\mu(x) \quad \text{for all } \mu \in M(X) \text{ and } f \in C(X), \\
\langle \nu, g \rangle_2 &= \int_Y g(x) d\nu(y) \quad \text{for all } \nu \in M(Y) \text{ and } g \in C(Y).
\end{aligned}
$$

For any $\varphi \in C(X \times Y)$, we define a linear operator $A : M(X) \to M(Y)$ by

$$
A\mu(B) = \int_B \phi(\mu, y) d\nu^*(y) \quad \text{for any } B \in B(Y) \text{ and } \mu \in M(X),
$$

where,

$$
\phi(\mu, y) = \int_X \varphi(x,y) d\mu(x) \in C(Y),
$$

is a continuous function on Y, and $\nu^* \in M(Y)$.

Applying Fubini theorem, we have

$$
\begin{aligned}
\langle A\mu, g\rangle_2 &= \int_Y g(y) \int_X \varphi(x,y)d\mu(x)d\nu^*(y) \\
&= \int_X \int_Y g(y)\varphi(x,y)d\nu^*(y)d\mu(x) \\
&= \langle \mu, A^*g\rangle_1,
\end{aligned}
$$

where $A^*g(\cdot) = \int_Y g(y)\varphi(x,y)d\nu^*(y)$ is the adjoint operator of A. It is clear that $M^+(X)$ is a $\sigma(M(X), C(Y))$-closed convex cone, and $M^+(Y)$ is a $\sigma(M(Y), C(X))$-closed convex cone. From Kretschmer (1961), we know that LPM has an associated dual problem:

$$
\begin{aligned}
DLPM: \quad &\text{maximize } \int_Y g(y)d\nu(y) \\
&\text{subject to } g \in C^+(Y) \quad \text{and} \\
&\int_Y g(y)\varphi(x,y)d\nu^*(y) \leq h(x) \quad \text{for all } x \in X,
\end{aligned}
$$

where $\nu, \nu^* \in M(Y)$, $\varphi \in C(X \times Y)$ and $h \in C(X)$ are given.

A feasible solution of an optimization problem (P) is a point satisfying the constraints of problem (P). The set of all feasible solutions for problem (P) is called the feasible set of problem (P), which we denote by $F(P)$. If $F(P)$ is not empty, then we say the problem (P) is consistent. We denote by $V(P)$ the optimal values for problem (P).

Wen and Wu (2001) proved that under some conditions there is no duality gap for LPM. Now we state the result as follows.

Theorem 1.1 *Suppose DLPM is consistent and $-\infty < V(DLPM) < \infty$. If there exists a $g^* \in C^+(Y)$ such that*

$$
\int_Y g^*(y)\varphi(x,y)d\nu^*(y) < h(x) \quad \text{for all} x \in X.
$$

Then LPM has no duality gap.

Corollary 1.1 . *Suppose DLPM (or LPM) is consistent with finite value. If $h(x) > 0$ or $h(x) < 0$ for all $x \in X$, then LPM has no duality gap.*

2   SOLVABILITY OF LPM

In this section, we shall prove that under some simple conditions, there exists
a solution for LPM.

Theorem 2.1 . *Suppose LPM is consistent with finite value. If $h(x) > 0$,
$\forall\, x \in X$, then LPM is solvable.*

Proof: It is obvious that $\int_X h(x)d\mu(x) \geq 0$, $\forall \mu \in F(LPM)$, since $h(x) > 0$,
$\forall\, x \in X$. If $\int_X h(x)d\mu(x) = 0$, $\forall \mu \in F(LPM)$, then LPM is solvable. Hence
we may assume that, without loss of generality, there exists a $\mu_0 \in F(LPM)$
such that

$$\int_X h(x)d\mu_0(x) > 0 \quad \text{and} \quad V(LPM) < \int_X h(x)d\mu_0(x).$$

Let $\{\mu_n : n \in \mathbb{N}\} \subseteq F(LPM)$ be such that $\int_X h(x)d\mu_n(x) \downarrow V(LPM)$. Since

$$V(LPM) < \int_X h(x)d\mu_0(x),$$

there exists $n^* \in \mathbb{N}$ such that

$$\int_X h(x)d\mu_n(x) < \int_X h(x)d\mu_0(x), \quad n \geq n^*.$$

Thus,

$$\min_{x\in X} h(x)\mu_n(X) \leq \int_X h(x)d\mu_n(x) < \int_X h(x)d\mu_0(x), \quad \forall n \geq n^*.$$

This implies that

$$\|\mu_n\| = \mu_n(X) < \frac{\int_X h(x)d\mu_0(x)}{\min\limits_{x\in X} h(x)}, \quad \forall n \geq n^*.$$

By the Banach-Alaoglu theorem, there exists $\mu^* \in M^+(X)$ and a subsequence
$\{\mu_{n_j} : j \in \mathbb{N}\}$ of $\{\mu_n : n \in \mathbb{N}\}$ such that

$$\mu_{n_j} \to \mu^* \quad \text{weakly as} \quad j \to \infty.$$

That is,

$$\int_X f(x)d\mu_{n_j}(x) \to \int_X f(x)d\mu^*(x), \quad \forall f \in C(X), \quad \text{as } j \to \infty.$$

Hence, for every $g \in C^+(Y)$,

$$\langle A\mu^*, g \rangle = \langle \mu^*, A^*g \rangle = \lim_{j \to \infty} \langle \mu_{n_j}, A^*g \rangle = \lim_{j \to \infty} \langle A\mu_{n_j}, g \rangle \geq \langle \nu, g \rangle,$$

since $\mu_{n_j} \in F(LPM)$, $\forall j \in \mathbb{N}$. This implies that $A\mu^* - \nu \in M^+(Y)$. Hence $\mu^* \in F(LPM)$.

Since,

$$\lim_{j \to \infty} \int_X h(x)d\mu_{n_j}(x) = \int_X h(x)d\mu^*(x) \quad \text{and}$$
$$\int_X h(x)d\mu_{n_j}(x) \downarrow V(LPM),$$

we obtain

$$V(LPM) = \int_X h(x)d\mu^*(x).$$

Therefore LPM is solvable.

3   AN APPROXIMATION SCHEME FOR LPM

In this section, let X and Y be compact sets on $\mathbb{R}$. We shall develop an scheme for solving LPM. This scheme is a discritization method. To derive the scheme, we now reformulate LPM in equality form as follows.

$$
\begin{aligned}
ELPM: \quad & \text{Minimize } \int_X h(x)d\nu(x) \\
& \text{s.t.} \quad \int_B \int_X \varphi(x,y)d\mu(x)d\nu^*(y) - \overline{\mu}(B) = \nu(B), \quad \forall B \in B(Y), \\
& \text{and} \quad \mu \in M^+(X), \overline{\mu} \in M^+(Y).
\end{aligned}
$$

For the given $\varphi \in C(X \times Y)$, $\nu^* \in M(Y)$, we define a linear operator

$$\widetilde{A} : M(X) \times M(Y) \to M(Y)$$

by

$$\widetilde{A}(\mu, \overline{\mu})(B) = \int_B \int_X \varphi(x,y)d\mu(x)d\nu^*(y) - \overline{\mu}(B) \text{ for any } B \in B(Y).$$

We also define the bilinear function $\langle \cdot, \cdot \rangle_3$ as follows:

$$\langle (\mu, \overline{\mu}), (f, g) \rangle_3 = \int_X f(x)d\mu(x) + \int_Y g(y)d\overline{\mu}(y)$$

for all $(\mu, \overline{\mu}) \in M(X) \times M(Y)$ and $(f, g) \in C(X) \times C(Y)$. Then ELPM can be rewritten as follows:

$$
\begin{aligned}
& \text{Minimize } \langle (\mu, \overline{\mu}), (h, 0) \rangle_3 \\
& \text{s.t.} \quad \widetilde{A}(\mu, \overline{\mu}) = \nu, \quad \text{and} \quad \mu \in M^+(X), \overline{\mu} \in M^+(Y).
\end{aligned}
$$

Applying Fubini theorem, we have

$$
\begin{aligned}
\langle \widetilde{A}(\mu,\overline{\mu}), g \rangle_2 &= \int_Y g(y) \int_X \varphi(x,y) d\mu(x) d\nu^*(y) - \int_Y g(y) d\overline{\mu}(y) \\
&= \int_X \int_Y g(y)\varphi(x,y) d\nu^*(y) d\mu(x) - \int_Y g(y) d\overline{\mu}(y) \\
&= \langle (\mu,\overline{\mu}), \widetilde{A}^*(g) \rangle_3,
\end{aligned}
$$

where $\widetilde{A}^*$, defined by

$$
\widetilde{A}^*(g) = \left(\int_Y \varphi(x,y) g(y) d\nu^*(y), -g(y) \right),
$$

is the adjoint of $\widetilde{A}$. Since $\widetilde{A}^*$ maps $C(Y)$ into $C(X) \times C(Y)$, $\widetilde{A}$ is weakly continuous by Proposition 4 in Anderson and Nash (1987).

For every $k \in \mathbb{N}$, we let

$$
P_k \equiv \{1, y, y^2, \dots y^k\}
$$

be a subset of polynomials. Instead of the original program ELPM, we consider the linear program

$$
\begin{aligned}
(ELPM)^k \quad &\text{minimize } \langle (\mu,\overline{\mu}), (h,0) \rangle_3 \\
&\text{subject to } \langle \widetilde{A}(\mu,\overline{\mu}) - \nu, g \rangle_2 = 0, \quad \forall g \in P_k, \\
&\text{and,} \quad \mu \in M^+(X), \overline{\mu} \in M^+(Y).
\end{aligned}
$$

The $(ELPM)^k$ program is a discritized version of LPM. As there will be no danger of confusion, from now on, the bilinear functionals $\langle \cdot, \cdot \rangle_1, \langle \cdot, \cdot \rangle_2$, and $\langle \cdot, \cdot \rangle_3$ are all represented by $\langle \cdot, \cdot \rangle$. Then we have the following result:

Theorem 3.1 *Suppose LPM is consistent with finite value. If the subprograms $(ELPM)^k$ satisfy the following conditions:*

(1) *For every $k \in \mathbb{N}$, $(ELPM)^k$ is solvable with an optimal solution $(\mu_k^*, \overline{\mu}_k^*)$, and*

(2) *there exists a positive number M such that $\|\mu_k^*\| + \|\overline{\mu}_k^*\| \leq M$.*

Then

(a) *ELPM is solvable.*

(b) *$V((ELPM)^k) = \langle (\mu_k^*, \overline{\mu}_k^*), (h,0) \rangle \uparrow V(ELPM)$.*

Proof: It is obvious that $F(ELPM) \subseteq F((ELPM)^{k+1}) \subseteq F((ELPM)^k)$, $\forall k \in \mathbb{N}$. Hence $V(ELPM) \geq V((ELPM)^{k+1}) \geq V((ELPM)^k)$, $\forall k \in \mathbb{N}$. That is,

$$\int_X h(x)d\mu_k^*(x) \leq \int_X h(x)d\mu_{k+1}^*(x) \leq V(ELPM), \quad \forall k \in \mathbb{N},$$

and so the sequence $\{\langle (\mu_k^*, \overline{\mu}_k^*), (h, 0) \rangle : k \in \mathbb{N}\}$ is increasing and has an upper bound. Hence

$$\lim_{k \to \infty} \langle (\mu_k^*, \overline{\mu}_k^*), (h, 0) \rangle = \lim_{k \to \infty} \int_X h(x)d\mu_k^*(x) \leq V(ELPM). \tag{3.1}$$

By condition (2) and the Banach-Alaoglu theorem, there exist $(\mu^*, \overline{\mu}^*) \in M^+(X) \times M^+(Y)$ and a subsequence $\{(\mu_{k_j}^*, \overline{\mu}_{k_j}^*) : j \in \mathbb{N}\} \subseteq \{(\mu_k^*, \overline{\mu}_k^*) : k \in \mathbb{N}\}$ such that

$$(\mu_{k_j}^*, \overline{\mu}_{k_j}^*) \to (\mu^*, \overline{\mu}^*) \quad \text{weakly as} \quad j \to \infty. \tag{3.2}$$

We claim that

$$(\mu^*, \overline{\mu}^*) \in F(ELPM). \tag{3.3}$$

To show this claim, it suffices to prove that

$$\langle \widetilde{A}(\mu^*, \overline{\mu}^*) - \nu, g \rangle = 0, \quad \forall g \in \bigcup_{k=1}^{\infty} span(P_k), \tag{3.4}$$

because, according to the fact that $\bigcup_{k=1}^{\infty} span(P_k)$ is dense in $C(Y)$, (3.4) implies that $(\mu^*, \overline{\mu}^*) \in F(ELPM)$.

Now to prove (3.4), let $g \in \bigcup_{k=1}^{\infty} span(P_k)$, and write

$$\begin{aligned} &\langle \widetilde{A}(\mu^*, \overline{\mu}^*), g \rangle \\ = \quad &\langle \widetilde{A}(\mu^*, \overline{\mu}^*) - \widetilde{A}(\mu_{k_j}^*, \overline{\mu}_{k_j}^*), g \rangle + \langle \widetilde{A}(\mu_{k_j}^*, \overline{\mu}_{k_j}^*) - \nu, g \rangle + \langle \nu, g \rangle. \end{aligned} \tag{3.5}$$

Note that

$$\langle \widetilde{A}(\mu^*, \overline{\mu}^*) - \widetilde{A}(\mu_{k_j}^*, \overline{\mu}_{k_j}^*), g \rangle \to 0 \quad \text{as} \quad j \to \infty, \tag{3.6}$$

since the weak convergence $(\mu_{k_j}^*, \overline{\mu}_{k_j}^*) \to (\mu^*, \overline{\mu}^*)$ and the weak continuity of $\widetilde{A}$. On the other hand, as g is in $\bigcup_{k=1}^{\infty} span(P_k)$, there is a $\overline{k}$ such that g is in $span(P_{\overline{k}})$. Since $g \in span(P_{\overline{k}})$, there exists $m \leq \overline{k} + 1$ such that

$$g = \sum_{i=1}^{m} \lambda_i g_i \quad \text{with} \quad g_i \in P_{\overline{k}} \quad \text{and} \quad \lambda_i \in R \ (i = 1, 2, 3, \ldots, m).$$

Moreover, as $F((ELPM)^{k+1}) \subset F((ELPM)^k)$ for every $k \in \mathbb{N}$,

$$(\mu_{k_j}^*, \overline{\mu}_{k_j}^*) \in F((ELPM)^m), \quad \forall\, k_j \geq m.$$

Hence,

$$\begin{aligned}
\langle \widetilde{A}(\mu_{k_j}^*, \overline{\mu}_{k_j}^*) - \nu, g \rangle &= \langle \widetilde{A}(\mu_{k_j}^*, \overline{\mu}_{k_j}^*) - \nu, \sum_{i=1}^{m} \lambda_i g_i \rangle \\
&= \sum_{i=1}^{m} \lambda_i \langle \widetilde{A}(\mu_{k_j}^*, \overline{\mu}_{k_j}^*) - \nu, g_i \rangle \\
&= 0, \quad \forall\, k_j \geq m.
\end{aligned} \tag{3.7}$$

Let $j \to \infty$ in (3.5), and from (3.6) and (3.6), we obtain (3.4). Thus we have (3.3).

Therefore, combining (3.1), (3.2) and, (3.3), we have

$$\begin{aligned}
V(ELPM) &\leq \langle (\mu^*, \overline{\mu}^*), (h, 0) \rangle \\
&= \lim_{j \to \infty} \langle (\mu_{k_j}^*, \overline{\mu}_{k_j}^*), (h, 0) \rangle \\
&= \lim_{j \to \infty} \int_X h(x) d\mu_{k_j}^*(x) \\
&\leq V(ELPM),
\end{aligned}$$

and complete the proof.

Corollary 3.1 *Suppose LPM is consistent with finite value. If the given data* h *and* ν^* *satisfy the following conditions:*

(1) $h(x) > 0$, $\forall\, x \in X$, *and*

(2) $\nu^* \in M^+(Y)$,

then

(a) $(ELPM)^k$ *is solvable for every* $k \in \mathbb{N}$.

(b) *For every* $k \in \mathbb{N}$, *let* $(\mu_k^*, \overline{\mu}_k^*)$ *be an optimal solution for* $(ELPM)^k$. *Then*

$$V((ELPM)^k) = \langle (\mu_k^*, \overline{\mu}_k^*), (h, 0) \rangle \uparrow V(ELPM).$$

Proof: (a) Given $k \in \mathbb{N}$. As in the proof of Theorem 2.1, we may assume that there exists $(\mu_0^{(k)}, \overline{\mu}_0^{(k)}) \in F((ELPM)^k)$ such that

$$\int_X h(x) d\mu_0^{(k)}(x) > 0 \quad \text{and} \quad V((ELPM)^k) < \int_X h(x) d\mu_0^{(k)}(x).$$

Let $\{(\mu_{k_n}, \overline{\mu}_{k_n}) : n \in \mathbb{N}\} \subseteq F((ELPM)^k)$ be such that

$$\langle (\mu_{k_n}, \overline{\mu}_{k_n}), (h, 0) \rangle = \int_X h(x) d\mu_{k_n}(x) \downarrow V((ELPM)^k). \qquad (3.8)$$

By the same argument as in the proof of Theorem 2.1, there exist $\mu_k^* \in M^+(X)$, $n^* \in \mathbb{N}$, and a subsequence $\{\mu_{k_{n_j}} : j \in \mathbb{N}\} \subseteq \{\mu_{k_n} : n \in \mathbb{N}\}$ such that

$$\|\mu_{k_{n_j}}\| < \frac{\int_X h(x) d\mu_0^{(k)}}{\min\limits_{x \in X} h(x)}, \quad \forall k_{n_j} \geq n^*, \qquad (3.9)$$

and,

$$\mu_{k_{n_j}} \to \mu_k^* \quad \text{weakly as} \quad j \to \infty. \qquad (3.10)$$

Since $(\mu_{k_{n_j}}, \overline{\mu}_{k_{n_j}}) \in F((ELPM)^k)$, $\langle \widetilde{A}(\mu_{k_{n_j}}, \overline{\mu}_{k_{n_j}}) - \nu, g \rangle = 0$, $\forall g \in P_k$. In particular, $\langle \widetilde{A}(\mu_{k_{n_j}}, \overline{\mu}_{k_{n_j}}) - \nu, 1 \rangle = 0$. That is,

$$\int_Y \int_X \varphi(x, y) d\mu_{k_{n_j}}(x) d\nu^*(y) - \int_Y d\overline{\mu}_{k_{n_j}}(y) - \int_Y d\nu(y) = 0.$$

Hence, by the nonegativity of ν^* and (3.9), we have

$$\begin{aligned}
\text{for } j \in \mathbb{N}, \ \|\overline{\mu}_{k_{n_j}}\| &= \int_Y \int_X \varphi(x, y) d\mu_{k_{n_j}}(x) d\nu^*(y) - \nu(Y) \\
&\leq \max_{\substack{x \in X \\ y \in Y}} \varphi(x, y) \|\mu_{k_{n_j}}\| \|\nu^*\| - \nu(Y) \\
&< \widetilde{k},
\end{aligned}$$

where, $\widetilde{k} = \max\limits_{\substack{x \in X \\ y \in Y}} \varphi(x, y) \dfrac{\int_X h(x) d\mu_0^{(k)}}{\min\limits_{x \in X} h(x)} \|\nu^*\| - \nu(Y)$ is a fixed positive number. Therefore, by the Banach-Alaoglu theorem, there exist $\overline{\mu}_k^* \in M^+(X)$ and a subsequence $\{\overline{\mu}_{k_{\ell_j}} : j \in \mathbb{N}\} \subseteq \{\overline{\mu}_{k_{n_j}} : j \in \mathbb{N}\}$ such that

$$\overline{\mu}_{k_{\ell_j}} \to \overline{\mu}_k^* \quad \text{weakly as} \quad j \to \infty. \qquad (3.11)$$

Combining (3.10) and (3.11), we obtain

$$(\mu_{k_{\ell_j}}, \overline{\mu}_{k_{\ell_j}}) \to (\mu_k^*, \overline{\mu}_k^*) \quad \text{weakly as} \quad j \to \infty. \qquad (3.12)$$

This implies that $(\mu_k^*, \overline{\mu}_k^*) \in F((ELPM)^k)$, since, for every $g \in P_k$,

$$\begin{aligned}
\langle \widetilde{A}(\mu_k^*, \overline{\mu}_k^*), g \rangle &= \langle (\mu_k^*, \overline{\mu}_k^*), \widetilde{A}^* g \rangle \\
&= \lim_{j \to \infty} \langle (\mu_{k_{\ell_j}}, \overline{\mu}_{k_{\ell_j}}), \widetilde{A}^* g \rangle \\
&= \lim_{j \to \infty} \langle \widetilde{A}(\mu_{k_{\ell_j}}, \overline{\mu}_{k_{\ell_j}}), g \rangle \\
&= \langle \nu, g \rangle, \quad \text{as } (\mu_{k_{\ell_j}}, \overline{\mu}_{k_{\ell_j}}) \in F((ELPM)^k), \quad \forall j \in \mathbb{N}.
\end{aligned}$$

Now (3.12) combining with (3.8) also yields that $(\mu_k^*, \overline{\mu}_k^*)$ is an optimal solution for $(ELPM)^k$, as

$$\langle(\mu_k^*, \overline{\mu}_k^*), (h, 0)\rangle = \lim_{j \to \infty} \langle(\mu_{k_{\ell_j}}, \overline{\mu}_{k_{\ell_j}}), (h, 0)\rangle = V((ELPM)^k).$$

(b) As in the proof of Theorem 3.1, the sequence $\{\langle(\mu_k^*, \overline{\mu}_k^*), (h, 0)\rangle\}_{k=1}^{\infty}$ is increasing and convergent. As in part (a), we may assume that there exists $(\mu_0, \overline{\mu}_0) \in F(ELPM)$ such that

$$\int_X h(x)d\mu_0(x) > 0 \quad \text{and} \quad V(ELPM) < \int_X h(x)d\mu_0(x).$$

Also by the same argument as in the proof of Theorem 2.1, there exists a subsequence $\{\mu_{k_j}^* : j \in \mathbb{N}\} \subseteq \{\mu_k^* : k \in \mathbb{N}\}$ such that

$$\|\mu_{k_j}^*\| < M_1, \ \forall j \in \mathbb{N}, \quad \text{where} \quad M_1 = \frac{\int_X h(x)d\mu_0(x)}{\min_{x \in X} h(x)}.$$

Since $(\mu_{k_j}^*, \overline{\mu}_{k_j}^*) \in F((ELPM)^{k_j})$ for every $j \in \mathbb{N}$, $\langle \widetilde{A}(\mu_{k_j}^*, \overline{\mu}_{k_j}^*) - \nu, 1\rangle = 0$. By the same argument as in part (a), we have

$$\|\overline{\mu}_{k_j}^*\| < M_2, \ \forall j \in, \quad \text{where} \quad M_2 = \max_{\substack{x \in X \\ y \in Y}} \varphi(x, y) M_1 \|\nu^*\| - \nu(Y).$$

So,

$$\|\mu_{k_j}^*\| + \|\overline{\mu}_{k_j}^*\| < M_1 + M_2, \quad \forall j \in \mathbb{N}.$$

Hence by Theorem 3.1, we obtain

$$\langle(\mu_{k_j}^*, \overline{\mu}_{k_j}^*), (h, 0)\rangle \uparrow V(ELPM),$$

and this implies

$$V((ELPM)^k) = \langle(\mu_k^*, \overline{\mu}_k^*), (h, 0)\rangle \uparrow V(ELPM).$$

Although the programs $(ELPM^k)$ are solvable, it is not always easy to solve them. To overcome the difficulty, we shall consider the dual problem of $(ELPM)^k$. For $k \in \mathbb{N}$, we define a linear operator

$$\widetilde{A}_k : M(X) \times M(Y) \to R^{k+1}$$

by

$$\widetilde{A}_k(\mu,\overline{\mu}) = \begin{bmatrix} \langle \widetilde{A}(\mu,\overline{\mu}), 1 \rangle \\ \langle \widetilde{A}(\mu,\overline{\mu}), y \rangle \\ \vdots \\ \langle \widetilde{A}(\mu,\overline{\mu}), y^k \rangle \end{bmatrix}.$$

Applying Fubini theorem, we have

$$\left\langle \widetilde{A}_k(\mu,\overline{\mu}), \begin{bmatrix} r_1 \\ r_2 \\ \vdots \\ r_{k+1} \end{bmatrix} \right\rangle = r_1[\int_Y \int_X \varphi(x,y)d\mu(x)d\nu^*(y) - \int_Y d\overline{\mu}(y)]$$

$$+ r_2[\int_Y y(\int_X \varphi(x,y)d\mu(x))d\nu^*(y) - \int_Y y d\overline{\mu}(y)] + \cdots$$

$$+ r_{k+1}[\int_Y y^k(\int_X \varphi(x,y)d\mu(x))d\nu^*(y) - \int_Y y^k d\overline{\mu}(y)]$$

$$= [\int_X r_1 \int_Y \varphi(x,y)d\nu^*(y)d\mu(x) - \int_Y r_1 d\overline{\mu}(y)] +$$

$$[\int_X r_2(\int_Y y\varphi(x,y)d\nu^*(y))d\mu(x) - \int_Y r_2 y d\overline{\mu}(y)] + \cdots +$$

$$[\int_X r_{k+1}(\int_Y y^{k+1}\varphi(x,y)d\nu^*(y))d\mu(x) - \int_Y r_{k+1} y^k d\overline{\mu}(y)]$$

$$= \int_X [r_1 \int_Y \varphi(x,y)d\nu^*(y) + r_2 \int_Y y\varphi(x,y)d\nu^*(y)$$

$$+ \cdots + r_{k+1} \int_Y y^{k+1}\varphi(x,y)d\nu^*(y)]d\mu(x)$$

$$+ \int_Y [-r_1 - r_2 y - \cdots - r_{k+1} y^k]d\overline{\mu}(y)$$

$$= \langle (\mu,\overline{\mu}), \widetilde{A}_k^* \begin{bmatrix} r_1 \\ r_2 \\ \vdots \\ r_{k+1} \end{bmatrix} \rangle,$$

where $\widetilde{A}_k^* : R^{k+1} \to C(X) \times C(Y)$, defined by

$$\widetilde{A}_k^* \begin{bmatrix} r_1 \\ r_2 \\ \vdots \\ r_{k+1} \end{bmatrix} = (r_1 \int_Y \varphi(x,y)d\nu^*(y) + r_2 \int_Y y\varphi(x,y)d\nu^*(y)$$

$$+ \cdots + r_{k+1} \int_Y y^k \varphi(x,y) d\nu^*(y), -r_1 - r_2 y - \cdots - r_{k+1} y^k),$$

is the adjoint operator of $\widetilde{A}_k$. Hence the dual problem for $(ELPM)^k$ is defined as follows:

$$(DELPM)^k : \quad \text{maximize } \sum_{i=0}^k r_{i+1} \langle \nu, y^i \rangle$$

$$\text{subject to } \widetilde{A}_k^* \begin{bmatrix} r_1 \\ r_2 \\ \vdots \\ r_{k+1} \end{bmatrix} \leq \begin{bmatrix} h \\ 0 \end{bmatrix}.$$

That is,

$$(DELPM)^k : \quad \text{maximize } \sum_{i=0}^k (\int_Y y^i d\nu(y)) r_{i+1}$$
$$\text{subject to } \sum_{i=0}^k (\int_Y y^i \varphi(x,y) d\nu^*(y)) r_{i+1} \leq h(x), \quad \forall x \in X$$
$$\text{and } \sum_{i=0}^k (-r_{i+1} y^i) \leq 0, \quad \forall y \in Y.$$

Note that $(DELPM)^k$ is basically the same as CSIP, the continuous semi-infinite program, except that the "minimization" in CSIP is replaced by "maximization" in $(DELPM)^k$. Moreover, for every $k \in \mathbb{N}$, there is no duality gap for $(ELPM)^k$ and $(DELPM)^k$ under the condition that $h(x) > 0, \forall x \in X$.

Theorem 3.2 *If $h(x) > 0, \forall x \in X$, then for every $k \in \mathbb{N}$, there is no duality gap for $(ELPM)^k$ and $(DELPM)^k$.*

Proof: Suppose $(\mu_0, \overline{\mu}_0) \in M^+(X) \times M^+(Y)$ such that

$$\widetilde{A}_k(\mu_0, \overline{\mu}_0) = 0 \quad \text{and} \quad \langle (\mu_0, \overline{\mu}_0), (h, 0) \rangle = 0.$$

Hence,

$$\int_Y \int_X \varphi(x,y) d\mu_0(x) d\nu^*(y) - \int_Y 1 d\overline{\mu}_0(y) = 0 \tag{3.13}$$

$$\vdots$$

$$\int_Y y^k \int_X \varphi(x,y) d\mu_0(x) d\nu^*(y) - \int_Y y^k d\overline{\mu}_0(y) = 0$$

and,

$$\int_X h(x) d\mu_0(x) = 0. \tag{3.14}$$

Since $h(x) > 0$, $\forall x \in X$ and owing to (3.14), μ_0 is the zero measure on X. Thus, from (3.13), we have

$$\int_Y 1 \, d\overline{\mu}_0(y) = 0,$$

and this implies that $\overline{\mu}_0$ is the zero measure on Y. Therefore, there is no $(\mu, \overline{\mu}) \in M^+(X) \times M^+(Y)$ other than zero measure with $\widetilde{A}_k(\mu, \overline{\mu}) = 0$ and $\langle (\mu, \overline{\mu}), (h, 0) \rangle = 0$. As the positive cone $M^+(X) \times M^+(Y)$ is weakly closed and the associated dual cone $C^+(X) \times C^+(Y)$ has a non-empty interior

$$\{(f, g) \in C(X) \times C(Y) : f(x) > 0, \, \forall x \in X \text{ and } g(y) > 0, \forall y \in Y\}$$

in the Mackey topology $\tau(C(X) \times C(Y), M(X) \times M(Y))$, by Corollary 3.18 in Anderson and Nash (1987), there is no duality gap for $(ELPM)^k$ and $(DELPM)^k$.

4   AN ALGORITHM FOR $(DELPM)^K$

In this section, we let X, Y be compact sets in $\mathbb{R}$ such that $X \cap Y = \phi$. We will develop an algorithm for solving the semi-infinite linear programming problem $(DELPM)^k$.

Given $k \in \mathbb{N}$, let $T = X \cup Y$, and for $i = 0, 1, 2, \ldots, k$, let

$$c_i = -\int_Y y^i d\nu(y),$$

$$f_i(t) = \begin{cases} -\int_Y y^i \varphi(t, y) d\nu^*(y), & \text{if } t \in X \\ t^i, & \text{if } t \in Y \end{cases},$$

and

$$g(t) = \begin{cases} -h(t), & \text{if } t \in X \\ 0, & \text{if } t \in Y \end{cases}.$$

Consider the semi-infinite linear programming problem SIP_k defined as follows:

$$(SIP_k): \quad \text{minimize } \sum_{i=0}^{k} c_i x_i$$
$$\text{such that } \sum_{i=0}^{k} f_i(t) x_i \geq g(t), \quad \forall t \in T.$$

Its dual problem can be formulated in the following form:

$$(DSIP_k): \quad \text{maximize } \int_T g(t) d\mu(t)$$
$$\text{such that } \int_T f_i(t) d\mu(t) = c_i, \, i = 0, 1, 2, \ldots, k,$$
$$\mu \in M^+(T),$$

where $M^+(T)$ is the space of all nonnegative bounded regular Borel measures on T. Note that $(ELPM)^k$ and SIP_k have the same optimal solution as well as

$$V((DELPM)^k) = -V(SIP_k).$$

Many algorithms in Glashoff and Gustafson (1982); Hettich and Kortanek (1993); Lai and Wu (1992a); Reemtsen and Görner (1998); Wu et al (1998) can be applied to solve SIP_k. Based on Wu et al (1998), we develop an explicit algorithm for sloving SIP_k.

To introduce the algorithm, let us start with some notations. Let $T' = \{t_1, t_2, \ldots, t_m\}$ be a subset with m elements in T. We denote by $(LP_k(T'))$ the following linear program with m constraints induced by T':

$$(LP_k(T')): \quad \text{minimize } \sum_{i=0}^{k} c_i x_i$$
$$\text{such that } \sum_{i=0}^{k} f_i(t_j) x_i \geq g(t_j), \quad j = 1, 2, \ldots, m.$$

The dual problem of $(LP_k(T'))$ can be formulated as the following problem:

$$(DLP_k(T')): \text{maximize } \sum_{j=1}^{m} g(t_j) y_j$$
$$\text{such that } \sum_{j=1}^{m} f_i(t_j) y_j = c_i, \quad i = 0, 1, 2, \ldots, k.$$
$$y_j \geq 0, \forall j = 1, 2, \ldots, m.$$

Given that $\delta > 0$ is a prescribed small number, we state our algorithm in the following steps:

Step 0 Let $n \leftarrow 1$, choose any $t_1^0 \in T$, set $T_1 = \{t_1^0\}$, and set $m_0 = 0$.

Step 1 Solve $LP_k(T_n)$. Let $X^n = (x_0^n, x_1^n, x_2^n, \ldots, x_k^n)$ be an optimal solution of $LP_k(T_n)$. Define

$$\phi_n(t) = \sum_{i=0}^{k} f_i(t) x_i^n - g(t). \tag{4.1}$$

Step 2 Solve $DLP_k(T_n)$. Let $Y^n = (y_1^n, y_2^n, \ldots, y_{m_{n-1}+1}^n)$ be an optimal solution $DLP_k(T_n)$. Define a discrete measure μ_n on T by letting

$$\mu_n(t) = \begin{cases} y_j^n \, (\geq 0), & \text{if } t = t_j^{n-1} \in T_n \\ 0, & \text{if } t \in T - T_n. \end{cases} \tag{4.2}$$

Set $E_n = \{t \in T_n \,|\, \mu_n(t) > 0\} = \{t_1^n, t_2^n, \ldots, t_{m_n}^n\}$.

Step 3 Find any $t^n_{m_n+1} \in T$ such that $\phi_n(t^n_{m_n+1}) < -\delta$. If such $t^n_{m_n+1}$ does not exist, stop and output X^n as a solution. Otherwise, set $T_{n+1} = E_n \cup \{t^n_{m_n+1}\}$.

Step 4 Update $n \leftarrow n + 1$, and go to step 1.

In the above algorithm, we make the following assumptions:

(A1) $LP_k(T_n)$ and $DLP_k(T_n)$ are both solvable for every $n \in \mathbb{N}$,

(A2) $LP_k(E_n)$ has the unique solution for every $n \in \mathbb{N}$.

Note that $t^n_{m_n+1} \in E_{n+1}$, that is, $\mu_{n+1}(t^n_{m_n+1}) > 0$. Otherwise, we have $\mu_{n+1}(t^n_{m_n+1}) = 0$, which implies that $V(LP_k(T_{n+1})) = V(DLP_k(T_{n+1})) = V(DLP_k(E_n))$. Since $V(DLP_k(E_n)) = V(DLP_k(T_n))$, we have $V(LP_k(T_{n+1})) = V(DLP_k(T_n)) = V(LP_k(T_n))$. Since X^n and X^{n+1} are feasible for $LP_k(E_n)$, they are both optimal solutions for $LP_k(E_n)$. Hence, by the assumption (A2), we have $X^n = X^{n+1}$, and this implies that $\phi_n(t^n_{m_n+1}) \geq 0$, which is a contradiction.

Theorem 4.1 $V(LP_k(T_{n+1})) - V(LP_k(T_n)) = -\phi_n(t^n_{m_n+1})\mu_{n+1}(t^n_{m_n+1})$.

Proof: Since $E_{n+1} - \{t^n_{m_n+1}\} \subset E_n$, we have, by the definition (4.2) of μ_n, $\mu_n(t) > 0$, $\forall t \in E_{n+1} - \{t^n_{m_n+1}\}$. Hence, from the complementary slackness theorem of linear programming, we get

$$\phi_n(t) = \sum_{i=0}^{k} f_i(t)x_i^n - g(t) = 0, \quad \forall t \in E_{n+1} - \{t^n_{m_n+1}\},$$

which implies

$$
\begin{aligned}
-\phi_n(t^n_{m_n+1})\mu_{n+1}(t^n_{m_n+1}) &= -\sum_{t \in E_{n+1}} \phi_n(t)\mu_{n+1}(t) \\
&= -\sum_{t \in E_{n+1}} [\sum_{i=0}^{k} f_i(t)x_i^n - g(t)]\mu_{n+1}(t) \\
&= -\sum_{t \in E_{n+1}} \sum_{i=0}^{k} f_i(t)x_i^n \mu_{n+1}(t) + \sum_{t \in E_{n+1}} g(t)\mu_{n+1}(t) \\
&= -\sum_{i=0}^{k} [\sum_{t \in E_{n+1}} f_i(t)\mu_{n+1}(t)]x_i^n + V(LP_k(T_{n+1}))
\end{aligned}
$$

$$= -\sum_{i=0}^{k} c_i x_i^n + V(LP_k(T_{n+1}))$$

$$= V(LP_k(T_{n+1})) - V(LP_k(T_n)),$$

and we complete the proof.

If, in each iteration, there exists a $\overline{\delta} > 0$ such that $\mu_n(t) \geq \overline{\delta}$, $\forall\, t \in E_n$, then by Theorem 4.1, we obtain

$$V(LP_k(T_{n+1})) - V(LP_k(T_n)) = -\phi(t_{m_n+1}^n)\mu_{n+1}(t_{m_n+1}^n) \geq \delta \cdot \overline{\delta} > 0.$$

Hence we have the following corollary.

Corollary 4.1 *Given any $\delta > 0$, in each iteration, if there exists a $\overline{\delta} > 0$ such that $\mu_n(t) \geq \overline{\delta}$, $\forall\, t \in E_n$, then $V(LP_k(T_{n+1})) > V(LP_k(T_n))$.*

Theorem 4.1 as well as Corollary 4.1 are fundamental results for the algorithm; with them, we show that, under proper conditions, for any given $\delta > 0$, the proposed algorithm actually terminates in a finite number iterations.

Theorem 4.2 *Given any $\delta > 0$, in each iteration, assume that:*

(A3) *There exists a $M > 0$ such that $\|X^n\| \leq M$;*

(A4) *There exists a $\overline{\delta} > 0$ such that $\mu_n(t) \geq \overline{\delta}$, $\forall\, t \in E_n$.*

Then, the algorithm terminates in a finite number of iterations.

Proof: Suppose the scheme does not stop in a finite number of iterations. By Corollary 4.1, we have

$$V(LP_k(T_1)) < V(LP_k(T_2)) < \cdots \leq V(SIP_k).$$

Thus,

$$\lim_{n \to \infty} V(LP_k(T_n)) = \alpha \leq V(SIP_k).$$

We claim that this is impossible. By (A3), the infinite sequence $\{X^n\}$ is confined in a compact set C in R^n. There exists a subsequence $\{X^{n_r}\}$ of $\{X^n\}$

such that X^{n_r} converges to X^*, and the subsequence $\{t^{n_r}_{m_{n_r}+1}\}$ converges to some point t_* as $r \to \infty$. Now we let

$$\phi_*(t) = \sum_{i=0}^{k} f_i(t) x_i^* - g(t).$$

Then, by (4.1), $\phi_{n_r}(t^{n_r}_{m_{n_r}+1})$ converges to $\phi_*(t_*)$. Since $\phi_{n_r}(t^{n_r}_{m_{n_r}+1}) < -\delta$, for each r, we have

$$0 \neq \phi_*(t_*) \leq -\delta. \tag{4.3}$$

Now, let ε be an arbitrary number, and we can find a large integer $n_N \in \{n_r\}_{r=1}^{\infty}$ such that

$$|V(LP_k(T_{n_N})) - \alpha| \leq \varepsilon, \quad |\phi_{n_N}(t^{n_N}_{m_{n_N}+1}) - \phi_*(t_*)| \leq \varepsilon.$$

By Theorem 4.1, we have

$$|V(LP_k(T_{n_N+1})) - V(LP_k(T_{n_N}))| = -\phi_{n_N}(t^{n_N}_{m_{n_N}+1})\mu_{n_N+1}(t^{n_N}_{m_{n_N}+1}) \leq \varepsilon.$$

By (A4), we obtain

$$|-\phi_{n_N}(t^{n_N}_{m_{n_N}+1})| \leq \frac{\varepsilon}{\delta},$$

wich implies

$$\phi_{n_N}(t^{n_N}_{m_{n_N}+1}) \to 0, \quad \text{as } \varepsilon \to 0. \tag{4.4}$$

But,

$$\phi_{n_N}(t^{n_N}_{m_{n_N}+1}) \to \phi_*(t_*) \neq 0, \quad \text{as } n_N \to \infty.$$

Hence, (4.4) cannot be true, and we have a contradiction. Therefore our claim is valid and the proof is complete.

Under conditions (A3) and (A4), Theorem 4.2 assures that the proposed scheme terminates in finitely many iterations, say n^* iterations, with an optimal solution

$$X^{n^*} = (x_0^{n^*}, x_1^{n^*}, x_2^{n^*}, \ldots, x_k^{n^*}).$$

In this case, X^{n^*} can be viewed as an approximate solution of (SIP_k). The next theorem tells us how good such an approximate solution can be.

Theorem 4.3 *For any given $\delta > 0$, if there exists $\overline{X} = (\overline{x}_0, \overline{x}_1, \overline{x}_2, \ldots, \overline{x}_k) \in \mathbb{R}^{k+1}$ such that*

$$\sum_{i=0}^{k} \overline{x}_i f_i(t) \geq 1, \quad \forall t \in T, \tag{4.5}$$

then

$$|V(LP_k(T_{n^*})) - V(SIP_k)| \leq \delta \Big| \sum_{i=0}^{k} c_i \overline{x}_i \Big|.$$

Proof: By the definition of X^{n^*}, we have

$$\sum_{i=0}^{k} f_i(t) x_i^{n^*} g(t) \geq -\delta, \quad \forall t \in T. \tag{4.6}$$

By (4.5), we have

$$\sum_{i=0}^{k} f_i(t) \delta \overline{x}_i \geq \delta, \quad \forall t \in T \tag{4.7}$$

It follows from (4.6) and (4.7) that

$$\sum_{i=0}^{k} f_i(t)(x_i^{n^*} + \delta \overline{x}_i) - g(t) \geq 0, \quad \forall t \in T.$$

Hence, $X^{n^*} + \delta \overline{X}$ is a feasible solution of (SIP_k). Therefore,

$$\begin{aligned}
|V(LP_k(T_{n^*})) - V(SIP_k)| &\leq \Big| \sum_{i=0}^{k} c_i x_i^{n^*} - \sum_{i=0}^{k} c_i(x_i^{n^*} + \delta \overline{x}_i) \Big| \\
&\leq \delta \Big| \sum_{i=0}^{k} c_i \overline{x}_i \Big|.
\end{aligned}$$

References

E.J. Anderson and P. Nash (1987), *Linear programming in infinite dimensional spaces*, Wiley, Chichester.

K. Glashoff and S.A. Gustafson (1982), *Linear optimization and approximation*, Springer–Verlarg, New York.

R. Hettich and K. Kortanek (1993), Semi–infinite programming: Theory, method and application, *SIAM Review*, 35, pp.380–429.

H.G. Kellerer (1988), Measure theoretic versions of linear programming, *Math. Zeitschrift*, 198, pp.367–400.

K.S. Kretschmer (1961), Programmes in paired spaces, *Canad. J. Math.*, 13, pp.221–238.

H.C. Lai and S.Y. Wu (1992), Extremal points and optimal solutions for general capacity problems, *Math. Programming (series A)*, 54, pp.87–113.

H.C. Lai and S.Y. Wu (1992a), On linear semi–infinite programming problems, *An algorithm, Numer. Funct. Anal. and Optimiz.*, 13, 3/4, pp.287–304.

H.C. Lai and S.Y. Wu (1994), Linear programming in measure spaces, *Optimization*, 29, pp.141–156.

R. Reemtsen and S. Görner (1998), Numerical methods for semi–infinite programming; a survey. In R. Reemtsen and J–J. Ruckmann, editors, *Semi–Infinite programming*, pp.195–275, Kluwer Academic Publishers, Boston.

C.F. Wen and S.Y. Wu (2001), Duality theorems and algorithms for linear programming in measure spaces, submitted to *Journal of Global Optimization.*

S.Y. Wu, S.C. Fang and C.J. Lin (1998), Relaxed cutting plane method for solving linear semi–infinite programming problems, *Journal of Optimization Theory and Applications*, 99, pp.759–779.

S.Y. Wu, C.J. Lin and S.C. Fang (2001), Relaxed cutting plane method for solving general capacity programming problems, to appear in *Ann of Operations Research.*

III OPTIMAL CONTROL

16 OPTIMAL CONTROL OF NONLINEAR SYSTEMS

S.P. Banks* and T. Cimen

Department of Automatic Control and Systems Engineering
The University of Sheffield
Mappin Street, Sheffield, S1 3JD, U.K.

Abstract: In this paper we study a nonlinear optimization problem with non-linear dynamics and replace it with a sequence of time-varying linear-quadratic problem, which can be solved classically.

Key words: Nonlinear optimal control, approximating systems.

*e-mail: s.banks@sheffield.ac.uk

1   INTRODUCTION

In recent papers Banks (2001), Banks and Dinesh (2000) have applied a sequence of linear time-varying approximations to find feedback controllers for nonlinear systems. Thus, for the optimal control problem

$$\min J = \frac{1}{2} x^T (t_f) F (x (t_f)) x (t_f)$$

$$+ \frac{1}{2} \int_0^{t_f} \left\{ x^T (t) Q (x (t)) x (t) + u^T (t) R (x (t)) u (t) \right\} dt \quad (1.1)$$

subject to the dynamics

$$\dot{x} (t) = A (x (t)) x (t) + B (x (t)) u (t), \qquad x (0) = x_0 \quad (1.2)$$

they have introduced the sequence of approximations

$$\min J^{[i]} = \tfrac{1}{2} x^{[i]T} (t_f) F \left(x^{[i-1]} (t_f) \right) x^{[i]} (t_f)$$
$$+ \tfrac{1}{2} \int_0^{t_f} \left\{ x^{[i]T} (t) Q \left(x^{[i-1]} (t) \right) x^{[i]} (t) + u^{[i]T} (t) R \left(x^{[i-1]} (t) \right) u^{[i]} (t) \right\} dt$$
$$(1.3)$$

subject to the dynamics

$$\dot{x}^{[i]} (t) = A \left(x^{[i-1]} (t) \right) x^{[i]} (t) + B \left(x^{[i-1]} (t) \right) u^{[i]} (t), \qquad x^{[i]} (0) = x_0 \quad (1.4)$$

for $i \geq 0$, where

$$\dot{x}^{[0]} (t) = A (x_0) x^{[0]} (t) + B (x_0) u^{[0]} (t), \qquad x^{[0]} (0) = x_0$$

and

$$\min J^{[0]} = \frac{1}{2} x^{[0]T} (t_f) F (x_0) x^{[0]} (t_f)$$

$$+ \frac{1}{2} \int_0^{t_f} \left\{ x^{[0]T} (t) Q (x_0) x^{[0]} (t) + u^{[0]T} (t) R (x_0) u^{[0]} (t) \right\} dt.$$

The approximations have been shown to converge under very mild conditions (that each operator is locally Lipschitz) and to provide very effective control in many examples. However, optimality has not been proved and, in fact, the limit control is unlikely to be optimal in general (and indeed, there may be no optimal control since the nonlinear systems considered are very general). Hence, in this paper, we consider the full necessary equations derived from Pontryagin's maximum principle and compare them with the original "approximate optimal control" method proposed in Banks and Dinesh (2000).

2   THE APPROXIMATING SYSTEMS

2.1   Classical Optimal Control Theory

Let us first look at classical optimal control theory where we consider the linear system

$$\dot{x}(t) = A(t)x(t) + B(t)u(t), \qquad x(0) = x_0 \tag{2.1}$$

with the finite-time cost functional

$$\min J = \frac{1}{2}x^T(t_f)Fx(t_f) + \frac{1}{2}\int_0^{t_f}\left\{x^T(t)Q(t)x(t) + u^T(t)R(t)u(t)\right\}dt. \tag{2.2}$$

It is well known (see, for example, Banks (1986)) that from the maximum principle, the solution to the linear-quadratic regulator problem is given by the coupled two-point boundary value problem

$$\begin{pmatrix} \dot{x} \\ \dot{\lambda} \end{pmatrix} = \begin{pmatrix} A(t) & -B(t)R^{-1}(t)B^T(t) \\ -Q(t) & -A^T(t) \end{pmatrix} \begin{pmatrix} x \\ \lambda \end{pmatrix} \tag{2.3}$$

with

$$\begin{aligned} x(0) &= x_0, \\ \lambda(t_f) &= Fx(t_f). \end{aligned} \tag{2.4}$$

Assuming that $\lambda(t) = P(t)x(t)$ for some positive-definite symmetric matrix $P(t)$, the necessary conditions are then satisfied by the Riccati equation

$$\dot{P}(t) = -Q(t) - P(t)A(t) - A^T(t)P(t) + P(t)B(t)R^{-1}(t)B^T(t)P(t),$$
$$P(t_f) = F \tag{2.5}$$

yielding the linear optimal control law given by

$$u(t) = -R^{-1}(t)B^T(t)P(t)x(t). \tag{2.6}$$

2.2   Global Optimal Control

Now consider applying the maximum principle to the nonlinear-quadratic regulator problem (1.1), (1.2) as in classical linear optimal control theory. We thus have the Hamiltonian

$$H = \frac{1}{2}\left(x^TQ(x)x + u^TR(x)u\right) + \lambda^T\left(A(x)x + B(x)u\right)$$

and from the equations

$$\dot{\lambda} = -\frac{\partial H}{\partial x}, \quad \frac{\partial H}{\partial u} = 0$$

we obtain

$$\begin{aligned}
\dot{\lambda} &= -\left(Q(x)\,x + \frac{1}{2}x^T \frac{\partial Q(x)}{\partial x} x + \frac{1}{2} u^T \frac{\partial R(x)}{\partial x} u + A^T(x)\lambda \right. \\
&\quad \left. + x^T \left(\frac{\partial A(x)}{\partial x} \right)^T \lambda + u^T \left(\frac{\partial B(x)}{\partial x} \right)^T \lambda \right),
\end{aligned}$$

$$R(x)\,u + B^T(x)\,\lambda = 0$$

giving

$$\begin{aligned}
\dot{\lambda} &= -\left(Q(x)\,x + \frac{1}{2}x^T \frac{\partial Q(x)}{\partial x} x + \frac{1}{2} u^T \frac{\partial R(x)}{\partial x} u + A^T(x)\lambda \right. \\
&\quad \left. + \left(\frac{\partial A(x)}{\partial x}x \right)^T \lambda + \left(\frac{\partial B(x)}{\partial x}u \right)^T \lambda \right),
\end{aligned}$$

$$u = -R^{-1}(x)\,B^T(x)\,\lambda$$

where

$$x^T \frac{\partial Q(x)}{\partial x} x = \left(x^T \frac{\partial Q(x)}{\partial x_1} x, \; x^T \frac{\partial Q(x)}{\partial x_2} x, \; \ldots, \; x^T \frac{\partial Q(x)}{\partial x_n} x \right)^T,$$

$$u^T \frac{\partial R(x)}{\partial x} u = \left(u^T \frac{\partial R(x)}{\partial x_1} u, \; u^T \frac{\partial R(x)}{\partial x_2} u, \; \ldots, \; u^T \frac{\partial R(x)}{\partial x_n} u \right)^T$$

are vectors of quadratic forms, and

$$\left(\frac{\partial A(x)}{\partial x} x \right)^T \lambda = \left(\frac{\partial A(x)}{\partial x_1} x, \; \frac{\partial A(x)}{\partial x_2} x, \; \ldots, \; \frac{\partial A(x)}{\partial x_n} x \right)^T \lambda,$$

$$\left(\frac{\partial B(x)}{\partial x} u \right)^T \lambda = \left(\frac{\partial B(x)}{\partial x_1} u, \; \frac{\partial B(x)}{\partial x_2} u, \; \ldots, \; \frac{\partial B(x)}{\partial x_n} u \right)^T \lambda.$$

Hence we obtain the equations

$$\dot{x} = A(x)\,x - B(x)\,R^{-1}(x)\,B^T(x)\,\lambda, \qquad x(0) = x_0$$

$$\begin{aligned}
\dot{\lambda} &= -Q(x)\,x - \frac{1}{2}x^T \frac{\partial Q(x)}{\partial x} x - \frac{1}{2}\lambda^T B(x)\,R^{-1}(x)\frac{\partial R(x)}{\partial x} R^{-1}(x)\,B^T(x)\,\lambda \\
&\quad - A^T(x)\,\lambda - \left(\frac{\partial A(x)}{\partial x} x \right)^T \lambda + \left(\frac{\partial B(x)}{\partial x} R^{-1}(x)\,B^T(x)\,\lambda \right)^T \lambda
\end{aligned}$$

together with the transversality condition $\lambda\left(t_f\right) = \frac{\partial}{\partial x}\left(\frac{1}{2}x^T\left(t_f\right)F\left(x\left(t_f\right)\right)x\left(t_f\right)\right)$, giving

$$\lambda\left(t_f\right) = F\left(x\left(t_f\right)\right)x\left(t_f\right) + \frac{1}{2}x^T\left(t_f\right)\frac{\partial F\left(x\left(t_f\right)\right)}{\partial x}x\left(t_f\right).$$

We write these differential equations in the form

$$\begin{pmatrix} \dot{x} \\ \dot{\lambda} \end{pmatrix} = $$

$$\begin{pmatrix} A\left(x\right) & -B\left(x\right)R^{-1}\left(x\right)B^T\left(x\right) \\ & -\frac{1}{2}\lambda^T B\left(x\right)R^{-1}\left(x\right)\frac{\partial R(x)}{\partial x}R^{-1}\left(x\right)B^T\left(x\right) \\ -Q\left(x\right)-\frac{1}{2}x^T\frac{\partial Q(x)}{\partial x} & -A^T\left(x\right)-\left(\frac{\partial A(x)}{\partial x}x\right)^T \\ & +\left(\frac{\partial B(x)}{\partial x}R^{-1}\left(x\right)B^T\left(x\right)\lambda\right)^T \end{pmatrix}\begin{pmatrix} x \\ \lambda \end{pmatrix}$$

or

$$\begin{pmatrix} \dot{x} \\ \dot{\lambda} \end{pmatrix} = \begin{pmatrix} A\left(x\right) & -B\left(x\right)R^{-1}\left(x\right)B^T\left(x\right) \\ -Q_1\left(x,\lambda\right) & -A^T\left(x\right) \end{pmatrix}\begin{pmatrix} x \\ \lambda \end{pmatrix} \tag{2.7}$$

where

$$\begin{aligned} Q_1\left(x,\lambda\right) = {}& Q\left(x\right)+\frac{1}{2}x^T\frac{\partial Q\left(x\right)}{\partial x} \\ & +\frac{1}{2}\lambda^T B\left(x\right)R^{-1}\left(x\right)\frac{\partial R\left(x\right)}{\partial x}R^{-1}\left(x\right)B^T\left(x\right)\frac{\lambda}{x} \\ & +\left(\frac{\partial A\left(x\right)}{\partial x}x\right)^T\frac{\lambda}{x}-\left(\frac{\partial B\left(x\right)}{\partial x}R^{-1}\left(x\right)B^T\left(x\right)\lambda\right)^T\frac{\lambda}{x}. \end{aligned}$$

Here $\frac{L\left(x,\lambda\right)}{x}$ means that we assume $L\left(x,\lambda\right)$ has a factor x so that it can be written as $L\left(x,\lambda\right) = M\left(x,\lambda\right)x$ for some function $M\left(x,\lambda\right)$ so that $\frac{L\left(x,\lambda\right)}{x} = M\left(x,\lambda\right)$. Using the approximation theory proposed in Banks and McCaffrey (1998), we can replace the nonlinear system of equations (2.7) by a sequence of linear time-varying approximations given by

$$\begin{pmatrix} \dot{x}^{[i]}\left(t\right) \\ \dot{\lambda}^{[i]}\left(t\right) \end{pmatrix} = $$

$$\begin{pmatrix} A\left(x^{[i-1]}\left(t\right)\right) & -B\left(x^{[i-1]}\left(t\right)\right)R^{-1}\left(x^{[i-1]}\left(t\right)\right)B^T\left(x^{[i-1]}\left(t\right)\right) \\ -Q_1\left(x^{[i-1]}\left(t\right),\lambda^{[i-1]}\left(t\right)\right) & -A^T\left(x^{[i-1]}\left(t\right)\right) \end{pmatrix}$$

$$\times\begin{pmatrix} x^{[i]}\left(t\right) \\ \lambda^{[i]}\left(t\right) \end{pmatrix}$$

or

$$\begin{pmatrix} \dot{x}^{[i]}\left(t\right) \\ \dot{\lambda}^{[i]}\left(t\right) \end{pmatrix} = \begin{pmatrix} \widetilde{A}\left(t\right) & -\widetilde{B}\left(t\right)\widetilde{R}^{-1}\left(t\right)\widetilde{B}^T\left(t\right) \\ -\widetilde{Q}\left(t\right) & -\widetilde{A}^T\left(t\right) \end{pmatrix}\begin{pmatrix} x^{[i]}\left(t\right) \\ \lambda^{[i]}\left(t\right) \end{pmatrix} \tag{2.8}$$

where

$$\widetilde{A}(t) = A\left(x^{[i-1]}(t)\right)$$
$$\widetilde{B}(t) = B\left(x^{[i-1]}(t)\right)$$
$$\widetilde{R}^{-1}(t) = R^{-1}\left(x^{[i-1]}(t)\right)$$
$$\widetilde{Q}(t) = Q_1\left(x^{[i-1]}(t),\ \lambda^{[i-1]}(t)\right)$$

(2.9)

with

$$x^{[i]}(0) = x_0,$$
$$\lambda^{[i]}(t_f) = F\left(x^{[i-1]}(t_f)\right)x^{[i]}(t_f) + \tfrac{1}{2}x^{[i-1]T}(t_f)\frac{\partial F\left(x^{[i-1]}(t_f)\right)}{\partial x}x^{[i]}(t_f) \quad (2.10)$$
$$= \widetilde{F}x^{[i]}(t_f)$$

where

$$\widetilde{F} = F\left(x^{[i-1]}(t_f)\right) + \frac{1}{2}x^{[i-1]T}(t_f)\frac{\partial F\left(x^{[i-1]}(t_f)\right)}{\partial x}. \qquad (2.11)$$

We know that the two-point boundary value problem (2.3) with conditions (2.4) represents the classical optimal control of a linear system with quadratic cost (2.2) subject to the dynamics (2.1), the solution of which is given by the Riccati equation (2.5) together with the optimal control law (2.6). Therefore, since each approximating problem in (2.8) is linear-quadratic, the system (2.8) with conditions (2.9), (2.10) and (2.11) is equivalent to the classical optimal control of a linear system with quadratic cost

$$\min \widetilde{J} = \frac{1}{2}x^T(t_f)\widetilde{F}x(t_f) + \frac{1}{2}\int_0^{t_f}\left\{x^T(t)\widetilde{Q}(t)x(t) + u^T(t)\widetilde{R}(t)u(t)\right\}dt$$

subject to the dynamics

$$\dot{x}(t) = \widetilde{A}(t)x(t) + \widetilde{B}(t)u(t), \qquad x(0) = x_0$$

and so the solution is given by the Riccati equation

$$\dot{P}(t) = -\widetilde{Q}(t) - P(t)\widetilde{A}(t) - \widetilde{A}^T(t)P(t) + P(t)\widetilde{B}(t)\widetilde{R}^{-1}(t)\widetilde{B}^T(t)P(t),$$
$$P(t_f) = \widetilde{F}$$

together with the optimal control law

$$u(t) = -\widetilde{R}^{-1}(t)\widetilde{B}^T(t)P(t)x(t)$$

where we have made the assumption that $\lambda(t) = P(t)x(t)$ as in classical optimal control theory. Hence the solution to the nonlinear optimal control problem (2.7) for these necessary conditions with quadratic cost (1.1) subject to the dynamics (1.2) is given by the approximate Riccati equation sequence

$$\dot{P}^{[i]}(t) = -\widetilde{Q}(t) - P^{[i]}(t) A\left(x^{[i-1]}(t)\right) - A^T\left(x^{[i-1]}(t)\right) P^{[i]}(t)$$
$$+ P^{[i]}(t) B\left(x^{[i-1]}(t)\right) R^{-1}\left(x^{[i-1]}(t)\right) B^T\left(x^{[i-1]}(t)\right) P^{[i]}(t),$$
$$P^{[i]}(t_f) = F\left(x^{[i-1]}(t_f)\right) + \tfrac{1}{2} x^{[i-1]T}(t_f) \frac{\partial F\left(x^{[i-1]}(t_f)\right)}{\partial x}$$

$$(2.12)$$

where

$$\widetilde{Q}(t) = Q\left(x^{[i-1]}(t)\right) + \tfrac{1}{2} x^{[i-1]T}(t) \frac{\partial Q\left(x^{[i-1]}(t)\right)}{\partial x}$$
$$+ \left\{ \begin{array}{l} \tfrac{1}{2} x^{[i-1]T}(t) P^{[i-1]}(t) B\left(x^{[i-1]}(t)\right) R^{-1}\left(x^{[i-1]}(t)\right) \frac{\partial R\left(x^{[i-1]}(t)\right)}{\partial x} \times \\ R^{-1}\left(x^{[i-1]}(t)\right) B^T\left(x^{[i-1]}(t)\right) P^{[i-1]}(t) \end{array} \right\}$$
$$+ \left(\frac{\partial A\left(x^{[i-1]}(t)\right)}{\partial x} x^{[i-1]}(t) \right)^T P^{[i-1]}(t)$$
$$- \left(\frac{\partial B\left(x^{[i-1]}(t)\right)}{\partial x} R^{-1}\left(x^{[i-1]}(t)\right) B^T\left(x^{[i-1]}(t)\right) P^{[i-1]}(t) x^{[i-1]}(t) \right)^T P^{[i-1]}(t)$$

together with the approximate optimal control law sequence

$$u^{[i]}(t) = -R^{-1}\left(x^{[i-1]}(t)\right) B^T\left(x^{[i-1]}(t)\right) P^{[i]}(t) x^{[i]}(t). \qquad (2.13)$$

The ith dynamical system then becomes

$$\dot{x}^{[i]}(t) = A\left(x^{[i-1]}(t)\right) x^{[i]}(t)$$
$$- B\left(x^{[i-1]}(t)\right) R^{-1}\left(x^{[i-1]}(t)\right) B^T\left(x^{[i-1]}(t)\right) P^{[i]}(t) x^{[i]}(t),$$
$$x^{[i]}(0) = x_0$$

$$(2.14)$$

for $i \geq 0$, where

$$\dot{x}^{[0]}(t) = \left(A(x_0) - B(x_0) R^{-1}(x_0) B^T(x_0) P^{[0]}(t) \right) x^{[0]}(t), \qquad x^{[0]}(0) = x_0.$$

2.3 Solving the Approximating Sequence

Similar to the "approximate optimal control" technique (see Banks and Dinesh (2000)), optimization is carried out for each sequence on the system trajectory for the "global optimal control" approach. In order to calculate the optimal solution, it is necessary to solve the approximate matrix Riccati equation sequence (2.12) storing the values of $P(t)$ from each sequence at every discrete time-step. In practice this will be done in a computer and it will be necessary to solve the Riccati equation using standard numerical integration procedures, starting at the final time $t = t_f$ and integrating backwards in time by taking negative time-steps. It should be noted that, even though the problem is to be solved in discrete-time, we do not solve the discrete-time Riccati equation as the dynamics are essentially continuous-time. The control that minimizes the finite-time quadratic cost functional at every time-step for each sequence is then given by (2.13). The approximating sequence for the states is obtained by solving (2.14), again storing the values $x(t)$ from each sequence at every time-step.

Let us consider solving the sequence of approximations. From (2.14) the first approximation is given by

$$\dot{x}^{[0]}(t) = \left(A(x_0) - B(x_0) R^{-1}(x_0) B^T(x_0) P^{[0]}(t) \right) x^{[0]}(t), \qquad x^{[0]}(0) = x_0$$

where $P^{[0]}(t)$ is the solution of equation (2.12) for $i = 0$ where we assume $x^{[i-1]}(t) = x_0$ and $P^{[i-1]}(t) = F(x_0)$. The second approximation is then given by

$$\dot{x}^{[1]}(t) = \overline{A}(t) x^{[1]}(t), \qquad x^{[1]}(0) = x_0$$

where

$$\overline{A}(t) = A\left(x^{[0]}(t) \right) - B\left(x^{[0]}(t) \right) R^{-1}\left(x^{[0]}(t) \right) B^T\left(x^{[0]}(t) \right) P^{[1]}(t)$$

such that $\overline{A} : \mathbb{R} \to \mathbb{R}^{n^2}$ and $\overline{A}(t)$ now represents a time-varying linear matrix. Here $P^{[1]}(t)$ is obtained by again solving (2.12) for $i = 1$, this time replacing $x^{[i-1]}(t)$ and $P^{[i-1]}(t)$ with the previous approximations $x^{[0]}(t)$ and $P^{[0]}(t)$ respectively. The subsequent approximations are obtained in a similar way. Thus, in solving (2.12)-(2.14), we obtain a sequence of time-varying linear equations where each sequence is solved as a standard numerical problem. For each sequence, optimization has to be carried out at every numerical integration time-step resulting in dynamic feedback control values. When both $x^{[i]}$ and $u^{[i]}$ have converged on the kth sequence, on applying the control $u^{[k]}$ of the converged sequence to the true nonlinear system (1.2), the solution obtained should be the same as that of the converged approximation. This is expected since $x^{[i]}(t) = x(t)$ when $x^{[i]}(t)$ has converged.

3   EXAMPLE

Balancing an inverted pendulum on a motor-driven cart, as shown in Figure 3.1, has become a popular controller design problem. The objective in the control

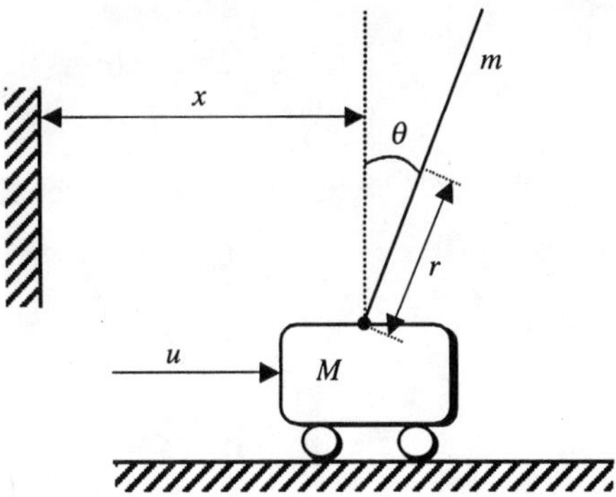

Figure 3.1   Inverted Pendulum System.

of this model is to move the cart to a specified position while maintaining the pendulum vertical. The inverted pendulum is unstable in that it may fall over any time in any direction unless a suitable control force u is applied and is often used to test new controller designs. Here only a two-dimensional problem is considered, that the pendulum moves only in the plane of the page.

Let us now apply the theories of "approximate optimal control" and "global optimal control" to the inverted pendulum system where the mathematical model is given by the equations (see Ogata (1997))

$$(M+m)\ddot{x} + mr\ddot{\theta}\cos\theta - mr\dot{\theta}^2\sin\theta = u$$
$$\ddot{x}\cos\theta + r\ddot{\theta} - g\sin\theta = 0. \tag{3.1}$$

Defining the four state-variables x_1, x_2, x_3, x_4 as x, θ, $\dot{x}$, $\dot{\theta}$ respectively, we can represent the system (3.1) in state-space in the form (1.2) as

$$
\begin{bmatrix} \dot{x}_1 \\ \dot{x}_2 \\ \dot{x}_3 \\ \dot{x}_4 \end{bmatrix} =
\begin{bmatrix}
0 & 0 & 1 & 0 \\
0 & 0 & 0 & 1 \\
0 & -\dfrac{mg\cos x_2}{M+m\sin^2 x_2}\mathrm{sinc}\,x_2 & 0 & \dfrac{mr x_4\sin x_2}{M+m\sin^2 x_2} \\
0 & \dfrac{(M+m)g}{(M+m\sin^2 x_2)r}\mathrm{sinc}\,x_2 & 0 & -\dfrac{m x_4\sin(2x_2)}{2(M+m\sin^2 x_2)}
\end{bmatrix}
\begin{bmatrix} x_1 \\ x_2 \\ x_3 \\ x_4 \end{bmatrix}
$$
$$
+ \begin{bmatrix} 0 \\ 0 \\ \dfrac{1}{M+m\sin^2 x_2} \\ -\dfrac{\cos x_2}{(M+m\sin^2 x_2)r} \end{bmatrix} u \tag{3.2}
$$

where

$$\operatorname{sinc} x_2 = \begin{cases} 1, & x_2 = 0 \\ \frac{\sin x_2}{x_2}, & x_2 \neq 0. \end{cases}$$

For the "global optimal control" technique we also require the Jacobians of $A(x)$ and $B(x)$, given by

$$\frac{\partial A(x)}{\partial x_1} = \begin{bmatrix} 0 & 0 & 0 & 0 \\ 0 & 0 & 0 & 0 \\ 0 & 0 & 0 & 0 \\ 0 & 0 & 0 & 0 \end{bmatrix}, \quad \frac{\partial A(x)}{\partial x_2} = \begin{bmatrix} 0 & 0 & 0 & 0 \\ 0 & 0 & 0 & 0 \\ 0 & \frac{\partial a_{32}}{\partial x_2} & 0 & \frac{\partial a_{34}}{\partial x_2} \\ 0 & \frac{\partial a_{42}}{\partial x_2} & 0 & \frac{\partial a_{44}}{\partial x_2} \end{bmatrix},$$

$$\frac{\partial A(x)}{\partial x_3} = \begin{bmatrix} 0 & 0 & 0 & 0 \\ 0 & 0 & 0 & 0 \\ 0 & 0 & 0 & 0 \\ 0 & 0 & 0 & 0 \end{bmatrix}, \quad \frac{\partial A(x)}{\partial x_2} = \begin{bmatrix} 0 & 0 & 0 & 0 \\ 0 & 0 & 0 & 0 \\ 0 & 0 & 0 & \frac{\partial a_{34}}{\partial x_4} \\ 0 & 0 & 0 & \frac{\partial a_{44}}{\partial x_4} \end{bmatrix},$$

$$\frac{\partial B(x)}{\partial x_1} = \begin{bmatrix} 0 \\ 0 \\ 0 \\ 0 \end{bmatrix}, \quad \frac{\partial B(x)}{\partial x_1} = \begin{bmatrix} 0 \\ 0 \\ \frac{\partial b_3}{\partial x_2} \\ \frac{\partial b_4}{\partial x_2} \end{bmatrix}, \quad \frac{\partial B(x)}{\partial x_1} = \begin{bmatrix} 0 \\ 0 \\ 0 \\ 0 \end{bmatrix}, \quad \frac{\partial B(x)}{\partial x_1} = \begin{bmatrix} 0 \\ 0 \\ 0 \\ 0 \end{bmatrix}$$

where

$$\frac{\partial a_{32}}{\partial x_2} = \frac{mg}{(M+m\sin^2 x_2)^2} \left\{ \begin{array}{l} \left(M + m\sin^2 x_2\right)\left[\sin x_2 \operatorname{sinc} x_2 - \cos x_2 \frac{\partial}{\partial x_2}\left(\operatorname{sinc} x_2\right)\right] \\ + m\sin\left(2x_2\right)\cos x_2 \operatorname{sinc} x_2 \end{array} \right\}$$

$$\frac{\partial a_{34}}{\partial x_2} = \frac{mr x_4}{(M+m\sin^2 x_2)^2} \left\{ \left(M + m\sin^2 x_2\right)\cos x_2 - m\sin x_2 \sin\left(2x_2\right) \right\}$$

$$\frac{\partial a_{42}}{\partial x_2} = \frac{(M+m)g}{(M+m\sin^2 x_2)^2 r} \left\{ \left(M + m\sin^2 x_2\right)\frac{\partial}{\partial x_2}\left(\operatorname{sinc} x_2\right) - m\sin\left(2x_2\right)\operatorname{sinc} x_2 \right\}$$

$$\frac{\partial a_{44}}{\partial x_2} = \frac{m x_4}{(M+m\sin^2 x_2)^2} \left\{ -\left(M + m\sin^2 x_2\right)\cos\left(2x_2\right) + \tfrac{1}{2}m\sin^2\left(2x_2\right) \right\}$$

$$\frac{\partial a_{34}}{\partial x_4} = \frac{mr\sin x_2}{M+m\sin^2 x_2}$$

$$\frac{\partial a_{44}}{\partial x_4} = -\frac{m\sin(2x_2)}{2(M+m\sin^2 x_2)}$$

$$\frac{\partial b_3}{\partial x_2} = -\frac{m\sin(2x_2)}{(M+m\sin^2 x_2)^2}$$

$$\frac{\partial b_4}{\partial x_2} = \frac{1}{(M+m\sin^2 x_2)^2 r} \left\{ \left(M + m\sin^2 x_2\right)\sin x_2 + m\sin\left(2x_2\right)\cos x_2 \right\}$$

with

$$\frac{\partial}{\partial x_2}\left(\operatorname{sinc} x_2\right) = \begin{cases} 0, & x_2 = 0 \\ \frac{x_2 \cos x_2 - \sin x_2}{x_2}, & x_2 \neq 0 \end{cases}$$

which can easily be shown by differentiating $\operatorname{sinc} x_2$ with respect to x_2 and using L'Hopital's rule.

By taking mass of the cart, $M = 1\ kg$, mass of the pendulum arm, $m = 0.1\ kg$, the distance from the pivot to the center of mass of the pendulum arm,

$r = 0.5\ m$, and the acceleration due to gravity, $g = 9.81\ m/\sec^2$, the state-space model (3.2) representing the inverted pendulum system was simulated using Euler's numerical integration technique with time-step 0.02 sec. The performance matrices have been chosen independent of the system states such that $Q = F = diag\{1,\ 1,\ 1,\ 1\}$ and $R = [1]$, so that the Jacobians of these become zero. In simulating the inverted pendulum system (using a simulation package such as MATLAB©) we only consider the regulator problem where our objective is to drive all the system states to zero. We also assume that the system starts from rest and simulate it for a given initial angle.

4   RESULTS

The "approximate optimal control" technique has been shown to provide very effective control in that the pendulum is stabilizable from any given initial angle. In fact taking the initial horizon time $t_f \leq 1.9$ sec and proceeding with the control of the approximating sequence from where left-off, by taking the final state values as initial conditions and defining a new horizon time, the pendulum can even be stabilized for its uncontrollable states ($\theta = \pm\pi/2\ rad$). This is because the "approximate optimal control" technique does not require a stabilizability condition to be satisfied - it only requires Lipschitz continuity. Although stability is not guaranteed in general, on a finite-time interval, it has been achieved for the uncontrollable states of the inverted pendulum system. A related "local freezing control" technique given in Banks and Mhana (1992) (where optimization is carried out point-wise on the system trajectory for the nonlinear system (1.2) has been shown to stabilize the inverted pendulum from any given initial angle except its uncontrollable states. The "global optimal control" strategy, however, provides control that stabilizes the inverted pendulum system for initial angles within the interval $\pm 1.1\ rad$ beyond which problems arise related to the convergence of the approximations for the necessary conditions. This may be related to the solution being a discontinuous feedback and a viscosity solution, causing the algorithm to eventually blow up. Note also that the "global optimal control" strategy is harder to implement, which requires the Jacobians of $A(x)$, $B(x)$, $F(x)$, $Q(x)$, and $R(x)$, hence taking a longer time to converge to the optimal solution.

Figures 4.1 and 4.2 illustrate and compare the converged solutions using both techniques when the initial angle is set to $\theta = 0.75\ rad$ and $\theta = 1.0\ rad$

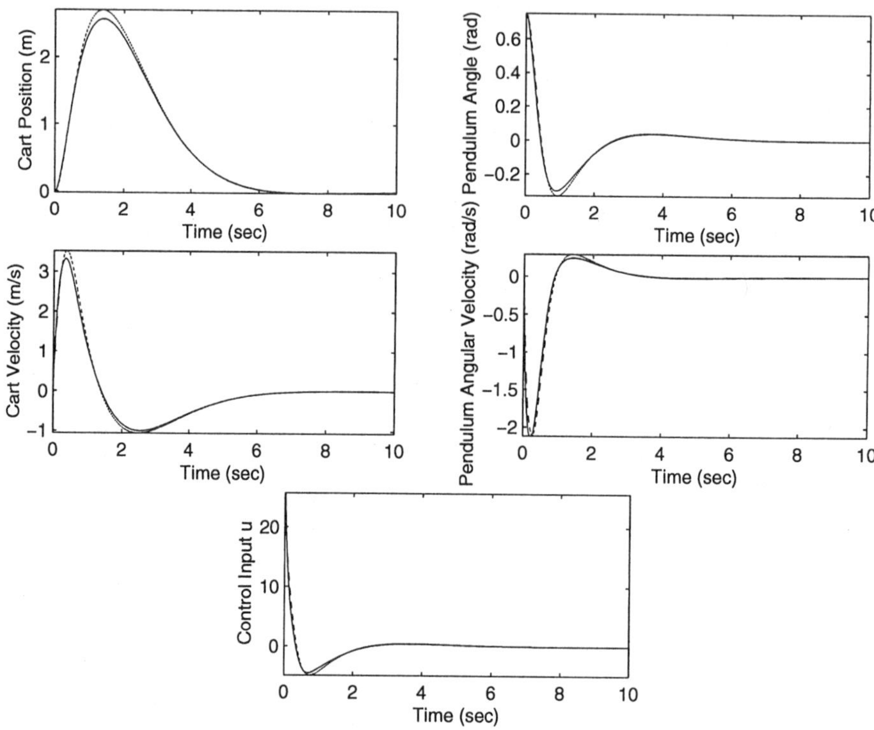

Figure 4.1 Response of the States and the Control Input of the Inverted Pendulum System when Subject to the Initial Angle $\theta = 0.75\ rad$ Using "Approximate Optimal Control" and "Global Optimal Control" Methods

respectively. The "approximate optimal control" solution is shown with a dotted line whereas the "global optimal control" solution is shown with a solid line.

From the plots it is obvious that the solutions obtained using these techniques are indeed very close. From Table 4.1, a similar statement can be made about the costs associated with each approximating strategy. However, note that there are problems in obtaining slightly larger costs using the "global optimal control" approach. This may be due to numerical problems since the costs obtained are only approximations. It may also be due to nonexistence of a global optimum since we have only considered the necessary (and not sufficient) conditions for an optimum, which may result in the approximating

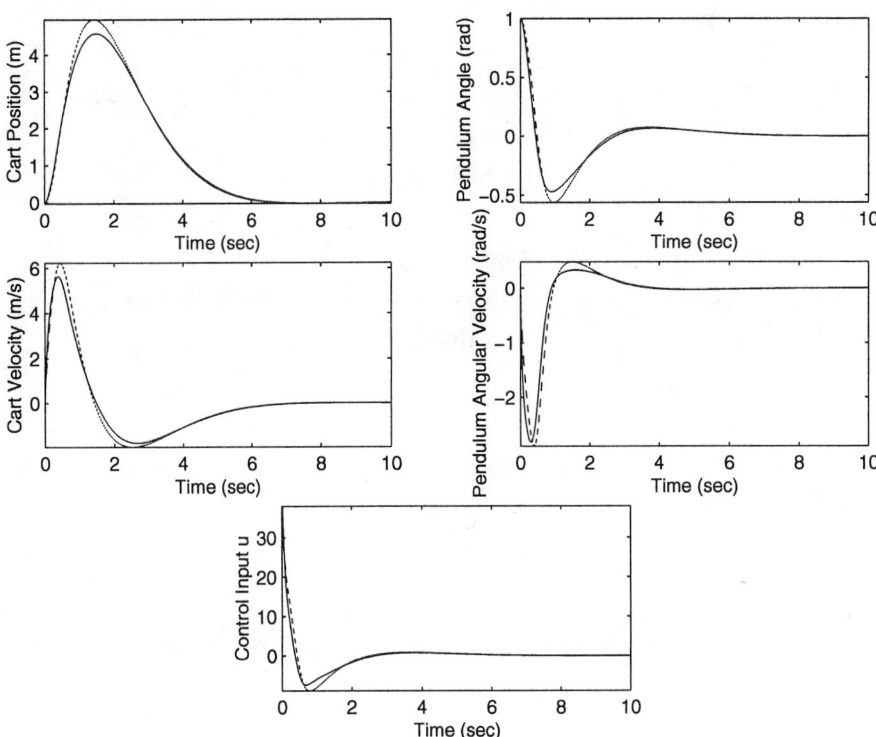

Figure 4.2   Response of the States and the Control Input of the Inverted Pendulum System when Subject to the Initial Angle $\theta = 1.0\ rad$ Using "Approximate Optimal Control" and "Global Optimal Control" Methods

systems converging to any of the local optima. Thus the proposed technique still remains "approximate optimal control", which provides solutions close to the optimal one, without having the need to compute any Jacobians of the system matrices and thus providing an easier implementation.

Table 4.1   Costs Associated with Each Optimization Technique Subject to Various Initial Angles.

Initial Angle (rad)	App. Optimal Control	Global Optimal Control
0.1	0.5501	0.5502
0.2	2.2532	2.2552
0.3	5.2765	5.2873
0.4	9.9306	9.9687
0.5	16.7286	16.8329
0.6	26.4951	26.7376
0.7	40.5709	41.0658
0.8	61.2150	62.1055
0.9	92.4568	93.8100
1.0	142.0922	143.4671
1.1	227.0319	225.3659

5   CONCLUSIONS

In this paper we have considered the full necessary conditions of a system with nonlinear dynamics, derived from Pontryagin's maximum principle, and compared them with the original "approximate optimal control" method. We have thus considered a nonlinear optimization problem with nonlinear dynamics and replaced it with a sequence of time-varying linear-quadratic regulator problem, where we have made the argument that the set of approximating systems are equivalent to the classical optimal control of a linear-quadratic regulator system and hence can be solved classically giving the solution to the "global optimal control", in the case where such a control exists. Even though the approximating systems using the "approximate optimal control" strategy may not converge to a global optimum of the nonlinear system, by considering a similar approximation sequence (given by the necessary conditions of the maximum principle), we have seen that the proposed method gives solutions very close to the optimal one in many cases for the inverted pendulum system. The methods used here are very general and apply to a very wide range of nonlinear systems. Future work will examine issues on discontinuity of the solution of the Hamilton-Jacobi equation and viscosity solutions.

References

Banks, S.P. (1986), *Control Systems Engineering*, Prentice-Hall International, Englwood Cliffs, New Jersey.

Banks, S.P. (2001), Exact Boundary Controllability and Optimal Control for a Generalized Korteweg de Vries Equation, *International Journal of Nonlinear Analysis, Methods and Applications*, Vol. 47, pp. 5537-5546.

Banks, S.P., and Dinesh, K. (2000), Approximate Optimal Control and Stability of Nonlinear Finite- and Infinite-Dimensional Systems, *Annals of Operations Research*, Vol. 98, pp. 19-44.

Banks, S.P., and McCaffrey, D. (1998), Lie Algebras, Structure of Nonlinear Systems and Chaotic Motion, *International Journal of Bifurcation and Chaos*, Vol. 8, No. 7, pp. 1437-1462, World Scientific Publishing Company.

Banks, S.P., and Mhana, K.J. (1992), Optimal Control and Stabilization for Nonlinear Systems, *IMA Journal of Control and Information*, Vol. 9, pp. 179-196.

Ogata, K. (1997), *Modern Control Engineering*, 3rd Edition, Prentice-Hall Inc., Upper Saddle River, New Jersey.

17 PROXIMAL-LIKE METHODS FOR CONVEX MINIMIZATION PROBLEMS

Christian Kanzow

University of Würzburg
Institute of Applied Mathematics and Statistics
Am Hubland
97074 Würzburg
Germany
e-mail: kanzow@mathematik.uni-wuerzburg.de

Abstract: This paper gives a brief survey of some proximal-like methods for the solution of convex minimization problems. Apart from the classical proximal-point method, it gives an introduction to several proximal-like methods using Bregman functions, φ-divergences etc. and discusses a couple of recent developments in this area. Some numerical results for optimal control problems are also included in order to illustrate the numerical behaviour of these proximal-like methods.

Key words: Convex minimization, proximal-point method, Bregman functions, φ-divergences, global convergence.

1 INTRODUCTION

This paper gives a brief survey of some proximal-like methods for the solution of convex minimization problems. To this end, let $f : \mathbb{R}^n \to \mathbb{R} \cup \{+\infty\}$ be a closed, proper, convex function, and consider the associated optimization problem

$$\min f(x), \quad x \in \mathbb{R}^n. \tag{1.1}$$

Formally, this is an unconstrained problem. However, since f is allowed to be extended-valued, any constrained problem of the form

$$\min \tilde{f}(x) \quad \text{subject to} \quad x \in X$$

for some convex function $\tilde{f} : \mathbb{R}^n \to \mathbb{R}$ and a closed, nonempty and convex set $X \subseteq \mathbb{R}^n$ can easily be transformed into a minimization problem of the form (1.1) by defining

$$f(x) := \begin{cases} \tilde{f}(x), & \text{if } x \in X, \\ +\infty, & \text{if } x \notin X. \end{cases}$$

Hence, theoretically, there is no loss of generality by considering the unconstrained problem (1.1). In fact, many theoretical results can be obtained in this (unifying) way for both unconstrained and constrained optimization problems. The interested reader is referred to the classical book by Rockafellar (1970) for further details.

Despite the fact that extended-valued functions allow such a unified treatment of both unconstrained and constrained problems, they are typically not tractable from a numerical point of view. Therefore, numerical algorithms for the solution of a problem like (1.1) have to take into account the constraints explicitly or, at least, some of these constraints. This can be done quite elegantly by so-called proximal-like methods. Similar to interior-point algorithms, these methods generate strictly feasible iterates and are usually applied to the problem with nonnegativity constraints

$$\min f(x) \quad \text{s.t.} \quad x \geq 0 \tag{1.2}$$

or to the linearly constrained problem

$$\min f(x) \quad \text{s.t.} \quad A^T x \leq b, \tag{1.3}$$

where, in the latter case, $A \in \mathbb{R}^{n \times m}$ and $b \in \mathbb{R}^m$ are the given data.

While we concentrate on the application of proximal-like methods to optimization problems, we should at least mention that proximal-like methods may also be applied to several other problem classes like nonlinear systems of equations, complementarity problems, variational inequalities and generalized equations. The interested reader is referred to Lemaire (1989); Lemaire (1992); Eckstein (1998); Censor et al. (1998); Auslender et al. (1999b); Solodov and Svaiter (2000b) and the corresponding references in these papers for more details.

This paper is organized in the following way: Section 2 first reviews the classical proximal-point method and then describes several proximal-like methods for the solution of the constrained optimization problems (1.2) and (1.3). In Section 3, we then present some numerical results obtained with some of these proximal-like methods when applied to some classes of optimal control problems. We then conclude with some final remarks in Section 4.

The notation used in this paper is quite standard: $\mathbb{R}^n$ denotes the n-dimensional Euclidean space, inequalities like $x \geq 0$ or $x > 0$ for a vector $x \in \mathbb{R}^n$ are defined componentwise, $\mathbb{R}^n_{++} := \{x \in \mathbb{R}^n \mid x > 0\}$ denotes the strictly positive orthant, and $\overline{S}$ is the closure of a subset $S \subseteq \mathbb{R}^n$.

2 PROXIMAL-LIKE METHODS

Throughout this section, we make the blanket assumption that $f : \mathbb{R}^n \to \mathbb{R} \cup \{+\infty\}$ is a closed, proper and convex function.

2.1 Classical Proximal-Point Method

The classical proximal-point method was introduced by Martinet (1970) and further developped by Rockafellar (1976) and others, see, e.g., the survey by Lemaire (1989). Being applied to the minimization problem (1.1), it generates a sequence $\{x^k\} \subseteq \mathbb{R}^n$ such that x^{k+1} is a solution of the subproblem

$$\min \ f(x) + \frac{1}{2\lambda_k} \|x - x^k\|^2 \tag{2.1}$$

for $k = 0, 1, \ldots$; here, λ_k denotes a positive number. The objective function of this subproblem is strictly convex since it is the sum of the original (convex) objective function f and a strictly convex quadratic term. This term is usually called the regularization term.

This strictly convex regularization term guarantees that the subproblem (2.1) has a unique minimizer for each $k \in \mathbb{N}$. Hence the classical proximal-point method is well-defined. Furthermore, it has the following global convergence properties, see, e.g., Güler (1991) for a proof of this result.

Theorem 2.1 *Let $\{x^k\}$ and $\{\lambda_k\}$ be two sequences generated by the classical proximal-point method (2.1), define $\sigma_k := \sum_{j=0}^{k} \lambda_j$, let $f_* := \inf\{f(x) \mid x \in \mathbb{R}^n\}$ be the optimal value and $\mathcal{S} := \{x^* \in \mathbb{R}^n \mid f(x^*) = f_*\}$ be the solution set of (1.1). Assume that $\sigma_k \to \infty$. Then the following statements hold:*

(a) The sequence of function values $\{f(x^k)\}$ converges to the optimal value f_*.

(b) If $\mathcal{S} \neq \emptyset$, then the entire sequence $\{x^k\}$ converges to an element of $\mathcal{S}$.

Theorem 2.1 states some very strong convergence properties under rather weak conditions. In particular, it guarantees the convergence of the entire sequence $\{x^k\}$ even if the solution set $\mathcal{S}$ contains more than one element; in fact, this statement also holds for an unbounded solution set $\mathcal{S}$. Note that the assumption $\sigma_k \to \infty$ holds, for example, if the sequence $\{\lambda_k\}$ is constant, i.e., if $\lambda_k = \lambda$ for all $k \in \mathbb{N}$ and some positive number λ.

We note that many variations of Theorem 2.1 are available in the literature. For example, it is not necessary to compute the exact minimizer of the subproblems (2.1) at each step, see Rockafellar (1976) for some criteria under which inexact solutions still provide similar global convergence properties. It should be noted, however, that the criteria of inexactness in Rockafellar (1976) are not implementable in general since they assume some knowledge regarding the exact solution of (2.1). On the other hand, Solodov and Svaiter (1999) recently gave a more constructive criterion in a slightly different framework.

Furthermore, some rate of convergence results can be shown for the classical proximal-point method under a certain error bound condition, cf. Luque (1984). This error bound condition holds, for example, for linearly constrained problems due to Hoffman's error bound Hoffman (1952). Moreover, being applied to linear programs, it is known that the classical proximal-point method has a finite termination property, see Ferris (1991) for details.

We also note that the classical proximal-point method can be extended to infinite-dimensional Hilbert spaces for which weak convergence of the iterates $\{x^k\}$ can be shown, see Rockafellar (1976); Güler (1991) once again. Strong convergence does not hold without any further modifications as noted by Güler (1991) who provides a counterexample. On the other hand, using a rather simple modification of the classical proximal-point method, Solodov and Svaiter (2000a) were able to present a strongly convergent version of the classical proximal-point method in Hilbert spaces.

In contrast to the classical proximal-point method, the proximal-like methods to be presented in our subsequent discussion have only been presented in the finite dimensional setting; this is mainly due to the fact that they typically involve some logarithmic functions.

2.2 Proximal-like Methods Using Bregman Functions

The simple idea which is behind each proximal-like method for the solution of convex minimization problems is, more or less, to replace the strictly quadratic term in the regularized subproblem (2.1) by a more general strictly convex function. Later on, we will see that this might be a very useful idea when solving constrained problems.

There are quite a few different possibilities to replace the term $\frac{1}{2}\|x - x^k\|^2$ by another strictly convex distance-like function. The one we discuss in this subsection is defined by

$$D_\psi(x, y) := \psi(x) - \psi(y) - \nabla\psi(y)^T(x - y),$$

where ψ is a so-called Bregman function. According to Solodov and Svaiter (2000a), a Bregman function may be defined in the following way.

Definition 2.1 *Let $S \subseteq \mathbb{R}^n$ be an open and convex set. A mapping $\psi : \overline{S} \to \mathbb{R}$ is called a* Bregman function *with zone S if it has the following properties:*

(i) ψ is strictly convex and continuous on $\overline{S}$;

(ii) ψ is continuously differentiable in S;

(iii) The partial level set

$$\mathcal{L}_\alpha(x) := \{y \in \overline{S} \mid D_\psi(x, y) \leq \alpha\}$$

is bounded for every $x \in \overline{S}$;

(iv) If $\{y^k\} \subseteq S$ converges to x, then $\lim_{k \to \infty} D_\psi(x, y^k) = 0$.

Earlier papers on Bregman functions require some additional properties, see De Pierro and Iusem (1986); Censor and Zenios (1992); Eckstein (1993); Güler (1994); Censor et al. (1998). However, as noted in Solodov and Svaiter (2000a), all these additional properties follow from those mentioned in Definition 2.1.

Two simple examples of Bregman functions are:

$$\psi_1(x) \quad := \quad \frac{1}{2}\|x\|^2 \quad \text{on } S = \mathbb{R}^n,$$

$$\psi_2(x) \quad := \quad \sum_{i=1}^{n} x_i \log x_i - x_i \quad \text{on } S = \mathbb{R}^n_{++} \quad (\text{convention: } 0\log 0 = 0).$$

Using $\psi = \psi_1$ in the definition of D_ψ, we reobtain the subproblem (2.1) of the classical proximal-point method. On the other hand, using $\psi = \psi_2$ in the definition of D_ψ gives the so-called *Kullback-Leibler relative entropy function*

$$D_{\psi_2}(x, y) = \sum_{i=1}^{n} x_i \log \frac{x_i}{y_i} + y_i - x_i. \tag{2.2}$$

This function may be used in order to solve the constrained optimization problem (1.2) by generating a sequence $\{x^k\}$ in such a way that x^{k+1} is a solution of the subproblem

$$\min \ f(x) + \frac{1}{\lambda_k} D_\psi(x, x^k), \quad x > 0 \tag{2.3}$$

for $k = 0, 1, 2, \ldots$, where $x^0 > 0$ is a strictly feasible starting point. Then a convergence result completely identical to Theorem 2.1 can be shown for this method, see Chen and Teboulle (1993) for details. Some related papers dealing with Bregman functions in the context of proximal-like methods are Eckstein (1993); Güler (1994); Eckstein (1998); Censor et al. (1998), where the interested reader will find rules which allow inexact solutions of the subproblems (2.3) and where he will also find some rate of convergence results. In contrast to the classical proximal-point method, however, the proximal-like methods do not possess any finite termination properties for linear programs.

On the other hand, a major advantage of the proximal-like methods is that a subproblem like (2.3) is essentially unconstrained and can therefore be solved by unconstrained minimization techniques.

2.3 Proximal-like Methods Using φ-Divergences

Another variant of the classical proximal-point method is a proximal-like method based on so-called φ-divergences. These φ-divergences will be used in order to replace the strictly convex quadratic term in the subproblem (2.1) of the classical proximal-point method.

Definition 2.2 *Let Φ denote the class of closed, proper and convex functions $\varphi : \mathbb{R} \to (-\infty, +\infty]$ with $dom(\varphi) \subseteq [0, +\infty)$ having the following properties:*

(i) φ is twice continuously differentiable on $int(dom\varphi) = (0, +\infty)$;

(ii) φ is strictly convex on its domain;

(iii) $\lim_{t \to 0+} \varphi'(t) = -\infty$;

(iv) $\varphi(1) = \varphi'(1) = 0$ and $\varphi''(1) > 0$;

(v) There exists $\nu \in (\frac{1}{2}\varphi''(1), \varphi''(1))$ such that

$$(1 - 1/t)(\varphi''(1) + \nu(t - 1)) \leq \varphi'(t) \leq \varphi''(1)(t - 1) \quad \forall t > 0.$$

Then the φ-divergence corresponding to a mapping $\varphi \in \Phi$ is defined by

$$d_\varphi(x, y) := \sum_{i=1}^{n} y_i \varphi\left(\frac{x_i}{y_i}\right)$$

for $x, y \in \mathbb{R}_{++}^n$.

Sometimes condition (v) is not required in Definition 2.2. However, the corresponding convergence results are weaker without this additional property.

Some examples of functions $\varphi \in \Phi$ are

$$\begin{aligned}
\varphi_1(t) &:= t \log t - t + 1, \\
\varphi_2(t) &:= -\log t + t - 1, \\
\varphi_3(t) &:= \left(\sqrt{t} - 1\right)^2.
\end{aligned}$$

The graphs of these three functions are shown in the following figure.

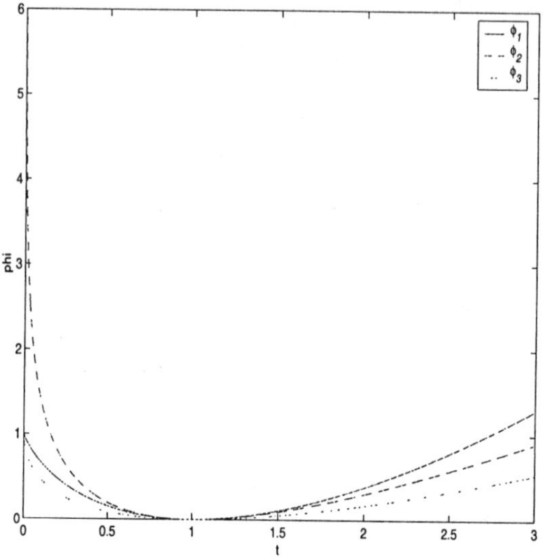

In particular, if we choose $\varphi = \varphi_1$ in the definition of d_φ, we obtain

$$d_{\varphi_1}(x, y) = \sum_{i=1}^{n} x_i \log \frac{x_i}{y_i} + y_i - x_i,$$

and this is precisely the Kullback-Leibler relative entropy function from (2.2).

Using any φ-divergence, we may try to solve the constrained minimization problem (1.2) by generating a sequence $\{x^k\}$ in such a way that x^{k+1} solves the subproblem

$$\min \; f(x) + \frac{1}{\lambda_k} d_\varphi(x, x^k), \quad x > 0 \tag{2.4}$$

for $k = 0, 1, \ldots$, where $x^0 > 0$ is any given starting point. For this method, Teboulle (1997) shows that it has the same global convergence properties as those mentioned in Theorem 2.1 for the classical proximal-point method. In addition, Teboulle (1997) also allows inexact solutions of the subproblems (2.4). The method may also be applied to the linearly constrained problem (1.3), and, once again, superlinear convergence can be shown under a certain error bound assumption, see Auslender and Haddou (1995) for details. Further references on φ-divergences include Csiszár (1967); Eggermont (1990); Teboulle (1992); Iusem et al. (1994); Iusem, Teboulle (1995).

2.4 Proximal-like Methods Using Quadratic Kernels and Regularization

The method we describe in this subsection is taken from Auslender et al. (1999a) and based on earlier work by Tseng and Bertsekas (1993) and Ben Tal and Zibulevsky (1997). It can be motivated by using the ideas from the previous subsection. To this end, assume for a while that the mapping f from the linearly constrained optimization problem (1.3) is twice continuously differentiable. Then it is most natural to use Newton's method in order to minimize the objective function from the (essentially unconstrained) subproblem in (2.4) in order to obtain the next iterate in this way. Using Newton's method, however, we need second order derivatives of these functions. In particular, we need second order derivatives of the function d_φ from Definition 2.2. Calculating the first derivative with respect to x, we obtain

$$\nabla_x d_\varphi(x, y) = \sum_{i=1}^{n} \varphi'\Big(\frac{x_i}{y_i}\Big) e_i,$$

and calculating the second derivative gives

$$\nabla_{xx}^2 d_\varphi(x, y) = \sum_{i=1}^{n} \frac{1}{y_i} \varphi''\Big(\frac{x_i}{y_i}\Big) e_i e_i^T,$$

where e_i denotes the i-th unit vector in $\mathbb{R}^n$. Hence, in each sum, we have the factor $\frac{1}{y_i}$ which increases to infinity during the iteration process for all indices i for which a constraint like $x_i \geq 0$ is active at a solution. Consequently, we therefore get Hessian matrices which are very ill-conditioned.

In order to avoid this drawback, it is quite natural to modify the idea from the previous subsection in the following way: Let Φ be the class of functions from Definition 2.2 and set

$$d_\varphi(x, y) := \sum_{i=1}^{n} y_i^2 \varphi\Big(\frac{x_i}{y_i}\Big) \tag{2.5}$$

The difference to the φ-divergence from the previous subsection is that we use y_i^2 instead of y_i. Calculating the second order derivative of the mapping d_φ from (2.5) gives

$$\nabla_{xx}^2 d_\varphi(x, y) = \sum_{i=1}^{n} \varphi''\Big(\frac{x_i}{y_i}\Big) e_i e_i^T,$$

i.e., the crucial factor $\frac{1}{y_i}$ vanishes completely.

Now consider the linearly constrained optimization problem (1.3), i.e., consider the problem

$$\min \; f(x) \quad \text{s.t.} \quad g(x) \geq 0, \tag{2.6}$$

with $g(x) := b - A^T x$. Auslender et al. (1999a) consider an algorithm which defines a sequence $\{x^k\}$ in such a way that, given a strictly feasible starting point x^0, the next iterate x^{k+1} is computed as a solution of the subproblem

$$\min \; f(x) + \frac{1}{\lambda_k} d_\varphi\big(g(x), g(x^k)\big), \quad g(x) > 0,$$

where, of course, d_φ denotes the mapping from (2.5). However, it turns out that this method has weaker global convergence properties than all other methods discussed so far, see Auslender et al. (1999a).

In order to overcome this problem, Auslender et al. (1999a) suggest to use a regularization technique. More precisely, they suggest to add a quadratic penalty term to a function $\varphi \in \Phi$ (with Φ being the set from Definition 2.2) in order to obtain a function $\tilde{\varphi}$ in this way. For example, using the three mappings $\varphi_1, \varphi_2, \varphi_3$ from Subsection 2.3, we obtain

$$\begin{aligned}
\tilde{\varphi}_1(t) &:= t \log t - t + 1 + \frac{\nu}{2}(t-1)^2, \\
\tilde{\varphi}_2(t) &:= -\log t + t - 1 + \frac{\nu}{2}(t-1)^2, \\
\tilde{\varphi}_3(t) &:= 2\big(\sqrt{t}-1\big)^2 + \frac{\nu}{2}(t-1)^2
\end{aligned}$$

for some constant $\nu > 1$. These functions are plotted in the following figure.

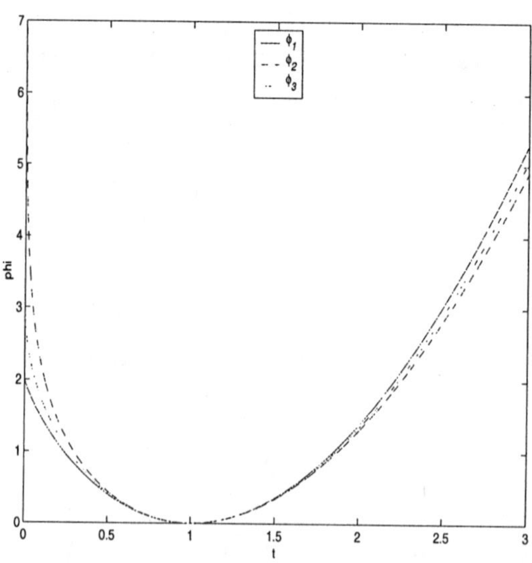

Now, let $\tilde{\varphi}$ denote any of these functions and set (similar to (2.5))

$$d_{\tilde{\varphi}}(x,y) := \sum_{i=1}^{n} y_i^2 \tilde{\varphi}\left(\frac{x_i}{y_i}\right).$$

The regularized method from Auslender et al. (1999a) then generates a sequence $\{x^k\}$ by starting with a strictly feasible point x^0 for problem (1.3) and by computing x^{k+1} as the solution of the subproblem

$$\min\ f(x) + \frac{1}{\lambda_k} d_{\tilde{\varphi}}\big(g(x), g(x^k)\big), \quad g(x) > 0. \tag{2.7}$$

Then the following result was shown in Auslender et al. (1999a).

Theorem 2.2 *Let $\{x^k\}$ be a sequence generated by the above method. Assume that the following assumptions are satisfied:*

(A.1) There exist constants $\lambda_{\max} \geq \lambda_{\min} > 0$ such that $\lambda_k \in [\lambda_{\min}, \lambda_{\max}]$ for all $k \in \mathbb{N}$.

(A.2) The optimal value $f_* := \inf\{f(x) \,|\, A^T x \leq b\}$ is finite.

(A.3) $\mathrm{dom}(f) \cap \mathrm{int}\{x \,|\, A^T x \leq b\}$ is nonempty.

(A.4) The matrix A has rank n.

Then the following statements hold:

(a) The sequence of function values $\{f(x^k)\}$ converges to the optimal value f_*.

(b) If $\mathcal{S} \neq \emptyset$, then the entire sequence $\{x^k\}$ converges to an element of $\mathcal{S}$.

The rank assumption (A.4) is satisfied, e.g., if $A = I$, i.e., if the feasible set is the nonnegative orthant. Moreover, this rank condition can be assumed to hold without loss of generality for linear programs if we view (1.3) as the dual of a standard form linear program.

2.5 Infeasible Proximal-like Methods

Consider again the linearly constrained minimization problem (1.3). The methods described in the previous subsections all assume, among other things, that the interior of the feasible set is nonempty. Moreover, it is assumed that we can find a strictly feasible point in order to start the algorithm. However, for linear constraints, it is usually not easy to find such a starting point. Furthermore, there do exist convex optimization problems which are solvable but whose interior of the feasible region is empty. In this case, it is not possible to apply one of the methods from the previous subsections.

Yamashita et al. (2001) therefore describe an infeasible proximal-like method which can be started from an arbitrary point and which avoids the assumption that the interior of the feasible set is nonempty. The idea behind the method from Yamashita et al. (2001) is to enlarge the feasible set

$$X := \left\{ x \in \mathbb{R}^n \,\middle|\, A^T x \le b \right\}$$

of the original minimization problem (1.3) by introducing a perturbation vector $\delta^k > 0$ and by replacing X at each iteration k by an enlarged region of the form

$$X_k := \left\{ x \in \mathbb{R}^n \,\middle|\, A^T x \le b + \delta^k \right\}.$$

Note that $X \ne \emptyset$ then implies $\mathrm{int}(X_k) \ne \emptyset$. Hence we can apply the previous method to the enlarged problem

$$\min f(x) \quad \text{s.t.} \quad x \in X_k.$$

The fact that the method from the previous subsection uses a quadratic penalty term actually fits perfectly into our situation where we allow infeasible iterates.

Obviously, we can hope that we obtain a solution of the original problem (1.3) by letting $\delta^k \to 0$. To be more precise, let us define

$$g_{\delta^k}(x) := b + \delta^k - A^T x,$$

and let x^{k+1} be a solution of the subproblem

$$\min f(x) + \frac{1}{\lambda_k} d_{\tilde{\varphi}}\big(g_{\delta^k}(x), g_{\delta^k}(x^k)\big), \quad g_{\delta^k}(x) > 0$$

for some perturbation vector $\delta^k > 0$. Then is has been shown in Yamashita et al. (2001) that the statements of Theorem 2.2 remain true under a certain set of assumptions. However, without going into the details, we stress that these assumptions include the condition $\mathrm{dom}(f) \cap \{x \,|\, A^T x \le b\} \ne \emptyset$ (in contrast to (A.3) which assumes that the domain of f intersected with the *interior* of the feasible set is nonempty). On the other hand, the main convergence result in

Yamashita et al. (2001) has to impose another condition which eventually guarantees that the iterates $\{x^k\}$ generated by the infeasible proximal-like method become feasible in the limit point.

3 NUMERICAL RESULTS FOR SOME OPTIMAL CONTROL PROBLEMS

3.1 Description of Test Problems

We consider two classes of optimal control problems. The first class contains control constraints, the second one involves state constraints.

The class of control constrained problems is as follows: Let $\Omega \subseteq \mathbb{R}^n$ be an open and bounded domain and consider the minimization problem

$$\begin{aligned}
\min \quad & J(u) := \tfrac{1}{2}\|y(u) - y_d\|^2_{L^2(\Omega)} + \tfrac{\alpha}{2}\|u - u_d\|^2_{L^2(\Omega)} \\
\text{s.t.} \quad & u \in \mathcal{F} := \{u \in L^2(\Omega) \,|\, u \leq \psi \text{ on } \Omega\},
\end{aligned} \tag{3.1}$$

where $y = y(u) \in H^1_0(\Omega)$ denotes the weak solution of Poisson's equation $-\Delta y = u$ on Ω and $y_d, u_d, \psi \in L^2(\Omega)$ are given square-integrable functions. Here y is the state and u is the control variable. The meaning of (3.1) is that we want to minimize the distance to a desired state y_d (hence the subscript 'd') subject to some constraints on the control u. The penalty term in the objective function J is a standard regularization term multiplied by a small constant $\alpha > 0$.

The following two particular instances of the class of problems (3.1) will be used in our numerical tests; they are taken from Bergounioux et al. (2001).

Example 3.1 Consider the following two instances:

(a) Consider problem (3.1) with the following data:

$$\begin{aligned}
\Omega \quad &:= \quad (0,1)^2 \subset \mathbb{R}^2, \\
y_d(x_1, x_2) \quad &:= \quad \tfrac{1}{6}\sin(2\pi x_1)\sin(2\pi x_2)\exp(2x_1), \\
u_d \quad &\equiv \quad 0, \\
\psi \quad &\equiv \quad 0, \\
\alpha \quad &:= \quad 10^{-2}.
\end{aligned}$$

(b) Consider problem (3.1) with the following data:

$$\Omega \ := \ (0,1)^2 \subset \mathbb{R}^2,$$

$$y_d(x_1, x_2) \ := \ \begin{cases} 200 x_1 x_2 (x_1 - \tfrac{1}{2})^2 (1 - x_2), & \text{if } 0 < x_1 \leq \tfrac{1}{2}, \\ 200 x_2 (x_1 - 1)(x_1 - \tfrac{1}{2})^2 (1 - x_2), & \text{if } \tfrac{1}{2} < x_1 \leq 1, \end{cases}$$

$$u_d \ \equiv \ 0,$$

$$\psi \ \equiv \ 1,$$

$$\alpha \ := \ 10^{-2}$$

In order to deal with this problem numerically, we discretize the domain $\Omega = (0,1)^2$ by using an equidistant $(N \times N)$-grid. The Laplacian was approximated by using the standard 5-point finite difference scheme. After shifting the variables in order to get nonnegativity constraints, we then obtain an optimization problem of the form (1.2).

The second class of problems we will deal with is the state constrained problem

$$\begin{aligned} \min \quad & J(u) := \tfrac{1}{2}\|y(u) - y_d\|_{L^2(\Omega)}^2 + \tfrac{\alpha}{2}\|u - u_d\|_{L^2(\Omega)}^2 \\ \text{s.t.} \quad & u \in \mathcal{F} := \{u \in L^2(\Omega) \mid y \leq \phi \text{ on } \Omega\}, \end{aligned} \qquad (3.2)$$

where we used the same notation as for the control constrained problem, i.e., $y = y(u) \in H_0^1(\Omega)$ denotes the weak solution of Poisson's equation $-\Delta y = u$ on Ω, $y_d, u_d, \phi \in L^2(\Omega)$ are given functions, and $\alpha > 0$ is a small regularization parameter. Note that the only difference between problems (3.1) and (3.2) is that we have different constraints.

The particular instance of problem (3.2) we are interested in is also taken from Bergounioux et al. (2001).

Example 3.2 Consider problem (3.2) with the following data:

$$\Omega \ := \ (0,1)^2 \subset \mathbb{R}^2,$$

$$y_d(x_1, x_2) \ := \ \sin(2\pi x_1 x_2),$$

$$u_d \ \equiv \ 0,$$

$$\phi \ \equiv \ 0.1,$$

$$\alpha \ := \ 10^{-3}$$

In order to deal with the state constrained problem (3.2), we use the same discretization scheme as for the control constrained problem (3.1). In this case, however, this results in a linearly constrained optimization problem of the form (1.3), and it is in general not easy to find a strictly feasible starting point.

3.2 Numerical Results for Control Constrained Problems

We begin with a word of caution: The numerical results presented in this and the next subsection are not intended to show that proximal-point methods are the best methods for solving the two classes of optimal control problems from the previous subsection. The only thing we want to do is to provide a brief comparison between some of the different proximal-like methods for the solution of these problems in order to get some hints which proximal-like methods seem to work best.

All methods were implemented in MATLAB and use the same parameter setting whenever this was possible. The particular methods we consider in this subsection for the solution of the control constrained problem (3.1) are the proximal-like methods from Subsections 2.3 (φ-divergences) and 2.4 (quadratic kernels with regularization term). The unconstrained minimization is always carried out by applying Newton's method. For reasons explained earlier, we took the function φ_2 in combination with the proximal-like method from Subsection 2.3, and the corresponding mapping $\tilde{\varphi}_2$ for the regularized method from Subsection 2.4.

Table 3.2 contains the numerical results for Example 3.1 (a) for different sizes of N (the dimension of the discretized problem is $n = N^2$). This table contains the cumulated number of inner iterations, i.e., we present the total number of Newton steps and therefore the total number of linear system solves for each test problem. Both methods seem to work reasonably well, and the number of iterations is more or less independent from the mesh size. However, the number of iterations using the φ-divergence approach is significantly higher than the number of iterations for the regularized approach. The resulting optimal control and state for Example 3.1 (a) are given in Figures 3.1 and 3.2, respectively.

The observation is similar for Example 3.1 (b) as shown in Table 3.2, although this time the number of iterations needed by the two methods is pretty much the same. The resulting optimal control and states for these two examples are given in Figures 3.3 and 3.4, respectively.

We also tested both methods on Example 3.1 (a) using smaller values of α. Due to the quadratic penalty term in the regularized method, we do expect a better behaviour for this method. This is indeed reflected by the numerical results shown in Table 3.4.

Table 3.1   Number of iterations for Example 3.1 (a) ($n = N^2$)

N	φ-divergence	regularized φ function
20	25	17
30	25	17
40	26	17
50	26	16

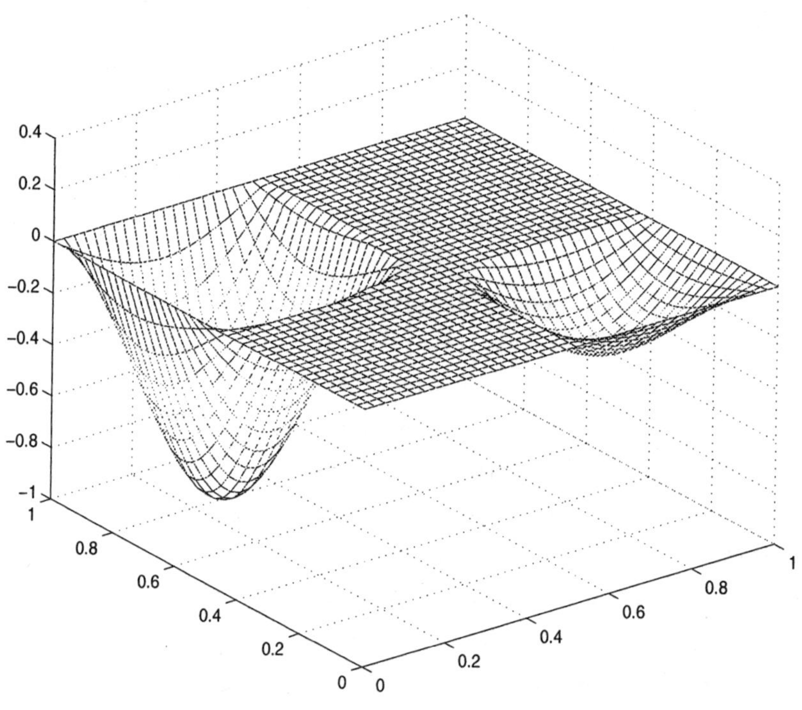

Figure 3.1   Resulting optimal control for Example 3.1 (a)

3.3   Numerical Results for State Constrained Problems

Since a strictly feasible starting point is usually not at hand for state constrained problems, we only applied the infeasible proximal-like method from Subsection 2.5 to the test problem from Example 3.2.

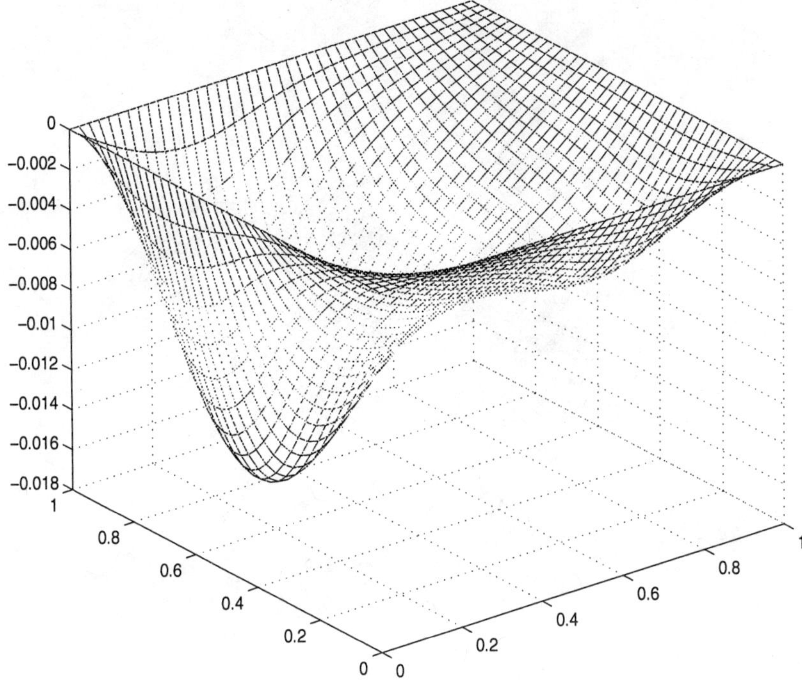

Figure 3.2   Resulting optimal state for Example 3.1 (b)

Table 3.2   Number of iterations for Example 3.1 (b) $(n = N^2)$

N	φ-divergence	regularized φ function
20	23	28
30	30	28
40	30	28
50	31	28

Similar to our description of the methods from the previous subsection, we use Newton's method in order to carry out the inner iterations. The cumulated number of inner iterations for different dimensions of this problem are reported in Table 3.3.

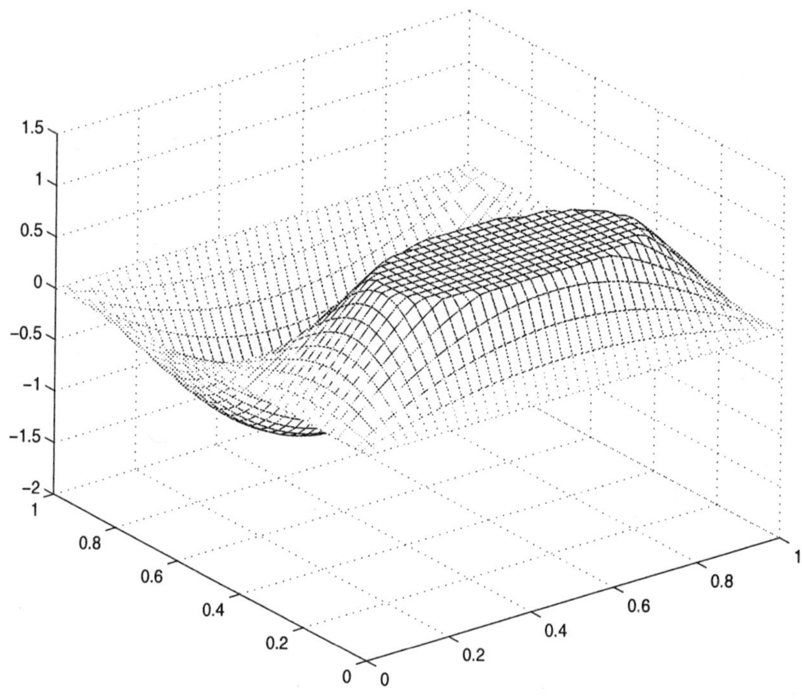

Figure 3.3   Resulting optimal control for Example 3.1 (b)

Table 3.3   Number of iterations for Example 3.1 (a) using different α ($N = 30$)

α	φ-divergence	regularized φ function
$\alpha = 10^{-2}$	25	17
$\alpha = 10^{-3}$	31	18
$\alpha = 10^{-4}$	34	18
$\alpha = 10^{-5}$	47	18
$\alpha = 10^{-6}$	60	18
$\alpha = 10^{-7}$	71	18
$\alpha = 10^{-8}$	62	18

The results indicate that the infeasible method works quite well. Similar to the results from the previous subsection, we can see from Table 3.4 that the number of iterations is again (more or less) independent of the mesh size.

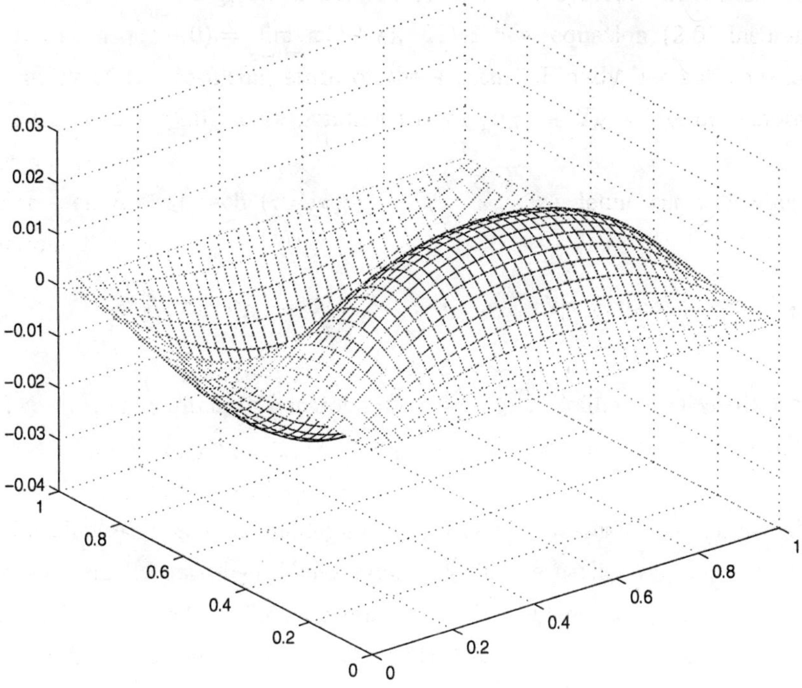

Figure 3.4   Resulting optimal state for Example 3.1 (b)

Table 3.4   Number of iterations for Example 3.2 ($n = N^2$)

N	iterations
20	16
30	14
40	16
50	14
60	14

The resulting optimal control and state for Example 3.2 are shown in Figures 3.5 and 3.6, respectively.

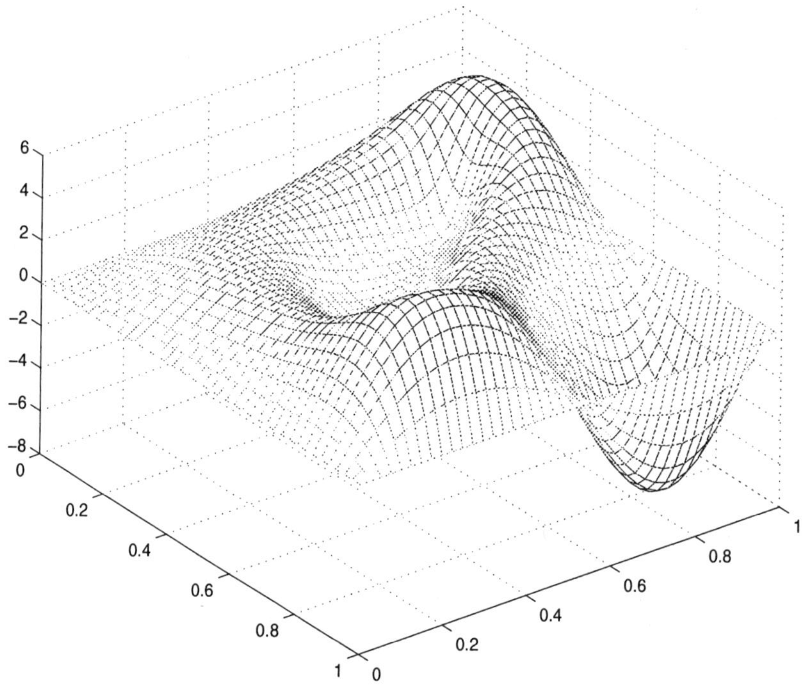

Figure 3.5   Resulting optimal control for Example 3.2

4   FINAL REMARKS

In this paper, we presented several proximal-like methods for the solution of convex minimization problems. A particular class of convex problems are the so-called semi-definite programs, see Nesterov and Nemirovskii (1994) for example. The main difference between semi-definite programs and a convex minimization problem of the form (1.1), say, is the fact that the variables in semi-definite programs are matrices, more precisely, symmetric positive semi-definite matrices. As far as the author is aware of, proximal-like methods have not been applied to semi-definite programs so far. However, due to the recent interest in semi-definite programs (mainly in the field of interior-point analysis), it might be a very useful research topic to see how proximal-like methods can be extended to semi-definite programs and how these methods behave for this class of convex problems.

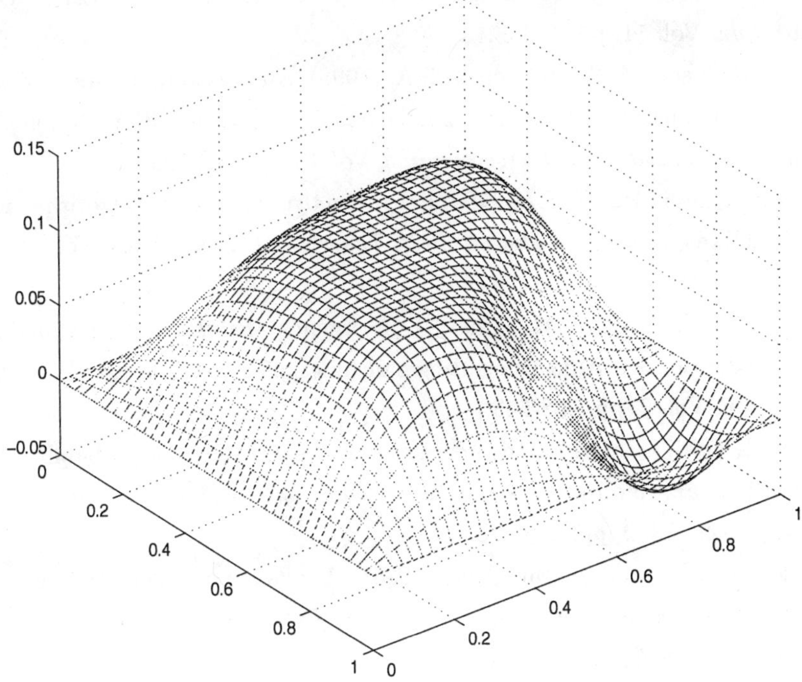

Figure 3.6   Resulting optimal state for Example 3.2

References

Auslender, A., and Haddou, M. (1995), An interior-proximal method for convex linearly constrained problems and its extension to variational inequalities, *Mathematical Programming*, Vol. 71, pp. 77-100.

Auslender, A., Teboulle, M., and Ben-Tiba, S. (1999a), Interior proximal and multiplier methods based on second order homogeneous kernels, *Mathematics of Operations Research*, Vol. 24, pp. 645-668.

Auslender, A., Teboulle, M., and Ben-Tiba, S. (1999b), A logarithmic-quadratic proximal method for variational inequalities, *Computational Optimization and Applications*, Vol. 12, pp. 31-40.

Ben-Tal, A., and Zibulevsky, M. (1997), Penalty-barrier methods for convex programming problems, *SIAM Journal on Optimization*, Vol. 7, pp. 347-366.

Bergounioux, M., Haddou, M., Hintermüller, M., and Kunisch, K. (2001): A comparison of a Moreau-Yosida-based active set strategy and interior point methods for constrained optimal control problems, *SIAM Journal on Optimization*, Vol. 11, pp. 495-521.

Censor, Y., Iusem, A.N., and Zenios, S.A. (1998), An interior point method with Bregman functions for the variational inequality problem with paramonotone functions, *Mathematical Programming*, Vol. 81, pp. 373-400.

Censor, Y., and Zenios, S.A. (1992), Proximal minimization algorithm with D-functions, *Journal of Optimization Theory and Applications*, Vol. 73, pp. 451-464.

Chen, G., and Teboulle, M. (1993), Convergence analysis of a proximal-like minimization algorithm using Bregman functions, *SIAM Journal on Optimization*, Vol. 3, pp. 538-543.

Csiszár, I. (1967), Information-type measures of difference of probability distributions and indirect observations, *Studia Scientiarium Mathematicarum Hungarica*, Vol. 2, pp. 299-318.

De Pierro, A.R., and Iusem, A.N. (1986), A relaxed version of Bregman's method for convex programming, *Journal of Optimization Theory and Applications*, Vol. 51, pp. 421-440.

Eckstein, J. (1993), Nonlinear proximal point algorithms using Bregman functions, with applications to convex programming, *Mathematics of Operations Research*, Vol. 18, pp. 202-226.

Eckstein, J. (1998), Approximate iterations in Bregman-function-based proximal algorithms. *Mathematical Programming*, Vol. 83, pp. 113-123.

Eggermont, P.P.B., Multiplicative iterative algorithms for convex programming, *Linear Algebra and its Applications*, Vol. 130, pp. 25-42.

Ferris, M.C. (1991), Finite termination of the proximal-point algorithm, *Mathematical Programming*, Vol. 50, pp. 359-366.

Güler, O. (1991), On the convergence of the proximal-point algorithm for convex minimization, *SIAM Journal on Control and Optimization*, Vol. 29, pp. 403-419.

Güler, O. (1994), Ergodic convergence in proximal point algorithms with Bregman functions, in Du, D.-Z. and Sun, J., *Advances in Optimization and Approximation*, Kluwer Academic Publishers, pp. 155-165.

Hoffman, A.J. (1952), On approximate solutions of systems of linear inequalities, *Journal of Research of the National Bureau of Standards*, Vol. 49, pp. 263-265.

Iusem, A.N., Svaiter, B.F., and Teboulle, M. (1994), Entropy-like proximal methods in convex programming. *Mathematics of Operations Research*, Vol. 19, pp. 790-814.

Iusem, A.N., and Teboulle, M. (1995), Convergence rate analysis of nonquadratic proximal methods for convex and linear programming, *Mathematics of Operations Research*, Vol. 20, pp. 657-677.

Lemaire, B. (1989), The proximal algorithm, in Penot. J.P. (ed.): *New Methods in Optimization and their Industrial Uses*, Birkhäuser-Verlag, Basel, pp. 73-87.

Lemaire, B. (1992), About the convergence of the proximal method, in *Advances in Optimization*. Lecture Notes in Economics and Mathematical Systems 382, Springer-Verlag, pp. 39-51.

Luque, F.J. (1984), Asymptotic convergence analysis of the proximal point algorithm. *SIAM Journal on Control and Optimization*, Vol. 22, pp. 277-293.

Martinet, B. (1970), Regularisation d'inéqations variationelles par approximations successives, *Revue Francaise d'Informatique et de Recherche Operationelle*, pp. 273-299.

Nesterov, Y., and Nemirovskii, A. (1994), *Interior-Point Polynomial Algorithms in Convex Programming*, SIAM, Philadelphia, PA.

Rockafellar, R.T. (1970), *Convex Analysis*. Princeton University Press.

Rockafellar, R.T. (1976), Monotone operators and the proximal point algorithm, *SIAM Journal on Control and Optimization*, Vol. 14, pp. 877-898.

Solodov, M.V., and Svaiter, B.F. (1999), A hybrid projection-proximal point algorithm, *Journal of Convex Analysis*, Vol. 6, pp. 59-70.

Solodov, M.V., and Svaiter, B.F. (2000a), Forcing strong convergence of proximal point iterations in a Hilbert space, *Mathematical Programming*, Vol. 87, pp. 189-202.

Solodov, M.V., and Svaiter, B.F. (2000b), A truly globally convergent Newton-type method for the monotone nonlinear complementarity problem, *SIAM Journal on Optimization*, Vol. 10, pp. 605-625.

Teboulle, M. (1992), Entropic proximal mappings with applications to nonlinear programming, *Mathematics of Operations Research*, Vol. 17, pp. 670-690.

Teboulle, M. (1997), Convergence of proximal-like algorithms, *SIAM Journal on Optimization* 7, pp. 1069-1083.

Tseng, P., and Bertsekas, D.P. (1993), On the convergence of the exponential multiplier method for convex programming, *Mathematical Programming*, Vol. 60, pp. 1-19.

Yamashita, N., Kanzow, C., Morimoto, T., and Fukushima, M. (2001), An infeasible interior proximal method for convex programming problems, *Journal of Nonlinear and Convex Analysis*, Vol. 2, 2001, pp. 139-156.

18   ANALYSIS OF TWO DIMENSIONAL NONCONVEX VARIATIONAL PROBLEMS

René Meziat

Departamento de Matemáticas
Universidad de Los Andes
Carrera 1 este No 18A-10
Bogotá, Colombia
rmeziat@uniandes.edu.co
and
OMEVA, Research Group on Optimization and Variational Methods
Departamento de Matemáticas
Universidad de Castilla La Mancha
13071, Ciudad Real, Spain
http://matematicas.uclm.es/omeva

Abstract: The purpose of this work is to carry out the analysis of two-dimensional scalar variational problems by the method of moments. This method is indeed shown to be useful for treating general cases in which the Lagrangian is a separable polynomial in the derivative variables. In these cases, it follows that the discretization of these problems can be reduced to a single large scale semidefinite program.

Key words: Calculus of variations, young measures, the method of moments, semidefinite programming, microstructure, non linear elasticity.

1 INTRODUCTION

The classical theory of variational calculus does not provide any satisfactory methods to analyze non-convex variational problems expressed in the form

$$\min_{u} I\left(u\right) = \int_{\Omega} f\left(\vec{\nabla} u\left(x,y\right)\right) dx\, dy \qquad s.t. \qquad u|_{\partial\Omega} = g \qquad (1.1)$$

where f is a coercive non-convex Lagrangian function, and u the family of all admissible scalar functions defined on Ω. For a review on recent methods in the calculus of variations see Dacorogna (1989).

In order to analyze this class of non-convex variational problems, we must appeal to a new formulation with respect to *Young measures*. For these problems we introduce the *generalized functional*

$$\widetilde{I}\left(\nu\right) = \int_{\Omega} \left(\int_{R^2} f\left(s,t\right) d\mu_{x,y}\left(s,t\right)\right) dx\, dy$$
$$with \quad \vec{\nabla} u\left(x,y\right) = \int_{R^2} \left(s,t\right) d\mu_{x,y}\left(s,t\right) \qquad (1.2)$$
$$and \quad the \quad boundary \quad condition \quad u|_{\partial\Omega} = g$$

where

$$\nu = \{\mu_{x,y} : \left(x,y\right) \in \Omega\}$$

is a parametrized family of probability measures supported on the plane. Each one of these sets ν is called a Young measure, hence the generalized functional $\widetilde{I}$ is defined in the family of all Young measures ν.

Young measures theory predicts that the generalized functional (1.2) has a Young measure minimizer

$$\nu^* = \{\bar{\mu}_{x,y} : \left(x,y\right) \in \Omega\}$$

which provides information about the limit behavior of the minimizing sequences of the functional I given in (1.1). Thus,

$$\vec{\nabla} u_n\left(x,y\right) \rightarrow d\bar{\mu}_{x,y} \qquad (1.3)$$

in measure, whenever u_n is a minimizing sequence for the functional I. One immediate conclusion is that functional I has a unique minimizer if and only if the generalized functional $\widetilde{I}$ has a minimizer ν^* composed only of Dirac measures. In this case

$$\bar{\mu}_{x,y} = \delta_{\vec{\nabla} \bar{u}(x,y)}$$

where $\bar{u}$ is a minimizer for I. For a thorough study on Young measures and calculus of variations see Pedregal (1997).

In the present work, we will study the particular case in which the Lagrangian function f takes the polynomial separable form

$$f(s,t) = \sum_{i=0}^{2n} a_i\, s^i + \sum_{j=0}^{2r} b_j\, t^j \quad with \quad c_{2n},\, b_{2r} > 0\,. \tag{1.4}$$

Under this assumption, Problem (1.2) may be reduced to a single semidefinite program using the theory of the classical problem of moments and elementary convex analysis.

The present paper is organized as follows: in Section 2 we will see a short review on the use of the Method of Moments for treating one dimensional non convex variational problems. In Section 3 we will see how the Method of Moments is used for analyzing the convex envelope of one-dimensional algebraic polynomials. Section 4 describes the general analysis of two dimensional non convex variational problems by the Method of Moments. Section 5 shows how transform the analytical formulation into a particular mathematical program. In Section 6 we will see some examples in detail and finally Section 7 gives some comments about the interplay of this work with pure and applied mathematics.

2   THE METHOD OF MOMENTS

The generalized formulation in Young measures is valid for one-dimensional non convex variational problems like

$$\min_u \int_0^1 f\left(u'\left(x\right)\right)\, dx \qquad s.t. \qquad u\left(0\right) = 0,\, u\left(1\right) = \alpha\,.$$

Assuming that f is a one-dimensional polynomial in the form

$$f\left(t\right) = \sum_{k=0}^{2n} c_k\, t^k \qquad c_{2n} > 0 \tag{2.1}$$

the generalized problem in Young measures

$$\begin{aligned} \min_\nu \widetilde{I}\left(\nu\right) &= \int_0^1 \int_R f\left(\lambda\right)\, d\mu_x\left(\lambda\right)\, dx \\ with \quad u'\left(x\right) &= \int_R \lambda\, d\mu_x\left(\lambda\right) \\ u\left(0\right) &= 0, \quad u\left(1\right) = \alpha \end{aligned} \tag{2.2}$$

can be recast as

$$\begin{aligned} \min_m \int_0^1 \sum_{k=0}^{2n} c_k\, m_k\left(x\right)\, dx \\ with \quad u'\left(x\right) = m_1\left(x\right) \\ u\left(0\right) = 0, \quad u\left(1\right) = \alpha \end{aligned} \tag{2.3}$$

where $m_k(x)$ are the algebraic moments of the parametrized measures μ_x which form the one-dimensional Young measure

$$\nu = \{\mu_x : 0 \leq x \leq 1\} .$$

The theory of moments provides a good characterization for the algebraic moments of positive measures supported on the real line. Therefore we can study the one-dimensional generalized problem (2.2) by solving the optimization problem (2.3). Here we will study two-dimensional problems defined by separable polynomials in the form (1.4).

For a short review on applications of the method of moments for one-dimensional non-convex variational problems see Meziat et al (2001). The essential facts about the characterization of one-dimensional algebraic moments are exposed in Akhiezer and Krein (1962). The difficulties about the characterization of two-dimensional algebraic moments are explained in Berg et al (1979).

3   CONVEX ENVELOPES

Given a one-dimensional polynomial (2.1), its convex envelope may be defined as

$$f_c(t) = \min_{\mu} \int_R f(\lambda) \, d\mu(\lambda) \tag{3.1}$$

where μ represents the family of all probability measures with mean t. In this approach, every probability measure represents a convex combination of points on the real line. Therefore, the measure

$$\bar{\mu} = \lambda_1 \delta_{t_1} + \lambda_2 \delta_{t_2} \tag{3.2}$$

which solves (3.1), represents the convex combination which satisfies

$$\lambda_1 (t_1, f(t_1)) + \lambda_2 (t_2, f(t_2)) = (t, f_c(t)) .$$

From this point of view, it is clear that optimal measure $\bar{\mu}$ has a very precise geometric meaning. Here we have assumed that $\bar{\mu}$ is supported in two points at the most, because of Caratheodory's theorem in convex analysis.

Since f is a polynomial function in the form (2.1), every integral in (3.1) can be written as

$$\sum_{k=0}^{2n} c_k \, m_k$$

where values $m_0, \ldots, m_{2n}$ are the algebraic moments of measure μ. So we can express the convex envelope of f using the next semidefinite program

$$f_c(t) = c_0 + c_1 t + \min_{m_k} \sum_{k=2}^{2n} c_k\, m_k \quad s.t.$$

$$\begin{pmatrix} 1 & t & m_2 & \cdots & m_n \\ t & m_2 & m_3 & \cdots & m_{n+1} \\ m_2 & m_3 & & \cdots & m_{n+2} \\ & & \cdots & & \\ m_n & m_{n+1} & m_{n+2} & \cdots & m_{2n} \end{pmatrix} \succeq 0 \qquad (3.3)$$

where we have used the classical representation of one-dimensional algebraic moments: *The convex cone of positive definite Hankel matrices $H = (m_{k+l})_{k,l=0}^{n}$ is the interior of the convex cone of algebraic moments $(m_0, \ldots, m_{2n})$ of positive measures supported on the real line.* For more details we refer the reader to Akhiezer and Krein (1962).

By using an elementary algebraic procedure, we can obtain the optimal measure $\bar{\mu}$ for problem (3.1) from the optimal values $\bar{m}_2, \ldots, \bar{m}_{2n}$ of the semidefinite program (3.3). Indeed, if $\bar{m}_2 = t^2$, take

$$\lambda_1 = 1, \ \lambda_2 = 0, \ t_1 = t_2 = t$$

so the optimal measure $\bar{\mu}$ is equal to the Dirac measure

$$\bar{\mu} = \delta_t.$$

Otherwise, take t_1 and t_2 as the roots of the polynomial

$$P(x) = \begin{vmatrix} 1 & t & \bar{m}_2 \\ t & \bar{m}_2 & \bar{m}_3 \\ 1 & x & x^2 \end{vmatrix}$$

and denote by λ_1, λ_2 the quantities

$$\lambda_1 = \frac{t_2 - t}{t_2 - t_1} \qquad \lambda_2 = \frac{t - t_1}{t_2 - t_1}$$

where $t_1 < t < t_2$. Using these values in the expression (3.2), we obtain the optimal measure $\bar{\mu}$. It is remarkable that only three moments are needed for recovering the optimal measure $\bar{\mu}$. Finally, we conclude that Problem (3.1) and Problem (3.3) are equivalent. For additional details see Pedregal et al (2003)

When f is a two-dimensional separable polynomial with the form (1.4), its convex envelope is defined as

$$f_c(s,t) = \min_{\mu} \int_{R^2} f(\sigma, \gamma) \, d\mu(\sigma, \gamma) \qquad (3.4)$$

where μ represents the family of all probability measures supported in the plane satisfying

$$(s,t) = \int_{R^2} (\sigma, \gamma) \, d\mu(\sigma, \gamma).$$

Note that it is analogous to the definition of convex envelopes for one-dimensional functions.

However, in order to estimate the convex envelope of the separable polynomial f, we must use another well known result of convex analysis: *the convex envelope of a separable function is the sum of the convex envelopes of its components.* See Dacorogna (1989). From this result and the explanation on convex envelopes of one-dimensional polynomials given above, one observes that the convex envelope of f is given by the semidefinite program

$$f_c(s,t) = a_0 + b_0 + a_1 s + b_1 t + \min_{m_i, p_j} \sum_{i=2}^{2n} a_i m_i + \sum_{j=2}^{2r} b_j p_j \quad s.t$$

$$\begin{pmatrix} 1 & s & m_2 & \cdots & m_n \\ s & m_2 & m_3 & \cdots & m_{n+1} \\ m_2 & m_3 & & \cdots & m_{n+2} \\ & & \cdots & & \\ m_n & m_{n+1} & m_{n+2} & \cdots & m_{2n} \end{pmatrix} \succeq 0$$

$$\begin{pmatrix} 1 & t & p_2 & \cdots & p_r \\ t & p_2 & p_3 & \cdots & p_{r+1} \\ p_2 & p_3 & & \cdots & p_{r+2} \\ & & \cdots & & \\ p_r & p_{r+1} & p_{r+2} & \cdots & p_{2r} \end{pmatrix} \succeq 0.$$

$$(3.5)$$

The optimal values $\bar{m}_2, \ldots, \bar{m}_{2n}, \bar{p}_2, \ldots, \bar{p}_{2r}$ for problem (3.5) allow us to determine the optimal probability measure $\bar{\mu}$ which satisfies (3.4).

From a practical point of view, $\bar{\mu}$ is the direct product of two independent one-dimensional distributions $\bar{\mu}_X$ and $\bar{\mu}_Y$, so we have

$$\bar{\mu} = \bar{\mu}_X \times \bar{\mu}_Y$$

where $\bar{\mu}_X$ represents the convex envelope of the first polynomial

$$\sum_{i=0}^{2n} a_i s^i$$

in (1.4) and respectively, $\bar{\mu}_Y$ represents the convex envelope of the second polynomial

$$\sum_{j=0}^{2r} b_j t^j$$

in (1.4). Thus, marginal distributions $\bar{\mu}_X$ and $\bar{\mu}_Y$ are obtained from values $s, \bar{m}_2, \bar{m}_3$ and $t, \bar{p}_2, \bar{p}_3$ respectively, in the same way that we did for the one-dimensional case. In other words, there is no essential difference between the one-dimensional polynomial case (2.1) and the two-dimensional separable polynomial case (1.4). Finally, it is very important to note that the optimal measure $\bar{\mu} = \bar{\mu}_X \times \bar{\mu}_Y$ determines the convex combination which defines the convex envelope of the separable polynomial f at the point (s, t).

4   PROBLEM ANALYSIS

Our concern here is the analysis of non-convex variational problems like

$$\min_u I(u) = \int_\Omega f\left(\vec{\nabla} u(x, y)\right) dx\, dy \qquad s.t. \qquad u|_{\partial\Omega} = g. \qquad (4.1)$$

We will study the case where f is a two-dimensional separable polynomial in the general form (1.4). We first notice that direct discretization of functional I in (4.1) provides a non-convex optimization problem which is not particularly adequate to be solved by standard numerical optimization software. The reason behind that is the lack of convexity on f, which can cause the search algorithm to stop at some wrong local minima instead of providing the right global minima for the functional I. In addition, we must consider the possibility that I lacks minimizers on the space of admissible functions. Normally, admissible functions belong to the Sobolev space $W_0^{1,p}(\Omega) + g$ where index p depends on the integrand function f in (4.1).

To overcome this difficulty we study the generalized problem

$$\begin{aligned} \min_\nu \widetilde{I}(\nu) &= \int_\Omega \left(\int_{R^2} f(s, t)\, d\mu_{x,y}(s, t)\right) dx\, dy \\ with \quad \vec{\nabla} u(x, y) &= \int_{R^2} (s, t)\, d\mu_{x,y}(s, t) \\ and \quad the \quad boundary \quad condition \quad & u|_{\partial\Omega} = g \end{aligned} \qquad (4.2)$$

whose solution in Young measures provides information about the minimizers of the original functional I. By using the separable polynomial structure of f, we can transform the generalized functional in (4.2) into the functional

$$J(m,p) = \int_\Omega \left(\sum_{i=0}^{2n} a_i m_i(x,y) + \sum_{j=0}^{2r} b_j p_j(x,y) \right) dx\, dy$$

where $m = (m_i(x,y))_{i=0}^{2n}$ and $p = (p_j(x,y))_{j=0}^{2r}$ represent the algebraic moments of the parametrized measures $\mu_{x,y}$ in the Young measure ν. In this way, we must solve the optimization problem

$$\min_{m,p} J(m,p) = \int_\Omega \left(\sum_{i=0}^{2n} a_i m_i(x,y) + \sum_{j=0}^{2r} b_j p_j(x,y) \right) dx\, dy$$
$$with \quad \vec{\nabla} u(x,y) = (m_1(x,y), p_1(x,y)) \tag{4.3}$$
$$and \quad the \quad boundary \quad condition \quad u|_{\partial\Omega} = g$$

where the new sets of variables m and p must be characterized as the algebraic moments of one-dimensional probability measure. In order to do so, we impose the linear matrix inequalities

$$\begin{pmatrix} 1 & m_1(x,y) & \cdots & m_n(x,y) \\ m_1(x,y) & m_2(x,y) & \cdots & m_{n+1}(x,y) \\ m_2(x,y) & m_3(x,y) & \cdots & m_{n+2}(x,y) \\ & & \cdots & \\ m_n(x,y) & m_{n+1}(x,y) & \cdots & m_{2n}(x,y) \end{pmatrix} \succeq 0$$
$$\begin{pmatrix} 1 & p_1(x,y) & \cdots & p_r(x,y) \\ p_1(x,y) & p_2(x,y) & \cdots & p_{r+1}(x,y) \\ p_2(x,y) & p_3(x,y) & \cdots & p_{r+2}(x,y) \\ & & \cdots & \\ p_r(x,y) & p_{r+1}(x,y) & \cdots & p_{2r}(x,y) \end{pmatrix} \succeq 0 \tag{4.4}$$

for every point $(x,y) \in \Omega$. After an appropriate discretization, this problem can be posed as a single semidefinite program. See Vandenverghe and Boyd (1996) and Boyd et al (1994) for an introduction to semidefinite programming.

5   DISCRETE AND FINITE MODEL

Here we will transform the optimization problem (4.3) subject to the constraints (4.4), into an equivalent discrete mathematical program. First, we take a finite set of N points on the domain Ω indexed by k, that is

$$(x_k, y_k) \in \Omega \quad for \quad k = 1, \ldots, N. \tag{5.1}$$

Next, for every discrete point (x_k, y_k) we take the algebraic moments

$$(m_i\,(x_k, y_k))_{i=1}^{2n} \qquad (p_j\,(x_k, y_k))_{j=1}^{2r} \tag{5.2}$$

of the respective parametrized measure μ_{x_k, y_k}. Using the $2N \times (n + r)$ variables listed in (5.2), we can express the functional J in the discrete form:

$$J_d\,(m, p) = \sum_{k=0}^{N} \left(\sum_{i=0}^{2n} a_i m_i\,(x_k, y_k) + \sum_{j=0}^{2r} b_j p_j\,(x_k, y_k) \right) \Delta x_k\, \Delta y_k\,. \tag{5.3}$$

The constraints (4.4) form a set of linear matrix inequalities for every point in Ω, hence they should keep the same for every point (x_k, y_k) in the mesh (5.1). So we have a set of $2N$ linear matrix inequalities expressed as

$$[m_{i+j}\,(x_k, y_k)]_{i,j=0}^{n} \succeq 0 \qquad [p_{i+j}\,(x_k, y_k)]_{i,j=0}^{r} \succeq 0 \tag{5.4}$$

where $m_0 = 1$ and $p_0 = 1$ for every $k = 1, \dots, N$.

In order to impose the boundary conditions

$$u\big|_{\partial\Omega} = g$$

and the constraints

$$\vec{\nabla} u\,(x, y) = (m_1\,(x, y)\,, p_1\,(x, y)) \tag{5.5}$$

in (4.3), we use the following fact: Given any Jordan curve $\mathcal{C}$ inside the domain Ω, the restriction (5.5) implies

$$\int_{\mathcal{C}} (m_1\,dx + p_1\,dy) = u\,(x_f, y_f) - u\,(x_0, y_0)$$

where (x_0, y_0) and (x_f, y_f) are two endpoints of curve $\mathcal{C}$.

We shall select a finite collection of M curves $\mathcal{C}_l$ with $l = 1, \dots, M$ which, in some sense, *sweep* the whole domain Ω. It will suffice that each point (x_k, y_k) on the mesh belongs to at least one curve $\mathcal{C}_l$. In order to impose the boundary conditions in (4.3), every curve $\mathcal{C}_l$ must link two boundary points of Ω. So we obtain a new set of M constraints in the form

$$\int_{\mathcal{C}_l} (m_1\,dx + p_1\,dy) = g\,(x_f^l, y_f^l) - g\,(x_0^l, y_0^l) \tag{5.6}$$

which can be incorporated as linear equalities in the discrete model.

We can see that optimization problem (4.3) can be transformed into a single semidefinite program after discretization. Note that objective function J_d in (5.3) is a linear function of the variables in (5.2). Those variables are restricted by the set of $2N$ linear matrix inequalities given in (5.4) and the set of M linear equations given in (5.6). Thus, we have obtained a very large single semidefinite program.

6 EXAMPLES

To illustrate the method proposed in this work, we will analyze the non-convex variational problem

$$\min_u \int_{[-1,1]^2} \left\{ \left(1 - \left(\frac{\partial u}{\partial x} \right)^2 \right)^2 + \left(\frac{\partial u}{\partial y} \right)^2 \right\} \, dx \, dy$$

under the following boundary conditions

$$a) \quad g(x,y) = 0 \qquad b) \quad g(x,y) = 1 - |x| \qquad c) \quad g(x,y) = x + 1 \,.$$

The corresponding generalized problem has the form

$$\min_\nu \int_{[-1,1]^2} \left\{ \int_{R^2} \left((1 - \sigma^2)^2 + \gamma^2 \right) d\mu_{x,y}(\sigma, \gamma) \right\} \, dx \, dy$$
$$under \quad the \quad constraint \quad \left(\frac{\partial u}{\partial x}, \frac{\partial u}{\partial y} \right) = \int_{R^2} (\sigma, \gamma) \, d\mu_{x,y}(\sigma, \gamma)$$
$$and \quad the \quad boundary \quad conditions \ u|_{\partial\Omega} = g(x) \quad with \quad \Omega = [-1, 1]^2$$

$$(6.1)$$

which transforms into the optimization problem

$$\min_{m,p} \int_{[-1,1]^2} \left\{ 1 - 2m_2(x,y) + m_4(x,y) + p_2(x,y) \right\} \, dx \, dy$$
$$under \quad the \quad constraints$$
$$\left(\frac{\partial u}{\partial x}, \frac{\partial u}{\partial y} \right) = (m_1(x,y), p_1(x,y)), \ [m_{i+j}(x,y)]_{i,j=0}^2 \succeq 0, \ [p_{i+j}(x,y)]_{i,j=0}^1 \succeq 0$$
$$and \quad the \quad boundary \quad conditions \ u|_{\partial\Omega} = g(x) \quad with \quad \Omega = [-1, 1]^2 \,.$$

$$(6.2)$$

In order to perform the discretization of this problem, we use the straight lines with slope 1 crossing the square $[-1,1]^2$. With them we can impose the boundary conditions in the finite model. After solving the discrete model, not to be exposed here, we obtain the optimal moments for (6.2), and the Young measure solution for the generalized problem (6.1).

For the three cases studied, we obtain the following optimal parametrized measures

$$a)\quad \bar{\mu}_{x,y} = \tfrac{1}{2}\delta_{(-1,0)} + \tfrac{1}{2}\delta_{(1,0)} \quad \forall\, x,y \in [-1,1]^2$$

$$b)\quad \bar{\mu}_{x,y} = \begin{cases} \delta_{(1,0)} & if \quad -1 \le x \le 0 \\ \delta_{(-1,0)} & if \quad\ \ 0 \le x \le 1 \end{cases}$$

$$c)\quad \bar{\mu}_{x,y} = \delta_{(1,0)} \quad\quad \forall\, x,y \in [-1,1]^2$$

hence we infer that Problem a) does not have minimizers, Problem b) has the minimizer $\bar{u}(x,y) = 1 - |x|$ and Problem c) has the minimizer $\bar{u}(x,y) = x + 1$. Although Problem a) lacks minimizers, the optimal Young measure obtained gives enough information about the limit behavior of the minimizing sequences. Indeed, if u_n is an arbitrary minimizing sequence for Problem a) we have

$$\vec{\nabla} u_n(x,y) \to (\pm 1, 0)$$

in measure, where gradient $(1,0)$ is preferred with 50% of possibilities and gradient $(-1,0)$ is preferred with the remaining 50% of possibilities in the minimizing process, for every point $(x,y) \in [-1,1]^2$.

7   CONCLUDING REMARKS

The major contribution of this work is that it settles the way for studying non-convex variational problems of the form (4.1). Indeed, the direct method of the calculus of variations does not provide any answer for them if the integrand f is not convex. See Dacorogna (1989). In addition, in this work we propose a method for solving generalized problems like (4.2) when the integrand f has the separable form described in (1.4). In fact to the best knowledge of the author, do not exist other proposals to analyze this kind of generalized problems in two dimensions.

An important remark about this work is that we have reduced the original non convex variational problem (4.1) to the optimization problem (4.3). In addition, the reader should note that Problem (4.3) is a convex problem because the objective function is linear and the feasible set convex. That is a remarkable qualitative difference since numerical implementation of problem (4.1) may provide wrong answers when the search algorithm stops in local minima, whereas a good implementation of Problem (4.3) should yield the global minima of the problem.

Since we can pose Problem (4.3) as a single large scale semidefinite program, we can use existing software for solving non convex variational problems in the form (4.1) whenever the integrand f has the separable form (1.4). This situation prompts further research on large scale semidefinite programming specially suited for generalized problems in the form (4.3).

We should also stress that, although the original non convex variational problem (4.1) may not have a solution, its new formulation (4.3) always has one. In general, this solution is unique and provides information about the existence of minimizers for problem (4.1). If Problem (4.1) has a unique minimizer $\overline{u}(x, y)$, then Problem (4.3) provides the moments of the Dirac measures

$$\left\{ \delta_{\overrightarrow{\nabla}\,\overline{u}(x,y)} : (x, y) \in \Omega \right\}.$$

Moreover, if Problem (4.3) provides the moments of a family of Dirac measures like

$$\left\{ \delta_{\overrightarrow{F}(x,y)} : (x, y) \in \Omega \right\}$$

then problem (4.1) has a unique minimizer $\overline{u}(x, y)$ which satisfies $\overrightarrow{\nabla}\overline{u}(x, y) = \overrightarrow{F}(x, y)$.

One fundamental question we feel important to raise is whether the discrete model (5.3) is an adequate representation of the convex problem (4.3). From an analytical point of view, we need to find a particular qualitative feature on the solution of Problem (4.3), that is the Dirac mass condition on all optimal measures. So we can hope that even rough numerical models can provide us with the right qualitative answer about the existence of minimizers for the non convex variational problem (4.1). This has actually been observed in many numerical experiments.

It is also extremely remarkable that we can get a numerical answer to an analytical question. Indeed, we are clarifying the existence of minimizers of one particular variational problem from a numerical procedure. This point is crucial because no analytical method exists which allows to solve this question when we are coping with general non convex variational problems.

On the other hand, we really need a fine numerical model because the solution of problem (4.3) contains the information about the oscillatory behavior of minimizing sequences of the non convex problem (4.1). In those cases where Problem (4.1) lacks solution, minimizing sequences show similar oscillatory behavior linked with important features in the physical realm. For example, in elasticity models of solid mechanics such behavior represents the distribution of several solid phases inside some particular body. This information provides the microstructure of the crystalline net of the material.

To discover such phenomena we need a good representation of the optimal Young measure of the generalized problem (4.2), which in turns, is embedded into the solution of the convex formulation in moments (4.3). In conclusion, we feel that it is important to devise a good numerical treatment of problem (4.3) by solving a semidefinite model like (5.3).

Acknowledgments

Author wishes to thank Serge Prudhomme and Juan C. Vera for their comments and suggestions on this paper.

References

Dacorogna, B. (1989), Direct Methods in the Calculus of Variations, Springer Verlag.

Pedregal, P. (1997), Parametrized Measures and Variational Principles, Birkhäuser.

Meziat, R., J.J. Egozcue and P. Pedregal (2001), The method of moments for non-convex variational problems, in Advances in Convex Analysis and Global Optimization, Kluwer Nonconvex Optimization and its Applications Series, vol 54, 371-382.

Akhiezer, N. and M. Krein (1962), Some Questions in the Theory of Moments, AMS.

Berg, C., J. Christensen and C. Jensen (1979), A remark on the multidimensional moment problem, Math. Ann. 243, 163-169.

Pedregal, P., R. Meziat and J.J. Egozcue (2003), From a non linear, non convex variational problem to a linear, convex formulation, accepted in Journal of Applied Mathematics, Springer Verlag.

Vandenverghe, L. and S. Boyd (1996), *Semidefinite programming*, SIAM Review, vol. 38, no. 1.

Boyd, S., L. El Ghaoui, E. Feron and V. Balakrishnan (1994), *Linear matrix inequalities in system and control theory*, SIAM, studies in applied mathematics series, vol. 15.

19 STABILITY OF EQUILIBRIUM POINTS OF PROJECTED DYNAMICAL SYSTEMS

Mauro Passacantando

Department of Applied Mathematics, University of Pisa
Via Bonanno 25/b, 56126 Pisa, Italy
e-mail: passacantando@dma.unipi.it

Abstract: We present a survey of the main results about asymptotic stability, exponential stability and monotone attractors of locally and globally projected dynamical systems, whose stationary points coincide with the solutions of a corresponding variational inequality. In particular, we show that the global monotone attractors of locally projected dynamical systems are characterized by the solutions of a corresponding Minty variational inequality. Finally, we discuss two special cases: when the domain is a polyhedron, the stability analysis for a locally projected dynamical system, at regular solutions to the associated variational inequality, is reduced to one of a standard dynamical system of lower dimension; when the vector field is linear, some global stability results, for locally and globally projected dynamical systems, are proved if the matrix is positive definite (or strictly copositive when the domain is a convex cone).

Key words: Variational inequality, projected dynamical system, equilibrium solution, stability analysis.

1 INTRODUCTION

Equilibrium is a central concept in the study of complex and competitive systems. Examples of well-known equilibrium problems include oligopolistic market equilibrium and traffic network equilibrium problems. For these problems many variational formulations have been introduced in the last years, nevertheless their analysis is focused on the static study of the equilibrium, while it is also of interest to analyse the time evolution of adjustment processes for these equilibrium problems. Recently, two models for studying dynamic behaviour of such systems have been proposed: they are based on constrained dynamical systems involving projection operators. The first one is the so-called locally projected dynamical system (first proposed in Dupuis et al (1993)), the other is known as globally projected dynamical system (introduced in Friesz et al (1994)). The most important connection between these dynamical models and variational models lies is the possibility of characterizing stationary points for dynamical models by solutions of a particular variational inequality. The locally projected dynamical systems have had recently important applications in economics (see Dong et al (1996), Nagurney et al (1996a) and Nagurney et al (1996b)) and in the traffic networks equilibrium problems (see Nagurney et al (1997a), Nagurney et al (1997b) and Nagurney et al (1998)); the globally projected dynamical systems have been introduced as a model for describing network disequilibria (see Friesz et al (1994)) and they have been applied to neural networks for solving a class of optimization problems (see Xia et al (2000)).

The main purpose of this paper is to give, to the best of my knowledge, a survey of the main results about the stability of these two types of projected dynamical systems (see Nagurney et al (1995) and Nagurney et al (1996c), Pappalardo et al (2002) and Xia et al (2000)). The paper is organized as follows. In Section 2, we recall the definitions of locally and globally projected dynamical systems and we show the equivalence between their equilibrium points and the solutions of a suitable associated variational inequality. In Section 3, we recall some stability definitions needed throughout the paper (monotone attractor, asymptotic stability, exponential stability); we show some stability results for a locally projected dynamical system under monotonicity assumptions on the vector field; we state the correspondence between the global monotone attractors of a locally projected dynamical system and the solutions to a related

Minty variational inequality; we give asymptotic and exponential stability results for a globally projected dynamical system when the jacobian matrix of the vector field is symmetric; finally we provide a stability result, for both the projected dynamical systems, similar to the nonlinear sink theorem for standard dynamical systems. Section 4 is dedicated to the stability analysis in two special cases. When the domain is a convex polyhedron, the stability of a locally projected dynamical system, at regular solutions to the associated variational inequality, is essentially the same as that of a standard dynamical system of lower dimension, the so-called minimal face flow. When the vector field is linear and the matrix is positive definite (or strictly copositive if the domain is a convex cone), the global exponential stability for locally and globally projected dynamical systems is proved. Finally, some suggestions for future research in the linear case are described.

2 VARIATIONAL AND DYNAMICAL MODELS

Throughout this paper K denotes a closed convex subset of R^n and $F : R^n \to R^n$ a vector field. The following monotonicity definitions will be needed for our later discussions.

Definition 2.1 *F is said to be locally pseudomonotone at x^* if there is a neighborhood $N(x^*)$ of x^* such that*

$$\langle F(x^*), x - x^* \rangle \geq 0, \implies \langle F(x), x - x^* \rangle \geq 0, \quad \forall \, x \in N(x^*),$$

where $\langle \cdot, \cdot \rangle$ denotes the inner product in R^n;
* F is said to be pseudomonotone on K if*

$$\langle F(y), x - y \rangle \geq 0, \implies \langle F(x), x - y \rangle \geq 0, \quad \forall \, x, y \in K;$$

F is said to be locally strictly pseudomonotone at x^* if there is a neighborhood $N(x^*)$ of x^* such that

$$\langle F(x^*), x - x^* \rangle \geq 0, \implies \langle F(x), x - x^* \rangle > 0, \quad \forall \, x \in N(x^*);$$

F is said to be strictly pseudomonotone on K if

$$\langle F(y), x - y \rangle \geq 0, \implies \langle F(x), x - y \rangle > 0, \quad \forall \, x, y \in K;$$

F is said to be monotone on K if

$$\langle F(x) - F(y), x - y \rangle \geq 0, \quad \forall \, x, y \in K;$$

F is said to be locally strongly monotone at x^* if there is a neighborhood $N(x^*)$ of x^* and $\eta > 0$ such that

$$\langle F(x) - F(x^*), x - x^* \rangle \geq \eta \|x - x^*\|^2, \quad \forall\, x \in N(x^*);$$

F is said to be strongly monotone on K if there is $\eta > 0$ such that

$$\langle F(x) - F(y), x - y \rangle \geq \eta \|x - y\|^2, \quad \forall\, x, y \in K.$$

We recall that a variational inequality of Stampacchia-type SVI(F,K) consists in determining a vector $x^* \in K$, such that

$$\langle F(x^*), x - x^* \rangle \geq 0, \quad \forall\, x \in K;$$

the associated Minty variational inequality MVI(F,K) consists in finding a vector $x^* \in K$, such that

$$\langle F(x), x^* - x \rangle \leq 0, \quad \forall\, x \in K.$$

It is well-known that if F is continuous on K, then each solution to MVI(F,K) is a solution to SVI(F,K); whereas if F is pseudomonotone on K, then each solution to SVI(F,K) is also a solution to MVI(F,K).

The first dynamical model we consider in this context is the so-called locally projected dynamical system (first introduced in Dupuis et al (1993)), denoted by LPDS(F,K), which is defined by the following ordinary differential equation

$$\dot{x} = P_{T_K(x)}(-F(x)),$$

where $T_K(x)$ denotes the tangent cone to K at x, P_S denotes the usual projection on a closed convex subset S:

$$P_S(x) = \arg\min_{z \in S} \|x - z\|,$$

and $\|\cdot\|$ is the euclidean norm on R^n.

We remark that if the vector field F is continuous on K, then the right-hand side of LPDS(F,K) is also continuous on the relative interior of K, and it can be discontinuous elsewhere. By a solution to the LPDS(F,K) we mean an absolutely continuous function $x : [0, +\infty) \to K$ such that

$$\dot{x}(t) = P_{T_K(x(t))}(-F(x(t))),$$

for all $t \geq 0$ save on a set of Lebesgue measure zero.

We are interested in the equilibrium (or stationary) points of LPDS(F,K), i.e. the vectors $x^* \in K$ such that

$$P_{T_K(x^*)}(-F(x^*)) = 0;$$

that is once a solution of the LPDS(F,K) is at x^*, it will remain at x^* for all future times. The first connection between the locally projected dynamical systems and the variational inequalities is that the stationary points of the LPDS(F,K) coincide with the solutions of SVI(F,K). This equivalence allows us to carry out a stability analysis of a solution to SVI(F,K) respect to the dynamical system LPDS(F,K), as we will see in Section 3.

Since the main purpose of this paper is to analyse the stability of the equilibrium points, we confine ourself to cite only the following result about the existence, uniqueness and continuous dependence on the initial value of solutions to LPDS(F,K), in the special case where K is a polyhedron (see Nagurney et al (1996c)).

Theorem 2.1 *Let K be a polyhedron. If there exists a constant $M > 0$ such that*

$$\|F(x)\| \leq M\,(1 + \|x\|), \quad \forall\, x \in K,$$

and also

$$\langle F(y) - F(x), x - y \rangle \leq M\,\|x - y\|^2, \quad \forall\, x, y \in K,$$

then

- *for any $x_0 \in K$, there exists a unique solution $x_0(t)$ to LPDS(F,K), such that $x_0(0) = x_0$;*

- *if $x_n \to x_0$ as $n \to +\infty$, then $x_n(t)$ converges to $x_0(t)$ uniformly on every compact set of $[0, +\infty)$.*

We note that Lipschitz continuity of F on K is a sufficient condition for the properties stated in Theorem 2.1.

The second dynamical model we consider is the so-called globally projected dynamical system, denoted by GPDS(F,K,α), which is defined as the following ordinary differential equation

$$\dot{x} = P_K(x - \alpha\,F(x)) - x,$$

where α is a positive constant. If the vector field F is continuous on K, then the right-hand side of GPDS(F,K,α) is also continuous on K, but it can be different from $-\alpha\, F(x)$ even if x is an interior point to K. Hence the solutions of GPDS(F,K,α) and LPDS(F,K) are different in general.

The following result about global existence and uniqueness of solutions to GPDS(F,K,α) follows on the ordinary differential equations theory (see Xia et al (2000)).

Theorem 2.2 *If F is locally Lipschitz continuous and there exists a constant $M > 0$ such that*

$$\|F(x)\| \leq M\,(1 + \|x\|), \qquad \forall\, x \in K,$$

then for any $x_0 \in K$ there exists a unique solution $x_0(t)$ for GPDS(F,K,α), such that $x_0(0) = x_0$, that is defined for all $t \in R$.

It can be proved that, as in the case of LPDS(F,K), a solution to GPDS(F,K,α) starting from a point in K has to remain in K (see Xia et al (2000) and Pappalardo et al (2002)).

The equilibrium (or stationary) points of GPDS(F,K,α) are naturally defined as the vectors $x^* \in K$ such that

$$P_K(x^* - \alpha\, F(x^*)) = x^*.$$

It is easy to check that they also coincide with the solutions to SVI(F,K). Hence, we can analyse the stability of the solutions to SVI(F,K) also respect to the dynamical system GPDS(F,K,α).

3   STABILITY ANALYSIS

We have seen that LPDS(F,K) and GPDS(F,K,α) have the same stationary points, but, in general, their solutions are different. In this section we analyse the stability of these equilibrium points, namely we wish to know the behaviour of the solutions, of LPDS(F,K) and GPDS(F,K,α) respectively, which start near an equilibrium point. Since we are mainly focused on the stability issue, we can assume the property of existence and uniqueness of solutions to the Cauchy problems corresponding to locally and globally projected dynamical systems. In the following, $B(x^*, r)$ denotes the open ball with center x^* and radius r.

Now we recall some definitions on stability.

Definition 3.1 *Let x^* be a stationary point of LPDS(F,K) and GPDS(F,K,α).*

x^* *is called* stable *if for any* $\epsilon > 0$, *there exists* $\delta > 0$ *such that for every solution* $x(t)$, *with* $x(0) \in B(x^*, \delta) \cap K$, *one has* $x(t) \in B(x^*, \epsilon)$ *for all* $t \geq 0$;

x^* *is said* asymptotically stable *if* x^* *is stable and* $\lim_{t \to +\infty} x(t) = x^*$ *for every solution* $x(t)$, *with* $x(0) \in B(x^*, \delta) \cap K$; x^* *is said* globally asymptotically stable *if it is stable and* $\lim_{t \to +\infty} x(t) = x^*$ *for every solution* $x(t)$ *with* $x(0) \in K$.

We recall also that x^* *is called* monotone attractor *if there exists* $\delta > 0$ *such that, for every solution* $x(t)$ *with* $x(0) \in B(x^*, \delta) \cap K$, *the euclidean distance between* $x(t)$ *and* x^*, *that is* $\|x(t) - x^*\|$, *is a nonincreasing function of* t; *whereas* x^* *is said* strictly monotone attractor *if* $\|x(t) - x^*\|$ *is decreasing to zero in* t. *Moreover* x^* *is a (strictly)* global monotone attractor *if the same properties hold for any solution* $x(t)$ *such that* $x(0) \in K$.

Finally, x^* *is a* finite-time attractor *if there is* $\delta > 0$ *such that, for every solution* $x(t)$, *with* $x(0) \in B(x^*, \delta) \cap K$, *there exists some* $T < +\infty$ *such that* $x(t) = x^*$ *for all* $t \geq T$.

It is trivial to remark that the monotone attractors are stable equilibrium points, whereas the strictly monotone attractors and the finite-time attractors are asymptotically stable ones.

It is easy to check that the stability of a locally (or globally) projected dynamical system can differ from the stability of a standard dynamical system in the same vector field (see examples in Nagurney et al (1995)).

The pseudo-monotonicity property of F is directly related to the monotone attractors of LPDS(F,K), as shown in the following theorem which is a direct generalization of a result proved in Nagurney et al (1995).

Theorem 3.1 *Let* x^* *be a stationary point of LPDS(F,K). If* F *is locally (strictly) pseudomonotone at* x^*, *then* x^* *is a (strictly) monotone attractor for LPDS(F,K); if* F *is (strictly) pseudomonotone on* K, *then* x^* *is a (strictly) global monotone attractor for LPDS(F,K).*

However, the monotonicity of F is not sufficient to prove the stability for an equilibrium point of a globally projected dynamical system, as the following example shows (for the details see Xia et al (2000)).

Example 3.1 *We consider* $K = \{x \in R^3 : -10 \leq x_i \leq 10\}$, $\alpha = 1$ *and* $F(x) = Ax + b$, *where*

$$A = \begin{pmatrix} 0.1 & 0.1 & -0.5 \\ 0.1 & 0.1 & 0.5 \\ 0.5 & -0.5 & 0 \end{pmatrix}, \quad b = \begin{pmatrix} -1 \\ 1 \\ -0.5 \end{pmatrix}.$$

The affine vector field F is monotone, because A is positive semidefinite, but the unique equilibrium point of GPDS(F,K,α), i.e. $x^* = (0.5, -0.5, -2)^T$, is not stable.

In addition to the monotonicity of F, the symmetry of the jacobian matrix of F, denoted by JF, is necessary in order to achieve the asymptotic stability of a stationary point of a globally projected dynamical system (see Xia et al (2000)).

Theorem 3.2 *Let x^* be the unique stationary point of GPDS(F,K,α). If F is monotone on K and the jacobian matrix JF is symmetric on an open convex set including K, then x^* is globally asymptotically stable for GPDS(F,K,α) for any $\alpha > 0$.*

We remark that under the assumptions of Theorem 3.2, the vector field F is the gradient map of a real convex function on K.

Now we go back to the monotone attractors. When the vector filed F is continuous, there is a further connection between locally projected dynamical systems and variational inequalities: the global monotone attractors of LPDS(F,K) are equivalent to the solutions of the Minty variational inequality MVI(F,K) (see Pappalardo et al (2002)).

Theorem 3.3 *Let F be continuous on K. Then $x^* \in K$ is a global monotone attractor for LPDS(F,K) if and only if it is solution to MVI(F,K).*

We remark that Theorem 3.3 does not hold for globally projected dynamical systems: the following example (see Pappalardo et al (2002)) shows that the solutions to MVI(F,K) are not necessarily monotone attractors of GPDS(F,K,α), even if the vector field F is continuous on K.

Example 3.2 *We consider $K = R_+^2$ and the vector field*

$$F(x) = \begin{pmatrix} x_2 + ((x_1 - 1)^2 + x_2^2 - 1)^2 \\ -x_1 + ((x_1 - 1)^2 + x_2^2 - 1)^2 \end{pmatrix}.$$

The point $x^* = (0,0)^T$ is solution to MVI(F,K), but it is not a monotone attractor for GPDS(F,K,α) for any fixed $\alpha > 0$.

Another stability type we consider is the so-called exponential stability.

Definition 3.2 *Let x^* be a stationary point of LPDS(F,K) and GPDS(F,K,α). It is said exponentially stable, if the solutions starting from points close to x^* are convergent to x^* with exponential rate, that is if there is $\delta > 0$ and two constants $a > 0$ and $C > 0$ such that for every solution $x(t)$, with $x(0) \in B(x^*,\delta) \cap K$, one has*

$$\|x(t) - x^*\| \leq C \, \|x(0) - x^*\| \, e^{-at} \quad \forall \, t \geq 0; \qquad (3.1)$$

x^* is globally exponentially stable if (3.1) holds for all solutions $x(t)$ such that $x(0) \in K$.

We remark that a strictly monotone attractor is not necessarily exponentially stable and vice versa (see examples in Pappalardo et al (2002)).

The exponential stability of a stationary point of LPDS(F,K) is proved under the strong monotonicity assumption of the vector field F (see Nagurney et al (1995) and Nagurney et al (1996c)).

Theorem 3.4 *Let x^* be a stationary point of LPDS(F,K). If F is locally strongly monotone at x^*, then x^* is a strictly monotone attractor and exponentially stable for LPDS(F,K); if F is strongly monotone on K, then x^* is a strictly global monotone attractor and globally exponentially stable for LPDS(F,K).*

The strong monotonicity and the Lipschitz continuity of F give the exponential stability of a stationary point of GPDS(F,K,α), provided that α is small enough (see Pappalardo et al (2002)).

Theorem 3.5 *Let x^* be a stationary point of GPDS(F,K,α). If F is locally strongly monotone at x^* with constant η, and locally Lipschitz continuous at x^* with constant L, then x^* is a strictly monotone attractor and exponentially stable for GPDS(F,K,α), provided that $\alpha < 2\eta/L^2$; if F is strongly monotone on K with constant η and locally Lipschitz continuous on K with constant L, then x^* is a strictly global monotone attractor and globally exponentially stable for GPDS(F,K,α), provided that $\alpha < 2\eta/L^2$.*

The global exponential stability for GPDS(F,K,α), where α is small enough, has been proved even when the jacobian matrix JF is symmetric and positive definite (see Xia et al (2000)).

Theorem 3.6 *Let x^* be a stationary point of GPDS(F,K,α). If JF is symmetric and uniformly positive definite in R^n, $\|JF\|$ has an upper bound, then x^* is globally exponentially stable for GPDS(F,K,α), provide that $\alpha < 2/ \max\limits_{x \in R^n} \|JF(x)\|$.*

A further result on the exponential stability for globally projected dynamical systems can be proved when the jacobian matrix JF is not symmetric, but the domain K is bounded (see Xia et al (2000)).

Theorem 3.7 *Let x^* be a stationary point of GPDS(F,K,α). If K is bounded, F is continuously differentiable on K and JF is positive definite on K, then there exists $\alpha_0 > 0$ such that x^* is globally exponentially stable for GPDS(F,K,α) for any $\alpha < \alpha_0$.*

Now we present a stability result for LPDS(F,K) analogous to the nonlinear sink theorem for classical (i.e. not projected) dynamical systems (see Hirsch et al (1974)); we will assume a stronger condition on F, that is the jacobian matrix of F in x^* is positive definite, instead of having its eigenvalues positive real parts, but we also obtain a stronger result, i.e. x^* is a strictly monotone attractor and exponentially stable (see Pappalardo et al (2002)), instead of only exponentially stable.

Theorem 3.8 *Let x^* be a stationary point of LPDS(F,K). If F is continuously differentiable on a neighborhood of x^* and the jacobian matrix $JF(x^*)$ is positive definite, then x^* is a strictly monotone attractor and exponentially stable for LPDS(F,K).*

A result similar to Theorem 3.8 can be proved also for GPDS(F,K,α), provided that α is small enough (see Pappalardo et al (2002)).

Theorem 3.9 *Let x^* be a stationary point of GPDS(F,K,α). If F is continuously differentiable on a neighborhood of x^* and the jacobian matrix $JF(x^*)$ is positive definite, then there exists $\alpha_0 > 0$ such that x^* is a strictly monotone attractor and exponentially stable for GPDS(F,K,α) for any $\alpha < \alpha_0$.*

4 SPECIAL CASES

This section is devoted to the stability analysis in two special cases: when the domain K is a convex polyhedron and when the vector field F is linear.

We remarked that the stability for a locally projected dynamical system is generally different from that of a standard dynamical system; however, when K is a convex polyhedron, many local stability properties for LPDS(F,K) follow on that of a classical dynamical system in lower dimension, under suitable assumption on the regularity of the stationary points of LPDS(F,K) (see Nagurney et al (1995)).

We assume that K is specified by

$$K = \{x \in R^n : \ Bx \le b\},$$

where B is an $m \times n$ matrix, with rows B_i, and $b \in R^n$. We recall that a face of K is the intersection of K and a number of hyperplanes that support K, and the minimal face of K containing a point x, denoted by $E(x)$, is the intersection of all the faces of K containing x. If we denote

$$I(x) = \{i : \ B_i x = b_i\} \quad \text{and} \quad S(x) = \{x \in R^n : \ B_i x = 0, \ \forall i \in I(x)\},$$

then $E(x) = (S(x) + x) \cap K$. We assume that $S(x) = R^n$, when $I(x) = \emptyset$.

Let x^* be a stationary point of LPDS(F,K), with $\dim S(x^*) \ge 1$, then there is a $\delta > 0$ such that

$$z + x^* \in E(x^*), \quad \forall \, z \in S(x^*) \cap B(0, \delta).$$

The following ordinary differential equation defined on the subspace $S(x^*)$:

$$\dot{z} = P_{S(x^*)}(-F(z + x^*)), \quad z \in S(x^*);$$

is called the minimal face flow and it is denoted by MFF(F,K,x^*). Note that if F is locally Lipschitz continuous, then so is the right hand side of MFF(F,K,x^*), hence, for any $z_0 \in S(x^*)$, there is a unique solution $z_0(t)$ to MFF(F,K,x^*), defined in a neighborhood of 0, such that $z_0(0) = z_0$. Moreover, it is clear that $0 \in S(x^*)$ is a stationary point of MFF(F,K,x^*). The stability of $0 \in S(x^*)$ for MFF(F,K,x^*) assures the stability of x^* for LPDS(F,K), under some regularity condition on x^*, which we now introduce. Since x^* solves the variational inequality SVI(F,K), we have

$$-F(x^*) \in N_K(x^*),$$

where $N_K(x^*) = \{y \in R^n : \langle y, x - x^* \rangle \leq 0, \ \forall \, x \in K\}$ is the normal cone of K at x^*. We say that x^* is a *regular solution* of SVI(F,K) if

$$-F(x^*) \in \mathrm{ri} N_K(x^*),$$

where $\mathrm{ri} N_K(x^*)$ denotes the relative interior of $N_K(x^*)$. Note that any interior solution of SVI(F,K) is regular if we assume $\mathrm{ri}\{0\} = \{0\}$; moreover, when x^* is a solution of SVI(F,K) that lies on an (n-1)-dimensional face of K, it is regular if and only if $F(x^*) \neq 0$.

Now we show two stability results proved in Nagurney et al (1995). First, a regular solution to SVI(F,K) has the strongest stability when it is an extreme point of K.

Theorem 4.1 *If x^* is a regular solution to SVI(F,K) and $S(x^*) = \{0\}$, then it is a finite-time attractor for LPDS(F,K) and there are $\gamma > 0$ and $\delta > 0$ such that, for any solution $x(t)$, with $x(0) \in B(x^*, \delta) \cap K$*

$$\frac{d}{dt}\|x(t) - x^*\| \leq -\gamma.$$

The stability results in the general case are summarized in the following theorem.

Theorem 4.2 *If x^* is a regular solution to SVI(F,K) and $\dim S(x^*) \geq 1$, then*

- *if 0 is stable for MFF(F,K,x^*), then x^* is stable for LPDS(F,K);*

- *if 0 is asymptotically stable for MFF(F,K,x^*), then x^* is asymptotically stable for LPDS(F,K);*

- *if 0 is a finite-time attractor for MFF(F,K,x^*), then x^* is a finite-time attractor for LPDS(F,K).*

We remark that the local stability of a stationary point x^* of LPDS(F,K) depends on the combination of the regularity of x^* and the local stability of MFF(F,K,x^*). In the extreme case $S(x^*) = \{0\}$ the stability for LPDS(F,K) is implied only by the regularity condition, in the other extreme case, $S(x^*) = R^n$, MFF(F,K,x^*) is just a translation of LPDS(F,K) from x^* to the origin, hence they enjoy the same stability.

We now proceed to consider the special case of F being a linear vector field, that is $F(x) = Ax$, where A is a real n-dimensional matrix. Under this assumption, the existence and uniqueness property for the solutions to the Cauchy problems associated to LPDS(F,K) and GPDS(F,K,α) holds for any closed convex domain K (see Dupuis et al (1993) and Xia et al (2000)).

We first remark that when the matrix A is positive definite, the local stability properties obtained for LPDS(F,K), by Theorem 3.8, and for GPDS(F,K,α), by Theorem 3.9, become global properties, as shown by the following result.

Proposition 4.1 *Assume that $F(x) = Ax$ and x^* is a stationary point of LPDS(F,K) and GPDS(F,K,α). If A is a positive semidefinite matrix, then x^* is a global monotone attractor for LPDS(F,K). If A is positive definite, then x^* is the unique stationary point for LPDS(F,K) and GPDS(F,K,α); moreover there is $\alpha_0 > 0$ such that x^* is a strictly global monotone attractor and globally exponentially stable for LPDS(F,K) and for GPDS(F,K,α) for any $\alpha < \alpha_0$.*

Proof: It easy to prove that F is strongly monotone on K with constant

$$\eta = \min_{\|x\|=1} \langle Ax, x \rangle > 0;$$

therefore the variational inequality SVI(F,K) has a unique solution x^*, which is also a stationary point for LPDS(F,K) and GPDS(F,K,α). By Theorem 3.4, x^* is a strictly global monotone attractor and globally exponentially stable for LPDS(F,K). Moreover F is Lipschitz continuous on K with constant $\|A\|$. By Theorem 3.5, x^* is a strictly global monotone attractor and globally exponentially stable for GPDS(F,K,α), provided that $\alpha < 2\eta/\|A\|^2$.

In addition, if we assume that the domain K is a closed convex cone, then it is well-known that SVI(F,K) is equivalent to a generalized complementarity problem. In this case the origin is a trivial stationary point for LPDS(F,K) and GPDS(F,K,α), and we can easily prove some stability properties for it, under the weaker assumption that the matrix A is (strictly) copositive with respect to the cone K.

Proposition 4.2 *Assume that K be a closed convex cone and $F(x) = Ax$. If A is a copositive matrix with respect to K, that is $\langle x, Ax \rangle \geq 0$ for all $x \in K$, then $x^* = 0$ is a global monotone attractor for LPDS(F,K). If A is strictly copositive with respect to K, that is $\langle x, Ax \rangle > 0$ for all $x \in K$, then $x^* = 0$ is the unique stationary point for LPDS(F,K) and GPDS(F,K,α), and there exists $\alpha_0 > 0$ such that $x^* = 0$ is a strictly global monotone attractor and* *globally exponentially stable for LPDS(F,K) and for GPDS(F,K,α), for any $\alpha < \alpha_0$.*

The stability analysis in the linear case is still open; future research might be carried on to study suitable conditions on the matrix A providing stability for any closed and convex domain. Also, it might be of interest to check if, when F is an affine vector field and $K = R^n_+$, the classes of matrices needed for the study of existence and uniqueness of the solutions to the linear complementarity problem are sufficient to guarantee some stability properties.

References

Aubin, J.P. and Cellina, A. (1984), *Differential inclusions*, Springer, Berlin, Germany.

Cottle, R.K., Pang, J.-S. and Stone, R.E. (1992), *The linear complementarity problem*, Academic Press, Inc., Boston, Massachussets.

Dong, J., Nagurney, A. and Zhang, D. (1996), A projected dynamical system model of general financial equilibrium with stability analysis, *Mathematical and Computer Modelling*, Vol. 24, pp. 35-44.

Dupuis, P. (1987), Large deviations analysis of reflected diffusions and constrained stochastic approximation algorithms in convex sets, *Stochastics*, Vol. 21, pp. 63-96.

Dupuis, P. and Nagurney, A. (1993), Dynamical systems and variational inequalities, *Annals of Operations Research*, Vol. 44, pp. 9-42.

Friesz, T.L., Bernstein, D.H., Metha, N.J., Tobin, R.L. and Ganjlizadeh, S. (1994), Day-to-day dynamic network disequilibria and idealized traveler information systems, *Operations Research*, Vol. 42, pp. 1120-1136.

Hirsch, M.W. and Smale, S. (1974), *Differential equations, dynamical systems, and linear algebra*, Academic Press, New York, New York.

Karamardian, S. and Schaible, S. (1990), Seven kinds of monotone maps, *Journal of Optimization Theory and Applications*, Vol. 66, pp. 37-46.

Kinderlehrer, D. and Stampacchia, G. (1980), *An introduction to variational inequality and their application*, Academic Press, New York, New York.

Nagurney, A. (1993), *Network economics: a variational inequality approach*, Kluwer Academic Publishers, Boston, Massachusetts.

Nagurney, A. (1997), Parallel computation of variational inequalities and projected dynamical systems with applications, *Parallel computing in optimization*, Kluwer Academic Publishers, Dordrecht, Holland, pp. 343-411.

Nagurney, A. and Zhang, D. (1995), On the stability of projected dynamical systems, *Journal of Optimization Theory and Applications*, Vol. 85, pp. 97-124.

Nagurney, A. and Zhang, D. (1996a), On the stability of an adjustment process for spatial price equilibrium modeled as a projected dynamical system, *Journal of Economic Dynamics and Control*, Vol. 20, pp. 43-62.

Nagurney, A. and Zhang, D. (1996b), A stability analysis of an adjustment process for oligopolistic market equilibrium modeled as a projected dynamical system, *Optimization*, Vol. 36, pp. 263-285.

Nagurney, A. and Zhang, D. (1996c), *Projected dynamical systems and variational inequalities with applications*, Kluwer Academic Publishers, Dordrecht, Holland.

Nagurney, A. and Zhang, D. (1997a), Massively parallel computation of dynamic traffic networks modeled as projected dynamical systems, *Network optimization*, Springer, Berlin, Germany, pp. 374-396.

Nagurney, A. and Zhang, D. (1997b), A formulation, stability and computation of traffic network equilibria as projected dynamical systems, *Journal of Optimization Theory and Applications*, Vol. 93, pp. 417-444.

Nagurney, A. and Zhang, D. (1998), A massively parallel implementation of a discrete-time algorithm for the computation of dynamic elastic demand traffic problems modeled as projected dynamical systems, *Journal of Economic Dynamics and Control*, Vol. 22, pp. 1467-1485.

Pappalardo, M. and Passacantando, M. (2002), Stability for equilibrium problems: from variational inequalities to dynamical systems, *Journal of Optimization Theory and Applications*, Vol. 113, pp. 567-582.

Xia, Y.S. and Wang, J. (2000), On the stability of globally projected dynamical systems, *Journal of Optimization Theory and Applications*, Vol. 106, pp. 129-150.

20   ON A QUASI-CONSISTENT APPROXIMATIONS APPROACH TO OPTIMIZATION PROBLEMS WITH TWO NUMERICAL PRECISION PARAMETERS

Olivier Pironneau

Laboratoire d'Analyse Numerique,
Universitéde Paris 6, Paris, France
Email: pironneau@ann.jussieu.fr

and Elijah Polak

Department of Electrical Engineering and Computer Sciences,
University of California, Berkeley, CA 94720, USA
Email: polak@eecs.berkeley.edu

Abstract: We present a theory of quasi-consistent approximations that combines the theory of consistent approximations with the theory of algorithm implementation, presented in Polak (1997), and enables us to solve infinite-dimensional optimization problems whose discretization involves two precision parameters. A typical example of such a problem is an optimal control problem with initial and final value constraints. The theory includes new algorithm models that can be used with two discretization parameters. We illustrate the applicability of these algorithm models by implementing them using an approximate steepest descent method and applying it them to a simple two point boundary value optimal control problem. Our numerical results (not only the ones in this paper) show that these new algorithms perform quite well and are fairly insensitive to the selection of user-set parameters. Also, they appear to be superior to some alternative, *ad hoc* schemes.

Key words: Optimization, partial differential equations, acceleration methods

1   INTRODUCTION

The research presented in this paper was motivated by optimal control problems with ODE or PDE dynamics, whose discretized dynamics cannot be solved explicitly. For example, consider a classical optimal control problem of the form

$$\min_{u \in L_{\infty,2}^m[0,1]} f(u),\tag{1.1}$$

where $L_{\infty,2}^m[0,1]$ is a linear space whose elements are in $L_\infty^m[0,1]$, but it uses the $L_2[0,1]$ norm, $f(u) = F(x^u(1))$, and $x^u(t) \in \mathbb{R}^n$ is the solution of the two point boundary value problem

$$\dot{x}(t) = h(x(t), u(t)), \quad t \in [0,1], \quad g_0(x(0)) = 0, \quad g_1(x(1)) = 0,\tag{1.2}$$

with the usual assumptions (see, e.g., Polak (1997), Ch. 4). If we discretize the dynamics (1.2) by means of Euler's method, using a step-size $1/N$, where $N > 0$ is an integer, we get

$$x_{k+1} - x_k = \frac{1}{N} h(x_k, u_k), \quad k = 0, 1, ..., N-1, \quad g_0(x_0) = 0, \quad g_1(x_1) = 0,\tag{1.3}$$

and the discretized problem assumes the form

$$\min_{u \in L_N} f_N(u),\tag{1.4}$$

where L_N is the space of functions taking values in $\mathbb{R}^m$, which are constant of the intervals $[k/N, (k+1)/N)$, $k = 0, 1, ..., N-1$, $f_N(u) = F(x_N^u)$, and x_N^u is the solution of (1.3). Generally, (1.3) cannot be solved explicitly, and hence must be solved by some recursive technique, that we will call a "solver". Since only a finite number of iterations of the solver can be contemplated, in solving the problem (1.1) numerically, we find ourselves dealing with two approximation parameters: N, which determines the Euler integration step-size, and, say K, the number of iterations of the solver used to approximate x_N^u and hence also $f_N(u)$. If we denote by $x_{N,K}^u$ the result of K iterations of the solver in solving (1.3), we get a second level approximating problem

$$\min_{u \in L_N} f_{N,K}(u),\tag{1.5}$$

where $f_{N,K}(u) = F(x_{N,K}^u)$.

Note that while the function $f_N(u)$ is continuously differentiable under standard assumptions, depending on the solver, the function $f_{N,K}(u)$ may fail to be even continuous. Hence we may not assume that (1.5) is solvable by means of standard nonlinear programming type algorithms.

An examination of the literature shows that efficient approaches to solving infinite dimensional problems use "dynamic discretization," i. e., they start out with low discretization precision and increase the precision progressively as the computation proceeds. Referring to Polak (1997), we see that there are essentially two distinct approaches to "dynamic" discretization, both of which have been used only in situations with a single discretization parameter.

The first and oldest is that of *algorithm implementation*, see, e.g., Becker et al (2000); Betts et al (1998); Carter (1991); Carter (1993); Deuflhard (1974); Deuflhard (1975); Deuflhard (1991); Dunn et al (1983); Kelley et al (1991); Kelley et al (1999); Polak et al (1976); Mayne et al (1977); Sachs (1986). In this approach, first one develops a *conceptual* algorithm for the original problem and then a *numerical implementation* of this algorithm. In each iteration, the numerical implementation adjusts the precision with which the function and derivative values used by the conceptual algorithm are approximated so as to ensure convergence to a stationary point of the original problem. When far from a solution the approximate algorithms perform well at low precision, but as a solution is approached, the demand for increased precision progressively increases. Potentially, this approach is extendable to the case where two discretization parameters must be used.

The second, and more recent approach to dynamic discretization uses sequences of finite dimensional approximating problems, and is currently restricted to problems with a single discretization parameter. It was formalized in Polak (1993); Polak (1997), in the form of a theory of consistent approximations. Applications to optimal control are described in Schwartz (1996a); Schwartz et al (1996), and a software package for optimal control, based on consistent approximations, can be obtained from Schwartz (1996b). Within this approach, an infinite dimensional problem, $\mathbf{P}$, such as an optimal control problem with either ODE or PDE dynamics, is replaced by an infinite sequence of "nested", epi-converging finite dimensional problems $\{\mathbf{P}_N\}$. Problem $\mathbf{P}$ is then solved by a recursive scheme which applies a nonlinear programming algorithm to problem $\mathbf{P}_N$ until a test is satisfied, at which point it proceeds to solve problem $\mathbf{P}_{k+1}$, using the last point obtained for $\mathbf{P}_N$ as the initial point for the new calculation. In Polak (1997) we find a number of Algorithm Models for organizing such a calculation. The advantages of the consistent approximations approach over the algorithm implementation approach are that (i) there is a much richer set of possibilities for constructing precision refinement tests, and hence for devising one that enhances computational efficiency, and (ii) one can use unmodified nonlinear programming code libraries as subroutines, see Schwartz (1996a); Schwartz (1996b).

In this paper we develop two-tier algorithms for solving infinite dimensional optimization problems for which two discretization parameters must be used. In the first tier, this algorithm constructs an infinite sequence of epi-converging finite dimensional approximating problems $\{\mathbf{P}_N\}$, such as (1.4), and in the second tier, it uses an algorithm implementation strategy in solving each $\mathbf{P}_N$. The main task that we had to address was that of constructing efficient tests for dynamically adjusting *two* precision parameters: N and K, where K determines the precision used in the implementation strategy. The end result can be viewed as a *quasi consistent approximations* approach. As we will see in Section 3, our new algorithm performs considerably better than an ad hoc algorithm implementation scheme on the problems tested.

Finally, to make this paper reasonably self contained we include an appendix in which we define and summarize the properties of consistent approximations.

2 AN ALGORITHM MODEL

We now proceed in an abstract setting.[1] Let S be a normed space, $f : S \to \mathbb{R}$ a continuously differentiable function, and $V \subset S$. We will consider the problem

$$\mathbf{P} \qquad\qquad \min_{v \in V} f(v) \qquad\qquad (2.1)$$

and we will assume that we have an optimality function $\theta : S \to \mathbb{R}$ for $\mathbf{P}$.

Next, we will assume that $\{S_N\}_{N=N_0}^{\infty}$ is a nested sequence of finite dimensional subspaces of S such that $\cup S_N$ is dense in S, and that the pair $(\mathbf{P}, \theta)$ can be approximated by an infinite sequence of consistent approximations $(\mathbf{P}_N, \theta_N)$, $N = N_o, \ldots$, where $\theta_N : S_N \to \mathbb{R}$ is an optimality function for $\mathbf{P}_N$, with $\mathbf{P}_N$ of the form

$$\mathbf{P}_N \qquad\qquad \min_{v \in V_N} f_N(v), \qquad\qquad (2.2)$$

where the functions $f_N : S_N \to \mathbb{R}$ are continuously differentiable, and $V_N \subset S_N$.

Finally, we assume that the exact evaluation of the functions $f_N(\cdot)$ and their gradients is not practical, and that an iterative "solver" must be used, with K iterations of the solver yielding an approximation $f_{N,K}(u)$ to $f_N(u)$, and similarly, approximations $\nabla_K f_N(u)$ to $\nabla f_N(u)$, and $\theta_{N,K}(u)$ to $\theta_N(u)$. We will make no continuity assumptions on $f_{N,K}(\cdot)$, $\nabla_K f_N(\cdot)$, or $\theta_{N,K}(\cdot)$.

In response of the above assumption, we will develop new algorithm models, with two precision parameters, for solving problems of the form $\mathbf{P}$, by mimicking the one precision parameter Algorithm Model 3.3.17 in Polak (1997). Algorithm Model 3.3.17 in Polak (1997) assumes that the functions $f_N(\cdot)$, in (2.2) and the associated optimality functions θ_N, are computable, and that for any bounded set $B \subset V$, there exists a function $\Delta : \mathcal{N} \to \mathbb{R}_+$ and a $\kappa \in (0, \infty)$, such that for all $v \in V_N \cap B$,

$$|f_N(v) - f(v)| \leq \kappa \Delta(N), \tag{2.3}$$

with $\Delta(N) \to 0$ as $N \to \infty$.

Algorithm Model 3.3.17 in Polak (1997) uses a parametrized iteration function $A_N : V_N \to V_N$, $N \in \{N_{-1}, N_0, N_1, ...\} \subset \mathcal{N} := \{0, 1, 2, 3, ...\}$. When $A_N(v)$ is derived from the Armijo gradient method, it will have the form

$$A_N(v) = v - \lambda(v)\nabla f_N(v), \tag{2.4}$$

where $\lambda(v) > 0$ is the Armijo step-size. For convenience, we reproduce Algorithm Model 3.3.17 in Polak (1997) below.

Algorithm Model 1a: Solves problem $\mathbf{P}$.

Parameters. $\omega \in (0, 1)$, $\sigma > 0$.

Data. $N_{-1} \in \mathcal{N}$, and $v_0 \in V_{N_{-1}}$.

Step 0. Set $i = 0$.

Step 1. Compute the smallest N_i, of the form $2^k N_{i-1}$, $k \in \mathcal{N}$, and $v_{i+1} \in V_{N_i}$, such that

$$v_{i+1} = A_{N_i}(v_i), \tag{2.5}$$

and

$$f_{N_i}(v_{i+1}) - f_{N_i}(v_i) \leq -\sigma \Delta(N_i)^\omega . \tag{2.6}$$

Step 2. Replace i by $i + 1$, and go to **Step 1.**   ◇

Algorithm Model 1a has the following convergence property:

Theorem 2.1 *Suppose that*

(i) *for every bounded set $B \subset V$, there exists $\kappa < \infty$ and a function $\Delta :$ $\mathcal{N} \to \mathbb{R}_+$ such that $\lim_{N \to \infty} \Delta(N) = 0$, and for all $N \in \mathcal{N}$, $N \geq N_{-1}$, $v \in V_N \cap B$,*

$$|f_N(v) - f(v)| \leq \kappa \Delta(N); \tag{2.7}$$

(ii) *For every $v^* \in V$ such that $\theta(v^*) \neq 0$, there exist $\rho^* > 0$, $\delta^* > 0$, $N^* < \infty$, such that*

$$f_N(A_N(v)) - f_N(v) \leq -\delta^*, \quad \forall v \in V_N \cap B(v_*, \rho^*), \quad \forall N \geq N^*. \tag{2.8}$$

Then, every accumulation point $\hat{v}$ of a sequence $\{v_i\}_{i=0}^{\infty}$, constructed by Algorithm Model 1a, satisfies $\theta(\hat{v}) = 0$.   ◇

Referring to Section 1.2 in Polak (1997), we note that Algorithm Model 1a can also be used in an algorithm implementation scheme. Thus, suppose N is fixed and that K iterations of the solver applied to function and gradient evaluations yield the approximations $f_{N,K}(v)$ and $A_{N,K}(v)$, to $f_N(v)$ and $A_N(v)$ respectively, and that for any bounded subset $B \subset V_N$, there is a $\kappa \in (0, \infty)$ and a function $\varphi : \mathcal{N} \times \mathcal{N} \to \mathbb{R}_+$, such that $\varphi(N, K) \to 0$ as $K \to \infty$ and

$$|f_{N,K}(v) - f_N(v)| \leq \kappa \varphi(N, K). \tag{2.9}$$

Making use of these definitions, we can now state the following scheme, based on the idea of algorithm implementation, for solving the problem $\mathbf{P}_N$.

Algorithm Model 1b: Solves problem $\mathbf{P}_N$.

Parameters. $\omega \in (0, 1)$, $\sigma > 0$ $K^* \in \mathcal{N}$.

Data. $K_{-1} \in \mathcal{N}$, and $v_0 \in V_N$.

Step 0. Set $i = 0$.

Step 1. Compute the smallest K_i, of the form $K_{i-1} + kK^*$, $k \in \mathcal{N}$, and $v_{i+1} \in V_N$, such that

$$v_{i+1} = A_{N,K_i}(v_i), \qquad (2.10)$$

and

$$f_{N,K_i}(v_{i+1}) - f_{N,K_i}(v_i) \leq -\sigma\varphi(N, K_i)^\omega . \qquad (2.11)$$

Step 2. Replace i by $i + 1$, and go to **Step 1.** $\Diamond$

Referring to Theorem 1.2.37 in Polak (1997), we see that Algorithm Model 1b has the following convergence property:

Theorem 2.2 *Suppose that*

(i) *for every bounded set $B \subset V_N$, there exists $\kappa < \infty$ and a function φ : $\mathcal{N} \times \mathcal{N} \to Re_+$ such that $\lim_{K \to \infty} \varphi(N, K) = 0$, and for all $K \in \mathcal{N}$, $K \geq K_{-1}$, $v \in V_N \cap B$,*

$$|f_{N,K}(v) - f_N(v)| \leq \kappa\varphi(N, K); \qquad (2.12)$$

(ii) *For every $v^* \in V_N$ such that $\theta_N(v^*) \neq 0$, there exist $\rho^* > 0$, $\delta^* > 0$, $K^* < \infty$, such that*

$$f_{N,K}(A_{N,K}(v)) - f_{N,K}(v) \leq -\delta^*, \quad \forall v \in V_N \cap B(v^*,\rho^*), \quad \forall K \geq K^*.$$
$$(2.13)$$

Then, every accumulation point $\hat{v}$ of a sequence $\{v_i\}_0^\infty$, constructed by Algorithm Model 1b, satisfies $\theta_N(\hat{v}) = 0$. $\Diamond$

In view of this, we propose to construct an algorithm model which behaves as follows. Given a discretization parameter N, it applies Algorithm Model 1b to $\mathbf{P}_N$ until a sufficiently good approximation to its solution is obtained, and then it increases N and repeats the process. We will use the failure of a test of the form (2.6) to determine that that one is close enough to a solution of $\mathbf{P}_N$. As a result, the problems $\mathbf{P}_N$ will be solved more and more accurately as the computation progresses. Parameters in the algorithm will allow the user to balance the precision with which the problems $\mathbf{P}_N$ are solved versus the speed with which N is advanced.

At this point we must introduce some realistic assumptions. In particular, we assume that for every $N, K \in \mathcal{N}$, we can construct an iteration map $A_{N,K} : V_N \to V_N$, where K is the number of iterations of a solver.

Assumption: *We will assume as follows:*

(i) *The function $f(\cdot)$ is continuous and bounded from below, and for all $N \in \mathcal{N}$, the functions $f_N(\cdot)$ are continuous and bounded from below.*

(ii) *For every bounded set $B \subset V$, there exists $\kappa < \infty$, a function $K^* : \mathcal{N} \to \mathcal{N}$, and functions $\varphi : \mathcal{N} \times \mathcal{N} \to \mathbb{R}_+$, $\Delta : \mathcal{N} \to \mathbb{R}_+$ with the properties*

$$\lim_{N \to \infty} K^*(N) = \infty, \tag{2.14}$$

$$\lim_{K \to \infty} \varphi(N, K) = 0, \quad \forall N \in \mathcal{N}, \tag{2.15}$$

$$\lim_{N \to \infty} \varphi(N, K_N) = 0, \quad \forall K_N \geq K^*(N), \tag{2.16}$$

$$\lim_{N \to \infty} \Delta(N) = 0, \tag{2.17}$$

such that for all $N \in \mathcal{N}$, $v \in V_N \cap B$,

$$|f_N(v) - f(v)| \leq \kappa \Delta(N), \tag{2.18}$$

and for all $N \in \mathcal{N}$, $K \in \mathcal{N}$, $v \in V_N \cap B$,

$$|f_{N,K}(v) - f_N(v)| \leq \kappa \varphi(N, K). \tag{2.19}$$

(iii) *For every $v^* \in V$ such that $\theta(v^*) < 0$, there exist $\rho^* > 0$, $\delta^* > 0$, $N^* > 0$, $K^{**} < \infty$, such that*

$$f_{N,K}(A_{N,K}(v)) - f_{N,K}(v) \leq -\delta^*, \quad \forall v \in V_N \cap B(v_*, \rho^*), \quad \forall N \geq N^*, \quad \forall K \geq K^{**}. \tag{2.20}$$

Algorithm Model 2: Solves problem $\mathbf{P}$.

Parameters. $\omega \in (0, 1)$, $K, k^* \in \mathcal{N}$, $K^*(\cdot)$, $\Delta(\cdot)$, $\varphi(\cdot, \cdot)$ verifying (2.14), (2.15), (2.16), (2.17).

Data. $N_0 \in \mathcal{N}$, $v_0 \in V_{N_0}$.

Begin Outer Loop

Step 0. Set $i = 0$.

> **Begin Inner Loop**
>
> **Step 1.** Set $K_i = N^*(K_i)$.
>
> **Step 2.** Compute $A_{N_i, K_i}(v_i)$.
>
> **Step 3. If** $K_i < k^* K^*(N_i)$ and
>
> $$f_{N_i, K_i}(A_{N_i, K_i}(v_i)) - f_{N_i, K_i}(v_i) > -\varphi(N_i, K_i)^\omega, \qquad (2.21)$$
>
> replace K_i by $K_i + K$ and go to **Step 2.**
>
> **Else,** set
>
> $$\hat{\Delta}(N_i, K_i) = \Delta(N_i) + \varphi(N_i, K_i), \qquad (2.22)$$
>
> and go to **Step 4.**
>
> **End Inner Loop**

Step 4. If

$$f_{N_i, K_i}(A_{N_i, K_i}(v_i)) - f_{N_i, K_i}(v_i) > -\hat{\Delta}(N_i, K_i)^\omega, \qquad (2.23)$$

replaceN_i by $2N_i$ and go to **Step 1.**

Else, set

$$v_{i+1} = A_{N_i, K_i}(v_i), \qquad (2.24)$$

replace i by $i + 1$ and go to **Step 2.**

End Outer Loop $\hfill \Diamond$

Remark 2.1

1. The main function of the test (2.21) is to increase N over the initial value of $N = N^*(N_i)$ if that is necessary. It gets reset to $N = N^*(N_i)$ whenever N_i is halved.

2. Note that the faster $\varphi(N, K) \to 0$ as $K \to \infty$, the easier it is to satisfy the test (2.21) at a particular value of N. Thus, when the solver is fast, the precision parameter K_i will be increased more slowly than when it is slow. A similar argument applies to the increase of N_i, on the basis of the test in (2.23). In the context of dynamics defined by differential equations, the integration mesh size,

$1/N_i$, will be refined much faster when the Euler method is used for integration than when a Runge-Kutta method is used for integration. $\diamond$

Lemma 2.1 *Suppose that Assumption is satisfied.*

(a) If $v_i \in V_{N_i}$ is such that $\theta(v_i) \neq 0$, then there exists an $N_i < \infty$ such that (2.23) fails, i.e., v_{i+1} is constructed.

(b) If Algorithm Model 2 constructs an infinite sequence $\{v_i\}_{i=0}^{\infty}$ that has at least one accumulation point, then $N_i \to \infty$, as $i \to \infty$, and hence, also, $K_i \to \infty$, as $i \to \infty$.

Proof. (a) Suppose that $v_i \in V_i$ is such that $\theta(v_i) \neq 0$. Then, by Assumption (iii), there exist an $\hat{N} < \infty$, $\hat{K} < \infty$, and $\hat{\delta} < 0$, such that for all $N \geq \hat{N}$, $K \geq \hat{K}$,

$$f_{N,K}(A_{N,K}(v_i) - f_{N,K}(v_i) \leq -\hat{\delta} \leq -\hat{\Delta}(N,K)^{\omega}. \tag{2.25}$$

Since by construction $K_i \geq K^*(N_i)$ and $K^*(N) \to \infty$, as $N \to \infty$, it follows that there exists an $N_i < \infty$ such that (2.23) fails.

(b) For the sake of contradiction, suppose that the monotone increasing sequence $\{N_i\}_{i=0}^{\infty}$ is bounded from above by $b < \infty$. Then there exists an i_0 such that $N_i = N_{i_0} = N* < \infty$ for all $i \geq i_0$. In this case, there is a $K* \leq k * K^*(N*) + K$ and an $i_1 \geq i_0$ such that $K_i = K*$ for all $i \geq i_1$. Hence, for all $i \geq i_1$, since (2.23) fails for each such i,

$$f_{N*,K*}(v_{i+1}) - f_{N*,K*}(v_i) \leq -\hat{\Delta}(N*,K*), \tag{2.26}$$

which shows that $f_{N*,K*}(v_i) \to -\infty$, as $i \to \infty$. Now, it follows from (2.18) and (2.19) that for all $i \geq i_1$,

$$|f_{N*,K*}(v_i) - f(v_i)| \leq \kappa\hat{\Delta}(N*,K*), \tag{2.27}$$

which implies that $f(v_i) \to -\infty$, as $i \to \infty$. However, by assumption, there exists an infinite subsequence $\{v_{i_j}\}$ and a $v* \in V_{h*}$, such that $v_{i_j} \to v*$, as $j \to \infty$. Since $f(\cdot)$ is continuous, by assumption, we conclude that $f(v_{i_j}) \to f(v*)$, as $j \to \infty$, which is a contradiction, and completes our proof. $\diamond$

Theorem 2.3 *Suppose that Assumption is satisfied.*

(a) If $\{v_i\}_{i=0}^{\infty}$ is a sequence constructed by Algorithm Model 2, in solving the problem $\mathbf{P}$, then every accumulation point $v*$ of $\{v_i\}_{i=0}^{\infty}$ satisfies $\theta(v*) = 0$.

(b) *If $f(\cdot)$ is strictly convex, with bounded level sets, and $\{v_i\}_{i=0}^{\infty}$ is a sequence constructed by Algorithm Model 2, in solving the problem* **P**, *then $\{v_i\}_{i=0}^{\infty}$ converges to the unique solution of* **P**.

Proof. (a) Suppose that $\{v_i\}_{i=0}^{\infty}$ is a sequence constructed by Algorithm Model 2 and that $\{v_{i_j}\}_{j=0}^{\infty}$ is a subsequence converging to a point $\hat{v}$ and that $\theta(\hat{v}) \neq 0$.

Now, by Lemma 2.1, $N_i \to 0$, as $i \to \infty$, and by Assumption (iii), since $K_i \geq K^*(N_i)$, there exist $\hat{\rho} > 0$, $\hat{\delta} > 0$, $\hat{h} > 0$, such that

$$f_{N_i,K_i}(A_{N_i,K_i}(v_i)) - f_{N_i,K_i}(v_i) \leq -\hat{\delta}, \quad \forall v_i \in B(\hat{v}, \hat{\rho}), \quad \forall N_i \geq \hat{N}. \tag{2.28}$$

Next we note that in view of (2.18) and (2.19), for any $v \in V_i$,

$$|f(v) - f_{N_i,K_i}(v)| \leq \kappa \hat{\Delta}(N_i, K_i) := \kappa[\Delta(N_i) + \varphi(N_i, K_i)]. \tag{2.29}$$

Let i_0 be such that for all $i_j \geq i_0$, $v_{i_j} \in B(\hat{v}, \hat{\rho})$ and

$$2\kappa \hat{\Delta}(N_{i_j}, K_{i_j}) \leq \tfrac{1}{2}\hat{\delta}, \tag{2.30}$$

$$2\kappa \hat{\Delta}(N_{i_j}, K_{i_j})^{1-\omega} \leq 1. \tag{2.31}$$

Finally, let $i_1 \geq i_0$ be such that $N_i \leq \hat{h}$ for all $i \geq i_1$. Then, for the subsequence $\{v_{i_j}\}_{j=0}^{\infty}$, with $i_j \geq i_1$,

$$\begin{aligned}
f(v_{i_j+1}) - f(v_{i_j}) &\leq f_{N_{i_j},K_{i_j}}(v_{i_j+1}) - f_{N_{i_j},K_{i_j}}(v_{i_j}) + 2\kappa\hat{\Delta}(N_{i_j}, K_{i_j}) \\
&\leq -\hat{\delta} + 2\kappa\hat{\Delta}(N_{i_j}, K_{i_j}) \leq -\tfrac{1}{2}\hat{\delta},
\end{aligned} \tag{2.32}$$

and in addition, in view of (2.29) and the test (2.23), for all $i \geq i_1$,

$$\begin{aligned}
f(v_{i+1}) - f(v_i) &\leq 2\kappa\hat{\Delta}(N_i, K_i) - \hat{\Delta}(N_i, K_i)^{\omega} \\
&= -\hat{\Delta}(N_i, v_i)^{\omega}[1 - 2\kappa\hat{\Delta}(N_i, K_i)^{1-\omega}] \leq 0.
\end{aligned} \tag{2.33}$$

Hence we see that the sequence $\{f(v_i)\}_{i=i_1}^{\infty}$ is monotone decreasing, and therefore, because $f(\cdot)$ is continuous, it must converge to $f(\hat{v})$. Since this is contradicted by (2.31), our proof is complete.

(b) Since a strictly convex function, with bounded level sets, has exactly one stationary point, the desired result follows from **(a)** and the fact that $\{f(v_i)\}_{i=i_1}^{\infty}$ is monotone decreasing. $\diamond$

Remark 2.2 The following Algorithm Model differs from Algorithm Model 2 in two respects: first the integer K is never reset and hence increases monotonically, and second the test for increasing N is based on the magnitude of the approximate optimality function value. As a result, the proof of its convergence is substantially simpler than that for Algorithm Model 2. However, convergence can be established only for the diagonal subsequence $\{v_{i_j}\}_{j=0}^{\infty}$ at which N_i is doubled. $\diamond$

Algorithm Model 3: Solves problem **P**.

Parameters. $\omega \in (0,1)$, $\epsilon_0 > 0$, $K \in \mathcal{N}$, $K^*(\cdot)$, $\varphi(\cdot,\cdot)$ verifying (2.14), (2.15), (2.16).

Data. $N_0 \in \mathcal{N}$, $v_0 \in V_{N_0}$.

Begin Outer Loop

Step 0. Set $i = 0$, $j = 0$, $K_0 = K^*(N_0)$.

    Begin Inner Loop

    Step 1. Compute a point $v_* = A_{N_i, K_i}(v_i)$.

    Step 2. If

$$\theta_{N_i, K_i}(v_*) \leq -\epsilon_i \tag{2.34}$$

    and

$$f_{N_i, K_i}(v_*) - f_{N_i, K_i}(v_i) > -\varphi(N_i, K_i)^\omega, \tag{2.35}$$

    replace K_i by $K_i + K$ and go to **Step 1. Else,** set $v_{i+1} = v_*$, and go to **Step 3**.

    End Inner Loop

Step 3. If

$$\theta_{N_i, K_i}(v_{i+1}) \geq -\epsilon_i \text{ and } K_i \geq K^*(N_i), \tag{2.36}$$

set $v_j^* = v_{i+1}$, $K_{j+1}^* = K_i$, $N_{j+1}^* = N_i$, replace j by $j + 1$, N_i by $2N_i$, ϵ_i by $\epsilon_i/2$, i by $i+1$, and go to **Step 1. Else,** replace i by $i+1$ and go to **Step 1**.

End Outer Loop $\diamond$

At this point we need an additional assumption:

Assumption: *We will assume as follows:*

(i) The optimality functions $\theta(\cdot)$ and $\theta_N(\cdot)$ are continuous for all $N \in \mathcal{N}$.

(ii) For every bounded set $B \subset V$, there exists $\kappa < \infty$, a function $K^* : \mathcal{N} \to \mathcal{N}$, and functions $\varphi : \mathcal{N} \times \mathcal{N} \to \mathbb{R}_+$, $\Delta : \mathcal{N} \to \mathbb{R}_+$ satisfying (2.15)-(2.17), such that for all $N \in \mathcal{N}$, $v \in V_N \cap B$,

$$|\theta_N(v) - \theta(v)| \leq \kappa \Delta(N), \tag{2.37}$$

and for all $N \in \mathcal{N}$, $K \in \mathcal{N}$, $v \in V_N \cap B$,

$$|\theta_{N,K}(v) - \theta_N(v)| \leq \kappa \varphi(N, K), \tag{2.38}$$

where $\theta_{N,K}(v)$ is the approximation to $\theta_N(v)$ obtained as a result of K iterations of a solver.

Theorem 2.4 *Suppose that Assumptions and are satisfied and that $\{v_j^*\}$ is a sequence constructed by Algorithm Model 3, in solving the problem **P**.*

(a) If $\{v_j^*\}$ is finite, then the sequence $\{v_i\}_{i=0}^{\infty}$ has no accumulation points.

(b) If $\{v_j^*\}$ is infinite, then every accumulation point $v*$ of $\{v_j^*\}_{j=0}^{\infty}$ satisfies $\theta(v*) = 0$.

(c) If $f(\cdot)$ is strictly convex, with bounded level sets, and $\{v_j^*\}_{j=0}^{\infty}$ is a bounded sequence constructed by Algorithm Model 3, in solving the problem **P***, then it converges to the unique solution of* **P***.*

Proof. (a) Suppose that the sequence $\{v_j^*\}$ is finite and that he sequence $\{v_i\}_{i=0}^{\infty}$ has an accumulation point $v*$. Then there exists an i_0, an $N* < \infty$, and an $\epsilon* > 0$, such that for all $i \geq i_0$, $N_i = N*$, $\epsilon_i = \epsilon*$, and $\theta_{N*,K_i}(v_i) < -\epsilon*$. But, in this case, for $i \geq i_0$, the Inner Loop of Algorithm Model 2 is recognized as being of the form of Master Algorithm Model 1.2.36, in Polak (1997). It now follows from Theorem 1.2.37 in Polak (1997) that $K_i \to \infty$, as $i \to \infty$, and that $\theta_{N*}(v*) = 0$. Next, it follows from (2.38), in Assumption and the continuity of $\theta_{N*}(\cdot)$, that for some infinite subsequence $\{v_{i_j}\}$, $\theta_{N*,K_{i_j}}(v_{i_j}) \to \theta_{N*}(v*) = 0$, which shows that (2.36) could not be violated an infinite number of times, a contradiction.

(b) When the sequence $\{v_*^j\}$ is infinite, it follows directly from Assumption 2 and the test (2.36) that if $v*$ is an accumulation point of $\{v_j^*\}$, then $\theta(v*) = 0$.

(c) When the function $f(\cdot)$ is strictly convex, with bounded level sets, it has a unique minimizer $v*$ which is the only point in V satisfying $\theta(v*) = 0$. Hence the desired result follows from **(b)**. $\diamondsuit$

3   A DISTRIBUTED PROBLEM WITH CONTROL IN THE COEFFICIENTS

An absorbant coating of thickness α on a multi-component airfoil S is to be optimized to cancel the reflected wave u from an accoustic incident wave u_∞ of frequency ω, in a sector angle Σ.

The Leontovich condition models the thin coating and the thickness of the coating layer is proportional to the Leontovich coefficient α. A first order absorbing boundary condition is applied on the outer boundary Γ_∞ which, for computational purposes is assumed to be at finite distance; therefore the sector angle is a portion of Γ_∞. The problem is

$$
\begin{cases}
\min_\alpha f(\alpha) \equiv \int_\Sigma |u|^2 & \text{subject to} \\
\omega^2 u + \Delta u = 0 & \dfrac{\partial u}{\partial n} - i\omega u = 0 \text{ on } \Gamma_\infty, \quad \dfrac{\partial u}{\partial n} + \alpha\omega(u + u_\infty) = g \text{ on } S
\end{cases}
\tag{3.1}
$$

In weak form the PDE is

$$
\int_\Omega (\omega^2 uv - \nabla u \nabla v) - i\omega \int_{\Gamma_\infty} uv + \int_S (\omega\alpha(u + u_\infty) - g)v = 0 \quad \forall v \in H^1(\Omega) \tag{3.2}
$$

It can be shown that it has one and only one solution which depends continuously upon the data α. Note that u is a nonlinear function of α.

The problem is discretized by the finite element method of degree one Ciarlet (1977) on triangles combined with a domain decomposition strategy with the purpose of having a finer mesh in desired regions without having to touch the rest of the domain. All linear systems are solved with the Gauss factorization method.

The airfoil is made of two parts, a main airfoil S_m and an auxiliary airfoil S_a, below and slightly behind the main one. To apply Domain Decomposition we need to partition the physical domain Ω as a union $\Omega_1 \cup \Omega_2$ of two sub-domains with a non empty intersection. This is done by surrounding the auxiliary airfoil by a domain Ω_2 outside S_m and with boundary $\partial\Omega_2 = \Gamma_2 \cup S_a$ and by taking

$\Omega_1 = \Omega \backslash D$ where D contains S_a but is contained in Ω_2. Note that $\Gamma_\infty \cup S_m \cup \Gamma_1$ is the boundary of Ω_1 if Γ_1 denotes ∂D.

Each domain is triangulated by an automatic mesh generator which is monitored by the mesh of the boundaries. So as the mesh sizes on the boundary tend to zero, so does the size of the triangles.

We denote by V_h^i, $i = 1,2$ the spaces of piecewise linear functions on the triangulations of Ω_i which are zero on the approximations of Γ_2 and C respectively.

Consider the two problems: given z_h^j, $j = 1,2$, piecewise linear continuous on the triangulation of Ω_j, find u_h^j such that $u_h^j - z_h^j \in V_h^j$ and

$$\int_{\Omega_1} (\omega^2 u_h^1 v - \nabla u_h^1 \nabla v) - i\omega \int_{\Gamma_\infty} u_h^1 v = 0 \quad \forall v \in V_h^1(\Omega_1) \qquad (3.3)$$

$$\int_{\Omega_2} (\omega^2 u_h^2 v - \nabla u_h^2 \nabla v) + \int_S (\omega \alpha (u_h^2 + u_\infty) - g)v = 0 \quad \forall v \in V_h^1(\Omega_2) \quad (3.4)$$

The Schwarz algorithm is as follows:

1. Choose z_h^1, z_h^2, $\epsilon > 0$

2. `do`

3.   Solve (3.3) and (3.4)

4.   set $z_h^1 = \Pi_h u_h^2$, $z_h^2 = \Pi_h u_h^1$

5. `while`$(\|z_h^1 - z_h^2\| > \epsilon)$

where Π_h is the interpolation operator from one mesh to the other. The aproximate solution after K iterations or the Schwarz algorithm is defined to be

$$u_{h,K} = u_h^i \text{ in } \Omega_i \backslash (\Omega_1 \cap \Omega_2) \text{ and } u_{h,K} = \frac{1}{2}(u_h^1 + u_h^2) \text{ in } \Omega_1 \cap \Omega_2. \qquad (3.5)$$

However, for compatibility with the theory in this paper, we determine the mesh size $h = 1/N$ and the number of Schwarz iterations K as required in Algorithm Model 2, and we do not use a $z^1 - z^2$ to determine the number of Schwarz iterations.

The convergence of the Schwarz algorithm is known only for compatible meshes, i.e. meshes of Ω_1 and Ω_2 identical in $\Omega_1 \cap \Omega_2$.

3.1   Computation of Gradients

With self explanatory notation we have

$$\delta f = 2 \int_\Sigma u \delta u \qquad (3.6)$$

where δu is the solution of

$$\int_\Omega (\omega^2 \delta u v - \nabla \delta u \nabla v) - i\omega \int_{\Gamma_\infty} \delta u v + \int_S \omega \alpha \delta u v + \int_S \omega \delta \alpha u v = 0 \qquad (3.7)$$

Let p be the solution of

$$\int_\Omega (\omega^2 p w - \nabla p \nabla w) - i\omega \int_{\Gamma_\infty} p w + \int_S \omega \alpha p w = 2 \int_\Sigma u v \quad \forall v \in H^1(\Omega) \qquad (3.8)$$

Then $v = p$ and $w = \delta u$ yield

$$2 \int_\Sigma u \delta u = \int_\Omega (\omega^2 p \delta u - \nabla p \nabla \delta u) - i\omega \int_{\Gamma_\infty} p \delta u + \int_S \omega \alpha p \delta u$$
$$= - \int_S \omega \delta \alpha u p \qquad (3.9)$$

Corollary 3.1 In $L^2(\Sigma)$ the gradient of f, with respect to α, is given by

$$\nabla f(\alpha) = -\omega u p \qquad (3.10)$$

where p is solution of (3.8)

The same calculation applies to the discrete problem and yields:

Corollary 3.2 In $L^2(\Sigma)$, the gradient of f_h, with respect to α, is given by

$$\nabla f_h(\alpha_h) = -\omega u_h p_h \qquad (3.11)$$

where $p_h \in V_h$ is the piecewise linear continuous solution of

$$\int_\Omega (\omega^2 p_h w_h - \nabla p_h \nabla w_h) - i\omega \int_{\Gamma_\infty} p_h w_h + \int_S \omega \alpha p_h w_h = 2 \int_\Sigma u_h v_h \quad \forall v_h \in V_h$$
$$(3.12)$$

The approximate gradient of f_h, with respect to α, is given by

$$\nabla_K f_h = -\omega u_{h,K} p_{h,K} \tag{3.13}$$

where $u_{h,K}$ and $p_{h,K}$ are computed by K iterations of the Schwarz algorithm.

Details of the validity of this calculus of variations calculation can be found in Lions (1958)

3.2 Verification of Hypotheses

(i) Continuity of $f(\cdot)$ with respect to the control is established in Cessenat (1998). Continuity of $f_h(\cdot)$ with respect to the control is obvious from the matrix representation of the problem and the fact that the matrices are non singular.

(ii) It follows from the finite element error estimates given in Ciarlet (1977) that for some $C < \infty$,

$$\|u_h - u\|_0 < Ch^2, \tag{3.14}$$

which implies that

$$|f_h(v_h) - f(v)| < Ch^2. \tag{3.15}$$

Hence we can set $\Delta(h) = h^2$.

(iii) The Schwarz algorithm converges linearly with rate constant $(1 - d/D)$, where d is the diameter of $\Omega_1 \cap \Omega_2$ and D is the diameter of $\Omega_1 \cup \Omega_2$, so instead of (3.14) we have the bound

$$\|u_{h,K} - u_h\| \le C(1 - \frac{d}{D})^K \quad \forall K \in \mathcal{N}, \tag{3.16}$$

for some $C \in (0, \infty)$, which implies that we can set $\varphi(h, K) = (1 - \frac{d}{D})^K$. Note that in this case $\varphi(h, K)$ is actually independent of h. In view of this, we can take $K^*(h) = C\texttt{ceil}(1/h)$, where $C > 0$ is arbitrary.

(iv) It follows from the properties of the method of steepest descent that, given any $v^* \in V = L^2(0, 1)$ such that $\nabla f(v^*) \ne 0$, there exist a $\rho^* > 0$, a $\delta^* > 0$, λ^*, and an $h^* > 0$, such that for all $v \in V \cap B(v^*, \rho)$, (i) $\nabla f_h(v) \ne 0$ and (ii)

$$f(v - \lambda(v)\nabla f(v)) - f(v) \le f(v - \lambda^* \nabla f(v)) - f(v) \le -\delta^*, \tag{3.17}$$

where $\lambda(v)$ is the exact step-size computed by the Steepest Descent Algorithm.

To show that there exist an $h^* > 0$ and an $N^{**} < \infty$, such that for all $h \leq h^*$, $N \geq N^{**}$, and $v \in V_h \cap B(v^*, \rho)$

$$f_{h,N}(v - \lambda_N(v)\nabla f_h(v)) - f_{h,N}(v) \leq f_{h,N}(v - \lambda(v)\nabla f_h(v)) - f_{h,N}(v) \leq -\delta^*/2, \tag{3.18}$$

we make use of the facts that (a)

$$\nabla f_h(\alpha) = -\omega u_h p_h, \quad \nabla_K f_h(\alpha) = -\omega u_{h,K} p_{h,K}. \tag{3.19}$$

(b) By inspection, the bound functions $\Delta(h)$, $\varphi(h, K)$, and $K^*(h)$ have the required properties.

3.3   Numerical Results

The following test was conducted in cooperation with G. Lemarchand. Other test cases can be found in Pironnea et al (2002). Compared with these, the novelty of this accoustic problem is that it is nonlinear.

The numerical software freefem+Bernardi et at (1999) was used to solve the problem where the obstacles are two NACA0012 airfoils of length 1 in a circular domain of radius 5. The accoustic wave comes horizontally with frequency $\omega = 1.7$. The optimization starts with $\alpha = 0$. The Schwarz iteration number K is augmented by 1 when the test fails until the criteria is met and the mesh size goes from c/n to $c/(n+1)$ when refinement is required by the algorithm.

Figure 3.1 and figure 3.2 show the solution and figure 3.3 shows the history of the convergence compared with a straight steepest descent method and a steepest descent with mesh refinement only and no DDM.

On this example Algorithm model 3 proved ro be fairly stable. NOte that the approximation of the objective function increases quite a lot whenever the mesh is subdivided, at least in the beginning, but that it a common phenomenon.

The speed-up over using fixed very small h and very large K is considerable because most of the optimization occurs before the mesh is too fine or the number of Schwarz iterations is too large.

4   CONCLUSIONS

We have developed algorithm models based on the consistent approximations approach for solving infinite dimensional problems with two independent precision parameters. We have applied it to a nonlinear optimal control problems

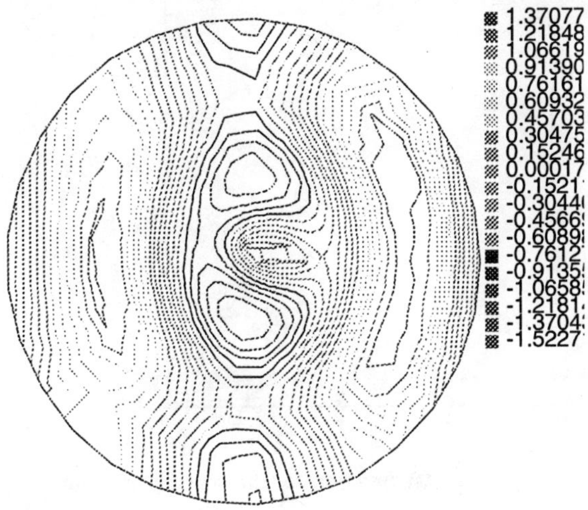

Figure 3.1   Real part of the solution of Helmholtz equation.

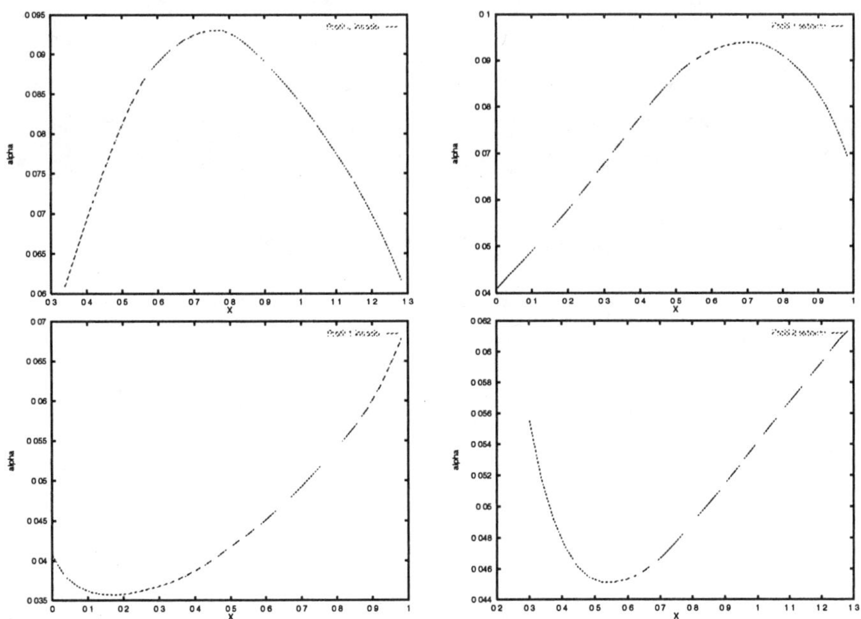

Figure 3.2 α versus distance to the leading edge on the two sides of each airfoil.

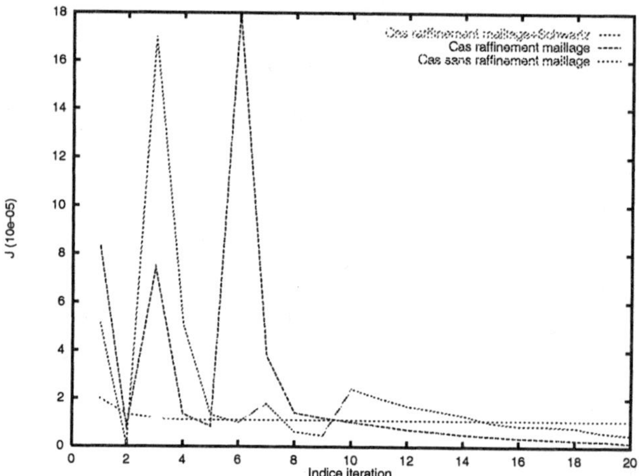

Figure 3.3   History of the convergence of the cost function for the coating problem. The method with mesh refinement and adapted Schwarz iteration number (green curve) is compared with a straight steepest descent method (red curve) and a steepest descent with mesh refinement only and DDM up to convergence (blue curve). The objective function augments whenever the mesh is refined.

with PDE dynamics having two precision parameters, the step size and an iteration loop count in the solver. Our numerical results show that our algorithms are effective. The numerical study was done using the method of steepest descent but the models and the proofs are general and are likely to work also with Newton methods, conjugate gradient methods, etc.

5   APPENDIX: CONSISTENT APPROXIMATIONS

To make this paper reasonably self contained we define consistent approximations and state their most important properties.

Definition 5.1 *Let S be a normed space, let $\{S_N\}_{N=N_0}^{\infty}$ be a nested sequence of finite dimensional subspaces of S such that $\cup S_N$ is dense in S, and consider the problems*

P

$$\min_{u \in U} f(u) \qquad (5.1)$$

where U is a subset of S and $f : S \rightarrow \mathbb{R}$ is continuous, together with the approximating problems

$$\mathbf{P}_N \qquad\qquad \min_{u \in U} f_N(v) \qquad\qquad (5.2)$$

where U_N is a subset of $\mathcal{S}_N$ and $f_N : \mathcal{S}_N \to \mathbb{R}$ is continuous.

(a) We say that the problems $\mathbf{P}_N$ epi-converge[2] to $\mathbf{P}$ if (i) for every $u \in U$ there exists a sequence $\{u_N\}$, with $u_N \in U_N$, such that $u_N \to u$, as $N \to \infty$, and $\limsup f_N(u_N) \leq f(u)$; and (ii) for every infinite sequence $\{u_N\}$, such that $u_N \in U_N$ and $u_N \to u$, $u \in U$ and $\liminf f_N(u_N) \geq f(u)$.

(b) We say that upper-semicontinuous, nonpositive-valued functions $\theta_N : U_N \to \mathbb{R}$ $(\theta : U \to \mathbb{R})$ are optimality functions for $\mathbf{P}_N$ $(\mathbf{P})$, if they vanish at local minimizers of $\mathbf{P}_N$ $(\mathbf{P})$[3].

(c) We say that the problem-optimality function pairs $\{\mathbf{P}_N, \theta_N\}$ are consistent approximations to the problem-optimality function pair $\{\mathbf{P}, \theta\}$, if the $\mathbf{P}_N$ epi-converge to $\mathbf{P}$ and for every infinite sequence $\{u_N\}$, such that $u_N \in U_N$ and $u_N \to u \in U$, $\limsup \theta_N(u_N) \leq \theta(u)$[4]. $\qquad \diamond$

Theorem 5.1 *Suppose that the problems $\mathbf{P}_N$ epi-converge to $\mathbf{P}$. .*

(a) If $\{\hat{u}_N\}$ is a sequence of global minimizers of the $\mathbf{P}_N$ and $\hat{u}$ is any accumulation point of this sequence, then $\hat{u}$ is a global minimizer of $\mathbf{P}$, and $f_N(\hat{u}_N) \to f(\hat{u})$.

(b) If $\{\hat{u}_N\}$ is a sequence of local minimizers of the $\mathbf{P}_N$, sharing a common radius of attraction, and $\hat{u}$ is accumulation point of this sequence, then $\hat{u}$ is a local minimizer of $\mathbf{P}$. $\qquad \diamond$

The reason for introducing optimality functions into the definition of consistency of approximation is that it enables us to ensure that not only *global* optimal solutions of the problems $\mathbf{P}_h$ converge to *global* optimal solutions of $\mathbf{P}$, but also *local* optimal solutions converge to either local solutions or stationary points.

Acknowledgments

This work was supported in part by the National Science Foundation under Grant No. ECS-9900985 and by the Institut Universitaire de France.

Notes

1. Please refer to the Appendix or Polak (1997) for the definitions of optimality functions and consistent approximations.

2. The epigraphs of f_N, restricted to U_N, converge to the epigraph of f, restricted to U, in the Painlevé-Kuratowski sense.

3. When optimality functions are properly constructed, their zeros are standard stationary points, for examples see Polak (1997).

4. Note that this property ensures that the limit point of a converging sequence of approximate stationary points for the $\mathbf{P}_h$ must be a stationary point for $\mathbf{P}$

References

Becker, R., Kapp, H., and Rannacher, R. (2000), Adaptive finite element methods for optimal control of partial differential equations: basic concept, *SIAM J. Control and Optimization*, Vol. 39, No. 1, pp. 113-132.

Bernardi D., Hecht, F., Otsuka K., Pironneau O. (1999) : freefem+, a finite element software to handle several meshes. Dowloadable from ftp://ftp.ann.jussieu.fr/pub/soft/pironneau/.

Cessenat M. (1998), Mathematical Methods in Electromagnetism, World Scientific, River Edge, NJ.

Betts, J. T. and Huffman, W. P. (1998), Mesh refinement in direct transcription methods for optimal control, *Optm. Control Appl.*, Vol. 19, pp. 1-21.

Carter, R. G. (1991), On the global convergence of trust region algorithms using inexact gradient information, *SIAM J. Numer. Anal.*, Vol. 28, pp. 251-265.

Carter, R. G. (1993), Numerical experience with a class of algorithms for nonlinear optimization using inexact function and gradient information, *SIAM J. Sci. Comput.*, Vol. 14, No. 2, pp.368-88.

Ciarlet, P.G. (1977), *The Finite Element Method*, Prentice Hall. Deuflhard, P. (1974), A modified Newton method for the solution of ill-conditioned systems of nonlinear equations with application to multiple shooting, *Numerische Mathematik*, Vol.22, No.4, p.289-315.

Deuflhard, P. (1975), A relaxation strategy for the modified Newton method. Optimization and Optimal Control, Proc. *Conference on Optimization and Optimal Control, Oberwolfach, West Germany, 17-23 Nov. 1974*, Eds. Bulirsch, R.; Oettli, W.; Stoer, J., Springer-Verlag, Berlin, p.59-73.

Deuflhard, P. (1991), Global inexact Newton methods for very large scale nonlinear problems, *Impact of Computing in Science and Engineering*, Vol.3, (No.4), p.366-93.

Dunn, J. C., and Sachs, E. W. (1983), The effect of perturbations on the convergence rates of optimization algorithms, *Applied Math. and Optimization*, pp. 143-147, Vol. 10.

Kelley, C. T. and Sachs, E. W. (1991), Fast algorithms for compact fixed point problems with inexact function evaluations, *SIAM J. Sci. Statist. Comput.*, Vol. 12, pp. 725–742.

Kelley, C. T. and Sachs, E. W. (1999), A Trust Region Method for Parabolic Boundary Control Problems, *SIAM J. Optim.*, Vol. 9, pp. 1064–1081.

Lions J.L. (1968),*Contrôle Optimal de systèmes gouvernés par des équations aux dérivées partielles*. Dunod-Gauthier Villars, 1968.

Mayne D. Q., and Polak E. (1977), A Feasible Directions Algorithm for Optimal Control Problems with Terminal Inequality Constraints, *IEEE Transactions on Automatic Control*, Vol. AC-22, No. 5, pp. 741-751.

Pironneau O., Polak E. (2002), Consistent Approximations and Approximate Functions and Gradients In Optimal Control, *J. SIAM Control and Optimization*, Vol 41, pp.487-510.

Polak E., and Mayne D. Q. (1976), An Algorithm for Optimization Problems with Functional Inequality Constraints, *IEEE Transactions on Automatic Control*, Vol. AC-21, No. 2.

Polak E. (1993), On the Use of consistent approximations in the solution of semi-Infinite optimization and optimal control problems", *Mathematical Programming*, Series B, Vol. 62, No.2, pp 385-414.

POLAK E. (1997), *Optimization: Algorithms and Consistent Approximations*, Springer-Verlag, New York.

Sachs, E. (1986), Rates of Convergence for adaptive Newton methods,*JOTA*, Vol. 48, No.1, pp. 175-190.

Schwartz, A. L. (1996a), *Theory and Implementation of Numerical Methods Based on Runge-Kutta Integration for Solving Optimal Control Problems*, Ph. D. Dissertation, University of California, Berkeley.

Schwartz, A. L. (1996b), *RIOTS The Most Powerful Optimal Control Problem Solver*. Available from http://www.accesscom.com/ adam/RIOTS/

Schwartz, A. L., and Polak, E. (1996), Consistent Approximations for Optimal Control Problems Based on Runge-Kutta Integration, *SIAM Journal on Control and Optimization*, Vol. 34, No.4, pp. 1235-69.

21 NUMERICAL SOLUTIONS OF OPTIMAL SWITCHING CONTROL PROBLEMS

T. Ruby and V. Rehbock

Department of Mathematics and Statistics
Curtin University of Technology
GPO Box U 1987, Perth 6845, Australia

Abstract: We develop a numerical solution strategy for a general class of optimal switching control problems. We view this class of problems as a natural extension of related classes considered previously, namely discrete valued optimal control problems and time optimal control problems. We show that techniques developed for these subclasses may be readily extended to the more general class considered in this paper. Numerical results are given to illustrate the proposed approach and to compare it with another recently developed solution technique.

Key words: Optimal control, computational methods, switching dynamics, control parametrization, transformations.

1 INTRODUCTION

In this paper, we consider a class of optimal switching control problems. These are characterized by the fact that the state variables of the problem are generated from a given initial state in conjunction with a successive sequence of dynamical systems which the user can select from amongst a given finite set. It is assumed that each system in this sequence is active for a certain duration within the time horizon and that the state of the system is continuous across a time point where systems are switched, *i.e.* the final state over the previous time duration is chosen as the initial state for the next time duration. The aim in these problems is to choose a sequence of dynamical systems such that the resulting state of the system optimizes a given performance index. At the same time, the aim is to also optimally choose the time durations over which each of the chosen dynamical systems is active. In this work, we limit our attention to problems involving dynamical systems describes by systems of ordinary differential equations (ODEs).

This class of problems has significant practical applications in economics, management, production planning and industrial engineering, as dynamical systems arising in these areas are typically subject to frequent and sudden changes.

A number of subclasses of the general switching control problem exist. These include time optimal control problems (where the optimal control is of bang-bang type) and discrete valued optimal control problems (where the control is only allowed to take values from a finite, discrete set). The common feature of all these problems is the need to accurately determine the optimal times for the control function to switch (between its upper and lower bounds in the case of time optimal control and between discrete values for the discrete valued optimal control). Regarding these switching times as parameters, it is known that the objective function gradients with respect to these parameters is discontinuous (see Teo et al. (1991)). Hence, direct solution via gradient based optimization methods does not usually work well for these problems. Recently, the Control Parametrization Enhancing Technique (CPET) has been proposed to overcome this difficulty. The basic idea is to introduce an auxiliary control function (known as the *enhancing control*) which is used to scale the original time horizon in such a way that the switching times are mapped to fixed points in the transformed time scale. CPET was originally introduced for time optimal control problems in Lee et al. (1997). The application of CPET to discrete valued optimal control problems is demonstrated in Lee et al. (1999); Lee et al. (2001); Lee et al. (2001a). An additional difficulty arises with discrete valued optimal control problems in that one needs to find not only the optimal switching times but also the corresponding optimal sequence of control values. Assuming a finite number of switching times, a heuristic method is proposed in Lee et al. (1999) to deal with this difficulty. Similar issues need to be addressed for the class of problems considered in this paper. A review of the CPET can be found in Rehbock et al. (1999).

Further, note that another efficient numerical approach for time optimal control and discrete valued optimal control problems has recently been proposed in Kaya and Noakes (2001a) and Kaya and Noakes (2001b). This is based on a combination of the Switching Time Computation (STC) method and the Time - Optimal Switchings (TOS) algorithms.

General optimal switching control problems have received limited attention in the literature. Theoretical and some computational aspects of these problems can be found in Yong (1989) and Li and Yong (1995) and in the references cited therein. A computational approach for solving optimal switching control problems with a fixed number of switching times has recently been proposed in Liu and Teo (2000). The approach in Liu and Teo (2000) involves a two transformations which turn the original problem into a form suitable for solution by a standard optimal control software. The first transformation introduces a set of scalar variables and associated constraints which determine whether a particular choice of dynamical system is active over a given time interval. This is similar to an approach proposed in Wang et al. (1996) to deal with general mixed programming problem. The resulting constraints are appended to the cost functional as penalty functions. The second transformation involves a rescaling of the original time horizon, along the lines of the CPET, to deal with the variable switching times.

The purpose of this paper is to demonstrate that a general class of optimal switching control problems can be transformed into a form suitable for solution by standard optimal control software (such as MISER3, see Jennings et al. (1991), Jennings et al. (2001)) through a more direct application of CPET. The resulting algorithm differs somewhat to that of Liu and Teo (2000) in that we do not assume a fixed number of switching times. As demonstrated by the numerical results, this can sometimes lead to more optimal objective function values.

Note, though, that neither of these algorithms can guarantee a globally optimal solution, due to the combinatorial aspect of having to choose an optimal sequence of dynamical systems.

2   PROBLEM FORMULATION

Suppose that we have a total of M given dynamical systems, $\Omega_1, \Omega_2, \ldots, \Omega_M$, defined on the time horizon $[0, T]$. Each of these may be invoked over any subinterval of the time horizon. For $i = 1, 2, \ldots, M$, let the i-th candidate system be defined by a set of first order ordinary differential equations, *i.e.*

$$\dot{\boldsymbol{x}}(t) = \boldsymbol{f}^i(t, \boldsymbol{x}(t)), \qquad t \in [0, T), \tag{2.1}$$

where $\boldsymbol{x} = [x_1, x_2, \ldots, x_n]^\top \in \mathbb{R}^n$ is the state of the system and $\boldsymbol{f}^i = [f_1^i, f_2^i, \ldots, f_n^i]^\top \in \mathbb{R}^n$ is continuously differentiable in all its arguments. We denote these systems collectively by the set

$$\Omega = \{\Omega_1, \Omega_2, \ldots, \Omega_M\}.$$

Let $K \geq 1$ be an integer. Then, we define

$$\mathcal{T}_K = \left\{ [t_1, t_2, \ldots, t_{K-1}]^\top \in \mathbb{R}^{K-1} : 0 \leq t_1 \leq t_2 \leq t_3 \leq \ldots \leq t_{K-1} \leq T \right\}, \tag{2.2}$$

$$\mathcal{V}_K = \left\{ [v_1, v_2, \ldots, v_K]^\top \in \mathbb{R}^K : v_i \in \{1, 2, \ldots, M\}, \, i = 1, 2, \ldots, K \right\}. \tag{2.3}$$

Here, each $\boldsymbol{\tau}_K = [t_1, t_2, \ldots, t_{K-1}]^\top \in \mathcal{T}_K$ is called a *feasible switching time sequence of length* K, and each $\boldsymbol{v}_K = [v_1, v_2, \ldots, v_K]^\top \in \mathcal{V}_K$ is called an *index sequence of length* K. Note how the index sequence $\boldsymbol{v}_K = [v_1, v_2, \ldots, v_K]^\top$ can be readily used to define a sequence of dynamical systems $\{\Omega_{v_1}, \Omega_{v_2}, \ldots, \Omega_{v_K}\}$. We refer to $\mathcal{T}_K \times \mathcal{V}_K$ as *the set of feasible switching sequences of length* K.

Corresponding to each K and each feasible switching sequence $(\boldsymbol{\tau}_K, \boldsymbol{v}_K) \in \mathcal{T}_K \times \mathcal{V}_K$, we can define a dynamical system over the entire time horizon as follows. Let $t_0 = 0$ and $t_K = T$, then consider

$$
\begin{aligned}
\dot{\boldsymbol{x}}(t) &= \boldsymbol{f}^{v_i}(t, \boldsymbol{x}(t)), & t \in [t_{i-1}, t_i), \quad i = 1, \ldots, K, & \tag{2.4} \\
\boldsymbol{x}(0) &= \boldsymbol{x}_0, & & \tag{2.5} \\
\boldsymbol{x}(t_i) &= \boldsymbol{x}(t_i - 0), & i = 1, \ldots, K-1, & \tag{2.6}
\end{aligned}
$$

where $x_0 \in I\!\!R^n$ is a given initial condition of the system and where we use the notation $x(t-0) = \lim_{\epsilon \to 0^-} x(t+\epsilon)$. Note how equation (2.6) insures the continuity of the resulting state of the system. Finally, let the solution to the system (2.4)-(2.6) corresponding to $(\tau_K, v_K) \in \mathcal{T}_K \times \mathcal{V}_K$ be denoted as $x(t|\tau_K, v_K)$.

For each K and each $(\tau_K, v_K) \in \mathcal{T}_K \times \mathcal{V}_K$, we define the following cost functional:

$$J(K, \tau_K, v_K) = \Phi(x(T)) + \int_0^T g(t, x(t|\tau_K, v_K))\, dt. \qquad (2.7)$$

The optimal switching control problem is then defined as:**Problem (P):** Find a K and $(\tau_K, v_K) \in \mathcal{T}_K \times \mathcal{V}_K$ to minimize the cost functional (2.7) subject to the dynamics (2.4)-(2.6).

Note how this general model encompasses the classes of time-optimal and discrete valued control problems addressed in the earlier references. For instance, in a discrete valued optimal control problem, we can consider each possible combination of the discrete control values as leading to a different dynamical system Ω_i which is obtained by simply substituting the control values. In problems where the control has multiple components with each one capable of taking on a large number of discrete values, this can lead to a large set Ω and we come back to this issue later.

We could include a range of other parameters, control functions and canonical constraints in Problem (P), but choose not to do so here for the sake of brevity. If ordinary (*i.e.* non-discrete valued) control functions were incorporated, issues regarding the partitioning of $[0, T]$ with respect to the parametrization of the controls would need to be addressed carefully to avoid discrepancies with the switching times.

Finally, note that the basic class of problems considered here is slightly different to that addressed in (Liu and Teo (2000)), because we allow K to be variable. Also, we do not consider terms directly measuring a cost of switching in the cost functional, although it is not clear whether the technique developed in Liu and Teo (2000) can actually deal with these in all cases.

Before proceeding, we make one assumption about the nature of Problem (P):

Assumption 1: There exists an optimal solution of Problem (P) where K is finite.

3 SOLUTION STRATEGY

The aim of this section is to describe a transformation which turns Problem (P) into a standard form suitable for solution by the ordinary control parametrization approach. As for the subclass of discrete valued optimal control problems dealt with in Lee et al. (1999), we use the control parametrization enhancing technique (CPET) in the second stage of this transformation.

In view of Assumption 1, let us initially suppose that the optimal number of switchings is at most N. In order to allow for all possible combinations of systems in Ω to occur, we let $L = M(N+1)$ and consider the fixed index sequence $\boldsymbol{w}_L = [w_1, w_2, \ldots, w_L]^\top \in \mathcal{V}_L$ where $w_j = ((j-1) \bmod M) + 1$, $j = 1, \ldots, L$, i.e.

$$\boldsymbol{w}_L = [1, 2, \ldots, M, 1, 2, \ldots, M, \ldots, 1, 2, \ldots, M],$$

i.e. the ordered sequence $[1, 2, \ldots, M]$ is repeated $N+1$ times within $\boldsymbol{w}_L$. Then, we consider the following system on $[0, T]$.

$$\dot{\boldsymbol{x}}(t) = \boldsymbol{f}^{w_j}(t, \boldsymbol{x}(t)), \qquad t \in [\tau_{j-1}, \tau_j), \quad j = 1, \ldots, L, \qquad (3.1)$$

$$\boldsymbol{x}(0) = \boldsymbol{x}_0, \qquad\qquad\qquad\qquad\qquad\qquad\qquad (3.2)$$

$$\boldsymbol{x}(\tau_j) = \boldsymbol{x}(\tau_j - 0), \qquad j = 1, \ldots, L-1, \qquad\qquad (3.3)$$

where $\boldsymbol{\tau}_L = [\tau_1, \tau_2, \ldots, \tau_{L-1}]^\top \in \mathcal{T}_L$ must, of course, satisfy

$$0 = \tau_0 \leq \tau_1 \leq \tau_2 \leq \ldots \leq \tau_{L-2} \leq \tau_{L-1} \leq \tau_L = T. \qquad (3.4)$$

We define the following problem:**Problem (P^N):** Given N, for $L = M(N+1)$, find a $\boldsymbol{\tau}_L \in \mathcal{T}_L$ (i.e. its components must satisfy (3.4)) such that the cost functional $J(L, \boldsymbol{\tau}_L, \boldsymbol{w}_L)$ defined by (2.7) is minimized subject to the dynamics (3.1)-(3.3).

Note how the only variables in Problem (P^N) are the components of $\boldsymbol{\tau}_L$, since both L and the index sequence $\boldsymbol{w}_L$ are assumed fixed. It should also be clear that Problem (P^N) may allow solutions with more then N switchings. In fact, it may allow solutions with up to $L-1$ switchings. In view of the algorithm described below, this is does not pose a problem, though. What is important to note is that all feasible switching sequences of length N for Problem (P) can be replicated as feasible switching sequences for Problem (P^N), simply by choosing an appropriate $\boldsymbol{\tau}_L \in \mathcal{T}_L$.

Note that a solution of Problem (P^N) may only be optimal with respect to allowing up to N switchings. Even if an optimal solution of Problem (P) has less than N switchings (in which case it is guaranteed to be amongst the feasible switching sequences of Problem (P^N)) we have no sure way of finding such a globally optimal solution. Instead, our proposed algorithm is only guaranteed to find a locally optimal solution. Indeed, if the number of candidate system, M, is large, such as is often the case in discrete valued optimal control problems, it is highly likely that only a locally optimal solution will be generated. The task of actually finding an optimal number of switchings and a corresponding optimal index sequence is theoretically challenging and has not been solved for the general class of problems under consideration here. However, in many practical applications, careful examination of the problem, based on the minimum principle, can give clues to the optimal switching sequence and thus simplify this task. See Pudney et al. (1992) and Howlett (1996) for an example involving the optimal control of a train.

We can, however, suggest a reasonable heuristic approach to approximate the globally optimal number of switchings as follows. Starting with a fixed N, we solve the Problem (P^N). We then increase N (*e.g.* $N = N + 1$) and solve Problem (P^N) again. If there is no decrease in the optimal cost, we assume that the optimal solution resulting from to the previous value of N contains the optimal number of switchings. Otherwise, we increase N further (by Assumption 1, we will only have to increase N a finite number of times). Note that, while this approach yields satisfactory results in many practical problems, it can not guarantee that the resulting N is indeed optimal.

For a given $\tau_L \in \mathcal{T}_L$, the right hand side of (3.1) may be a discontinuous function of t at the switching points τ_j, $j = 1, 2, ..., L-1$. Hence, we encounter a number of difficulties when trying to calculate the optimal τ_L in a conventional manner:

(a) We need to perform piecewise integration over intervals whose endpoints are variable;

(b) The gradient of the cost functional with respect to τ_L is not continuous(see Teo et al. (1991));

(c) When τ_{i-1} and τ_i coalesce, the number of decision variables changes.

All of these difficulties are overcome by the CPET originally developed in Lee et al. (1997) for time optimal control problems and also employed for discrete valued optimal control problems in Lee et al. (1999).

As in the cited references, we introduce the new time variable $s \in [0, L]$. Let $\mathcal{U}$ denote the class of non-negative piecewise constant scalar functions defined on $[0, L)$ with fixed switching points located at $\{1, 2, 3, ..., L-1\}$. The transformation (CPET) from $t \in [0, T]$ to $s \in [0, L]$ is defined by

$$\frac{dt}{ds} = u(s), \quad t(0) = 0, \tag{3.5}$$

where the scalar function $u \in \mathcal{U}$, $u(s) = \tau_j - \tau_{j-1}$ for $s \in [j-1, j)$, is called the *enhancing control*. Clearly, it must satisfy

$$\int_0^M u(s)ds = T. \tag{3.6}$$

Alternatively, we could replace (3.6) by the constraint

$$t(L) = T. \tag{3.7}$$

Remark 3.1 *As noted earlier, we are introducing many artificial switchings into the transformed problem by the above technique. This is necessary to allow all the possible orderings of the sequence of discrete control values to be considered when solving the transformed problem. These artificial switchings do not cause any difficulties, though, since, if $u(s) = 0$ on $[j-1, j)$, then $\tau_{j-1} = \tau_j$. Consequently, the choice of dynamical system Ω_{w_j}, valid for $s \in [j-1, j)$, has no bearing on the the solution of the original problem.*

To complete the application of the CPET to the Problem (P^N), note that, in the new time scale, the system dynamics may be conveniently rewritten as

$$\dot{\bar{x}}(s) = u(s)\bar{f}^j(s, \bar{x}(s)), \quad s \in [j-1, j), \; j = 1, \ldots, L \tag{3.8}$$

$$\bar{x}(0) = x_0 \tag{3.9}$$

$$\bar{x}(j) = \bar{x}(j-0), \quad j = 1, \ldots, L-1, \tag{3.10}$$

where we define $\bar{x}(s) = x(t(s))$, $\bar{f}^j(s, \bar{x}(s)) = f^j(t(s), x(t(s)))$ and $t(s)$ is the solution of (3.5). Furthermore, the objective functional is transformed to

$$\bar{J}(u) = \Phi(\bar{x}(L)) + \int_0^L u(s)\,\bar{g}(s, \bar{x}(s))\,ds, \tag{3.11}$$

where $\bar{g}(s, \bar{x}(s)) = g(t(s), x(t(s)))$. Finally, we define **Problem (P_e^N)**: Given N, find a $u \in \mathcal{U}$ such that the cost functional (3.11) is minimized subject to the dynamics (3.5), (3.8)-(3.10) and subject to the constraint (3.7).

Note that Problems (P^N) and (P_e^N) are equivalent. In Problem (P_e^N), rather than finding $\tau_1, \tau_2, ..., \tau_{L-1}$, we look for $u \in \mathcal{U}$. All switching points of the original problem are mapped onto the set of integers in chronological order. Piecewise integration can now be performed easily since all points of discontinuity of the dynamics in the s-domain are known and fixed. Moreover, $u \in \mathcal{U}$ is a piecewise constant function and hence Problem (P_e^N) is readily solvable by the optimal control software MISER3 (see Jennings et al. (1991), Jennings et al. (2001)), which is an implementation of the control parametrization technique (see Teo et al. (1991)). Note that the piecewise integration of (3.8) is performed automatically in MISER3. The continuity constraints (3.10) are also satisfied automatically when executing the code in standard mode.

The solution of (3.5) yields $t(s)$, so the the state trajectory $x(t)$, of the original problem defined on $[0, T]$ can be reconstructed easily. It is clear from Assumption 1 that there exists an integer N such that an optimal solution to Problem (P_e^N) is also an optimal solution of the original Problem (P). We must note, though, that MISER3 uses a gradient based optimization approach and we are therefore not guaranteed of finding a globally optimal solution to Problem (P_e^N) and therefore Problem (P).

Remark 3.2 *A large number, M, of candidate dynamical systems will result in a large number of control parameters once Problem (P_e^N) is parametrized. This is particularly true in discrete valued control problems where the control has multiple components each capable of taking on a large number of discrete values and it reflects the combinatorial nature of the class of problems we consider here. A similarly large number of parameters arises in the technique proposed in Liu and Teo (2000).*

Remark 3.3 *Note that an optimal solution of Problem (P_e^N) found by MISER3 does not necessarily represent a unique parametrization of the corresponding optimal solution of Problem (P^N), although this is of no practical consequence. The particular parametrization found depends on the initial guess supplied to MISER3.*

4 NUMERICAL EXAMPLES AND DISCUSSION

We consider the numerical example given in Liu and Teo (2000). In this problem, there are 3 candidate dynamical systems and the time horizon is $[0, 2]$. We have

$$
\begin{aligned}
\Omega_1 : \quad \dot{x}(t) &= f^1(t, x) = 1, \quad t \in [0, 2), \\
\Omega_2 : \quad \dot{x}(t) &= f^2(t, x) = -1, \quad t \in [0, 2), \\
\Omega_3 : \quad \dot{x}(t) &= f^3(t, x) = 2t, \quad t \in [0, 2),
\end{aligned}
$$

The given initial condition is $x(0) = 0$. The cost functional is

$$
J(\boldsymbol{\tau}_K, \boldsymbol{v}_K) = (x(2))^2 + \int_0^2 \left(\sin(\frac{\pi t}{2}) - x(t) \right)^2 dt.
$$

We have $M = 3$ in this case. Following the approach in the previous section, we have $L = 3(N + 1)$ and

$$
\boldsymbol{w}_L = [w_1, w_2, \ldots, w_L]^\top = [1, 2, 3, 1, 2, 3, \ldots, 1, 2, 3]^\top.
$$

The transformed problem may then be written asMinimize the cost functional

$$
J(u) = (\bar{x}(L))^2 + \int_0^L \left(\sin(\frac{\pi t(s)}{2}) - \bar{x}(s) \right)^2 ds
$$

subject to the dynamics

$$
\begin{aligned}
\dot{\bar{x}}(s) &= u(s) f^j(t(s), \bar{x}(s)), \quad s \in [j - 1, j), \ j = 1, \ldots, L, \\
\bar{x}(0) &= 0, \\
\bar{x}(j) &= \bar{x}(j - 0), \quad j = 1, \ldots, L - 1.
\end{aligned}
$$

We solve the problem for various values of N. For $N = 1$, we get different solutions depending on the initial guess we provide for u. The results are summarized in Table 4.1. Note that we do not express these results in terms of u but in terms of the original statement of the problem for the sake of clarity.

We note from Table 4.1 that the proposed method will generate local solutions depending on the choice of initial guess. For the first initial guess listed in Table 4.1, we obtain a locally optimal solution which involves one switching and is identical to that produced in Liu and Teo (2000) (where at most one switching

	Case 1	Case 2
Initial Sequence	$\{\Omega_1\}$	$\{\Omega_1, \Omega_2, \Omega_3, \Omega_1, \Omega_2, \Omega_3\}$
Initial Switching Times	no switch	$0.1, 0.3, 0.8, 1.1, 1.5$
Optimal Sequence	$\{\Omega_1, \Omega_2\}$	$\{\Omega_1, \Omega_3, \Omega_1, \Omega_2\}$
Optimal Switching Times	1.03463	$0.501, 0.797, 0.990$
Optimal Cost	0.003611	0.003204

Table 4.1 Result for $N = 1$.

was allowed). However, as the second line in the table shows, a different initial guess produces quite a different solution with 3 switches and a significantly lower cost. Virtually all other initial guesses we tried resulted in one of these two solutions.

Next we tried increasing N to $N = 3$ in order to see if there are more optimal solutions if we allow more switchings. Again, many initial guesses were tested, with most of these leading to a solution with optimal switching sequence $\{\Omega_1, \Omega_3, \Omega_1, \Omega_2, \Omega_1, \Omega_2\}$ and corresponding switching times 0.50156, 0.82533, 0.91906, 0.97595, and 1.03862. The optimal cost value was 0.003170. Again, we did obtain a slightly worse local optimal solution with one of the initial guesses tested.

A further increase to $N = 7$ did not yield any other solutions with a lower cost value, so the one obtained with $N = 3$ appears to be optimal, *i.e.* our estimate for the optimal number of switches is 5.

5 CONCLUSIONS

We have presented a new approach to solve a large class of optimal switching control problems. It is a natural extension of the approaches used for certain subclasses of these problems in some of our earlier work. Furthermore, it is more transparent than a recent alternative approach and it does not require the introduction of a set of equality constraints handled via penalty methods. On the other hand, the new approach is not always able to solve a problem when a fixed number of switching points is prescribed.

Neither of these methods can necessary guarantee a globally optimal solution and future work will need to address this issue.

References

P. Howlett (1996), Optimal strategies for the control of a train. *Automatica*, Vol. 32, pp. 519-532.

L.S. Jennings, M.E. Fisher, K.L. Teo and C.J. Goh (1991), Miser3: Solving optimal control problems - an update. *Advances in Engineering Software and Workstations*, Vol. 13, pp. 190-196.

L.S. Jennings, M.E. Fisher, K.L. Teo and C.J. Goh (2001), *MISER3 Optimal Control Software, Version 3: Theory and User Manual*, http://www.cado.uwa.edu.au/miser/.

C.Y. Kaya and J.L. Noakes, Computations and time - optimal controls, *Optimal Control Applications and Methods*, to appear.

C.Y. Kaya and J.L. Noakes, Computational method for time - optimal switching control, in press.

H.W.J. Lee, X.Q. Cai and K.L. Teo (2001), An Optimal Control Approach to Manpower Planning Problem, *Mathematical Problems in Engineering*, Vol. 7, pp.155-175.

H.W.J. Lee, K.L. Teo, L.S. Jennings and V. Rehbock (1997), Control parametrization enhancing technique for time optimal control problems, *Dynamical Systems and Applications*, Vol. 6(2), pp. 243-261.

H.W.J. Lee, K.L. Teo and A.E.B. Lim (2001), Sensor Scheduling in Continuous Time, *Automatica*, Vol. 37, pp. 2017-2023.

H.W.J. Lee, K.L. Teo, V. Rehbock and L.S. Jennings (1999), Control parametrization enhancing technique for discrete-valued control problems, *Automatica*, Vol. 35(8), pp. 1401-1407.

X. Li and J. Yong (1995), *Optimal Control Theory for Infinite Dimensional Systems*, Birkhäuser, Boston.

Y. Liu and K.L Teo (2000), Computational method for a class of optimal switching control problems, in X.Q. Yang *et al* (Eds.), *Progress in Optimization: Contributions from Australasia*, Kluwer Academic Publishers, Dordrecht, pp. 221-237.

P. Pudney P. Howlett and B. Benjamin (1992), Determination of optimal driving strategies for the control of a train. In B.R. Benjamin B.J. Noye and L.H. Colgan, editors, *Proc. Computational Techniques and Applications, CTAC 91,* pages 241-248. Computational Mathematical Group, Australian Mathematical Society.

V. Rehbock, K.L. Teo, L.S. Jennings and H.W.J. Lee (1999), A survey of the control parametrization and control parametrization enhancing methods for constrained optimal control problems, A. Eberhard *et al* (Eds.), *Progress in Optimization: Contributions from Australia,* Kluwer Academic Publishers, Dordrecht, pp. 247-275.

K.L. Teo, C.J. Goh and K.H. Wong (1991), *A Unified Computational Approach to Optimal Control Problems,* Longman Scientific and Technical, Essex.

S. Wang, K.L. Teo and H.W.J. Lee (1996), A new approach to nonlinear mixed programming problems, Preprint.

J. Yong (1989), Systems governed by ordinary differential equations with continuous, switching and impulse controls, *Appl. Math. Optim.,* Vol. 20, pp. 223-235.

22   A SOLUTION TO HAMILTON-JACOBI EQUATION BY NEURAL NETWORKS AND OPTIMAL STATE FEEDBACK CONTROL

Kiyotaka Shimizu

Faculty of Science and Technology,   Keio University

Abstract: This paper is concerned with state feedback controller design using neural networks for nonlinear optimal regulator problem. Nonlinear optimal feedback control law can be synthesized by solving the Hamilton-Jacobi equation with three layered neural networks. The Hamilton-Jacobi equation generates the value function by which the optimal feedback law is synthesized. To obtain an approximate solution of the Hamilton-Jacobi equation, we solve an optimization problem by the gradient method, which determines connection weights and thresholds in the neural networks. Gradient functions are calculated explicitly by the Lagrange multiplier method and used in the learning algorithm of the networks. We propose also a device such that an approximate solution to the Hamilton-Jacobi equation converges to the true value function. The effectiveness of the proposed method was confirmed with simulations for various plants.

Key words: Nonlinear optimal control, Hamilton-Jacobi equation, neural network, state feedback.

1   INTRODUCTION

This paper is concerned with optimal state feedback control of nonlinear systems. To solve nonlinear optimal regulator problem, we solve the Hamilton-Jacobi equation using neural networks and then synthesize the optimal state feedback control law with its approximate solution.

Most studies on optimal control of nonlinear systems have been made by application of calculus of variation. They are aimed at calculating optimal control input $\boldsymbol{u}^o(t)$, $t \in [0, t_1]$ and the corresponding optimal trajectory $\boldsymbol{x}^o(t)$, $t \in [0, t_1]$ starting from an initial condition $\boldsymbol{x}(0)$. This yields the so called open-loop control, but a practically interesting matter is to obtain optimal state feedback control law $\boldsymbol{u}^o(t) = \boldsymbol{\alpha}(\boldsymbol{x}(t))$, $t \in [0, t_1]$ which brings us a closed loop system.

As is well known, optimal regulator is a typical control problem in which a linear system and a quadratic performance functional are considered and the Riccati equation plays an important role. The reason of attaching importance to the optimal regulator is that it offers a systematic method to design a state feedback control law and consequently one can construct a closed-loop control system.

In contrast it is very hard for nonlinear systems to design such an optimal state feedback controller. To synthesize the optimal state feedback control law resulting in the stable closed loop system, one must solve the Hamilton-Jacobi partial differential equation (H-J equation).

However, nonlinear optimal regulator is of restrictive use since it is extremely difficult to solve H-J equation analytically. Hence we need to develop approximate solution of the H-J equation.

In the past several approaches to solve H-J equation were proposed as follows.

(1)   application of Taylor series expansion [Lukes (1969)]

(2)   application of neural networks [Goh (1993)]

(3)   application of Galerkin method [Beard (1997)]

(4)   application of Spline function [Lee (1996)]

With the Taylor series expansion, one can obtain an accurate approximate solution around an operating point. However, it is difficult to approximate uniformly in broad range.

The principle of neural network approximation for H-J equation is all explained in [Goh (1993)]. However, gradient functions necessary for neural network learning was not explicitly represented there. In the Galerkin method the problem is how to select basis functions, and besides its application is limited only to linear partial differential equations. Hence it was required to introduce the generalized H-J equation [Saridis (1986),Beard (1997)].

In this paper we propose a method to obtain an approximate solution to the H-J equation and to realize optimal state feedback control law using a three layered neural network. The method is relatively easy when applied to affine nonlinear systems, since we have only to approximate the value function in the H-J equation with the network such that an error of H-J equation goes to zero.

When applied to general nonlinear systems, however, its computation becomes very complex, because we must approximate both the value function and control inputs with neural networks such that necessary optimality conditions for the control inputs are satisfied. We can calculate gradient functions with respect to connection weights and thresholds in the network explicitly by applying Lagrange multiple method. Hence learning of the networks is carried out very efficiently and systematically.

Note that the H-J equation does not necessarily possess a unique solution and so an approximate solution obtained by the neural network is not guaranteed to be the true value function from which the optimal feedback law is synthesized.

Thus we make a device for learning of the networks, that is, we propose an idea to let any approximate solution converge to the true value function, using a stabilizing solution of the Riccati equation for the linearized LQ regulator problem.

The usefulness of the proposed method was confirmed from simulation results for both affine and general nonlinear systems.

2   NONLINEAR OPTIMAL REGULATOR AND HAMILTON-JACOBI EQUATION

Consider nonlinear optimal regulator problem

$$\min_{\boldsymbol{u}} \int_0^\infty \{q(\boldsymbol{x}(t)) + \boldsymbol{u}(t)^T R \boldsymbol{u}(t)\}dt \qquad (2.1a)$$

$$\text{subj. to } \dot{\boldsymbol{x}}(t) = \boldsymbol{f}(\boldsymbol{x}(t), \boldsymbol{u}(t)), \quad \boldsymbol{x}(0) = \boldsymbol{x}_0 \qquad (2.1b)$$

where $x(t) \in R^n$ and $u(t) \in R^r$ are the state vector and the control vector, respectively.

We assume the following:

Assumption 2.1 *System (2.1b) is stabilizable in the sence that for any initial condition $x(0)$ there exists a control law $u(\cdot)$ such that $x(t) \to 0$ as $t \to \infty$.*

Assumption 2.2 $f : R^n \times R^r \to R^n$ *is continuously differentiable in (x, u) and $f(0, 0) = 0$.*

Assumption 2.3 *It holds that $q(0) = 0$, $q(x) \geq 0$, and $R > 0$.*

We design a nonlinear state feed back controller $u(t) = \alpha(x(t))$ for problem (2.1), where $\alpha : R^n \to R^r$ is C^1 and $\alpha(0) = 0$. Then the nonlinear optimal regulator problem (2.1) results in the following stationary Hamilton-Jacobi equation

$$\begin{aligned}
0 &= \min_{u} H\left(x, u, V_x(x)\right) \\
&= \min_{u} \left\{ q(x) + u^T R u + V_x(x) f(x, u) \right\}
\end{aligned} \tag{2.2}$$

where, $H(x, u, V_x(x))$ is the Hamiltonian function

$$H(x, u, V_x(x)) = q(x) + u^T R u + V_x(x) f(x, u) \tag{2.3}$$

and $V(x)$ denotes the value function that is semi-positive definite.

$$V(x) = \min_{u} \int_t^{\infty} \{q(x) + u^T R u\} dt, \quad x(t) = x \tag{2.4}$$

Further, the stationary H-J equation (2.2) satisfies a boundary condition,

$$V(0) = 0 \tag{2.5}$$

Meanwhile, a necessary optimality condition

$$\nabla_u H(x, u, V_x(x)) = 2R^T u + f_u(x, u)^T V_x(x)^T = 0 \tag{2.6}$$

holds for u satisfying (2.2). Therefore the optimal control must satisfy the following partial differential equations.

$$q(x) + u^T R u + V_x(x) f(x, u) = 0 \tag{2.7}$$

$$V(0) = 0 \tag{2.8}$$

$$2R^T u + f_u(x, u)^T V_x(x)^T = 0 \tag{2.9}$$

As mentioned above, the nonlinear optimal regulator problem is equivalent to find $V(x)$ and $u=\alpha(x)$ which satisfy conditions $(2.7)\sim(2.9)$ at the same time.

Here let us consider affine nonlinear system

$$\dot{x} = f(x) + G(x)u \tag{2.10}$$

From (2.6)

$$\nabla_u H(x, u, V_x(x)) = 2R^T u + G(x)^T V_x(x)^T = 0 \tag{2.11}$$

Solve (2.11) with regard to u to obtain

$$u(x) = -\frac{1}{2}R^{-1}G(x)^T V_x(x)^T \tag{2.12}$$

Substituting this into (2.7) and taking account of symmetricity of R, we have

$$q(x) + V_x(x)f(x) - \frac{1}{4}V_x(x)G(x)R^{-1}G(x)^T V_x(x)^T = 0 \tag{2.13}$$

If we can obtain $V(x)$ satisfying this equation, then the optimal state feedback control law $u(x)$ is given by (2.12).

3 APPROXIMATE SOLUTION TO HAMILTON-JACOBI EQUATION AND OPTIMAL STATE FEEDBACK CONTROL LAW

[A] Affine Nonlinear System Case

Firstly, we try to generate an approximate solution to the H-J equation (2.13) by using a neural networks. As the neural networks a three layered neural network is used :

$$z = W_1 x + \theta \tag{3.1}$$

$$y = W_2 \sigma(z) + a \tag{3.2}$$

where $y \in R^n$ and $z \in R^q$ are the output and internal state of the neural network, respectively and $W_1 \in R^{q \times n}$, $W_2 \in R^{n \times q}$, $\theta \in R^q$ and $a \in R^n$ are the connection weight matrices, the threshold and the constant, respectively.

Further $\sigma : R^n \to R^q$ denotes the sigmoid function, and we use hyperbolic tangent functions as the sigmoid function $\sigma(z)$, i.e.,

$$\sigma_i(z_i) = \tanh z_i = \frac{\exp(z_i) - \exp(-z_i)}{\exp(z_i) + \exp(-z_i)}, \quad i = 1, \cdots, q$$

Then the value function $V(x)$ is approximated with y as

$$V^N(x) = y(x)^T y(x) \tag{3.3}$$

The boundary conditon $V(0) = 0$ is easily satisfied by taking $a = -W_2\sigma(\theta)$. Note that when we set the value function and a like this, $V^N(0) = 0$, $V^N(x) \geq 0 \ \forall x$ are always satisfied regardless of values of W_1, W_2, θ.

Next let us consider learning of the neural network so that $V^N(x)$ satisfies (2.13). As an error of (2.13) put

$$e(x) = q(x) + V_x^N(x)f(x) - \frac{1}{4}V_x^N(x)G(x)R^{-1}G(x)^T V_x^N(x)^T \tag{3.4}$$

and define a performance function $E[W_1, W_2, \theta]$ for learning as follows.

$$E[W_1, W_2, \theta] = \sum_{p=1}^{P} |e(x^p)|^2 \tag{3.5}$$

Here x^p denotes the element of a set $\Delta \triangleq \{x^p | x^p \in \Omega, \ p = 1, 2, \cdots, P\}$ where $\Omega \subset R^n$ is a subregion of state space and Δ is the discretized set of Ω.

Learning problem of the neural network is formulated as the following optimization problem.

$$\min_{W_1, W_2, \theta} E[W_1, W_2, \theta] = \sum_{p=1}^{P} |e(x^p)|^2 \tag{3.6a}$$

$$\text{subj. to } z^p = W_1 x^p + \theta \tag{3.6b}$$

$$y(x^p) = W_2\sigma(z^p) - W_2\sigma(\theta) \tag{3.6c}$$

$$V^N(x^p) = y(x^p)^T y(x^p) \tag{3.6d}$$

$$e(x^p) = q(x^p) + V_x^N(x^p)f(x^p)$$
$$- \frac{1}{4}V_x^N(x^p)G(x^p)R^{-1}G(x^p)^T V_x^N(x^p)^T \tag{3.6e}$$
$$p = 1, 2, \cdots, P$$

Here $V_x^N(x)$ included in $e(x)$ becomes from (3.6b)$\sim$(3.6d) as follows.

$$V_x^N(x) = \frac{\partial V^N}{\partial y}\frac{\partial y}{\partial \sigma}\frac{\partial \sigma}{\partial z}\frac{\partial z}{\partial x} = 2y^T W_2 \nabla\sigma(z) W_1$$
$$= 2\{\sigma(z) - \sigma(\theta)\}^T W_2^T W_2 \nabla\sigma(z) W_1 \tag{3.7}$$

For learning of the network we need concrete expressions of gradients of the performance function with respect to connection weights, etc., i.e., $\nabla_{W_1} E[W_1, W_2, \theta]$,

$\nabla_{W_2} E[W_1, W_2, \boldsymbol{\theta}], \nabla_{\boldsymbol{\theta}} E[W_1, W_2, \boldsymbol{\theta}]$. Since

$$\nabla_{W_1} E[W_1, W_2, \boldsymbol{\theta}] = \sum_{p=1}^{P} \nabla_{W_1} |e(\boldsymbol{x}^p)|^2 \tag{3.8}$$

$$\nabla_{W_2} E[W_1, W_2, \boldsymbol{\theta}] = \sum_{p=1}^{P} \nabla_{W_2} |e(\boldsymbol{x}^p)|^2 \tag{3.9}$$

$$\nabla_{\boldsymbol{\theta}} E[W_1, W_2, \boldsymbol{\theta}] = \sum_{p=1}^{P} \nabla_{\boldsymbol{\theta}} |e(\boldsymbol{x}^p)|^2 \tag{3.10}$$

we have only to derive expressions of $\nabla_{W_1} |e(\boldsymbol{x})|^2$, $\nabla_{W_2} |e(\boldsymbol{x})|^2$, and $\nabla_{\boldsymbol{\theta}} |e(\boldsymbol{x})|^2$. Hence we derive these gradients, applying the Lagrange multiplier method as below.

Define first the following variable $\boldsymbol{v} \in R^n$:

$$\boldsymbol{v} = 2W_1^T \nabla \boldsymbol{\sigma}(\boldsymbol{z}) W_2^T \boldsymbol{y} = V_{\boldsymbol{x}}^N(\boldsymbol{x})^T$$

Then we have

$$e(\boldsymbol{x}) = q(\boldsymbol{x}) + \boldsymbol{v}^T \boldsymbol{f}(\boldsymbol{x}) - \frac{1}{4} \boldsymbol{v}^T G(\boldsymbol{x}) R^{-1} G(\boldsymbol{x})^T \boldsymbol{v} \tag{3.11}$$

Next, to calculate the gradients of $|e(\boldsymbol{x})|^2$, let us define the Lagrangian L with Lagrange multipliers $\boldsymbol{\lambda} \in R^n, \boldsymbol{\beta} \in R^q, \boldsymbol{\gamma} \in R^n$.

$$\begin{aligned}
&L(\boldsymbol{x}; W_1, W_2, \boldsymbol{\theta}; \boldsymbol{v}, \boldsymbol{z}, \boldsymbol{y}; \boldsymbol{\lambda}, \boldsymbol{\beta}, \boldsymbol{\gamma}) \\
&= \left(q(\boldsymbol{x}) + \boldsymbol{v}^T \boldsymbol{f}(\boldsymbol{x}) - \frac{1}{4} \boldsymbol{v}^T G(\boldsymbol{x}) R^{-1} G(\boldsymbol{x})^T \boldsymbol{v} \right)^2 \\
&\quad + \boldsymbol{\lambda}^T (2W_1^T \nabla \boldsymbol{\sigma}(\boldsymbol{z}) W_2^T \boldsymbol{y} - \boldsymbol{v}) + \boldsymbol{\beta}^T (W_1 \boldsymbol{x} + \boldsymbol{\theta} - \boldsymbol{z}) \\
&\quad + \boldsymbol{\gamma}^T (W_2 \boldsymbol{\sigma}(\boldsymbol{z}) - W_2 \boldsymbol{\sigma}(\boldsymbol{\theta}) - \boldsymbol{y})
\end{aligned} \tag{3.12}$$

By the chain rule of derivatives and formulae[1] for gradients and symmetricity of $\nabla \boldsymbol{\sigma}(\boldsymbol{z})$, partial derivatives of the Lagrangian L with respect to each variable are calculated as follows.

$$\nabla_{W_1} L = 2 \nabla \boldsymbol{\sigma}(\boldsymbol{z}) W_2^T \boldsymbol{y} \boldsymbol{\lambda}^T + \boldsymbol{\beta} \boldsymbol{x}^T \tag{3.13}$$

$$\nabla_{W_2} L = 2 \boldsymbol{y} \boldsymbol{\lambda}^T W_1^T \nabla \boldsymbol{\sigma}(\boldsymbol{z}) + \boldsymbol{\gamma}(\boldsymbol{\sigma}(\boldsymbol{z})^T - \boldsymbol{\sigma}(\boldsymbol{\theta})^T) \tag{3.14}$$

$$\nabla_{\boldsymbol{\theta}} L = \boldsymbol{\beta} - \nabla \boldsymbol{\sigma}(\boldsymbol{\theta}) W_2^T \boldsymbol{\gamma} \tag{3.15}$$

$$\nabla_{\boldsymbol{v}} L = 2 \left(\boldsymbol{f}(\boldsymbol{x}) - \frac{1}{2} G(\boldsymbol{x}) R^{-1} G(\boldsymbol{x})^T \boldsymbol{v} \right) e(\boldsymbol{x}) - \boldsymbol{\lambda} = 0 \tag{3.16}$$

$$\nabla_y L = 2W_2 \nabla \sigma(z) W_1 \lambda - \gamma = 0 \tag{3.17}$$

$$\nabla_z L = \nabla^2 \sigma(z) \bullet (2W_1 \lambda \otimes y^T W_2) - \beta + \nabla \sigma(z) W_2^T \gamma = 0 \tag{3.18}$$

$$\nabla_\lambda L = 2W_1^T \nabla \sigma(z) W_2^T y - v = 0 \tag{3.19}$$

$$\nabla_\beta L = W_1 x + \theta - z = 0 \tag{3.20}$$

$$\nabla_\gamma L = W_2 \sigma(z) - W_2 \sigma(\theta) - y = 0 \tag{3.21}$$

where $x \otimes y$ and $x \bullet y$ denote the tensor product and the inner product of array $x \in X$ and $y \in Y$, respectively, and $\nabla^2 \sigma(z) \in R^{q \times q \times q}$ is the second order derivative array. From $(3.16) \sim (3.21)$ variables λ, β, γ are obtained as

$$\lambda = 2 \left(f_0(x) - \frac{1}{2} G(x) R^{-1} G(x)^T v \right) e(x) \tag{3.22}$$

$$\gamma = 2W_2 \nabla \sigma(z) W_1 \lambda \tag{3.23}$$

$$\beta = \nabla^2 \sigma(z) \bullet (2W_1 \lambda \otimes y^T W_2) + \nabla \sigma(z) W_2^T \gamma \tag{3.24}$$

By substituting these λ, β, γ into $(3.13) \sim (3.15)$ we obtain $\nabla_{W_1} L, \nabla_{W_2} L, \nabla_\theta L$. Then it holds that

$$\nabla_{W_1} L = \nabla_{W_1} |e(x)|^2 \tag{3.25}$$

$$\nabla_{W_2} L = \nabla_{W_2} |e(x)|^2 \tag{3.26}$$

$$\nabla_\theta L = \nabla_\theta |e(x)|^2 \tag{3.27}$$

In the above calculation of partial derivatives, not vector-matrix expression but array expression is used in only one place. Defining matrices

$$Z = \mathrm{diag}[\sigma_1''(z_1)h_1, \sigma_2''(z_2)h_2, \cdots, \sigma_q''(z_q)h_q]$$
$$\text{where} \quad h = W_2^T y = W_2^T W_2 \{\sigma(z) - \sigma(\theta)\}$$

however, its array expression part $\nabla^2 \sigma(z) \bullet \{2W_1 \lambda \otimes y^T W_2\}$ can be rewritten by vector-matrix one as follows.

$$\nabla^2 \sigma(z) \bullet \{2W_1 \lambda \otimes y^T W_2\} = 2ZW_1 \lambda$$

After all, from $(3.13) \sim (3.15)$, $(3.22) \sim (3.24)$ the gradients of performance function $(3.6a)$ $\nabla_{W_1} E, \nabla_{W_2} E$, and $\nabla_\theta E$ are acquired as follows.

$$\nabla_{W_1} E[W_1, W_2, \theta] = \sum_{p=1}^{P} 2\nabla \sigma(z^p) W_2^T y^p \lambda^{pT} + \beta^p x^{pT} \tag{3.28}$$

$$\nabla_{W_2} E[W_1, W_2, \boldsymbol{\theta}] = \sum_{p=1}^{P} 2y^p \lambda^{pT} W_1^T \nabla \boldsymbol{\sigma}(z^p) + \gamma^p (\boldsymbol{\sigma}(z^p)^T - \boldsymbol{\sigma}(\boldsymbol{\theta})^T) \quad (3.29)$$

$$\nabla_{\boldsymbol{\theta}} E[W_1, W_2, \boldsymbol{\theta}] = \sum_{p=1}^{P} \beta^p - \nabla \boldsymbol{\sigma}(\boldsymbol{\theta}) W_2^T \gamma^p \quad (3.30)$$

where $\lambda^p, \beta^p, \gamma^p$ are given by (3.22)$\sim$(3.24) coresponding to x^p.

Using these gradients, we can apply the steepest descent method to obtain optimal connection weights W_1, W_2 and threshold $\boldsymbol{\theta}$:

$$W_1^{k+1} = W_1^k - \alpha \nabla_{W_1} E[W_1^k, W_2^k, \boldsymbol{\theta}^k] \quad (3.31)$$

$$W_2^{k+1} = W_2^k - \alpha \nabla_{W_2} E[W_1^k, W_2^k, \boldsymbol{\theta}^k] \quad (3.32)$$

$$\boldsymbol{\theta}^{k+1} = \boldsymbol{\theta}^k - \alpha \nabla_{\boldsymbol{\theta}} E[W_1^k, W_2^k, \boldsymbol{\theta}^k] \quad (3.33)$$

where $\alpha > 0$ is a proportional coefficient and k denotes the iteration number.

Consequently, $V_{\boldsymbol{x}}^N(\boldsymbol{x})$ in (3.7) is obtained, which yields the optimal state feedback control law given by (2.12), as follows.

$$\boldsymbol{u}_N(\boldsymbol{x}) = -R^{-1} G(\boldsymbol{x})^T W_1^T \nabla \boldsymbol{\sigma}(z) W_2^T W_2 \{ \boldsymbol{\sigma}(z) - \boldsymbol{\sigma}(\boldsymbol{\theta}) \} \quad (3.34)$$

[B] General Nonlinear System Case

In order to get the optimal feedback law satisfying (2.7)$\sim$(2.9) by solving the optimal regulator problem (2.1), we must approximate the value function $V(\boldsymbol{x})$ and the state feedback control law $\boldsymbol{u}(\boldsymbol{x})$ with separate neural networks. Hence we use for $V(\boldsymbol{x})$ the same neural network used in the affine nonlinear case.

$$z_1 = W_1 \boldsymbol{x} + \boldsymbol{\theta}_1 \quad (3.35)$$

$$y = W_2 \boldsymbol{\sigma}(z_1) + \boldsymbol{a} \quad (3.36)$$

where $W_1 \in R^{q \times n}$, $W_2 \in R^{n \times q}$, $\boldsymbol{\theta}_1 \in R^q$, and $\boldsymbol{a} \in R^n (\boldsymbol{a} = -W_2 \boldsymbol{\sigma}(\boldsymbol{\theta}_1))$. In like manner as the case [A] the value function $V(\boldsymbol{x})$ is approximated with y as

$$V^N(\boldsymbol{x}) = y(\boldsymbol{x})^T y(\boldsymbol{x}) \quad (3.37)$$

It is noted that the boundary condition $V(\boldsymbol{0}) = 0$ is easily satisfied by setting $\boldsymbol{a} = -W_2 \boldsymbol{\sigma}(\boldsymbol{\theta}_1)$. Next for $\boldsymbol{u}(\boldsymbol{x})$ we use the neural network

$$z_2 = W_3 \boldsymbol{x} + \boldsymbol{\theta}_2 \quad (3.38)$$

$$\boldsymbol{u}_N = W_4 \boldsymbol{\sigma}(z_2) + \boldsymbol{b} \quad (3.39)$$

where $W_3 \in R^{m \times n}$, $W_4 \in R^{r \times m}$, $\boldsymbol{\theta}_2 \in R^m$, and $\boldsymbol{b} \in R^r$. $\boldsymbol{z}_2 \in R^m$ and $\boldsymbol{u}_N \in R^r$ denote the internal state and output vector of the neural network, respectively. The condition $\boldsymbol{u}_N(\mathbf{0}) = \mathbf{0}$ is always satisfied by setting $\boldsymbol{b} = -W_4 \boldsymbol{\sigma}(\boldsymbol{\theta}_2)$.

Define errors of equation (2.7) and (2.9):

$$e_1(\boldsymbol{x}) = q(\boldsymbol{x}) + \boldsymbol{u}_N(\boldsymbol{x})^T R \boldsymbol{u}_N(\boldsymbol{x}) + V_{\boldsymbol{x}}^N(\boldsymbol{x}) \boldsymbol{f}(\boldsymbol{x}, \boldsymbol{u}_N(\boldsymbol{x})) \tag{3.40}$$

$$e_2(\boldsymbol{x}) = 2R^T \boldsymbol{u}_N(\boldsymbol{x}) + \boldsymbol{f}_{\boldsymbol{u}}(\boldsymbol{x}, \boldsymbol{u}_N(\boldsymbol{x}))^T V_{\boldsymbol{x}}^N(\boldsymbol{x})^T \tag{3.41}$$

Note here that $e_2 \in R^r$. For simplicity letting $\overline{W} \overset{\triangle}{=} \{W_1, W_2, W_3, W_4\}$ and $\overline{\boldsymbol{\theta}} \overset{\triangle}{=} \{\boldsymbol{\theta}_1, \boldsymbol{\theta}_2\}$, we define the performance function $E[\overline{W}, \overline{\boldsymbol{\theta}}]$ for learning as follows.

$$E[\overline{W}, \overline{\boldsymbol{\theta}}] = \sum_{p=1}^{P} \left\{ |e_1(\boldsymbol{x}^p)|^2 + \|e_2(\boldsymbol{x}^p)\|^2 \right\} \tag{3.42}$$

The learning problem is formulated as the following optimizaton problem.

$$\min_{\overline{W}, \overline{\boldsymbol{\theta}}} E[\overline{W}, \overline{\boldsymbol{\theta}}] = \sum_{p=1}^{P} \left\{ |e_1(\boldsymbol{x}^p)|^2 + \|e_2(\boldsymbol{x}^p)\|^2 \right\} \tag{3.43a}$$

$$\text{subj. to} \quad \boldsymbol{z}_1^p = W_1 \boldsymbol{x}^p + \boldsymbol{\theta}_1 \tag{3.43b}$$

$$\boldsymbol{y}(\boldsymbol{x}^p) = W_2 \boldsymbol{\sigma}(\boldsymbol{z}_1^p) - W_2 \boldsymbol{\sigma}(\boldsymbol{\theta}_1) \tag{3.43c}$$

$$V^N(\boldsymbol{x}^p) = \boldsymbol{y}(\boldsymbol{x}^p)^T \boldsymbol{y}(\boldsymbol{x}^p) \tag{3.43d}$$

$$\boldsymbol{z}_2^p = W_3 \boldsymbol{x}^p + \boldsymbol{\theta}_2 \tag{3.43e}$$

$$\boldsymbol{u}_N(\boldsymbol{x}^p) = W_4 \boldsymbol{\sigma}(\boldsymbol{z}_2^p) - W_4 \boldsymbol{\sigma}(\boldsymbol{\theta}_2) \tag{3.43f}$$

$$e_1(\boldsymbol{x}^p) = q(\boldsymbol{x}^p) + \boldsymbol{u}_N(\boldsymbol{x}^p)^T R \boldsymbol{u}_N(\boldsymbol{x}^p) + V_{\boldsymbol{x}}^N(\boldsymbol{x}^p) \boldsymbol{f}(\boldsymbol{x}^p, \boldsymbol{u}_N(\boldsymbol{x}^p))$$

$$\tag{3.43g}$$

$$e_2(\boldsymbol{x}^p) = 2R^T \boldsymbol{u}_N(\boldsymbol{x}^p) + \boldsymbol{f}_{\boldsymbol{u}}(\boldsymbol{x}^p, \boldsymbol{u}_N(\boldsymbol{x}^p))^T V_{\boldsymbol{x}}^N(\boldsymbol{x}^p)^T \tag{3.43h}$$

$$p = 1, 2, \cdots, P$$

Here $V_{\boldsymbol{x}}^N(\boldsymbol{x})$ becomes

$$
\begin{aligned}
V_{\boldsymbol{x}}^N(\boldsymbol{x}) &= 2\boldsymbol{y}^T W_2 \nabla \boldsymbol{\sigma}(\boldsymbol{z}_1) W_1 \\
&= 2\{\boldsymbol{\sigma}(\boldsymbol{z}_1) - \boldsymbol{\sigma}(\boldsymbol{\theta}_1)\}^T W_2^T W_2 \nabla \boldsymbol{\sigma}(\boldsymbol{z}_1) W_1
\end{aligned}
\tag{3.44}
$$

as the same as (3.7).

Concrete expressions of $\nabla_{W_1} E$, $\nabla_{W_2} E$, $\nabla_{W_3} E$, $\nabla_{W_4} E$, $\nabla_{\boldsymbol{\theta}_1} E$, $\nabla_{\boldsymbol{\theta}_2} E$ are necessary for the learning, but since

$$\nabla_{W_i} E[\overline{W}, \overline{\theta}] = \sum_{p=1}^{P} \nabla_{W_i} \left\{ |e_1(x^p)|^2 + \|e_2(x^p)\|^2 \right\} \tag{3.45}$$

$$\nabla_{\theta_i} E[\overline{W}, \overline{\theta}] = \sum_{p=1}^{P} \nabla_{\theta_i} \left\{ |e_1(x^p)|^2 + \|e_2(x^p)\|^2 \right\} \tag{3.46}$$

we have to know $\nabla_{W_i}\{|e_1(x)|^2 + \|e_2(x^p)\|^2\}$, $i = 1, 2, 3, 4$ $\nabla_{\theta_i}\{|e_1(x)|^2 + \|e_2(x^p)\|^2\}$, $i = 1, 2$. So we derive these gradients, applying the Lagrange multiplier method as below.

First define the variable $v \in R^n$ as

$$v = 2W_1^T \nabla \sigma(z_1) W_2^T y = V_x^N(x)^T$$

then

$$|e_1(x)|^2 + \|e_2(x)\|^2 = \left(q(x) + u_N^T R u_N + v^T f(x, u_N)\right)^2$$
$$+ (2R^T u_N + f_u(x, u_N)^T v)^T (2R^T u_N + f_u(x, u_N)^T v) \tag{3.47}$$

Introducing the Lagrange multipliers $\lambda \in R^n, \beta \in R^q, \gamma \in R^n, \eta \in R^m, \delta \in R^r$, define the Lagrangian

$$L(x; \overline{W}, \overline{\theta}; z_1, z_2, v, y, u_N; \lambda, \beta, \gamma, \eta, \delta)$$
$$= \left(q(x) + u_N^T R u_N + v^T f(x, u_N)\right)^2$$
$$+ (2R^T u_N + f_u(x, u_N)^T v)^T (2R^T u_N + f_u(x, u_N)^T v)$$
$$+ \lambda^T (2W_1^T \nabla \sigma(z_1) W_2^T y - v) + \beta^T (W_1 x + \theta_1 - z_1)$$
$$+ \gamma^T (W_2 \sigma(z_1) - W_2 \sigma(\theta_1) - y) + \eta^T (W_3 x + \theta_2 - z_2)$$
$$+ \delta^T (W_4 \sigma(z_2) - W_4 \sigma(\theta_2) - u_N) \tag{3.48}$$

Then in the similar manner as the previous case [A], concrete expressions of gradients are obtained.

Calculating partial derivatives of Lagrangian L with respect to each variable and noticing $\nabla_{W_i}\{|e_1(x)|^2 + \|e_2(x)\|^2\} = \nabla_{W_i} L$, $\nabla_{\theta_i}\{|e_1(x)|^2 + \|e_2(x)\|^2\} = \nabla_{\theta_i} L$, we can obtain the gradients of performance function (3.42) with respect to connection weight matrices W_i and threshold θ_i as follows.

$$\nabla_{W_1} E[\overline{W}, \overline{\theta}] = \sum_{p=1}^{P} \left(2\nabla \sigma(z_1^p) W_2^T y^p \lambda^{pT} + \beta^p x^{pT} \right) \tag{3.49}$$

$$\nabla_{W_2} E[\overline{W}, \overline{\theta}] = \sum_{p=1}^{P} \left(2y^p \lambda^{pT} W_1^T \nabla \sigma(z_1^p) + \gamma^p \left(\sigma(z_1^p)^T - \sigma(\theta_1)^T \right) \right) \qquad (3.50)$$

$$\nabla_{W_3} E[\overline{W}, \overline{\theta}] = \sum_{p=1}^{P} \eta^p x^{pT} \qquad (3.51)$$

$$\nabla_{W_4} E[\overline{W}, \overline{\theta}] = \sum_{p=1}^{P} \delta^p \left(\sigma(z_2^p)^T - \sigma(\theta_2)^T \right) \qquad (3.52)$$

$$\nabla_{\theta_1} E[\overline{W}, \overline{\theta}] = \sum_{p=1}^{P} \left(\beta^p - \nabla \sigma(\theta_1) W_2^T \gamma^p \right) \qquad (3.53)$$

$$\nabla_{\theta_2} E[\overline{W}, \overline{\theta}] = \sum_{p=1}^{P} \left(\eta^p - \nabla \sigma(\theta_2) W_4^T \delta^p \right) \qquad (3.54)$$

where $\lambda, \beta, \gamma, \eta$, and δ are given as

$$\lambda = 2f(x, u_N)e_1(x) + 2f_u(x, u_N)e_2(x) \qquad (3.55)$$

$$\gamma = 2W_2 \nabla \sigma(z_1) W_1 \lambda \qquad (3.56)$$

$$\beta = \nabla^2 \sigma(z_1) \bullet (2W_1 \lambda \otimes y^T W_2) + \nabla \sigma(z_1) W_2^T \gamma \qquad (3.57)$$

$$\delta = 2e_1(x)e_2(x) + 4Re_2(x) + 2\nabla_{uu}^2 f(x, u_N) \bullet \left\{ e_2(x) \otimes v^T \right\} \qquad (3.58)$$

$$\eta = \nabla \sigma(z_2) W_4^T \delta \qquad (3.59)$$

Using these gradients we can execute the steepest descent method $(\alpha > 0)$

$$W_i^{k+1} = W_i^k - \alpha \nabla_{W_i} E[\overline{W}^k, \overline{\theta}^k], \quad i = 1, 2, 3, 4 \qquad (3.60)$$

$$\theta_i^{k+1} = \theta_i^k - \alpha \nabla_{\theta_i} E[\overline{W}^k, \overline{\theta}^k], \quad i = 1, 2 \qquad (3.61)$$

to get optimal connection weight matrices W_i, $i = 1, 2, 3, 4$ and thresholds θ_i, $i = 1, 2$.

As the results both the solution $V^N(x)$ to the generalized H-J equations $(2.7)\sim$ (2.9) and the optimal state feedback law $u_N(x)$ are obtained. This optimal state feedback control law

$$u_N(x) = W_4 \sigma(W_3 x + \theta_2) - W_4 \sigma(\theta_2) \qquad (3.62)$$

becomes the solution for nonlinear optimal regulator problem (2.1)

4   IMPROVEMENT OF LEARNING ALGORITHM OF NEURAL NETWORK

Since it does not necessarily follow that partial differential equations $(2.7)\sim(2.9)$ possess a unique solution, an arbitary solution of the H-J equation is not always the true value function. This difficulty is caused by the fact that the H-J equation is only a necessary condition for optimality. In general it is very hard to prove that any approximate solution of the H-J equation converges to the true value function $V(x)$. However, we can improve the possibility that the approximate solution converges to the value function, making a device on learning of networks.

It is not guaranteed that $V^N(x)$ obtained from the learning problem (3.6) or (3.43) coincides with the value function $V(x)$ of the performance functional $(2.1a)$. In fact it sometime happens that $V^N(x) \neq V(x)$. This is caused by that in general the solution to the H-J equation is not unique. Accordingly, we need a device of learning such that $V^N(x)$ converges to the true value function $V(x)$.

For simplicity let us assume there exits $u = u^o(x, V_x(x))$ satisfying (2.9) around $(x, u) = (0, 0)$ globally. Substitute this into (2.7) to get

$$q(x) + u^o(x, V_x(x))^T R u^o(x, V_x(x)) + V_x(x) f(x, u^o(x, V_x(x))) = 0 \quad (4.1)$$

The solution to the H-J equation (4.1) is not unique because the H-J equation is only a necessary condition for optimality. The value function of problem (2.1) certainly satisfies the H-J equation (4.1) but there may exist any other solutions. Let us here denote the value function (2.4) especially by $V^o(x)$ and distinguish it from any other solutions $V(x)$.

The minimum solution among semi-positive definite solutions to the H-J equation (4.1) coincides with $V^o(x)$. But asymptotical stability of the closed-loop system is not guaranteed by implementing the optimal control law $u^o(x, V_x^o(x))$.

If there exists the solution $V(x)$ of the H-J equation such that $\dot{x} = f(x, u^o(x, V_x(x)))$ becomes asymptotically stable, then we call it the stabilizing solution denoted by $V^-(x)$. The following lemma gives a condition that $V^-(x)$ exists uniquely [Kucera (1972),Schaft (1996)].

Lemma 4.1 *Assume that the Hamilton matrix*

$$\mathcal{H} = \begin{bmatrix} A & -B(2R)^{-1}B^T \\ -2Q & -A^T \end{bmatrix}$$

does not possess pure imaginary eigenvalues and $\{A, B\}$ is stabilizable, where $A = f_x(0,0)$, $B = f_u(0,0)$, $2Q = q_{xx}(0)$. Then the H-J equation possesses the unique stabilizing solution $V^-(x)$.

As this lemma holds, it is known from the uniqueness of the stabilizing solution that $V^-(x)$ becomes equal to the minimum performance function for the stabilizing optimal control problem, that is

$$\min_{u} \left\{ \int_t^\infty \left(q(x) + u^T Ru \right) dt \,\middle|\, \dot{x} = f(x, u), x(t) = x, \lim_{t\to\infty} x(t) = 0 \right\} = V^-(x) \tag{4.2}$$

At this time the stabilizing optimal control law is given by $u^o(x, V_x^-(x))$. Further it can be easily shown [Schaft (1996)] that $V^-(x)$ is the maximum solution of the H-J equation.

Now $V^o(x)$ was the minimum among the semi-positive definite solutions of the H-J equation (4.1), while $V^-(x)$ is the maximum one. Hence if there exists more than one semi-positive definite solution, then $V^o(x) \neq V^-(x)$. However if the semi-positive definite solution is unique, then it becomes $V^o(x) = V^-(x)$. The uniqueness of semipositive definite solution is guaranteed by assuming detectability.

Assumption 4.1 *$\{f(x, u), q(x) + u^T Ru\}$ is detectable. That is, it holds along solutions of (2.1b) that $\lim_{t\to\infty} x(t) = 0$ as $\lim_{t\to\infty} \{q(x(t)) + u(t)^T Ru(t)\} = 0$.*

In order that $V^o(x)$ exists, it must hold that $\lim_{t\to\infty} \{q(x(t)) + u(t)^T Ru(t)\} = 0$. Meanwhile, if we assume the detectability, we have $\lim_{t\to\infty} x(t) = 0$. Therefore the optimal regulator problem coincides with the stabilizing optimal regulator problem, and so it holds that $V^o(x) = V^-(x)$. Equality of the minimum and the maximum solutions indicates the uniqueness of semi-positive definite solution.

In below assume the detectability. Then our aim is to approximate the unique semi-positive definite solution to the H-J equation by the neural network.

To make matters worth, however, it happens sometime that $V^N(x)$ converges to something other than $V^-(x)$ partially on the way that $V^N(x)$ is learnt to converge to $V^-(x)$. Yet at least in the neighborhood of $x = 0$ it is easy to let $V^N(x)$ coincide with $V^-(x)$. Moreover, once $V^N(x)$ begins to converge to $V^-(x)$, it does not happen to converge to another solution thereafter.

The following well-known fact is important for letting $V^N(x)$ converge to $V^-(x)$;

Under the same assumption of Lemma 1, there exists a unique stabilizing solution P^- to the Riccati equation

$$PA + A^T P - PB(2R)^{-1}B^T P + 2Q = O \tag{4.3}$$

And it holds that

$$\nabla^2 V^-(0) = P^- \tag{4.4}$$

for the stabilizing solution $V^-(x)$ of the H-J equation (4.1). Further, since $V^o(x) = V^-(x)$, it holds $\nabla^2 V^o(0) = P^-$ also.

Now from (4.2) $V^-(x)$ takes the minimum $V^-(0) = 0$ and it holds $V_x^-(0) = 0$. Meanwhile $V^N(x)$ satisfies $V^N(0) = 0$ and $V_x^N(0) = 0$. Thus letting the relation

$$\nabla^2 V^N(0) = P^- \tag{4.5}$$

hold for $V^N(x)$ as well as (4.4), we can make $V^N(x)$ coincide with $V^-(x) = V^o(x)$ in the neighborhood of $x = 0$.

Here, we show the improvement of learning algorithm as to affine nonlinear system case. Note that we can apply the same approach for the general nonlinear system case.

$\nabla^2 V^N(0)$ can be calculated from (3.7) as

$$\nabla^2 V^N(0) = 2W_1^T \nabla\sigma(\theta)W_2^T W_2 \nabla\sigma(\theta)W_1 \tag{4.6}$$

Put $D \triangleq \nabla^2 V^N(0) - P^-$, then the norm of D, i.e. $\|D\|$ must be required to be zero. Therefore we modify the performance function (3.5) for learning as follows.

$$E[W_1, W_2, \theta] = \sum_{p=1}^{P} |e(x^p)|^2 + \|D\|^2 \tag{4.7}$$

Hence the learning problem is to solve the following optimization problem.

$$\min_{W_1, W_2, \theta} E[W_1, W_2, \theta] = \sum_{p=1}^{P} |e(x^p)|^2 + \|D\|^2 \tag{4.8a}$$

$$\text{subj. to } z^p = W_1 x^p + \boldsymbol{\theta} \tag{4.8b}$$

$$y(x^p) = W_2 \boldsymbol{\sigma}(z^p) - W_2 \boldsymbol{\sigma}(\boldsymbol{\theta}) \tag{4.8c}$$

$$V^N(x^p) = y(x^p)^T y(x^p) \tag{4.8d}$$

$$e(x^p) = q(x^p) + V_{\boldsymbol{x}}^N(x^p) f(x^p)$$
$$-\frac{1}{4} V_{\boldsymbol{x}}^N(x^p) G(x^p) R^{-1} G(x^p)^T V_{\boldsymbol{x}}^N(x^p)^T \tag{4.8e}$$

$$D = 2W_1^T \nabla \boldsymbol{\sigma}(\boldsymbol{\theta}) W_2^T W_2 \nabla \boldsymbol{\sigma}(\boldsymbol{\theta}) W_1 - P^- \tag{4.8f}$$

$$p - 1, 2, \cdots, P$$

Although gradients of (4.8a) w.r.t. $W_1, W_2, \boldsymbol{\theta}$ are necessary to solve (4.8), the gradients of the first term in (4.8a) have been obtained by (3.28)$\sim$(3.30) already. Thus we have only to obtain gradients of $\|D\|^2$ (w.r.t. $W_1, W_2, \boldsymbol{\theta}$). But they are easily calculated by using a formulae in the notes as follows.

$$\nabla_{W_1} \|D\|^2 = 8 \nabla \boldsymbol{\sigma}(\boldsymbol{\theta}) W_2^T W_2 \nabla \boldsymbol{\sigma}(\boldsymbol{\theta}) W_1 D \tag{4.9a}$$

$$\nabla_{W_2} \|D\|^2 = 8 W_2 \nabla \boldsymbol{\sigma}(\boldsymbol{\theta}) W_1 D W_1^T \nabla \boldsymbol{\sigma}(\boldsymbol{\theta}) \tag{4.9b}$$

$$\nabla_{\boldsymbol{\theta}} \|D\|^2 = 8 \nabla^2 \boldsymbol{\sigma}(\boldsymbol{\theta}) \bullet \left(W_1 D W_1^T \nabla \boldsymbol{\sigma}(\boldsymbol{\theta}) W_2^T W_2 \right) \tag{4.9c}$$

By using $\nabla_{W_1} E, \nabla_{W_2} E, \nabla_{\boldsymbol{\theta}} E$ newly obtained, the learning of neural network is executed by the steepest descent method (3.31)$\sim$(3.33).

5   SIMULATION RESULTS

We made computer simulations for the following example [Isidori (1989)]. Here let us consider the case where the method in [Goh (1993)] is difficult to apply for learning.

$$\min_{u} \int_0^\infty x_1^2 + x_2^2 + x_3^2 + u^2 \, dt \tag{5.1a}$$

$$\text{subj. to } \dot{x}_1 = -x_1 + e^{2x_2} u \tag{5.1b}$$

$$\dot{x}_2 = 2x_1 x_2 + \sin(x_2) + 0.5u \tag{5.1c}$$

$$\dot{x}_3 = 2x_2 \tag{5.1d}$$

$$x(0) = x_0$$

From (2.12) and (2.13) the H-J equation and the optimal control law $u^o(x, V_x(x))$ become:

$$x_1^2 + x_2^2 + x_3^2 - x_1 V_{x_1}(x) + \{2x_1 x_2 + \sin(x_2)\} V_{x_2}(x) + 2x_2 V_{x_3}(x)$$
$$-0.25\{e^{4x_2} V_{x_1}(x)^2 + e^{2x_2} V_{x_1}(x) V_{x_2}(x) + 0.25 V_{x_2}(x)^2\} = 0 \tag{5.2}$$

$$u^o(x, V_x(x)) = -0.5\{e^{2x_2}V_{x_1}(x) + 0.5V_{x_2}(x)\} \tag{5.3}$$

In order to obtain the desired value function $V^o(x)$, we improve the learning algorithm of neural network as stated in Section 4, using the condition $\nabla^2 V^N(0) = P^-$. For that purpose we consider LQ regulator problem generated by linearizing (5.1b)$\sim$(5.1d) and by approximating (5.1a) in a quadratic form.

$$\min_u \int_0^\infty x^T Q x + r u^2 \, dt \tag{5.4a}$$

$$\text{subj.to} \quad \dot{x} = Ax + bu, \quad x(0) = x_0 \tag{5.4b}$$

where A, b, Q, and r are given as

$$A = \begin{bmatrix} -1 & 0 & 0 \\ 0 & 1 & 0 \\ 0 & 2 & 0 \end{bmatrix}, \, b = \begin{bmatrix} 1 \\ 0.5 \\ 0 \end{bmatrix}, \quad \begin{matrix} Q = \text{diag}(1,1,1) \\ r = 1 \end{matrix}$$

Since $\{A, b\}$ and $\{\sqrt{Q}, A\}$ are controllable and observable, respectively, the assumption in Lemma 1 is satisfied. The Riccati equation for problem (5.4) becomes

$$PA + A^T P - Pb(2r)^{-1}b^T P + 2Q = O, \tag{5.5}$$

and the stabilizing solution P^- is calculated as follows.

$$P^- = \begin{bmatrix} 1 & -2 & 0 \\ -2 & 28.492 & 4 \\ 0 & 4 & 4.123 \end{bmatrix}$$

Since $q(x) = x_1^2 + x_2^2 + x_3^2$ is positive definite, Assumption 4.1 is satisfied. Hence the state variables become asymptotically stable by the optimal control law.

A number of middle layer of the network was taken 20. The learning domain was set as $\Omega = \{(x_1, x_2, x_3)| -2 \leq x_1 \leq 2, -1 \leq x_2 \leq 1, -2 \leq x_3 \leq 2\}$ and was discretized by an orthogonal lattice. The distance between the adjoining lattice points was set as $\Delta x_1 = 0.2$, $\Delta x_2 = 0.1$, $\Delta x_3 = 0.2$. Initial values of W_1, W_2 and θ were given by random numbers between -0.5 and 0.5.

The results of optimal feedback control is presented in Figure 5.1 and 5.2 in case of initial state $x(0) = (1.5, 1, 1.5)$. Then the value of performance function is 19.797. For comparison optimal control in the open-loop style was computed by a usual optimization algorithm [Shimizu (1994)] in case of the same initial

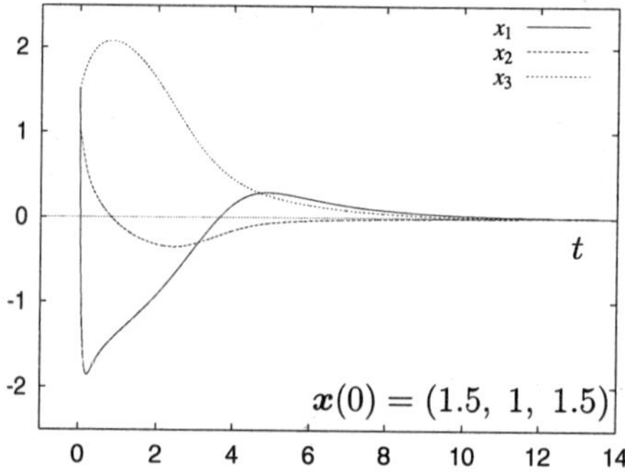

Figure 5.1   Optimal state feedback control by neural network (state variables)

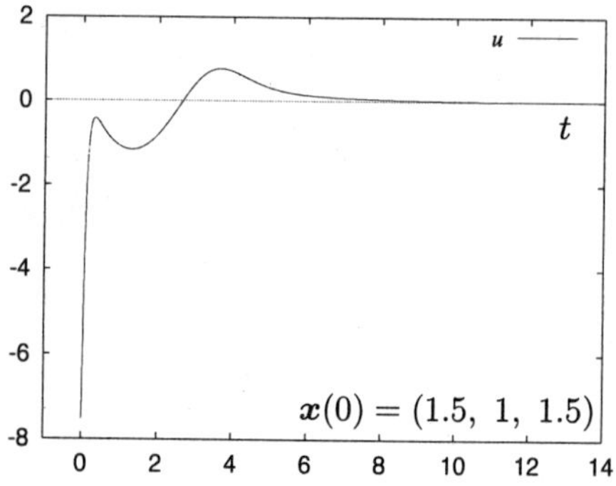

Figure 5.2   Optimal state feedback control by neural network (control input)

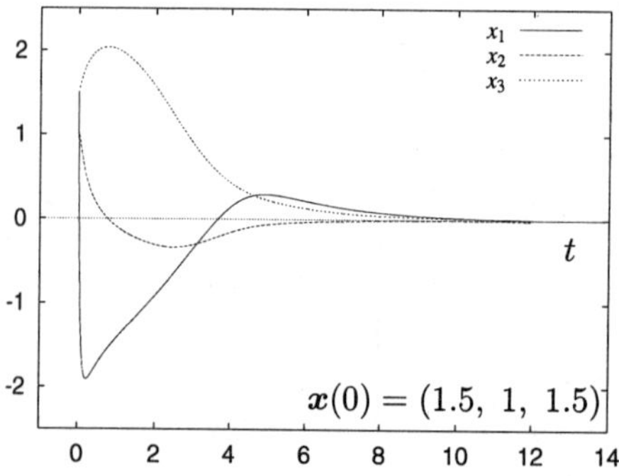

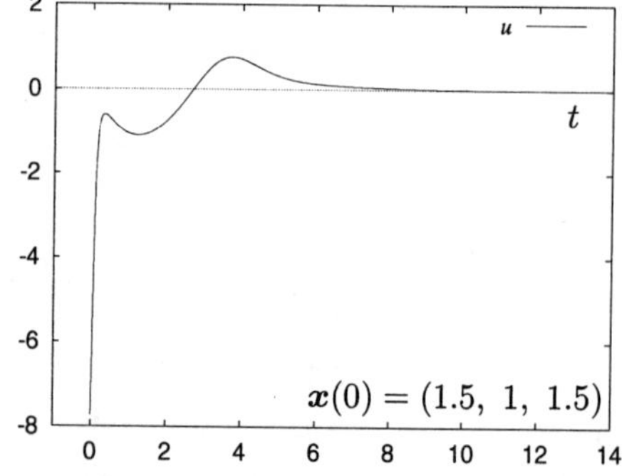

Figure 5.3   Optimal control in open loop style (state variables)

Figure 5.4   Optimal control in open loop style (control input)

state $x(0)$ (see Figure 5.3&5.4). Then the value of performance function is 19.705.

Comparing both results, we can see that the optimal feedback neural controller generates the true optimal control in sufficiently satisfactory accuracy.

6 CONCLUSIONS

We proposed a method using a neural network to obtain an approximate solution of the H-J equation for the nonlinear optimal regulator problem. It was confirmed from the simulation results of various examples that the proposed method is effective to synthesize the optimal feedback control law of nonlinear systems.

With regard to learning methods of the neural network, however, there is room for further improvements. To obtain a global optimum of connection weights and thresholds, the so called pattern mode search [Saridis (1986)] is considered to apply. How to initialize the neural network is also our future problem.

Notes

1.   (i) Let $f(x) = a^T \bullet x,\quad a^T \in Z \otimes X^*$, ($*$ denotes the conjugate space).
Then $\nabla f(x) = a \in X \otimes Z^*$
(ii) Let $f(D) = x^T \bullet D \bullet y,\quad x \in X, y \in Y, D \in X \otimes Y^*$,
then $\nabla f(D) = x \otimes y^T \in X \otimes Y^*$
(see the proof in [Suzuki (1990)])

References

Beard, R.W., Saridis, G.N. and Wen, J.T. (1997), Galerkin Approximations of the Generalized Hamilton-Jacobi-Bellman Equation, *Automatica*, Vol. 33, No. 12, pp. 2159-2177.

Doya, K. (2000), Reinforcement Learning in Continuous Time and Space, *Neural Computation*, Vol.12, pp. 219-245.

Goh, C.J. (1993), On the Nonlinear Optimal Regulator Problem, *Automatica*, Vol. 29, No. 3, pp. 751-756.

Isidori, A. (1989), *Nonlinear Control Systems: An Introduction*, Springer-Verlag.

Kucera, V. (1972), A Contribution to Matrix Quadratic Equations, *IEEE Trans. Automatic Control*, pp. 344-347.

Lee, H.W.J., Teo, K.L. and Yan, W.Y. (1996), Nonlinear Optimal Feedback Control Law for a Class of Nonlinear Systems, *Neural Parallel & Scientific Computations* 4, pp. 157-178.

Lukes, D.L. (1969), Optimal Regulation of Nonlinear Dynamical Systems, *SIAM. J. Control*, Vol. 7, No. 1, pp. 75-100.

Saridis, G.N. and Balaram, J. (1986), Suboptimal Control for Nonlinear System, *Control Theory and Advanced Technology*, Vol. 2, No. 3, pp. 547-562.

van der Schaft, A. (1996), *L₂ Gain and Passivity Techniques in Nonlinear Control*, Lecture Notes in Control and Information Science 218, Springer.

Shimizu, K. (1994), *Optimal Control—Theory and Algorithms*, Chap. 10.

Suzuki, M. and Shimizu, K. (1990), Analysis of Distributed Systems by Array Algebra, *Int. J. of Systems Science*, Vol. 21, No. 1, pp. 129-155.

23 H_∞ CONTROL BASED ON STATE OBSERVER FOR DESCRIPTOR SYSTEMS

Wei Xing, Q.L. Zhang,

College of Science, Northeastern University,
Shenyang, Liaoning 110006, P. R. China, Email: qlzhang@mail.neu.edu.cn

W.Q. Liu

School of Computing, Curtin University of Technology, WA 6102,
Australia. Email: wanquan@cs.curtin.edu.au

and Qiyi Wang

College of Science, Northeastern University,
Shenyang, Liaoning 110006, P. R. China, Email: qlzhang@mail.neu.edu.cn

Abstract: In this paper, the H_∞ control problem based on a state observer for descriptor systems is investigated. The motivation of this paper is two-fold. One is to extend the corresponding H_∞ results for linear time invariant systems to the case for descriptor systems. The other is to obtain more explicit results compared to those in the existing literature. From the results obtained here, one can figure out some special features for singular systems, which are different from normal linear systems. Also the approach adopted here is pure algebraic, which is easy to understand. As to the results, a necessary and sufficient condition for the solvability of the H_∞ control problem based observer design is obtained in terms of a Generalized Algebraic Riccati Inequality (GARI). Moreover, the desired controller is also explicitly constructed in this case.

Key words: Descriptor system, observer, state-feedback, admissible, H_∞ norm, GARI.

1 INTRODUCTION

H_∞ control theory in normal linear time-invariant systems has made significant progress since the fundamental work was published in 1981 Zames (1981). Within the state-space approach, the existing results can be roughly classified into two categories. One is the state feedback H_∞ control problems, such as Petersen (1987); Khargonekar et al (1988); Zhou et al (1988); Stoorvogel (1990); Scherer (1992) and the other category is the output feedback problems, see Doyel et al (1989); Sampei rt al (1990); Gahinet et al (1994); Barabanov (1998).

Descriptor systems can be seen as a generalization of the normal linear time-invariant systems and they can describe many practical systems that the normal systems can not model properly (e.g., see Verhese et al (1981); Dai (1989); Brenan et al (1996); Zhang (1997)). Though a lot of concepts and results in the normal linear time invariant systems have been generalized to the case of descriptor systems Cobb (1984); Lewis (1986); Dai (1989), the research on the H_∞ control problems for descriptor systems is still far behind compared to that for normal linear time invariant systems. Recently, output feedback H_∞ control problems for the descriptor systems using state-space method have been investigated Masubuchi et al (1997); Wang et al (1998). Further, static state-feedback H_∞ control problems for the descriptor systems is also examined in Gao et al (1999).

The authors in Masubuchi et al (1997) solved the H_∞ control problem based output feedback. The main result was given in terms of two generalized algebraic Riccatti equations (GAREs) with constraints. One possible algorithm was proposed for solving the two GAREs. It can be seen that it is not trivial to solve these GAREs with constraints. The results in Wang et al (1998); Gao et al (1999) were based on bounded real lemma for descriptor systems.

In this paper, we also study the H_∞ control problem based on state feed back for the descriptor systems. However, the state-feedback here is based on a state observer rather than static state feedback as in Gao et al (1999). In terms of one GARI, a necessary and sufficient condition for the solvability of our H_∞ control problem is obtained. Moreover, based on the solution of one GARI, an admissible controller is constructed explicitly. It should be noted that the results obtained here are much more explicit compared to those reported in Masubuchi et al (1997) since it only needs one GARI instead of two GARIs. In addition, we also investigate this result further when the orthogonal condition defined in this paper is satisfied.

The paper is organized as follows: In section 2, we will present some preliminary results. The main result will be given in section 3 and conclusions are given in section 4.

2 PRELIMINARIES

Consider the descriptor system

$$\begin{aligned}
\mathbf{E}\dot{\mathbf{x}}(t) &= \mathbf{A}\mathbf{x}(t) + \mathbf{B}_1\mathbf{w}(t) + \mathbf{B}_2\mathbf{u}(t) \\
\mathbf{z}(t) &= \mathbf{C}_1\mathbf{x}(t) + \mathbf{D}_{12}\mathbf{u}(t) \\
\mathbf{y}(t) &= \mathbf{C}_2\mathbf{x}(t) + \mathbf{D}_{21}\mathbf{w}(t)
\end{aligned} \tag{2.1}$$

where $\mathbf{x} \in \mathcal{R}^n$ is the descriptor state variable, $\mathbf{w} \in \mathcal{R}^m$ is the exogenous input variable (such as reference signal, command etc.), $\mathbf{u} \in \mathcal{R}^k$ is the control input variable, $\mathbf{z} \in \mathcal{R}^p$ is the controlled output variable, $\mathbf{y} \in \mathcal{R}^q$ is the measured output variable; $\mathbf{E}, \mathbf{A} \in \mathcal{R}^{n \times n}$, $\mathbf{B}_1, \mathbf{B}_2, \mathbf{C}_1, \mathbf{C}_2, \mathbf{D}_{12}$ and $\mathbf{D}_{21}$ are all constant real matrices with compatible dimensions. Moreover, it is generally assumed that $rank(\mathbf{E}) < n$. The following concepts will be used and they are mainly from Dai (1989).

1. The system (2.1) is said to be *regular*, if the polynomial (with respect to λ) $det(\lambda\mathbf{E} - \mathbf{A}) \neq 0$.

2. The *finite dynamic modes* of the system (2.1) are the finite eigenvalues of $(\mathbf{E}, \mathbf{A})$.

3. If all the finite dynamic modes lie in the open left half plane, Then the system (2.1) is said to be *stable*.

4. The infinite eigenvalues of $(\mathbf{E}, \mathbf{A})$ corresponding to such relative eigenvectors that their ranks are larger than 1 are said to be *impulsive modes*.

5. If the system (2.1) has no impulsive mode, Then it is said to be *impulse-free;*.

6. The triple $(\mathbf{E}, \mathbf{A}, \mathbf{C})$ is said to be *finite dynamics detectable* if there exists a constant matrix $\mathbf{M}$ such that the pair $(\mathbf{E}, \mathbf{A} + \mathbf{MC})$ is stable.

7. The triple $(\mathbf{E}, \mathbf{A}, \mathbf{C})$ is said to be *impulse observable* if there exists a constant matrix $\mathbf{N}$ such that the pair $(\mathbf{E}, \mathbf{A} + \mathbf{NC})$ is impulse-free.

Now one can present the following lemmas and definition for singular systems and they are useful to understand the results in this paper.

Lemma 2.1 *Dai (1989) The regular pair* $(\mathbf{E}, \mathbf{A})$ *is impulse-free if and only if*

$$\deg \det(s\mathbf{E} - \mathbf{A}) = rank\mathbf{E}$$

Definition 2.1 *The descriptor system (2.1) is said to be* **admissible** *if it is regular, stable and impulse-free.*

Lemma 2.2 *Dai (1989) The triple* $(\mathbf{E}, \mathbf{A}, \mathbf{C})$ *is finite dynamics detectable and impulse observable if and only if there exists a constant matrix* $\mathbf{L}$ *such that* $(\mathbf{E}, \mathbf{A} + \mathbf{LC})$ *is stable and impulse-free or equivalently* **admissible***.*

For the descriptor system (2.1), many researchers have devoted their effort to the output feedback H_∞ control problems Takaba et al (1994); Masubuchi et al (1997); Wang et al (1998). One important motivation is that the output feedback can be realized easily. However, the corresponding solutions to H_∞ control problems are much more complicated compared to the case of state feedback control. Since the state of a system contains all the essential information for the system, a controller based on state-feedback can lead to a more effective control. Particularly, the corresponding solution to the H_∞ control problems may become much more explicit Gao et al (1999); Wang et al (1998). The problem for state feedback control is that all the state variables are not available in practice, so we will choose to use a state observer in this paper and investigate the corresponding state-feedback H_∞ control problem for the descriptor system within this framework. This problem is between the output feedback and static feed back control. As can be seen in the sequel, it is not a trivial special case of output feedback control. With solving this problem for singular systems, one can visulize the difference more clearly between descriptor systems and normal systems.

As in the Full Information case Doyel et al (1989) for linear time-invariant system, it is also assumed here that the exogenous input signal $\mathbf{w}$ is always available.

The next lemma gives a result on the design of state observer for singular systems.

Lemma 2.3 *Dai (1989) Assume that* $(\mathbf{E}, \mathbf{A}, \mathbf{C}_2)$ *is finite dynamics detectable. Then the following dynamic system is a state observer for the system (2.1)*

$$\mathbf{E}\dot{\xi} = \mathbf{A}\xi + \mathbf{B}_1\mathbf{w} + \mathbf{B}_2\mathbf{u} - \mathbf{L}[\mathbf{y} - \mathbf{D}_{21}\mathbf{w} - \mathbf{C}_2\xi] \tag{2.2}$$

where matrix $\mathbf{L}$ *is such that* $(\mathbf{E}, \mathbf{A} + \mathbf{L}\mathbf{C}_2)$ *is stable and impulsive free.*

In this paper, a controller based on the observer (2.2) is assumed to be in the following form.

$$\begin{aligned} \mathbf{E}\dot{\xi} &= \mathbf{A}\xi + \mathbf{B}_1\mathbf{w} + \mathbf{B}_2\mathbf{u} - \mathbf{L}[\mathbf{y} - \mathbf{D}_{21}\mathbf{w} - \mathbf{C}_2\xi] \\ \mathbf{u} &= \mathbf{K}\xi \end{aligned} \tag{2.3}$$

Remark 2.1 *It should be noted that the state feedback controller given above is different from the output feedback controller given by Masubuchi et al (1997)*

$$\begin{aligned} \tilde{\mathbf{E}}\dot{\xi} &= \tilde{\mathbf{A}}\xi + \tilde{\mathbf{B}}\mathbf{y} \\ \mathbf{u} &= \tilde{\mathbf{K}}\xi \end{aligned} \tag{2.4}$$

In the controller (2.3), there are only two feedback parameters rather than the four parameters present in (2.4). It can be seen late in this paper that the results based on the controller (2.3) are much more explicit.

The next result is the basic result for H_∞ control for singular systems. It gives a necessary and sufficient condition for a singular system to be H_∞ norm bounded.

Lemma 2.4 *Masubuchi et al (1997) The descriptor system*

$$\begin{aligned} \mathbf{E}\dot{\mathbf{x}} &= \mathbf{A}\mathbf{x} + \mathbf{B}\mathbf{u} \\ \mathbf{y} &= \mathbf{C}\mathbf{x} \end{aligned} \tag{2.5}$$

is admissible and $\| \mathbf{C}(s\mathbf{E} - \mathbf{A})^{-1}\mathbf{B} \|_\infty < \gamma$ *if and only if there exists matrix* $\mathbf{X}$ *such that*

$$\mathbf{A}^T\mathbf{X} + \mathbf{X}^T\mathbf{A} + \mathbf{C}^T\mathbf{C} + \gamma^{-2}\mathbf{X}^T\mathbf{B}\mathbf{B}^T\mathbf{X} < \mathbf{0} \tag{2.6}$$

$$\mathbf{E}^T\mathbf{X} = \mathbf{X}^T\mathbf{E} \geq \mathbf{0} \tag{2.7}$$

The H_∞ control problem investigated here is to design a state-feedback controller (2.3) for the system (2.1) such that the resulting closed-loop system is admissible and the H_∞ norm of the transfer function matrix from $\mathbf{w}$ to $\mathbf{z}$ is strictly less than a prescribed positive number γ. In the next section, we will give solution of this problem.

3   MAIN RESULTS

In order to consider the H_∞ problem for the descriptor system (2.1), the following assumptions are made.

(A1) $(\mathbf{E}, \mathbf{A}, \mathbf{C}_2)$ is finite dynamics detectable and impulse observable.

(A2) $rank\mathbf{D}_{12} = k$.

The next lemma is an important result for the proof of our main result.

Lemma 3.1 *Assume that all the following matrices have appropriate dimensions. Let*

$$\mathbf{\Phi}(\mathbf{X} : \mathbf{A}, \mathbf{B}, \mathbf{C}) \;\triangleq\; \mathbf{A}^T\mathbf{X} + \mathbf{X}^T\mathbf{A} + \mathbf{C}^T\mathbf{C} + \gamma^{-2}\mathbf{X}^T\mathbf{B}\mathbf{B}^T\mathbf{X}$$

$$\mathbf{\Psi}(\mathbf{X} : \mathbf{A}, \mathbf{B}_1, \mathbf{B}_2, \mathbf{C}, \mathbf{D}) \;\triangleq\; \mathbf{A}^T\mathbf{X} + \mathbf{X}^T\mathbf{A} + \mathbf{C}^T\mathbf{C} + \gamma^{-2}\mathbf{X}^T\mathbf{B}_1\mathbf{B}_1^T\mathbf{X}$$
$$- (\mathbf{B}_2^T\mathbf{X} + \mathbf{D}^T\mathbf{C})^T(\mathbf{D}^T\mathbf{D})^{-1}(\mathbf{B}_2^T\mathbf{X} + \mathbf{D}^T\mathbf{C}).$$

where $\mathbf{D}$ *is of full column rank. Then*

$$\mathbf{\Phi}(\mathbf{X} : \mathbf{A} + \mathbf{B}_2\mathbf{K}, \mathbf{B}_1, \mathbf{C} + \mathbf{D}\mathbf{K}) = \mathbf{\Psi}(\mathbf{X} : \mathbf{A}, \mathbf{B}_1, \mathbf{B}_2, \mathbf{C}, \mathbf{D})$$
$$+\{\mathbf{D}[\mathbf{K} + (\mathbf{D}^T\mathbf{D})^{-1}(\mathbf{B}_2^T\mathbf{X} + \mathbf{D}^T\mathbf{C})]\}^T\{\mathbf{D}[\mathbf{K} + (\mathbf{D}^T\mathbf{D})^{-1}(\mathbf{B}_2^T\mathbf{X} + \mathbf{D}^T\mathbf{C})]\}$$

Proof

$$\mathbf{\Phi}(\mathbf{X} : \mathbf{A} + \mathbf{B}_2\mathbf{K}, \mathbf{B}_1, \mathbf{C} + \mathbf{D}\mathbf{K} = \mathbf{\Psi}(\mathbf{X} : \mathbf{A}, \mathbf{B}_1, \mathbf{B}_2, \mathbf{C}, \mathbf{D}) + \mathbf{\Pi}$$

where

$$\begin{aligned}
\mathbf{\Pi} \;=\;& \mathbf{K}^T\mathbf{B}_2^T\mathbf{X} + \mathbf{X}^T\mathbf{B}_2\mathbf{K} + \mathbf{C}^T\mathbf{D}\mathbf{K} + \mathbf{K}^T\mathbf{D}^T\mathbf{C} + \mathbf{K}^T\mathbf{D}^T\mathbf{D}\mathbf{K} \\
\;=\;& [\mathbf{K}^T + (\mathbf{X}^T\mathbf{B}_2 + \mathbf{C}^T\mathbf{D})(\mathbf{D}^T\mathbf{D})^{-1}](\mathbf{D}^T\mathbf{D})\mathbf{K} + \mathbf{K}^T(\mathbf{B}_2^T\mathbf{X} + \mathbf{D}^T\mathbf{C}) \\
& + \;(\mathbf{X}^T\mathbf{B}_2 + \mathbf{C}^T\mathbf{D})(\mathbf{D}^T\mathbf{D})^{-1}(\mathbf{B}_2^T\mathbf{X} + \mathbf{D}^T\mathbf{C}) - \\
& - \;(\mathbf{X}^T\mathbf{B}_2 + \mathbf{C}^T\mathbf{D})(\mathbf{D}^T\mathbf{D})^{-1}(\mathbf{B}_2^T X + \mathbf{D}^T\mathbf{C}) \\
\;=\;& [\mathbf{K}^T + (\mathbf{X}^T\mathbf{B}_2 + \mathbf{C}^T\mathbf{D})(\mathbf{D}^T\mathbf{D})^{-1}](\mathbf{D}^T\mathbf{D})[K + (\mathbf{D}^T\mathbf{D})^{-1}(\mathbf{B}_2^T\mathbf{X} + \mathbf{D}^T\mathbf{C})] \\
& - \;(\mathbf{X}^T\mathbf{B}_2 + \mathbf{C}^T\mathbf{D})(\mathbf{D}^T\mathbf{D})^{-1}(\mathbf{B}_2^T\mathbf{X} + \mathbf{D}^T\mathbf{C}) \\
\;=\;& \{\mathbf{D}[\mathbf{K} + (\mathbf{D}^T\mathbf{D})^{-1}(\mathbf{B}_2^T X + \mathbf{D}^T\mathbf{C})]\}^T\{\mathbf{D}[\mathbf{K} + (\mathbf{D}^T\mathbf{D})^{-1}(\mathbf{B}_2^T\mathbf{X} + \mathbf{D}^T\mathbf{C})]\} \\
& - \;(\mathbf{B}_2^T\mathbf{X} + \mathbf{D}^T\mathbf{C})^T(\mathbf{D}^T\mathbf{D})^{-1}(\mathbf{B}_2^T\mathbf{X} + \mathbf{D}^T\mathbf{C})
\end{aligned}$$

Therefore,

$$\Psi(\mathbf{X} : \mathbf{A}, \mathbf{B}_1, \mathbf{B}_2, \mathbf{C}, \mathbf{D}) = \Phi(\mathbf{X} : \mathbf{A}, \mathbf{B}_1, \mathbf{C}) -$$
$$(\mathbf{B}_2^T\mathbf{X} + \mathbf{D}^T\mathbf{C})^T(\mathbf{D}^T\mathbf{D})^{-1}(\mathbf{B}_2^T\mathbf{X} + \mathbf{D}^T\mathbf{C}) = \Phi(\mathbf{X} : \mathbf{A} + \mathbf{B}_2\mathbf{K}, \mathbf{B}_1, \mathbf{C} + \mathbf{D}\mathbf{K})$$
$$-\{\mathbf{D}[\mathbf{K} + (\mathbf{D}^T\mathbf{D})^{-1}(\mathbf{B}_2^T\mathbf{X} + \mathbf{D}^T\mathbf{C})]\}^T\{\mathbf{D}[\mathbf{K} + (\mathbf{D}^T\mathbf{D})^{-1}(\mathbf{B}_2^T\mathbf{X} + \mathbf{D}^T\mathbf{C})]\}$$

Now the main result of this paper can be stated as below.

Theorem 3.1 *Suppose that assumptions (A1) and (A2) hold. Then for the descriptor system (2.1), the following statements are equivalent.*

(i) *There exists a state-feedback controller (2.3) such that the resulting closed-loop system is admissible and the H_∞-norm of transfer function matrix $\mathbf{T}_{\mathbf{zw}}(s)$ from $\mathbf{w}$ to $\mathbf{z}$ is strictly less than a prescribed positive number γ, i.e.,*

$$\| \mathbf{T}_{\mathbf{zw}}(s) \|_\infty < \gamma$$

(ii) *There exists a solution to the following GARI*

$$\begin{aligned}
&\mathbf{A}^T\mathbf{X} + \mathbf{X}^T\mathbf{A} + \mathbf{C}_1^T\mathbf{C}_1 + \gamma^{-2}\mathbf{X}^T\mathbf{B}_1\mathbf{B}_1^T\mathbf{X} - \\
&-(\mathbf{X}^T\mathbf{B}_2 + \mathbf{C}_1^T\mathbf{D}_{12})(\mathbf{D}_{12}^T\mathbf{D}_{12})^{-1}(\mathbf{B}_2^T\mathbf{X} + \mathbf{D}_{12}^T\mathbf{C}_1) < 0
\end{aligned} \tag{3.1}$$

with a constraint

$$\mathbf{E}^T\mathbf{X} = \mathbf{X}^T\mathbf{E} \geq 0 \tag{3.2}$$

When (ii) holds, the matrix $\mathbf{K}$ in the controller (2.3) can be constructed as

$$\mathbf{K} \overset{\triangle}{=} -(\mathbf{D}_{12}^T\mathbf{D}_{12})^{-1}(\mathbf{B}_2^T\mathbf{X} + \mathbf{D}_{12}^T\mathbf{C}_1) \tag{3.3}$$

and $\mathbf{L}$ in the controller (2.3) satisfies the requirement that $(\mathbf{E}, \mathbf{A} + \mathbf{L}\mathbf{C}_2)$ is admissible.

Proof (i)$\Rightarrow$(ii): It can be seen that the closed-loop system is

$$\begin{bmatrix} \mathbf{E} & \mathbf{0} \\ \mathbf{0} & \mathbf{E} \end{bmatrix}\begin{bmatrix} \dot{\mathbf{x}} \\ \dot{\xi} \end{bmatrix} = \begin{bmatrix} \mathbf{A} & \mathbf{B}_2\mathbf{K} \\ -\mathbf{L}\mathbf{C}_2 & \mathbf{A} + \mathbf{B}_2\mathbf{K} + \mathbf{L}\mathbf{C}_2 \end{bmatrix}\begin{bmatrix} \mathbf{x} \\ \xi \end{bmatrix} + \begin{bmatrix} \mathbf{B}_1 \\ \mathbf{B}_1 \end{bmatrix}\mathbf{w}$$
$$\mathbf{z} = \begin{bmatrix} \mathbf{C}_1 & \mathbf{D}_{12}\mathbf{K} \end{bmatrix}\begin{bmatrix} \mathbf{x} \\ \xi \end{bmatrix} \tag{3.4}$$

Let

$$\begin{bmatrix} \mathbf{x} \\ \xi \end{bmatrix} = \begin{bmatrix} \mathbf{I} & \mathbf{I} \\ \mathbf{0} & \mathbf{I} \end{bmatrix}\begin{bmatrix} \lambda \\ \eta \end{bmatrix}$$

then the system (3.4) becomes

$$\begin{bmatrix} \mathbf{E} & 0 \\ 0 & \mathbf{E} \end{bmatrix} \begin{bmatrix} \dot{\lambda} \\ \dot{\eta} \end{bmatrix} = \begin{bmatrix} \mathbf{A} + \mathbf{LC}_2 & 0 \\ -\mathbf{LC}_2 & \mathbf{A} + \mathbf{B}_2\mathbf{K} \end{bmatrix} \begin{bmatrix} \lambda \\ \eta \end{bmatrix} + \begin{bmatrix} 0 \\ \mathbf{B}_1 \end{bmatrix} \mathbf{w}$$

$$\mathbf{z} = \begin{bmatrix} \mathbf{C}_1 & \mathbf{C}_1 + D_{12}\mathbf{K} \end{bmatrix} \begin{bmatrix} \lambda \\ \eta \end{bmatrix} \tag{3.5}$$

Denote

$$\widetilde{\mathbf{E}} = \begin{bmatrix} \mathbf{E} & 0 \\ 0 & \mathbf{E} \end{bmatrix}, \widetilde{\mathbf{A}} = \begin{bmatrix} \mathbf{A} + \mathbf{LC}_2 & 0 \\ -\mathbf{LC}_2 & \mathbf{A} + \mathbf{B}_2\mathbf{K} \end{bmatrix}, \widetilde{\mathbf{B}} = \begin{bmatrix} 0 \\ \mathbf{B}_1 \end{bmatrix},$$

$$\widetilde{\mathbf{C}} = \begin{bmatrix} \mathbf{C}_1 & \mathbf{C}_1 + \mathbf{D}_{12}\mathbf{K} \end{bmatrix}$$

then the closed-loop system (3.5) can be written as

$$\widetilde{\mathbf{E}}\,\dot{\theta} = \widetilde{\mathbf{A}}\theta + \widetilde{\mathbf{B}}\mathbf{w}$$

$$\mathbf{z} = \widetilde{\mathbf{C}}\theta \tag{3.6}$$

where

$$\theta = \begin{bmatrix} \lambda \\ \eta \end{bmatrix}$$

From Lemma 2.4, there exists a solution $\mathbf{X}$ satisfying the following inequality

$$\Phi(\mathbf{X}; \widetilde{\mathbf{A}}, \widetilde{\mathbf{B}}, \widetilde{\mathbf{C}}) < 0 \tag{3.7}$$

with the constraint

$$\widetilde{\mathbf{E}}^T\mathbf{X} = \mathbf{X}^T\widetilde{\mathbf{E}} \geq 0 \tag{3.8}$$

Partition $\mathbf{X}$ as in $\widetilde{\mathbf{A}}$

$$\mathbf{X} = \begin{bmatrix} \mathbf{X}_1 & \mathbf{X}_2 \\ \mathbf{X}_3 & \mathbf{X}_4 \end{bmatrix}$$

then the (2, 2)-block of the matrix in the left side of the inequalities (3.7) and (3.8) respectively are

$$\Phi(\mathbf{X}_4; \mathbf{A} + \mathbf{B}_2\mathbf{K}, \mathbf{B}_1, \mathbf{C}_1 + \mathbf{D}_{12}\mathbf{K}) < 0 \tag{3.9}$$

$$\mathbf{E}^T\mathbf{X}_4 = \mathbf{X}_4^T\mathbf{E} \geq 0 \tag{3.10}$$

According to Lemma 3.1, one can obtain

$$\Psi(\mathbf{X}_4; \mathbf{A}, \mathbf{B}_1, \mathbf{B}_2, \mathbf{C}_1, \mathbf{D}_{12}) \leq \Phi(\mathbf{X}_4; \mathbf{A} + \mathbf{B}_2\mathbf{K}, \mathbf{B}_1, \mathbf{C}_1 + \mathbf{D}_{12}\mathbf{K}) < 0$$

Then the statement (ii) is proved.

(ii)$\Rightarrow$(i): Since the descriptor system (2.1) satisfies assumption (A1), by Lemma 2.1, there exists a constant matrix $\mathbf{L}$ such that $(\mathbf{E}, \mathbf{A} + \mathbf{L}\mathbf{C}_2)$ is admissible. Then the observer (2.2) exists by Lemma 2.3. Now let

$$\mathbf{K} = -(\mathbf{D}_{12}^T \mathbf{D}_{12})^{-1}(\mathbf{B}_2^T \mathbf{X} + \mathbf{D}_{12}^T \mathbf{C}_1)$$

Then by Lemma 3.1,

$$\Phi(\mathbf{X}; \mathbf{A} + \mathbf{B}_2\mathbf{K}, \mathbf{B}_1, \mathbf{C}_1 + \mathbf{D}_{12}\mathbf{K}) = \Psi(\mathbf{X}; \mathbf{A}, \mathbf{B}_1, \mathbf{B}_2, \mathbf{C}_1, \mathbf{D}_{12}) < 0$$

¿From Lemma 2.4, the pair $(\mathbf{E}, \mathbf{A} + \mathbf{B}_2\mathbf{K})$ is admissible and

$$\| (\mathbf{C}_1 + \mathbf{D}_{12}\mathbf{K})(s\mathbf{E} - (\mathbf{A} + \mathbf{B}_2\mathbf{K}))^{-1}\mathbf{B}_1 \|_\infty < \gamma$$

With chosen $\mathbf{L}$ and $\mathbf{K}$ above in the controller (2.3), the resulting closed-loop system is the system (3.4) and its transfer function matrix from $\mathbf{w}$ to $\mathbf{z}$ is

$$\mathbf{T}_{\mathbf{zw}}(s) = (\mathbf{C}_1 + \mathbf{D}_{12}\mathbf{K})(s\mathbf{E} - (\mathbf{A} + \mathbf{B}_2\mathbf{K}))^{-1}\mathbf{B}_1$$

and so

$$\| \mathbf{T}_{\mathbf{zw}}(s) \|_\infty < \gamma$$

In order to complete the proof, it suffices to prove that the system (3.4) is admissible. Notice that

$$\det(s \begin{bmatrix} \mathbf{E} & 0 \\ 0 & \mathbf{E} \end{bmatrix} - \begin{bmatrix} \mathbf{A} & \mathbf{B}_2\mathbf{K} \\ -\mathbf{L}\mathbf{C}_2 & \mathbf{A} + \mathbf{B}_2\mathbf{K} + \mathbf{L}\mathbf{C}_2 \end{bmatrix})$$

$$= \det(s \begin{bmatrix} \mathbf{E} & 0 \\ 0 & \mathbf{E} \end{bmatrix} - \begin{bmatrix} \mathbf{A} + \mathbf{L}\mathbf{C}_2 & 0 \\ -\mathbf{L}\mathbf{C}_2 & \mathbf{A} + \mathbf{B}_2\mathbf{K} \end{bmatrix})$$

$$= \det \begin{bmatrix} s\mathbf{E} - (\mathbf{A} + \mathbf{L}\mathbf{C}_2) & 0 \\ \mathbf{L}\mathbf{C}_2 & s\mathbf{E} - (\mathbf{A} + \mathbf{B}_2\mathbf{K}) \end{bmatrix}$$

$$= \det(s\mathbf{E} - (\mathbf{A} + \mathbf{L}\mathbf{C}_2)) \det(s\mathbf{E} - (\mathbf{A} + \mathbf{B}_2\mathbf{K}))$$

and $(\mathbf{E}, \mathbf{A} + \mathbf{L}\mathbf{C}_2)$ and $(\mathbf{E}, \mathbf{A} + \mathbf{B}_2\mathbf{K})$ are stable. Then the closed-loop system is stable. Further,

$$\deg(\det(s \begin{bmatrix} \mathbf{E} & 0 \\ 0 & \mathbf{E} \end{bmatrix} - \begin{bmatrix} \mathbf{A} & \mathbf{B}_2\mathbf{K} \\ -\mathbf{L}\mathbf{C}_2 & \mathbf{A} + \mathbf{B}_2\mathbf{K} + \mathbf{L}\mathbf{C}_2 \end{bmatrix}))$$

$$
\begin{aligned}
&= \ \deg\!\left(\det\!\left(s\begin{bmatrix} \mathbf{E} & 0 \\ 0 & \mathbf{E} \end{bmatrix} - \begin{bmatrix} \mathbf{A}+\mathbf{LC_2} & 0 \\ -\mathbf{LC_2} & \mathbf{A}+\mathbf{B_2K} \end{bmatrix}\right)\right) \\
&= \ \deg\det(s\mathbf{E}-(\mathbf{A}+\mathbf{LC_2})) + \deg\det(s\mathbf{E}-(\mathbf{A}+\mathbf{B_2K})) \\
&= \ rank(\mathbf{E}) + rank(\mathbf{E}) \\
&= \ rank\begin{bmatrix} \mathbf{E} & 0 \\ 0 & \mathbf{E} \end{bmatrix}
\end{aligned}
$$

According to Lemma 2.1, one can see that the closed-loop system is impulse-free. Hence the closed-loop system is admissible.

It can be seen that the H_∞ control problem based on state feedback can be solved via one GARE with constraint. It should be noted that this result is not a special case for output feedback control, which involves two GAREs with constraints. Another important feature for this result is that there is only one parameter in the GARE obtained here instead of two parameters proposed in output case Masubuchi et al (1997). Moreover, if assumption (A2) is replaced by the following orthogonal condition:

(A2′) $\mathbf{D}_{12}^{T}[\mathbf{C_1} \ \ \mathbf{D_{12}}] = [0 \ \ \mathbf{I}]$ Then Theorem 3.1 can be simplified.

Theorem 3.2 *Suppose that assumptions (A1) and (A2′) hold. Then the following statements are equivalent for system (2.1).*

(i) There exists a controller (2.3) such that the resulting closed-loop system is admissible and

$$\| \mathbf{T_{zw}}(s) \|_\infty < \gamma$$

(ii) There exists a solution $\mathbf{X}$ satisfying the following GARI

$$\mathbf{A}^T\mathbf{X} + \mathbf{X}^T\mathbf{A} + \mathbf{C_1}^T\mathbf{C_1} + \mathbf{X}^T(\gamma^{-2}\mathbf{B_1}\mathbf{B_1}^T - \mathbf{B_2}\mathbf{B_2}^T)\mathbf{X} < 0$$

with constraint

$$\mathbf{E}^T\mathbf{X} = \mathbf{X}^T\mathbf{E} \geq 0. \qquad (3.11)$$

When (ii) hold, the matrix $\mathbf{K}$ in the controller (2.3) can be constructed as

$$\mathbf{K} \overset{\triangle}{=} -\mathbf{B_2}^T\mathbf{X} \qquad (3.12)$$

and $\mathbf{L}$ in the controller (2.3) is such that $(\mathbf{E}, \mathbf{A}+\mathbf{LC_2})$ is admissible.

In this section, the H_∞ control problem based on state feedback is investigated and a sufficient and necessary condition in terms of GARI are obtained.

It should be noted the GARI obtained here is much more simple than those obtained in Masubuchi et al (1997) in two ways.

(i) Only one parameter is involved here. This greatly simplifies the complexity of solving the GARI.

(ii) Only one GARI is required.

This motivates us that state feed back controller design is much easier if the estimated state information is available. In this case, the algorithm for solving GARI with constraints proposed in Masubuchi et al (1997) can also be significantly simplified. We will not discuss this simplification here since it is technically trivial.

4   CONCLUSIONS

For descriptor system (2.1), much effort of research on H_∞ control problems in the existing literature are devoted to output feedback case. In this paper, we obtained results on H_∞ control based on a state observer for the system (2.1). With mild assumptions, a necessary and sufficient condition is obtained for the solvability of our H_∞ control problem in terms of a GARI. Furthermore, the construction of the desired controller presented here is much more explicit.

It should be noted that the results obtained here is not a trivial case for output case as revealed in previous section. The algorithm for solving the GARI can be developed similarly to that in Masubuchi et al (1997).

References

N. E. Barabanov, (1998) On the static H_∞ control problem, *Systems Control Lett,* Vol. 35, pp. 13-18.

K. E. Brenan, S. L. Campbell and L. R. Petzold, (1996) *The numerical solution of initial value problems in differential-algebraic equations* , SIAM, Philadelphia.

D. Cobb, (1984) Controllability, observability, and duality in singular systems, *IEEE Trans. Automat. Contr.,* vol. AC-29, pp. 1076-1082.

L. Dai, (1989) *Singular control systems, Lecture notes in control and information sciences,* Springer, Berlin.

J. C. Doyle, K. Glover, P. Khargonekar and B. Francis, (1989) State-space solutions to standard H_2 and H_∞ control problems, *IEEE Trans. Automat. Contr.*, vol. AC-34, pp. 831-847.

P. Gahinet and P. Apkarian, (1994) A linear matrix inequality approach to H_∞ control, *Int. J. of Robust and Nonlinear Contr.*, vol. 4, pp. 421-448.

F. Gao, W. Q. Liu, V. Sreeram, K. L. Teo, (1999) Bound real lemma for descriptor systems and its application, *Proc. of the 14th world congress of IFAC*, Beijing, pp. 57-62.

P. P. Khargonekar, I. R. Petersen and M. Rotea, (1988) H^∞-optimal control with state-feedback, *IEEE Trans. Automat. Contr.*, AC-33, pp. 786-788.

F. L. Lewis, (1986) A survey of linear singular system, *Circ. Syst. Sig. Proc.*, Vol. 5, pp. 3-36.

I. Masubuchi, Y. Kamitane, A. Ohara and N. Suda, (1997) H_∞ control for descriptor systems: a matrix inequalities approach, *Automatica*, vol. 33, pp. 669-673.

I. Petersen, (1987) Disturbance attenuation and H^∞ optimization: a design method based on the algebraic Riccati equation, *IEEE Trans. Automat. Contr.* AC-32, pp. 427-429.

M. Sampei, T. Mita and M. Nakamichi, (1990) An algebraic approach to H_∞ output feedback control problems, *systems Control Lett.*, Vol. 14, pp. 13-24.

C. Scherer, (1992) H^∞-control by state-feedback for plants with zeros on the imaginary axis, *SIMA J. Control and Optimization*, Vol. 30, pp. 123-142.

A. A. Stoorvogel and H. L. Trentelman, (1990) The quadratic matrix inequality in singular H_∞ control with state-feedback, *SIMA J. Control and Optimization*, Vol. 28, pp. 1190-1208.

K. Takaba, N. Morihira and T. Katayama, (1994) H^∞ control for descriptor systems-a J-spectral factorization approach, *Proc. of the 33rd conference on decision and control*, pp. 2251-2256.

G. Verhese, B. C. Levy and T. Kailath, (1981) A generalized state-space for singular systems, *IEEE Trans. Automat. Contr.*, AC-26. pp. 811-831.

H. S. Wang, C. F. Yung, F. R. Chang, (1988) Bounded real lemma and H_∞ control for descriptor systems, *IEE Proc. -Control Theory Appl.* Vol. 145, (3) pp. 316-322.

G. Zames, (1981) Feedback and optimal sensitivity: Model reference transformations. Multiplicative seminorms and approximate inverses. *IEEE Tans. Automat. Contr.*, AC-26, pp. 585-601.

Q. L. Zhang, (1997) *Robust and decentralized control for large-scale descriptor systems*, Northwestern Polytechnical University Press.

K. Zhou and P. P. Kargonekar, (1988) An algebraic Riccati equation approach to H^∞ optimization, *systems Control Lett.*, Vol. 11, pp. 85-91.

IV VARIATIONAL INEQUALITY AND EQUILIBRIUM PROBLEMS

24   DECOMPOSABLE GENERALIZED VECTOR VARIATIONAL INEQUALITIES

E. Allevi, A. Gnudi and

Department of Mathematics, Statistics, Computer Science and Applications,
Bergamo University,
Piazza Rosate, 2, Bergamo 24129, Italy

I. V. Konnov

Department of Applied Mathematics,
Kazan University, ul. Kremlevskaya,18,
Kazan 420008, Russia

Abstract: In this paper, we consider vector variational inequalities with set-valued mappings over product sets in a real linear topological space setting. By employing concepts of relative pseudomonotonicity with variable weights, we establish several existence results for generalized vector variational inequalities and for systems of generalized vector variational inequalities. These results strengthen previous existence results which were based on the usual monotonicity type assumptions.

Key words: Vector variational inequalities, set-valued mappings, product sets, relative pseudomonotonicity, existence results.

1   INTRODUCTION

Vector variational inequalities were introduced by Giannessi e.g., Giannessi (1980) in a finite-dimensional Euclidean space as a natural extension of scalar variational inequalities. Since then, various classes of vector variational inequalities were studied both in finite- and in infinite-dimensional spaces; see e.g., Hadjisavvas et al (1998). It is well known that a number of equilibrium type problems arising in Economics, Game Theory and Transportation have a decomposable structure, namely, they can be formulated as vector variational inequalities over Cartesian product sets; see e.g. Yuan (1998)–Yang et al(1997) and the references therein. At the same time, most existence results for such problems are based on the known fixed point techniques, which require either the feasible set (otherwise, the corresponding subset associated to a coercivity condition) be compact in the strong topology or the cost mapping possess certain continuity type properties with respect to the weak topology; see e.g. Yuan (1998); Ansari et al (1999). Usually, to essentially weaken these assumptions one make use of the Ky Fan Lemma (Ky Fan (1961)) together with certain monotonicity type properties regardless of the decomposable structure of VI; see e.g. Hadjisavvas et al (1998); Oettli et al (1998). However, it was noticed by Bianchi (1993) that infinite-dimensional extensions of the concepts of M- and P-mappings are not sufficient to apply the Ky Fan Lemma for deriving exstence results.

In this paper, we develop some other approach, which is based on the invariance of the solution sets of decomposable equilibrium and variational inequality problems with respect to certain linear trasformations. This property enables one to extend the usual (generalized) monotonicity properties. Rosen (1965) introduced such an extension of strict monotonicity to establish the uniqueness of solutions for non-cooperative games with scalar payoffs. Being based on the same property, Konnov (2001) introduced new (generalized) monotonicity concepts, which are adjusted to a decomposable structure of the initial problem, and proved new existence results for scalar variational inequalities in a Banach space. These new relative monotonicity concepts can be regarded as intermediate ones between the usual monotonicity and order monotonicity ones. Allevi et al (2001) extended the results from Konnov (2001) to vector variational inequalities with set-valued mappings over the Cartesian product of a finite number of sets, where the parameters associated to relative (generalized) monotonicity are constant. Now we consider the case where the parameters (or weights) are variables, thus extending the results from Konnov (2001); Allevi et al (2001). Namely, we establish existence results for generalized vector variational inequalities and for systems of generalized vector variational inequalities in a topological vector space by employing new relative (pseudo)monotonicity concepts for set-valued mappings.

2 PROBLEM FORMULATIONS AND BASIC FACTS

Let I be the set of indexes $\{1, \ldots, m\}$. For each $s \in I$, let E_s be a real linear topological space and U_s be a nonempty subset of E_s. Set

$$U = \prod_{s \in I} U_s. \tag{2.1}$$

Let F be a real linear topological space with a partial order induced by a convex, closed and solid cone C.

Set $R_{>}^m = \{\mu \in R^m \mid \mu_i > 0, 1 \le i \le m\}$.

For each $s \in I$, let $G_s : U \to 2^{L(E_s, F)}$ be a mapping so that if we set

$$G = (G_s \mid s \in I), \tag{2.2}$$

then $G : U \to 2^{L(E, F)}$, where $E = \prod_{s \in I} E_s$. That is, if we take arbitrary points $u \in U$ and $v_s \in E_s$, then $G_s(u)v_s$ is a subset of F, hence, for each $v \in E$, the sum

$$\sum_{s \in I} G_s(u)v_s$$

is a subset of F. The *generalized vector variational inequality problem* is to find an element $u^* = (u_s^*)_{s \in I} \in U$ such that

$$\sum_{s \in I} G_s(u^*)(u_s - u_s^*) \not\subseteq -\mathrm{int}C \quad \forall u_s \in U_s, s \in I; \tag{2.3}$$

for brevity, we denote by $GVVI(I, U, G)$ problem (2.3).

Together with $GVVI(I, U, G)$ we shall consider its dual formulation, denoted by $DGVVI(I, U, G)$, which is to find an element $u^* = (u_s^*)_{s \in I} \in U$ such that

$$\sum_{s \in I} G_s(u)(u_s - u_s^*) \not\subseteq -\mathrm{int}C \quad \forall u_s \in U_s, \ s \in I.$$

Next, we shall also consider the *system of generalized vector variational inequalities*, denoted by $SGVVI(I, U, G)$, which is to find an element $u^* =$

$(u_s^*)_{s\in I} \in U$ such that

$$G_s(u^*)(u_s - u_s^*) \not\subseteq -\mathrm{int}C \quad \forall u_s \in U_s, s \in I.$$

Definition 1 The mapping $G : U \to 2^{L(E,F)}$, defined by (2.2), is said to be

(a) *u-hemicontinuous* if for any $u, v \in U$ and $\lambda \in [0, 1]$, the mapping $\lambda \to G(u + \lambda z)z$ with $z = v - u$ is upper semicontinuous at 0^+;

(b) *pseudo (w, P)-monotone* if for all $u, v \subset U$ we have

$$G_s(v)(u_s - v_s) \not\subseteq -\mathrm{int}C \; \forall s \in I \implies \sum_{s\in I} G_s(u)(u_s - v_s) \not\subseteq -\mathrm{int}C.$$

In Allevi et al (2001), the following result has been proved.

Lemma 2.1 *Suppose that the set U, defined by (2.1), is convex and that the mapping $G : U \to 2^{L(E,F)}$, defined by (2.2), is u-hemicontinuous. Then $DGVVI(I, U, G)$ implies $GVVI(I, U, G)$.*

Clearly, $SGVVI(I, U, G)$ implies $DGVVI(I, U, G)$ if G is pseudo (w, P)-monotone and, also, $GVVI(I, U, G)$ always implies $SGVVI(I, U, G)$. Combining the statements above, we obtain the following equivalence result.

Proposition 2.1 *Suppose that the set U, defined by (2.1), is convex and that the mapping $G : U \to 2^{L(E,F)}$, defined by (2.2), is u-hemicontinuous and pseudo (w, P)-monotone. Then $GVVI(I, U, G)$, $DGVVI(I, U, G)$, and $SGVVI(I, U, G)$ are equivalent.*

3   RELATIVE MONOTONICITY TYPE PROPERTIES

Let $G : U \to 2^{L(E,F)}$ be a mapping of form (2.2) and let $\gamma : U \to R_>^m$ be a single-valued mapping. Then we can define the composite mapping $[\gamma \circ G]$ of form (2.2) as follows

$$[\gamma \circ G](u) = (\gamma_1(u)G_1(u), \ldots, \gamma_m(u)G_m(u)).$$

The following result enables us to reduce $GVVI$ with "weighted" mapping to a usual $SGVVI$.

Lemma 3.1 *$GVVI(I, U, \gamma \circ G)$ implies $SGVVI(I, U, G)$.*

Proof: Clearly $GVVI(I,U,\gamma \circ G)$ implies $SGVVI(I,U,\gamma \circ G)$ which is equivalent to $SGVVI(I,U,G)$. $\qquad\qquad\qquad\qquad\qquad\qquad\qquad\qquad$ $\square$

Under certain additional assumptions, we also can obtain an equivalence result.

Proposition 3.1 *Suppose that the set U, defined by (2.1), is convex and that $\gamma \circ G$ is u-hemicontinous and pseudo (w,P)-monotone. Then $GVVI(I,U,\gamma \circ G)$ is equivalent to $SGVVI(I,U,G)$.*

Proof: $GVVI(I,U,\gamma \circ G)$ is now equivalent to $SGVVI(I,U,\gamma \circ G)$ due to Proposition 2.1 with using $\gamma \circ G$ instead of G. Obviously, $SGVVI(I,U,\gamma \circ G)$ is equivalent to $SGVVI(I,U,G)$. $\qquad\qquad\qquad\qquad\qquad\qquad\qquad$ $\square$

In Konnov (2001), new monotonicity type concepts for single-valued scalar mappings, which extend the usual ones, were proposed. Their extensions to the the multi-valued vector case involving constant weights, were suggested in Allevi et al (2001). Now we propose relative monotonicity concepts for vector multivalued mappings with non constant weights.

Definition 2 The mapping $G : U \to 2^{L(E,F)}$, defined by (2.2), is said to be

(a) *relatively monotone* if there exists a mapping $\alpha : U \to R^m_>$ such that for all $u,v \in U$, we have

$$\sum_{s \in I} \left[\alpha_s(u)G_s(u) - \alpha_s(v)G_s(v)\right](u_s - v_s) \subseteq C;$$

(b) *relatively pseudomonotone* if there exist mappings $\alpha, \beta : U \to R^m_>$ such that for all $u,v \in U$, we have

$$\sum_{s \in I} \beta_s(v)G_s(v)(v_s - u_s) \not\subseteq -\mathrm{int}C \implies \sum_{s \in I} \alpha_s(u)G_s(u)(u_s - v_s) \not\subseteq -\mathrm{int}C.$$

In what follows, we reserve the symbols α and β for the parameters associated with relative (pseudo)monotonicity.

The following well-known Ky Fan Lemma Ky Fan (1961), Lemma 1 (see also Yuan (1998), p.6) will play a crucial role in deriving existence results for GVVI's and SGVVI's. First this result was established in Ky Fan (1961),

Lemma 1 under a Hausdorff space setting. We give a somewhat strengthened version of this assertion (see e.g. Yuan (1998), p.6).

Proposition 3.2 *Let X and Y be non-empty sets in a topological vector space E and $Q : X \to 2^Y$ be such that*

(i) for each $x \in X$, $Q(x)$ is closed in Y;

(ii) for each finite subset $\{x^1, \cdots, x^n\}$ of X, its convex hull is contained in the corresponding union $\bigcup_{i=1}^n Q(x^i)$;

(iii) there exists a point $\tilde{x} \in X$ such that $Q(\tilde{x})$ is compact.

Then

$$\bigcap_{x \in X} Q(x) \neq \emptyset.$$

4   EXISTENCE RESULTS

In this section, in addition to the general assumptions, we shall suppose that the space $L(E, F)$ is topologized in such a manner that the mapping $T : Z \times U \to F$, defined by $T(g, u) = g(u)$, is continuous, if Z is a compact subset of $L(E, F)$. As usual, for each set $B \subseteq E$, we denote by $\overline{B}$ its closure. First we establish existence results for $SGVVI(I, U, G)$.

Theorem 4.1 *Let U be convex and compact. Suppose that G has nonempty and compact values and is relatively pseudomonotone, and that $\alpha \circ G$ is u-hemicontinuous. Then $SGVVI(I, U, G)$ is solvable.*

Proof: Define set-valued mappings $A, B : U \to 2^U$ by

$$B(v) = \{u \in U \mid \sum_{s \in I} \beta_s(u) G_s(u)(v_s - u_s) \not\subseteq -\mathrm{int} C\}$$

and

$$A(v) = \{u \in U \mid \sum_{s \in I} \alpha_s(v) G_s(v)(v_s - u_s) \not\subseteq -\mathrm{int} C\}.$$

We divide the proof into the following three steps.

(i) $\bigcap_{v \in U} \overline{B(v)} \neq \emptyset$.

Let z be in the convex hull of any finite subset $\{v^1, \ldots, v^n\}$ of K. Then $z = \sum_{j=1}^n \mu_j v^j$ for some $\mu_j \geq 0, j = 1, \ldots, n$; $\sum_{j=1}^n \mu_j = 1$. If $z \notin \bigcup_{j=1}^n B(v^j)$, then for all $g_s \in G_s(z)$, $s \in I$, we have

$$\sum_{s \in I} \beta_s(z) g_s(z)(v_s^j - z_s) \in -\mathrm{int} C \quad \forall j = 1, \ldots, n.$$

Since $-\mathrm{int}C$ is convex, we obtain

$$\sum_{j=1}^{n} \mu_j \left(\sum_{s \in I} \beta_s(z) g_s(z)(v_s^j - z_s) \right) \in -\mathrm{int}C.$$

It follows that

$$\begin{aligned}
0 &= \sum_{s \in I} \beta_s(z) g_s(z)(z_s - z_s) \\
&= \sum_{s \in I} \beta_s(z) g_s(z) \left(\sum_{j=1}^{n} \mu_j v_s^j - \sum_{j=1}^{n} \mu_j z_s \right) \\
&= \sum_{j=1}^{n} \mu_j \left(\sum_{s \in I} \beta_s(z) g_s(z)(v_s^j - z_s) \right) \in -\mathrm{int}C,
\end{aligned}$$

a contradiction. Therefore, the mapping $\overline{B} : U \to 2^U$ satisfies all the assumptions of Proposition 3.2 and we get

$$\bigcap_{v \in U} \overline{B(v)} \neq \emptyset.$$

(ii) $\bigcap_{v \in U} A(v) \neq \emptyset$.

From relative pseudomonotonicity of G it follows that $B(v) \subseteq A(v)$, but, for each $v \in U$, $A(v)$ is closed. In fact, let $\{u^\theta\}$ be a net in $A(v)$ such that u^θ converges to $\bar{u} \in U$. Then, for each θ, there exist elements $g_s^\theta \in G_s(v)$, $s \in I$, such that

$$\sum_{s \in I} \alpha_s(v) g_s^\theta(v_s - u_s^\theta) \notin -\mathrm{int}C.$$

Since $G(v)$ is compact, without loss of generality we can suppose that $g_s^\theta \to \bar{g}_s \in G_s(v)$ for each $s \in I$. It follows that $g_s^\theta(u_s^\theta) \to \bar{g}_s(\bar{u}_s)$ for each $s \in I$ and

$$\sum_{s \in I} \alpha_s(v) g_s^\theta(v_s - u_s^\theta) \to \sum_{s \in I} \alpha_s(v) \bar{g}_s(v_s - \bar{u}_s) \notin -\mathrm{int}C.$$

We conclude that $\bar{u} \in A(v)$, i.e. $A(v)$ is closed. Therefore, $\overline{B(v)} \subseteq A(v)$ and (i) now implies (ii).

(iii) $SGVVI(I, U, G)$ is solvable.

From (ii) it follows that $DGVVI(I, U, \alpha \circ G)$ is solvable. Applying now Lemmas 2.1 and 3.1, we conclude that $SGVVI(I, U, G)$ is solvable. The proof is complete. $\qquad\square$

Using the additional pseudo(w, P)-monotonicity and u-hemicontinuity assumptions on G, we obtain existence results of solutions for $GVVI(I, U, G)$.

Theorem 4.2 *Let G be relatively pseudomonotone and pseudo (w, P)-monotone. Suppose that G has nonempty and compact values and that $\alpha \circ G$ and G are u-hemicontinuous. Suppose that U is convex and compact. Then $GVVI(I, U, G)$ is solvable.*

The proof follows from Theorem 4.1 and Proposition 2.1.

By employing the corresponding coercivity condition, we obtain existence results on unbounded sets.

Corollary 4.1 *Suppose that G has nonempty and compact values and is relatively pseudomonotone, and that $\alpha \circ G$ is u-hemicontinuous. Suppose that U is convex and closed and that there exist a compact subset V of E and a point $\tilde{v} \in V \cap U$ such that*

$$\sum_{s \in I} \beta_s(u) G_s(u)(\tilde{v}_s - u_s) \subseteq -\mathrm{int}\, C \quad \textit{for all} \quad u \in U \backslash V. \qquad (4.1)$$

Then $SGVVI(I, U, G)$ is solvable.

Proof: In this case it suffices to follow the proof of Theorem 4.1 and observe that $B(\tilde{v}) \subseteq V$ under the above assumptions. Indeed, it follows that $\overline{B(\tilde{v})}$ is compact, hence the assertion of Step (i) will be true due to Proposition 3.2 as well.   □

We can now obtain an exitence result for $GVVI$ on unbounded sets.

Corollary 4.2 *Let G be relatively pseudomonotone and pseudo (w, P)-monotone. Suppose that G has nonempty and compact values and that $\alpha \circ G$ and G are u-hemicontinuous. Suppose that U is convex and closed and that there exist a compact subset V of E and a point $\tilde{v} \in V \cap U$ such that (4.1) holds. Then $GVVI(I, U, G)$ is solvable.*

The proof follows from Corollary 4.1 and the definition of pseudo (w, P)-monotonicity.

5   EXISTENCE RESULTS IN BANACH SPACES

By choosing different topologies, one can specify the above existence results for less general classes of topological vector spaces. For example, in this section, we specialize these results for a Banach space setting.

Namely, we suppose that E and F are real Banach spaces and that C is a convex, closed and solid cone C in F. We shall apply the weak topology in E, the strong topology in F, and the strong operator topology in $L(E, F)$. For this reason, we need the concept of a completely continuous mapping.

Definition 4 A mapping is said to be *completely continuous* if it maps each weakly convergent sequence into a strongly convergent sequence.

If a mapping $Q : E \to 2^{L(E,F)}$ has nonempty and compact values and each element $q \in Q(v)$ is completely continuous, then the mapping $T : Q(v) \times U \to F$, defined by $T(q, u) = q(u)$, will be continuous in the (strong) $\times$ (weak) topology. Taking this argument into account and following the proof of Theorem 4.1 and Corollary 4.1, we obtain an existence result of solutions for $SGVVI(I, U, G)$.

Corollary 5.1 *Let G be relatively pseudomonotone. Suppose that $\alpha \circ G$ is u-hemicontinuous, G has nonempty and compact values and that, for each $v \in U$, each element of $G(v)$ is completely continuous. Suppose that U is convex and that at least one of the following assumptions holds:*

(i) U is weakly compact;

(ii) U is closed and there exist a weakly compact subset V of E and a point $\tilde{v} \in V \cap U$ such that (4.1) holds.

Then $SGVVI(I, U, G)$ is solvable.

Adding the pseudo(w, P)-monotonicity and u-hemicontinuity assumptions on G, we obtain an analogue of Theorem 4.2 in Banach spaces.

Corollary 5.2 *Let all the assumptions of Corollary 5.1 hold and let G be pseudo (w, P)-monotone and u-hemicontinuous. Then $GVVI(I, U, G)$ is solvable.*

References

Allevi, E., Gnudi, A., and Konnov, I.V. (2001), Generalized vector variational inequalities over product sets, *Nonlinear Analysis Theory, Methods & Ap-*

plications, Proceedings of the Third Word Congress of Nonlinear Analysis, Vol. 47, Part 1, Editor-in-chief: V.Lamshmikantham, 573–582.

Ansari, Q.H. and Yao, J.-C. (1999), A fixed point theorem and its applications to a system of variational inequalities, *Bull. Austral. Math. Soc.*, Vol. 59, 433 – 442.

Ansari, Q.H. and Yao, J.-C., System of generalized variational inequalities and their applications, *Applicable Analysis*, Vol.76(3-4), 203-217, 2000.

Bianchi, M. (1993), Pseudo P-monotone operators and variational inequalities, *Research Report No.6*, Istituto di Econometria e Matematica per le Decisioni Economiche, Università Cattolica del Sacro Cuore, Milan.

Giannessi, F. (1980), Theory of alternative, quadratic programs and complementarity problems, *Variational Inequalities and Complementarity Problems*, R.W.Cottle, F.Giannessi, and J.L.Lions, eds., Wiley, New York, 151-186.

Hadjisavvas, H. and Schaible, S. (1998), Quasimonotonicity and pseudomonotonicity in variational inequalities and equilibrium problems, *Generalized Convexity, Generalized Monotonicity*, J.-P. Crouzeix, J.E. Martinez-Legaz and M. Volle, eds., Kluwer Academic Publishers, Dordrecht - Boston - London, 257–275.

Ky Fan (1961), A generalization of Tychonoff's fixed-point theorem, *Math. Annalen*, Vol. 142, 305–310.

Konnov, I.V. (1995), Combined relaxation methods for solving vector equilibrium problems, *Russ. Math. (Iz. VUZ)*, Vol. 39, no.12, 51–59.

Konnov, I.V. (2001), Relatively monotone variational inequalities over product sets, *Operations Research Letters*, Vol. 28, 21–26.

Oettli, W. and Schläger, D. (1998), Generalized vectorial equilibria and generalized monotonicity,*Functional Analysis with Current Applications in Science, Technology and Industry*, M. Brokate and A.H. Siddiqi, eds., Pitman Research Notes in Mathematical Series, No.377, Addison Wesley Longman Ltd., Essex, 145–154.

Rosen, J.B. (1965), Existence and uniqueness of equilibrium points for concave n-person games, *Econometrica*, Vol. 33, 520–534.

Yang, X.Q. and Goh, C.J. (1997), On vector variational inequalities: application to vector equilibria, *J. Optimiz. Theory and Appl.*, Vol. 95, 431–443.

Yuan,G.X.Z. (1998), The Study of Minimax Inequalities and Applications to Economies and Variational Inequalities, *Memoires of the AMS*, Vol.132, Number 625.

25 ON A GEOMETRIC LEMMA AND SET-VALUED VECTOR EQUILIBRIUM PROBLEM

Shui-Hung Hou

Department of Applied Mathematics
The Hong Kong Polytechnic University

Abstract: Based on a variation of Fan's geometric lemma, an existence theorem of solutions for set-valued vector equilibrium problem is given in this paper.

Key words: Set-valued map, geometric lemma, vector equilibrium.

1 INTRODUCTION

The following geometrical lemma was established in Fan (1961):

Lemma 1.1 *Let X be a nonempty compact convex set in a Hausdorff topological vector space E. Let A be a closed subset of $X \times X$ with the following properties:*

(a) *$(x, x) \in A$ for every $x \in X$;*

(b) *$\forall x \in X$, the set $\{y \in X : (x, y) \notin A\}$ is convex or empty.*

Then there exists a point $x_0 \in X$ such that $x_0 \times X \subset A$.

Subsequently, a slight generalization of Fan's lemma was given in Takahashi (1976) in the following form:

Lemma 1.2 *Let X be a nonempty compact convex set in a Hausdorff topological vector space. Let A be a subset of $X \times X$ having the following properties:*

(i) *For any $y \in X$, the set $\{x \in X : (x, y) \in A\}$ is closed;*

(ii) *$(x, x) \in A$ for every $x \in X$;*

(iii) *For any $x \in X$, the set $\{y \in X : (x, y) \notin A\}$ is convex.*

Then there exists a point $x_0 \in X$ such that $x_0 \times X \subset A$.

Later, Ha gave another generalization of Fan's geometric lemma by relaxing, among the others, the compactness condition (see Ha (1980)).

Lemma 1.3 *Let E, F be Hausdorff topological vector spaces, $X \subset E, Y \subset F$, be nonempty convex subsets and let $A \subset X \times Y$ be a subset such that*

(a) *$\forall y \in Y$, the set $\{x \in X : (x, y) \in A\}$ is closed in X;*

(b) *$\forall x \in X$, the set $\{y \in Y : (x, y) \notin A\}$ is convex or empty.*

Suppose that there exists a subset Γ of A and a compact convex subset K of X such that Γ is closed in $X \times Y$ and

(c) *$\forall y \in Y$, the set $\{x \in K : (x, y) \in \Gamma\}$ is nonempty and convex.*

Then there exists a point $x_0 \in K$ such that $x_0 \times Y \subset A$.

Remark 1.1 *Both Lemma 1.1 and 1.2 are special cases of Lemma 1.3 with* $X = Y = K$ *and* $\Gamma = \{(x, x) : x \in X\}$.

In this paper, we first give a variation of the geometric lemma of Fan et al, and then by applying the result, we prove an existence theorem of solutions for set-valued vector equilibrium problem.

2 PRELIMINARIES

We shall use the following notations and definitions in this paper.

Let A be a nonempty set. We denote by 2^A the family of all subsets of A. If A is a subset of a topological vector space E, we shall denote by $\operatorname{cl} A$ the closure of A in E, and by $\operatorname{co} A$ the convex hull of A. Also in E, the convex hull of its finite subset will be called a polytope.

Let X, Y be topological spaces and $F : X \to 2^Y$ be a set-valued map. The graph of F is the set $\operatorname{Gr}(F) := \{(x, y) \in X \times Y : y \in F(x)\}$, and F is said to have a closed graph if $\operatorname{Gr}(F)$ is a closed subset of $X \times Y$. For $B \subset Y$, we set $F^-(B) := \{x \in X : F(x) \cap B \neq \emptyset\}$ which is the lower inverse of B under F.

Recall also the following definitions from Tian (1992).

Definition 2.1 *A set-valued map* $F : X \to 2^Y$ *is said to be transfer closed-valued on* X *if for any* $x \in X$ *and* $y \notin F(x)$*, there exists an* $z \in X$ *such that* $y \notin \operatorname{cl} F(z)$.

Let us also mention the well known fixed point theorem of Kakutani (1941).

Theorem 2.1 *Let* X *be a nonempty compact convex subset in some finite dimensional space* $\mathbb{R}^n$ *and let the set-valued map* $F : X \to 2^X$ *have closed graph and nonempty convex vaules. Then* F *has a fixed point* $x^* \in F(x^*)$.

3 A VARIATION OF FAN'S GEOMETRIC LEMMA

Theorem 3.1 *Let* E, F *be Hausdorff topological vector spaces and* $X \subset E$, $Y \subset F$ *be nonempty subsets with* Y *convex. Let* $B \subset A \subset X \times Y$, $K \subset X$ *a compact convex subset, and* $\Omega, \Gamma : K \to 2^Y$ *be set-valued maps such that*

$$\operatorname{Gr}(\Gamma) \subset \operatorname{Gr}(\Omega) \subset B \subset A \subset X \times Y.$$

Suppose that

(a) *the set-valued map $y \to A^-(y) := \{x \in X : (x, y) \in A\}$ is transfer closed-valued on Y;*

(b) *$\forall x \in K$, the set $\{y \in Y : (x, y) \notin B\}$ is convex or empty;*

(c) *for every polytope P in Y, there exists a finite subset $\{x_1, \ldots, x_n\}$ of K such that $P \subset \bigcup_{i=1}^{n} \Gamma(x_i)$;*

(d) *for every polytope P in Y, $\Gamma(x) \cap P$ is closed in P for each x in K; and*

(e) *$\forall y \in Y$, the set $\Omega^-(y)$ is convex.*

Then there exists a point $x_0 \in K$ such that $x_0 \times Y \subset A$.

Proof: Suppose to the contrary that for each $x \in K$, there exists $y \in Y$ such that $x \notin A^-(y)$. Then, by (a) $x \notin \operatorname{cl} A^-(y')$ for some $y' \in Y$. Hence we have $K \subset \bigcup_{y \in Y} V_y$ where $V_y := \{z \in X : z \notin \operatorname{cl} A^-(y)\}$. By the compactness of K, there exists a finite family $\{y_1, \ldots, y_n\}$ of Y such that $K \subset \bigcup_{i=1}^{n} V_{y_i}$. Let $\{\beta_1, \ldots, \beta_n\}$ be a partition of unity on K subordinated to the finite covering $\{V_{y_i} : i = 1, \ldots, n\}$, that is, $\beta_1, \ldots, \beta_n$ are nonnegative real-valued continuous functions on K such that each β_i vanishes on $K \backslash V_{y_i}$, while $\sum_{i=1}^{n} \beta_i(x) = 1$ for all $x \in K$. Let $P := \operatorname{co}\{y_1, \ldots, y_n\} \subset Y$ and define a continuous mapping $p : K \to P$ by setting

$$p(x) = \sum_{i=1}^{n} \beta_i(x) y_i, \quad \forall x \in K.$$

Thus, for each i such that $\beta_i(x) > 0$, x lies in $V_{y_i} \cap K$, so that $(x, y_i) \notin B$ and whence by (b) we have

$$\left(x, \sum_{i=1}^{n} \beta_i(x) y_i \right) \notin B.$$

Hence,

$$(x, p(x)) \notin B, \quad \forall x \in K. \tag{3.1}$$

On the other hand, by (c) there exists a fintie subset $\{x_1, \ldots, x_n\}$ of K such that $P \subset \bigcup_{i=1}^{n} \Gamma(x_i)$. Let $\Delta := \operatorname{co}\{x_1, \ldots, x_n\} \subset K$. Define a set-valued map $H : \Delta \to 2^{\Delta}$ by $H(x) := \operatorname{co}\{x_i : p(x) \in \Gamma(x_i)\}$ for each x in Δ. Then each $H(x)$ is a nonempty closed convex subset of Δ. Moreover, H has a closed graph in $\Delta \times \Delta$. Indeed, let $(v, w) \in \Delta \times \Delta \backslash \mathrm{Gr}\,(H)$, i.e. $w \notin H(v)$. Then there exists an open neighborhood V of w in Δ which is disjoint from $H(v)$. Suppose $H(v) = \mathrm{co}\,\{x_i : i \in J\}$ for some $J \subset \{1, \ldots, n\}$. Then $p(v) \notin \Gamma(x_j)$ for $j \notin J$. Therefore by (d), $U := p^{-1}\left(\underset{j \notin J}{\cap}\, P \backslash \Gamma(x_j)\right)$ is an open neighborhood of v in Δ. If $z \in U$, then $p(z) \notin \Gamma(x_j)$ for $j \notin J$, and so $H(z) \subset H(v)$. This implies $V \cap H(z) = \emptyset$ for all $z \in U$. Hence we have an open neighborhood $U \times V$ of (v, w) which does not intersect the graph of H, that is, the graph $\mathrm{Gr}\,(H)$ is closed.

Applying the Kakutani's fixed point theorem, we have a point $\bar{x} \in \Delta$ such that $\bar{x} \in H(\bar{x})$. If $H(\bar{x}) = \mathrm{co}\,\{x_j : j \in J_0\}$ for some $J_0 \subset \{1, \ldots, n\}$, then $p(\bar{x}) \in \Gamma(x_j) \subset \Omega(x_j)$ for every $j \in J_0$, i.e. $x_j \in \Omega^-[p(\bar{x})], \forall j \in J_0$. Since $\Omega^-[p(\bar{x})]$ is convex by (e) and $\bar{x}$ is a convex combination of $\{x_j : j \in J_0\}$, we have $\bar{x} \in \Omega^-[p(\bar{x})]$. And so $(\bar{x}, p(\bar{x})) \in \mathrm{Gr}\,(\Omega) \subset B$, contradicting (3.1). The theorem is proven.

Remark 3.1 *The condition (b) in Theorem 3.1 may be replaced by the equivalent condition (b') below.*

(b') For any finite subset $\{y_1, \ldots, y_n\} \subset Y$,

$$K \cap B^-(\mathrm{co}\,\{y_1, \ldots, y_n\}) \subset \bigcup_{i=1}^{n} B^-(y_i) \cap K.$$

Indeed, (b) $\Rightarrow$ (b'): let $x \in K \cap B^-(\mathrm{co}\,\{y_1, \ldots, y_n\})$. Then, $x \in K$ and $(x, \lambda_1 y_1 + \cdots + \lambda_n y_n) \in B$ for some $0 \leq \lambda_1, \ldots, \lambda_n \leq 1$ with $\sum_{i=1}^{n} \lambda_i = 1$. It suffices to show that x belongs to one of the $B^-(y_i)$'s. Assume to the contrary that $x \notin B^-(y_i)$ for all $i = 1, \ldots, n$. It follows from (b) that $(x, \lambda_1 y_1 + \cdots + \lambda_n y_n) \notin B$ which is a contradiction. Therefore, $x \in \overset{n}{\underset{i=1}{\cup}} B^-(y_i) \cap K$. This establishes the implication (b') $\Rightarrow$ (b).

(b') $\Rightarrow$ (b): let $(x, y_i) \notin B$ for $i = 1, \ldots, n$ with $x \in K$ and let $z = \lambda_1 y_1 + \cdots + \lambda_n y_n$ be an arbitrary convex combination of y_i's. We need to show that $(x, z) \notin B$. Suppose otherwise that $(x, z) \in B$. Then, $x \in K \cap B^-(z)$, and so by (b') we have $x \in \overset{n}{\underset{i=1}{\cup}} B^-(y_i) \cap K$. This implies that for some $1 \leq j \leq n$, $(x, y_j) \in B$, which is a contradiction. The implication (b') $\Rightarrow$ (b) is established.

Thus, the theorem below is equivalent to Theorem 3.1.

Theorem 3.2 *Let E, F be Hausdorff topological vector spaces and $X \subset E$, $Y \subset F$ be nonempty subsets with Y convex. Let $B \subset A \subset X \times Y$, $K \subset X$ a compact convex subset, and $\Omega, \Gamma : K \to 2^Y$ be set-valued maps such that*

$$\mathrm{Gr}\,(\Gamma) \subset \mathrm{Gr}\,(\Omega) \subset B \subset A \subset X \times Y.$$

Suppose that

(a) *the set-valued map $y \to A_y := \{x \in X : (x, y) \in A\}$ is transfer closed-valued on Y;*

(b') *For any finite subset $\{y_1, \ldots, y_n\} \subset Y$,*

$$K \cap B^-(\mathrm{co}\,\{y_1, \ldots, y_n\}) \subset \bigcup_{i=1}^{n} B^-(y_i) \cap K;$$

(c) *for every polytope P in Y, there exists a finite subset $\{x_1, \ldots, x_n\}$ of K such that $P \subset \bigcup_{i=1}^{n} \Gamma(x_i)$;*

(d) *for every polytope P in Y, $\Gamma(x) \cap P$ is closed in P for each x in K; and*

(e) *$\forall y \in Y$, the set $\Omega^-(y)$ is convex.*

Then there exists a point $x_0 \in K$ such that $x_0 \times Y \subset A$.

4   SET-VALUED VECTOR EQUILIBRIUM PROBLEM

Let W be a topological vector space ordered by a pointed closed convex cone P. An order relation in W can be defined by $x \leq_P$ iff $y - x \in P$. Let E, F be Hausdorff topological vector spaces and $X \subset E$, $Y \subset F$ be nonempty subsets with Y convex.

Let $K \subset X$ be a compact convex subset and let $\Phi : X \times Y \to 2^W$ be a set-valued map. We are interested in the problem of finding

$$x_0 \in K \quad \text{such that} \quad \Phi(x_0, y) \subset P$$

holds for all $y \in Y$. This problem is called a set-valued vector equilibrium problem, and such a point x_0 is said to be an equilibrium point.

If Φ is a single-valued map, then the above problem becomes the vector equilibrium problem; see for example Blum and Oettli (1994) and Giannessi (2000) and the references therein.

Based on the result of Theorm 3.1, we are now ready to prove an existence theorem for the set-valued vector equilibrium problem.

Theorem 4.1 *Let $\Omega, \Gamma : K \to 2^Y$ be set-valued maps such that $\mathrm{Gr}\,(\Gamma) \subset \mathrm{Gr}\,(\Omega)$. Suppose that*

(i)   *the set-valued map $y \to \{x \in X : \Phi(x,y) \subset P\}$ is transfer closed-valued;*

(ii) *$\forall x \in K$, the set $\{y \in Y : \Phi(x,y) \not\subset P\}$ is convex or empty;*

(iii) *$\forall (x,y) \in \mathrm{Gr}\,(\Omega)$, $\Phi(x,y) \subset P$;*

(iv) *$\forall y \in Y$, the set $\Omega^-(y)$ is convex;*

(v)   *$\forall x \in K$, the set $\Gamma(x)$ is closed; and*

(vi) *$\forall y \in Y$, there exists an open neighborhood $N(y)$ of y such that*

$$\bigcap_{v \in N(y)} \Gamma^-(v) \neq \emptyset.$$

Then there exists $x_0 \in K$ such that $\Phi(x_0, y) \subset P$ for all $y \in Y$.

Proof: If we can show that all the conditions (a)-(e) of Theorem 3.1 are satisfied, then we may invoke the theorem to conclude the existence of an equilibrium point $x_0 \in K$. To this end, We define $A := \{(x,y) \in X \times Y : \Phi(x,y) \subset P\}$ and $B := A$. It follows from (i) and (ii) that conditions (a) and (b) in Thoerem 3.1 are satisfied. Also it is clear that (iii) implies $\mathrm{Gr}\,(\Omega) \subset B \subset A$, (iv) implies (e), and (v) implies (d) of Theorem 3.1. It remains to show that condition (c) of Theorem 3.1 is satisfied. Indeed, let Q be a polytope of Y. For each $y \in Q$, by condition (vi) we know that there exists an open neighborhood $N(y)$ of y such that $M(y) := \bigcap_{v \in N(y)} \Gamma^-(v) \neq \emptyset$. Since Q is compact, there exists a finite family $\{y_1, \ldots, y_r\} \subset Y$ such that $Q \subset \bigcup_{i=1}^r N(y_i)$. With each $M(y_i) \neq \emptyset$, we may take a point $x_i \in M(y_i)$; consequently $N(y_i) \subset \Gamma(x_i)$ and so $Q \subset \bigcup_{i=1}^r \Gamma(x_i)$. Thus, condition (c) is satisfied. We can now conclude from Theorem 3.1 the existence of an equilibrium point $x_0 \in K$. The theorem is proven.

Remark 4.1 *Condition (ii) is satisfied when for every $x \in K$ the set-valued map $\Phi(x, \cdot) : Y \to 2^W$ has the following P-properly quasiconvexity property (see Kuroiwa (1996)).*

Definition 4.1 *A set-valued map $\phi : Y \to 2^W$ is said to be P-properly quasi-convex if for every $y_1, y_2 \in Y$, $w_1 \in \phi(y_1)$, $w_2 \in \phi(y_2)$, and $\lambda \in (0, 1)$, there exists $w \in \phi(\lambda y_1 + (1 - \lambda)y_2)$ such that either $w \leq_P w_1$ or $w \leq_P w_2$.*

Indeed, let $M_x := \{y \in Y : \Phi(x, y) \not\subset P\}$ and $\lambda \in (0, 1)$, $y_1 \in M_x, y_2 \in M_x$. Then for $i = 1, 2$ there exist $z_i \in \Phi(x, y_i)$, while $z_i \notin P$. Since $\Phi(x, y)$ is P-properly quasi-convex in y, there exists a $z \in \Phi(x, \lambda y_1 + (1 - \lambda)y_2)$ such that either $z \leq_P z_1$ or $z \leq_P z_2$. If we can show $z \not\subset P$, then $\Phi(x, \lambda y_1 + (1-\lambda)y_2) \not\subset P$ implying $\lambda y_1 + (1 - \lambda)y_2 \in M_x$, so that M_x is convex. To this end, suppose otherwise $z \in P$. If $z \leq_P z_1$ holds, then this together with $z \in P$ imply $z_1 = (z_1 - z) + z \in P + P \subseteq P$ which is a contradiction. This establishes $z \notin P$. Similarly $z \notin P$ if $z \leq_P z_2$. Hence M_x is convex and condition (ii) of Theorem 4.1 is satisfied.

Acknowledgments

This work was supported by the Research Committee of The Hong Kong Polytechnic University.

References

Kakutani, S. (1941), A generalization of Brouwer's fixed point theorem, *Duke Math. Journal*, Vol. 8, pp. 457-459.

Fan, K. (1961), A generalization of Tychonoff's fixed point theorem, *Math. Ann.*, Vol. 142, pp. 305-310.

Takahashi, W. (1976), Nonlinear variational inequalities and fixed point theorems, *J. of Math. Soc. of Japan*, Vol. 28, No. 1, pp. 168-181.

Ha, C.W. (1980), Minimax and fixed point theorems, *Math. Ann.*, Vol. 248, pp. 73-77.

Tian, G. (1992), Generalizations of the FKKM theorem and the Ky Fan minimax inequality, with applications to maximal elements, price equilibrium and complementarity, *Journal of Mathematical Analysis and Applications*, Vol. 170, No. 2, pp. 457-471.

Blum, E. and Oettli, W. (1994), ¿From optimization and variational inequalities to equilibrium problems, *The Math. Student*, Vol. 63, pp.123-145.

Kuroiwa, D. (1996), Convexity for set-valued maps, *Applied Mathematics Letter*, Vol. 9, No. 2, pp. 97-101.

Giannessi, F., Editor (2000), *Vector Variational Inequalities and Vector Equilibria: Mathematical Theories*, Kluwer Academic Publishers, Dordrecht, The Netherlands.

26   EQUILIBRIUM PROBLEMS

Giovanna Idone

D.I.M.E.T.
Università di Reggio Calabria
Via Graziella, Loc. Feo di Vito
89100 Reggio Calabria - ITALIA

idone@ing.unirc.it

and Antonino Maugeri

Dipartimento di Matematica e Informatica
Università di Catania
Viale A.Doria, 6
95125 Catania - ITALIA

maugeri@dmi.unict.it

Abstract: We show that many equilibrium problems fulfill a common law expressed by a set of complementarity conditions and that the equilibrium solution is obtained as a solution to a Variational Inequality. In particular we show that various models of elastoplastic torsion are included in the framework above.

Key words: Variational inequality, elastoplastic torsion, quasi-relative interior, Lagrange multipliers.

1   INTRODUCTION

Many equilibrium problems arising from various fields of science may be expressed, under general conditions, in a unified way:

$$\begin{cases} Bu\mathcal{L}u = 0 \\ Bu \geq 0 \\ u \in S \end{cases} \qquad (1.1)$$

where B and $\mathcal{L}$ are suitable operators defined in a suitable functional class S. For example, many important problems of Mathematical Phisic follow the structure:

$$\begin{cases} [\mathcal{L}(u)u - F(u, \Delta u)][A(u, \Delta u) - \psi] = 0 \\ A(u, \Delta u) \leq \psi \\ u \geq \phi \\ u = \phi \text{ on } \partial\Omega. \end{cases}$$

A particular case is considered in section 5.4 of the book Troianiello (1987), where the following problem is considered:

$$\begin{cases} [\mathcal{L}(u) - F(u, \Delta u)][A(u) - \psi] = 0 \\ A(u, \Delta u) \leq \psi \\ u = \phi \text{ on } \partial\Omega. \end{cases}$$

The kind of EQUILIBRIUM described by the structure (1.1) is, in general, different from the one obtained by minimizing of a cost functional or of an energy integral. Moreover, the structure (1.1) leads, in general, to a VARIATIONAL INEQUALITY on a convex, closed subset of S (usually called K). This happens, for example, for: unilateral problems in continuum mechanics (the celebrated Signorini problem) the obstacle problem, the discrete and continuum traffic equilibrium problem, the spatial price problem, the financial problem, the Walras problem, see Idone (2003), Maugeri (2001), Maugeri (1998).

The equilibrium given by the structure (1.1) may be considered as an equilibrium from a "local point of view". It is different, in general, from the one, which we call "global", obtained by minimizing the usual functionals, and represents a complementary aspect which makes clear unknown features of equilibrium problems. The reason for which the past decades have witnessed an exceptional interest in the equilibrium problems of the type (1.1) rests on the fact that the Variational Inequality theory *which in general expresses the equilib-rium conditions* (1.1), provides a powerful methodology, that in these last years has been improved by studying the connections with the Separation Theory, Gap Functions, the Lagrangean Theory and Duality and many related compu-tational procedures.

In the present paper we show that various models of elasto-plastic torsion satisfy the structure (1.1) and that the equilibrium conditions (1.1) can be expressed in terms of a Variational Inequality.

2   A MODEL OF ELASTIC-PLASTIC TORSION

Let Ω be an open bounded Lipschitz domain with its boundary $\Gamma = \partial\Omega$; for the sake of simplicity, we confine ourselves to the case $\Omega \subset \mathbf{R}^2$.

Let K be the closed convex non empty subset of $H_0^1(\Omega)$:

$$K = \left\{ v \in H_0^1(\Omega) : v \geq 0, \ \left(\frac{\partial v}{\partial x_1} + \frac{\partial v}{\partial x_2} \right)^2 \leq 1 \ \text{a.e.} \ x \in \Omega \right\}. \tag{2.1}$$

Then for each $f \in L^2(\Omega)$ the V.I.

$$\text{Find } u \in K : a(u, v - u) \geq \int_\Omega f(v - u)dx \quad \forall v \in K \tag{2.2}$$

where:

$$a(u, v) := \sum_{i,j=1}^2 \int_\Omega a_{ij} \frac{\partial u}{\partial x_i} \frac{\partial v}{\partial x_j} dx + \int_\Omega \sum_{i=1}^2 b_i \frac{\partial u}{\partial x_i} v dx +$$

$$- \int_\Omega \sum_{i=1}^n c_i u \frac{\partial v}{\partial x_i} dx + \int_\Omega d\, uv dx + \lambda \int_\Omega uv dx$$

is the usual sesquilinear form with:

$a_{ij} \in L^\infty(\Omega), b_i \in L^{2+\epsilon}, \epsilon > 0$

$c_i \in L^{2+\epsilon}, \epsilon > 0$

$d \in L^{1+\epsilon}, \epsilon > 0$

$$\sum_{i,j=1}^2 a_{ij}(x)\xi_i\xi_j \geq \alpha \sum_{i=1}^2 \xi_i^2 \ \forall \xi \in R^2, \alpha > 0$$

$\lambda \geq \overline{\lambda}$ in such a way that:

$$a(u, u) \geq \nu \|u\|_{H_0^1}^2, \quad \nu > 0$$

admits a unique solution.

Let u be the unique solution to the *V.I.* and let us consider the Lagrangean Function:

$$\mathcal{L}(v, \mu, \lambda_1, \lambda_2) = \psi(v) - \int_\Omega \mu v dx - \int_\Omega \lambda_1(x) \left(1 - \frac{\partial v}{\partial x_1} - \frac{\partial v}{\partial x_2}\right) dx -$$

$$- \int_\Omega \lambda_2(x) \left(1 + \frac{\partial v}{\partial x_1} + \frac{\partial v}{\partial x_2}\right) dx \tag{2.3}$$

where:

$$\psi(v) = a(u, v - u) - \int_\Omega f(v - u) dx = \langle \mathcal{L}u, v - u \rangle \quad v \in H_0^1(\Omega)$$

$$(\mu, \lambda_1, \lambda_2) \in C^* = \{(\mu, \lambda_1, \lambda_2) : \mu, \lambda_1, \lambda_2 \in L^2(\Omega), \mu, \ \lambda_1, \ \lambda_2 \geq 0 \text{ a. e. in } \Omega\}.$$

Taking into account that the convex K satisfies the constraint qualification condition introduced in Borwein et al. (1991), namely the "quasi-relative interior of K is non empty", which replaces the standard Slater condition in the infinite dimensional case, following Daniele (1999) (see also Maugeri (1998)), it is possible to show the following Lemma:

Lemma 2.1 *There exist* $(\overline{\mu}, \overline{\lambda}_1, \overline{\lambda}_2) \in C^*$ *such that*

$$\mathcal{L}(u, \mu, \lambda_1, \lambda_2) \leq \mathcal{L}(u, \overline{\mu}, \overline{\lambda}_1, \overline{\lambda}_2) \leq \mathcal{L}(v, \overline{\mu}, \overline{\lambda}_1, \overline{\lambda}_2) \tag{2.4}$$

$\forall v \in H_0^1(\Omega), \forall(\mu, \lambda_1, \lambda_2) \in C^*$. *Moreover,* $\mathcal{L}(u, \overline{\mu}, \overline{\lambda}_1, \overline{\lambda}_2) = 0$.

Now let us denote by $\mathcal{L}u$ the operator defined by the relationship:

$$\langle \mathcal{L}u, v \rangle = a(u, v) - \int_\Omega f v dx \quad \forall v \in H_0^1(\Omega). \tag{2.5}$$

By means of Lemma 2.1, we can prove the following result:

Theorem 2.1 *Let u be solution to the problem (2.2) and $\mathcal{L}u$ the operator defined by (2.5). Then u fulfills the conditions:*

$$
\begin{cases}
\mathcal{L}(u)u\left[\left(\dfrac{\partial u}{\partial x_1}+\dfrac{\partial u}{\partial x_2}\right)^2-1\right]=0 \;\; in\; \Omega \\[3em]
u\geq 0 \;\; in\; \Omega \\[2em]
\left(\dfrac{\partial u}{\partial x_1}+\dfrac{\partial u}{\partial x_2}\right)^2\leq 1 \; in\; \Omega \\[3em]
u=0 \; on\; \partial\Omega.
\end{cases}
\tag{2.6}
$$

Proof.: Being: $\psi(u)=0,\; \overline{\mu}\geq 0 \;\; \overline{\lambda}_i\geq 0,\;\; i=1,2$ and

$$
-1\leq \frac{\partial u}{\partial x_1}+\frac{\partial u}{\partial x_2}\leq 1,\;\; u(x)\geq 0
$$

from $\mathcal{L}(u,\overline{\mu},\overline{\lambda}_1,\overline{\lambda}_2)=0$, it follows:

$$
\int_\Omega \overline{\mu}u\,dx=0 \Leftrightarrow \overline{\mu}u=0 \quad \text{a.e. } x\in\Omega
\tag{2.7}
$$

$$
\int_\Omega \overline{\lambda}_1(x)\left(1-\frac{\partial u}{\partial x_1}-\frac{\partial u}{\partial x_2}\right)dx=0
$$

$$
\Leftrightarrow \int_\Omega \overline{\lambda}_1(x)\,dx=\int_\Omega \overline{\lambda}_1(x)\left(\frac{\partial u}{\partial x_1}+\frac{\partial u}{\partial x_2}\right)dx
\tag{2.8}
$$

$$
\int_\Omega \overline{\lambda}_2(x)\left(1+\frac{\partial u}{\partial x_1}+\frac{\partial u}{\partial x_2}\right)dx=0
$$

$$
\Leftrightarrow \int_\Omega \overline{\lambda}_2(x)\,dx=-\int_\Omega \overline{\lambda}_2(x)\left(\frac{\partial u}{\partial x_1}+\frac{\partial u}{\partial x_2}\right)dx.
\tag{2.9}
$$

From (2.4), taking into account (2.7), (2.8), (2.9), we derive:

$$
\mathcal{L}(v,\overline{\mu},\overline{\lambda}_1,\overline{\lambda}_2)=\langle \mathcal{L}u,v-u\rangle-\int_\Omega \overline{\mu}(v-u)\,dx
$$

$$
-\int_\Omega \overline{\lambda}_1(x)\left(\frac{\partial(v-u)}{\partial x_1}+\frac{\partial(v-u)}{\partial x_2}\right)dx
$$

$$
+\int_\Omega \overline{\lambda}_2(x)\left(\frac{\partial(v-u)}{\partial x_1}+\frac{\partial(v-u)}{\partial x_2}\right)dx =
$$

$$\langle \mathcal{L}u, v - u \rangle - \int_\Omega \overline{\mu}(v - u)dx-$$

$$\int_\Omega \left(\frac{\partial \overline{\lambda}_1}{\partial x_1} + \frac{\partial \overline{\lambda}_1}{\partial x_2} \right)(v - u)dx - \int_\Omega \left(\frac{\partial \overline{\lambda}_2}{\partial x_1} + \frac{\partial \overline{\lambda}_2}{\partial x_2} \right)(v - u)\,dx =$$

$$\left\langle \mathcal{L}u - \overline{\mu} - \left(\frac{\partial \overline{\lambda}_1}{\partial x_1} + \frac{\partial \overline{\lambda}_1}{\partial x_2} \right) + \left(\frac{\partial \overline{\lambda}_2}{\partial x_1} + \frac{\partial \overline{\lambda}_2}{\partial x_2} \right), \, v - u \right\rangle \geq 0$$

$$\forall v \in H_0^1(\Omega). \tag{2.10}$$

From (2.10), choosing $v = u \pm \varphi, \forall \varphi \in H_0^1(\Omega)$ we obtain:

$$\mathcal{L}u - \overline{\mu} - \left(\frac{\partial \overline{\lambda}_1}{\partial x_1} + \frac{\partial \overline{\lambda}_1}{\partial x_2} \right) + \left(\frac{\partial \overline{\lambda}_2}{\partial x_1} + \frac{\partial \overline{\lambda}_2}{\partial x_2} \right) = 0 \tag{2.11}.$$

Now let us set:

$$E = \left\{ x \in \Omega : \left(\frac{\partial u}{\partial x_1} + \frac{\partial u}{\partial x_2} \right)^2 < 1 \right\} \tag{2.12}$$

E is called the *elastic region.*

From (2.8) and (2.9) it follows $\overline{\lambda}_i(x) = 0 \ i = 1, 2, \ x \in E$ and hence (2.11) becomes

$$\mathcal{L} - \overline{\mu} = 0 \ \text{ in } E.$$

Now if $x \in \Omega$ is such that

$$\mathcal{L} - \overline{\mu} \neq 0$$

also

$$\frac{\partial \overline{\lambda}_1(x)}{\partial x_1} + \frac{\partial \overline{\lambda}_1(x)}{\partial x_2} - \frac{\partial \overline{\lambda}_2(x)}{\partial x_1} - \frac{\partial \overline{\lambda}_2(x)}{\partial x_2} \neq 0$$

and in virtue of the previous arguments, it is not possible that:

$$\left(\frac{\partial u}{\partial x_1} + \frac{\partial u}{\partial x_2} \right)^2 < 1.$$

Then if $\mathcal{L} - \overline{\mu} \neq 0$, it follows that

$$x \in T = \left\{ x \in \Omega : \left(\frac{\partial u}{\partial x_1} + \frac{\partial u}{\partial x_2} \right)^2 = 1 \right\} \tag{2.13}$$

T is the *torsion region.*

Thus, we have obtained that:

$$(\mathcal{L}u - \overline{\mu})\left[\left(\frac{\partial u}{\partial x_1} + \frac{\partial u}{\partial x_2} \right)^2 - 1 \right] = 0 \ \text{in } \Omega \tag{2.14}.$$

Taking into account (2.7), we find that the solution to the Variational Inequality (2.2) verifies the conditions (2.6):

$$
\begin{cases}
(\mathcal{L}(u)u\left[\left(\dfrac{\partial u}{\partial x_1}+\dfrac{\partial u}{\partial x_2}\right)^2-1\right]=0 \ \text{in}\ \Omega \\[2em]
\left(\dfrac{\partial u}{\partial x_1}+\dfrac{\partial u}{\partial x_2}\right)^2 \leq 1\ \text{in}\ \Omega \\[2em]
u \geq 0\ \text{in}\ \Omega \\[1.5em]
u = 0\ \text{on}\ \partial\Omega
\end{cases}
\tag{2.15}
$$

(see Brézis (1972), Chiadò et al. (1994), Lanchon (1969), Ting (1969) for similar results).

This is a particular case of the general scheme (1.1).

In the paper Brézis et al. (1977) the author claims for the convenience to study problems of the general scheme (1.1) in a convex $\overline{K}$ in which an upper bound for the gradient of u is given:

$$
\overline{K}=\left\{v\in H^1_0(\Omega):v\geq 0\ \left(\dfrac{\partial v}{\partial x_1}\right)^2+\left(\dfrac{\partial v}{\partial x_2}\right)^2\leq 1\ \text{a.e.}\ x\in\Omega\right\}.
\tag{2.16}
$$

This case is studied in paper Idone et al. (2002), in which a similar characterization of (1.1) is proved. In particular the convex set

$$
\widetilde{K}=\{v\in H^1_0(\Omega):v\geq 0\ \ v(x)\leq\delta(x)\}
$$

$$
\delta(x)=\text{dist}(x,\partial\Omega)
$$

is considered and, under further assumptions, the following characterization is shown:

$$
\begin{cases}
(\mathcal{L}(u)-f-\overline{\mu})(\delta(x)-u(x))=0 \\[1.5em]
\mathcal{L}(u)-f-\overline{\mu}\leq 0 \\[1.5em]
u(x)\leq\delta(x).
\end{cases}
$$

It is also possible to prove

Theorem 2.2 *The solution u to problem:*

$$
\text{Find}\ u\in\overline{K}:a(u,v-u)\geq\int_\Omega f(v-u)dx\ \ \forall v\in\overline{K}
\tag{2.17}
$$

verifies the conditions:

$$
\begin{cases}
\mathcal{L}u - \overline{\mu} - \operatorname{div}(\overline{\lambda}\,\operatorname{grad} u) = 0 \\[2ex]
\mathcal{L}(u)u\left[\left(\dfrac{\partial u}{\partial x_1}\right)^2 + \left(\dfrac{\partial u}{\partial x_2}\right)^2 - 1\right] = 0 \;\; in\; \Omega \\[2ex]
u \geq 0 \;\; in\; \Omega \\[2ex]
\left(\dfrac{\partial u}{\partial x_1}\right)^2 + \left(\dfrac{\partial u}{\partial x_2}\right)^2 \leq 1 \; in\; \Omega \\[2ex]
u = 0 \; on\; \partial\Omega
\end{cases}
\tag{2.18}
$$

Moreover, using these results, it is possible to exhibit a computational procedure.

References

Borwein, J.M. and Lewis, A.S. (1991), Practical conditions for Fenchel duality in Infinite Dimension, *Pitman Research Notes in Mathematics Series*, 252, pp. 83-89.

Brézis, H. (1972), Multiplicateur de Lagrange en torsion élasto-plastique, *Arch. Rational Mech. Anal.*, 49, pp. 32-40.

Brézis, H. and Stampacchia, G. (1977), Remarks on some fourth order variational inequality, *Ann. Scuola Norm. Sup. Pisa* (4), pp. 363-371.

Brézis, H. (1972), Problèmes Unilatéraux, *J. Mat. pures et appl.* 51, pp. 1-168.

Chiadò, V. and Percivale, D. (1994), Generalized Lagrange Multipliers in Elasto-plastic torsion, *Journal of Differential Equations*, 114, pp. 570-579.

Daniele, P. (1999), Lagrangean function for dynamic Variational Inequalities, *Rendiconti del Circolo Matematico di Palermo*, 58, pp. 101-119.

Idone, G., Variational inequalities and application to a continuum model of transportation network with capacity constraints, to appear.

Idone, G., Maugeri, A. and Vitanza, C. (2002), Equilibrium problems in Elastic-Palstic Torsion, *Boundary Elements* 24th, Brebbia C.A., Tadeu A., Popov V. Eds., WIT Press, Southampton, Boston, pp. 611-616.

Lanchon, H. (1969), Solution du problème de torsion élasto-plastique, d'une barre cylindrique de section quelconque, *C.R. Acad. Sci. Paris*, 269, pp. 791-794.

Maugeri, A. (2001), Equilibrium problems and variational inequalities, in: *Equilibrium Problems: nonsmooth optimization and variational inequalities models*, Maugeri A., Giannessi F. and Pardalos P. Eds., Kluwer Academic Publishers, pp. 187-205.

Maugeri A. (1998), Dynamic models and generalized equilibrium problems, in: *New Trends in Mathematical Programming*, Giannessi F. et al. (eds.), Kluwer Academic Publishers, pp. 191-202.

Nagurney A. (1993), *Network economics. A Variational Inequality Approach*, Kluwer Academic Publishers.

Ting, J.W. (1969), Elasto-plastic torsion of convex cylindrical bars, *J. Math. Mech.*, 19, pp. 531-551.

Troianiello, G.M. (1987), *Elliptic Differential Equations and obstacle problems*, Plenum Press.

27   GAP FUNCTIONS AND DESCENT METHODS FOR MINTY VARIATIONAL INEQUALITY

Giandomenico Mastroeni

Department of Mathematics, University of Pisa
Via Buonarroti 2, 56127 Pisa, Italy

Abstract: A new class of gap functions associated to the variational inequality introduced by Minty is defined. Descent methods for the minimization of the gap functions are analysed in order to develop exact and inexact line-search algorithms for solving strictly and strongly monotone variational inequalities, respectively.

Key words: Variational inequality, gap function, descent methods.

1 INTRODUCTION

The gap function approach for Variational Inequalities (for short, VI) has allowed to develop a wide class of descent methods for solving the classic VI defined by the following problem:

$$\text{find } y^* \in K \text{ s.t. } \langle F(y^*), x - y^* \rangle \geq 0, \quad \forall x \in K, \qquad (VI)$$

where $F : \mathbb{R}^n \longrightarrow \mathbb{R}^n$, $K \subseteq \mathbb{R}^n$ and $\langle \cdot, \cdot \rangle$ is the inner product in $\mathbb{R}^n$.

We recall that a gap function $p : \mathbb{R}^n \longrightarrow \mathbb{R}$ is a non-negative function on K, such that $p(y) = 0$ with $y \in K$ if and only if y is a solution of VI. Therefore solving a VI is equivalent to the (global) minimization of the gap function on K.

In the last years the efforts of the scholars have been directed to the study of differentiable gap functions in order to simplify the computational aspects of the problem. See Harker et al (1990), for a survey on the theory and algorithms developed for VI.

The problem of defining a continuously differentiable gap function was first solved by Fukushima (1992) whose approach was generalized by Zhu et al (1994); they proved that

$$g(y) := \max_{x \in K}[\langle F(y), y - x \rangle - G(x, y)]$$

is a continuously differentiable gap function for VI under the following conditions: $G(x, y) : \mathbb{R}^n \times \mathbb{R}^n \longrightarrow \mathbb{R}$, is a non–negative, continuously differentiable, strongly convex function on the convex set K with respect to x, such that

$$G(y, y) = 0 \quad \text{and} \quad \nabla_x G(y, y) = 0, \quad \forall y \in K.$$

In the particular case where $G(x, y) := \frac{1}{2}\langle x - y, M(x - y) \rangle$, where M is a symmetric and positive definite matrix of order n, it is recovered the gap function introduced by Fukushima (1992).

Mastroeni (1999) showed that the gap function approach for VI developed by Fukushima (1992), Zhu et al (1994), can be extended to the variational inequality introduced by Minty (1962):

$$\text{find } x^* \in K \text{ s.t. } \langle F(y), x^* - y \rangle \leq 0, \quad \forall y \in K. \qquad (VI^*)$$

The interest in the study of Minty Variational Inequality had, at first, theoretical reasons, mainly in the analysis of existence results concerning the classic VI.

In fact, under the hypotheses of continuity and pseudomonotonicity of the operator F, VI^* is equivalent to VI (Karamardian (1976)). Recently, John (1998) has shown that VI^* provides a sufficient condition for the stability of equilibrium solutions of autonomous dynamical systems:

$$\frac{dx}{dt} + F(x) = 0, \qquad x \in K,$$

where $x = x(t)$, $t \geq 0$.

Moreover some algorithmic applications have been developed in the field of bundle methods for solving VI (see e.g. Lemarechal et al (1995)).

In this paper, we will deepen the analysis of descent methods for VI^* initiated by Mastroeni (1999). In particular, we will define an inexact line-search algorithm for the minimization of a gap function associated to the problem VI^*.

In Section 2 we will recall the main properties of the gap functions related to VI^*. In Section 3 we will develop an inexact descent method for VI^*, in the hypothesis of strong monotonicity of the operator F. Section 4 will be devoted to a brief outline of the applications of Minty Variational Inequality and to the, recently introduced, extension to the vector case (Giannessi (1998)).

We recall the main notations and definitions that will be used in the sequel.

A function $f : \mathbb{R}^n \longrightarrow \mathbb{R}$ is said quasi–convex on the convex set K iff:

$$f(\lambda x_1 + (1 - \lambda)x_2) \leq max\{f(x_1), f(x_2)\}, \tag{1.1}$$

$\forall x_1, x_2 \in K, \forall \lambda \in [0, 1]$.

If f is differentiable on K, then f is quasi–convex on K iff:

$$f(x_1) \leq f(x_2) \quad \Longrightarrow \quad \langle \nabla f(x_2), x_1 - x_2 \rangle \leq 0, \quad \forall x_1, x_2 \in K. \tag{1.2}$$

A function $f : K \longrightarrow \mathbb{R}$ is said strictly quasi–convex iff strict inequality holds in (1.1), for every $x_1 \neq x_2$ and every $\lambda \in (0, 1)$.

This last definition has been given by Ponstein (1967). Different definitions of strict quasi–convexity can be found in the literature (see e.g. Karamardian (1967)): for a deeper analysis on this topic see Avriel et al (1981) and references therein.

A strictly quasi–convex function has the following properties (Thomson et al (1973)):

(i)f is quasi–convex on K,

(ii) every local minimum point of f on K is also a global minimum point on K,

(iii) if f attains a global minimum point x^* on K then x^* is the unique minimum point for f on K.

Let X, Y be metric spaces. A point to set map $A : X \longrightarrow 2^Y$ is upper semicontinuous (for short, u.s.c.) according to Berge at a point $\lambda^* \in X$ if, for each open set $B \supset A\lambda^*$, there exists a neighborhood V of λ^* such that

$$A\lambda \subset B, \quad \forall \lambda \in V.$$

A is lower semicontinuous (for short, l.s.c.) according to Berge at a point $\lambda^* \in X$ if, for each open set B satisfying $B \cap A\lambda^* \neq \emptyset$, there exists a neighborhood V of λ^* such that

$$A\lambda \cap B, \quad \forall \lambda \in V.$$

A is called closed at $\lambda^* \in X$ iff

$$\lambda^k \longrightarrow \lambda^* \in X, \; y^k \longrightarrow y \in Y, \text{with } y^k \in A\lambda^k \; \forall k, \text{implies that } y \in A\lambda^*.$$

A point to set map is called closed on $S \subset X$ if it is closed at every point of S.

We will say that the mapping $F : \mathbb{R}^n \longrightarrow \mathbb{R}^n$ is monotone on K iff:

$$\langle F(y) - F(x), y - x \rangle \geq 0, \quad \forall x, y \in K;$$

it is strictly monotone if strict inequality holds $\forall x \neq y$.

We will say that the mapping F is pseudomonotone on K iff:

$$\langle F(y), x - y \rangle \geq 0 \quad \text{implies} \quad \langle F(x), x - y \rangle \geq 0, \quad \forall x, y \in K.$$

We will say that F is strongly monotone on K (with modulus $\mu > 0$) iff:

$$\langle F(y) - F(x), y - x \rangle \geq \mu \|y - x\|^2, \quad \forall x, y \in K.$$

It is known (Ortega et al (1970)) that, if F is continuously differentiable on K, then F is strongly monotone on K iff

$$\langle \nabla F(y)d, d \rangle \geq \mu \|d\|^2, \quad \forall d \in \mathbb{R}^n, \; \forall y \in K,$$

where ∇F denotes the Jacobian matrix associated to F.

2   A GAP FUNCTION ASSOCIATED TO MINTY VARIATIONAL INEQUALITY

In this section, we will briefly recall the main results concerning the gap function theory for VI^* (Mastroeni (1999)). Following the analysis developed for the classic VI, we introduce the gap function associated to VI^*.

Definition 2.1 *Let $K \subseteq \mathbb{R}^n$. The function $p : \mathbb{R}^n \longrightarrow \mathbb{R}$ is a gap function for VI^* iff:*
i) $p(y) \geq 0, \quad \forall y \in K$;
ii) $p(y) = 0$ and $y \in K$ iff y is a solution for VI^*.

By means of a suitable regularization of the variational inequality, it is possible to define a continuously differentiable gap function for VI^* (Mastroeni (1999)). Let $H(x, y) : \mathbb{R}^n \times \mathbb{R}^n \longrightarrow \mathbb{R}$ be a non-negative, differentiable function, such that

$$H(x, x) = 0, \quad \forall x \in K; \tag{2.1}$$

$$\nabla_y H(x, x) = 0, \quad \forall x \in K. \tag{2.2}$$

Proposition 2.1 *Let K be a convex set in $\mathbb{R}^n$. Suppose that $H : \mathbb{R}^n \times \mathbb{R}^n \longrightarrow \mathbb{R}$, is a non negative, differentiable function on K that fulfils (2.1) and (2.2) and $F : \mathbb{R}^n \longrightarrow \mathbb{R}^n$ is a differentiable and pseudomonotone operator on K. Then*

$$h(x) := \sup_{y \in K}[\langle F(y), x - y \rangle - H(x, y)]$$

is a gap function for VI^*.

Proof: We observe that $h(x) \geq 0, \forall x \in K$. Suppose that $h(x^*) = 0$ with $x^* \in K$. This is equivalent to say that x^* is a global minimum point of the problem

$$\min_{y \in K}[\langle F(y), y - x^* \rangle + H(x^*, y)].$$

The convexity of K implies that x^* is a solution of the variational inequality

$$\langle \nabla_y[q(x^*, x^*) + H(x^*, x^*)], y - x^* \rangle \geq 0, \quad \forall y \in K,$$

where $q(x, y) := \langle F(y), y - x \rangle$. From (2.2) we obtain

$$\langle \nabla_y q(x^*, x^*), y - x^* \rangle \geq 0.$$

Since $\nabla_y q(x, y) = F(y) + \nabla F(y)(y - x)$ then $\nabla_y q(x^*, x^*) = F(x^*)$, which implies that x^* is a solution of VI. By the pseudomonotonicity of F, we obtain that x^* is also a solution of VI^*.

Now suppose that x^* is a solution of VI^*. Since $H(x, y)$ is non negative, we have that

$$\langle F(y), y - x^* \rangle + H(x^*, y) \geq 0, \quad \forall y \in K,$$

which is equivalent to the condition

$$\max_{y \in K}[\langle F(y), x^* - y \rangle - H(x^*, y)] = 0.$$

Since $h(x) \geq 0$, $\forall x \in K$, we obtain

$$h(x^*) = \min_{x \in K} \max_{y \in K}[\langle F(y), x - y \rangle - H(x, y)] = 0.$$

$$\square$$

Let us consider the differentiability properties of the function $h(x)$.

Proposition 2.2 *Let K be a nonempty compact convex set in $\mathbb{R}^n$. Suppose that F is continuous on an open set $A \supset K$, $H : \mathbb{R}^n \times \mathbb{R}^n \longrightarrow \mathbb{R}$ is continuously differentiable on $A \times A$ and the function $\phi(x, y) := \langle F(y), y - x \rangle + H(x, y)$ is strictly quasi convex with respect to y, $\forall x \in K$, then $h(x)$ is continuously differentiable on K and its gradient is given by*

$$\nabla h(x) = F(y(x)) - \nabla_x H(x, y(x))$$

where $y(x)$ is the solution of the problem $\min_{y \in K} \phi(x, y)$.

Proof: We observe that

$$h(x) = -\inf_{y \in K} \phi(x, y) \tag{2.3}$$

Since $\phi(x, y)$ is strictly quasi convex with respect to y then there exists a unique minimum point $y(x)$ of the problem (2.3) . Applying Theorem 4.3.3 of Bank et al (1983) (see the Appendix), we obtain that $y(x)$ is u.s.c. according to Berge at x and, being $y(x)$ single-valued, it follows that $y(x)$ is continuous at x.

Since F is continuous and H is continuously differentiable then $\nabla_x \phi$ is continuous. Therefore, from theorem 1.7 Chapter 4 of Auslender (1976) (see the Appendix), taking into account that (2.3) has a unique minimum point, it follows that h is differentiable in the sense of Gateaux at x and

$$h'(x) = -\nabla_x \phi(x, y(x)).$$

From the continuity of F, $y(x)$ and $\nabla_x H$, it follows that $h'(x)$ is continuous at x so that h is continuously differentiable and

$$\nabla h(x) = h'(x) = F(y(x)) - \nabla_x H(x, y(x)).$$

$$\square$$

3   EXACT AND INEXACT DESCENT METHODS

In the previous section we have shown, that under suitable assumptions on the operator F and the function H, the gap function associated to the variational inequality VI^*:

$$h(x) := \sup_{y \in K} [\langle F(y), x - y \rangle - H(x, y)]$$

is continuously differentiable on K.

This considerable property allows us to define descent direction methods for solving the problem

$$\min_{x \in K} \; h(x). \tag{3.1}$$

After recalling an exact descent method proposed by Mastroeni (1999), we will analyse an inexact line search method. We will assume that

1. K is a nonempty compact and convex set in $\mathbb{R}^n$;

2. $\phi(x, y) := \langle F(y), y - x \rangle + H(x, y)$ is strictly quasi convex with respect to y, $\forall x \in K$;

3. F is a continuously differentiable operator on an open set $A \supset K$;

4. $H(x, y) : \mathbb{R}^n \times \mathbb{R}^n \longrightarrow \mathbb{R}$ is a non negative function on K, which is continuously differentiable on $A \times A$. Moreover, we suppose that it fulfils

conditions (2.1) and (2.2) and the further assumption:

$$\nabla_x H(x,y) + \nabla_y H(x,y) = 0, \quad \forall x, y \in K. \tag{3.2}$$

Remark 3.1 The hypothesis 4 is fulfilled by the function $H(x,y) := \frac{1}{2}\langle M(x - y), x - y\rangle$ where M is a symmetric matrix of order n. With this choice of the function H, the hypothesis 2 is fulfilled when $\langle F(y), y - x\rangle$ is convex with respect to y, $\forall x \in K$, and M is positive definite; for example when $F(y) = Cy + b$ where C is a positive semidefinite matrix of order n and $b \in \mathbb{R}^n$. A characterization of strict quasi convexity, in the differentiable case, is given in Theorem 3.26 of Avriel et al (1981).

In order to obtain a function H which fulfils (2.1),(2.2) and the condition (3.2), as noted by Yamashita et al (1997), it must necessarily be

$$H(x,y) = \psi(x - y),$$

where $\psi : \mathbb{R}^n \longrightarrow \mathbb{R}$ is nonnegative, continuously differentiable and such that $\psi(0) = 0$.

We recall that, from Proposition 2.2, h is a continuously differentiable function and $\nabla h(x) = F(y(x)) - \nabla_x H(x, y(x))$, where $y(x)$ is the solution of the problem

$$\min_{y \in K} \phi(x,y). \tag{$P(x)$}$$

Lemma 3.1 *Suppose that the hypotheses 1–4 hold and, furthermore, $\nabla F(y)$ is a positive definite matrix, $\forall y \in K$. Let $y(x)$ be the solution of $P(x)$. Then x^* is a solution of VI^* iff $x^* = y(x^*)$.*

Proof: Since $\nabla F(y)$ is a positive definite matrix, $\forall y \in K$, and F is continuously differentiable, then F is a strictly monotone operator (Ortega et al (1970), Theorem 5.4.3). Therefore x^* is a solution of VI^* iff

$$0 = h(x^*) = -\min_{y \in K} \phi(x^*, y)$$

and, by the uniqueness of the solution, iff $y(x^*) = x^*$. $\square$

Next result proves that $y(x) - x$ provides a descent direction for h at the point x, when $x \neq x^*$.

Proposition 3.1 *Suppose that the hypotheses 1–4 hold and F is strongly monotone on K (with modulus $\mu > 0$). Let $y(x)$ be the solution of the problem $P(x)$ and $d(x) := y(x) - x$. Then*

$$\langle \nabla h(x), d(x) \rangle \leq -\mu \|d(x)\|^2.$$

Proof: Since K is a convex set $y(x)$ fulfils the condition

$$\langle \nabla_y \phi(x, y(x)), z - y(x) \rangle \geq 0, \forall z \in K,$$

that is, putting $q(x, y) := \langle F(y), y - x \rangle$,

$$\langle \nabla_y q(x, y(x)), z - y(x) \rangle + \langle \nabla_y H(x, y(x)), z - y(x) \rangle \geq 0, \quad \forall z \in K.$$

In particular for $z := x$ we obtain

$$\langle \nabla_y q(x, y(x)), x - y(x) \rangle \geq -\langle \nabla_y H(x, y(x)), x - y(x) \rangle. \tag{3.3}$$

Since $\nabla_y q(x, y) = F(y) + \nabla F(y)(y - x)$, taking into account assumption 4 and (3.3), we have

$$\langle \nabla_x h(x), y(x) - x \rangle = \langle F(y(x)), y(x) - x \rangle - \langle \nabla_x H(x, y(x)), y(x) - x \rangle \leq$$
$$\langle F(y(x)), y(x) - x \rangle + \langle \nabla_y q(x, y(x)), x - y(x) \rangle = \langle F(y(x)), y(x) - x \rangle +$$
$$\langle F(y(x)), x - y(x) \rangle + \langle \nabla F(y(x))(y(x) - x), x - y(x) \rangle =$$

$$\langle \nabla F(y(x))(y(x) - x), x - y(x) \rangle \leq -\mu \|d(x)\|^2,$$

and the proposition is proved.

Remark 3.2 If we replace the hypothesis of strong monotonicity of the operator F, with the one of strict monotonicity, we obtain the weaker descent condition:

$$\langle \nabla h(x), d(x) \rangle < 0,$$

provided that $y(x) \neq x$.

The following exact line search algorithm has been proposed by Mastroeni (1999):

Algorithm 1

Step 1. Let $x_0 \in K$, ϵ be a tolerance factor and $k = 0$. If $h(x_0) = 0$, then STOP, otherwise go to step 2.

Step 2. Let $d_k := y(x_k) - x_k$.

Step 3. Let $t_k \in [0,1]$ be the solution of the problem

$$min\{h(x_k + td_k) : \ 0 \leq t \leq 1\}; \tag{3.4}$$

put $x_{k+1} = x_k + t_k d_k$.

If $\|x_{k+1} - x_k\| < \epsilon$, then STOP, otherwise let $k = k+1$ and go to step 2.

The following convergence result holds (Mastroeni (1999)):

Theorem 3.1 *Suppose that the hypotheses 1–4 hold and $\nabla F(y)$ is positive definite, $\forall y \in K$. Then, for any $x_0 \in K$ the sequence $\{x_k\}$ defined by Algorithm 1 belongs to the set K and converges to the solution of the variational inequality VI^*.*

Proof: Since $\nabla F(y)$ is positive definite $\forall y \in K$, and F is continuously differentiable then F is a strictly monotone operator (Ortega et al (1970), Theorem 5.4.3) and therefore both problems VI and VI^* have the same unique solution. The convexity of K implies that the sequence $\{x_k\} \subset K$ since $t_k \in [0,1]$. It is proved in the Proposition 2.2 that the function $y(x)$ is continuous, which implies the continuity of $d(x)$. It is known (see e.g. Minoux (1986), Theorem 3.1) that the map

$$U(x,d) := \{y : y = x + td, 0 \leq t \leq 1, h(y) = \min_{0 \leq t \leq 1} h(x + td)\}$$

is closed whenever h is a continuous function.

Therefore the algorithmic map $x_{k+1} = U(x_k, d(x_k))$ is closed, (see e.g. Minoux (1986), Proposition 1.3). Zangwill's convergence theorem (Zangwill (1969)) (see the Appendix) implies that any accumulation point of the sequence $\{x_k\}$ is a solution of VI^*. Since VI^* has a unique solution, the sequence $\{x_k\}$ converges to the solution of VI^*. $\square$

Algorithm 1 is based on an exact line search rule: it is possible to consider the inexact version of the previous method.

Algorithm 2

Step 1. Let x_0 be a feasible point, ϵ be a tolerance factor and β, σ parameters in the open interval $(0,1)$. Let $k = 0$.

Step 2. If $h(x_k) = 0$, then STOP, otherwise go to step 3.

Step 3. Let $d_k := y(x_k) - x_k$. Select the smallest nonnegative integer m such that

$$h(x_k) - h(x_k + \beta^m d_k) \geq \sigma \beta^m \|d_k\|^2,$$

set $\alpha_k = \beta^m$ and $x_{k+1} = x_k + \alpha_k d_k$.

If $\|x_{k+1} - x_k\| < \epsilon$, then STOP, otherwise let $k = k+1$ and go to step 2.

Theorem 3.2 *Suppose that the hypotheses 1–4 hold, F is a strongly monotone operator on K with modulus μ, $\sigma < \mu/2$, and $\{x_k\}$ is the sequence defined in the Algorithm 2. Then, for any $x_0 \in K$, the sequence $\{x_k\}$ belongs to the set K and converges to the solution of the variational inequality VI^*.*

Proof: The convexity of K implies that the sequence $\{x_k\} \subset K$, since $\alpha_k \in [0, 1]$. The compactness of K ensures that $\{x_k\}$ has at least one accumulation point. Let $\{\tilde{x}_k\}$ be any convergent subsequence of $\{x_k\}$ and x^* be its limit point.

We will prove that $y(x^*) = x^*$ so that, by Lemma 3.1, x^* is the solution of VI^*.

Since $y(x)$ is continuous (see the proof of Proposition 2.2) it follows that $d(x)$ is continuous; therefore we obtain that $d(\tilde{x}_k) \longrightarrow d(x^*) =: d^*$ and $h(\tilde{x}_k) \longrightarrow h(x^*) =: h^*$. By the line search rule we have

$$h(\tilde{x}_k) - h(\tilde{x}_{k+1}) \geq \sigma \tilde{\alpha}_k \|d(\tilde{x}_k)\|^2, \quad \forall k \in N, \tag{3.5}$$

for a suitable subsequence $\{\tilde{\alpha}_k\} \subseteq \{\alpha_k\}$.

Let us prove the relation (3.5). We observe that, by the line search rule, the sequence $\{h(x_k)\}$ is strictly decreasing. Let $k \in N$ and $x_{\bar{k}} := \tilde{x}_k$, for some $\bar{k} \in N$;

we have

$$h(\tilde{x}_k) - h(\tilde{x}_{k+1}) \geq h(x_{\bar{k}}) - h(x_{\bar{k}+1}) \geq \sigma \alpha_{\bar{k}} \|d(x_{\bar{k}})\|^2.$$

Putting $\tilde{\alpha}_k := \alpha_{\bar{k}}$, we obtain (3.5). Therefore,

$$\tilde{\alpha}_k \|d(\tilde{x}_k)\|^2 \longrightarrow 0.$$

If $\tilde{\alpha}_k > \beta^{m_0} > 0$, for some m_0, $\forall k > \bar{k} \in N$, then $\|d(\tilde{x}_k)\| \longrightarrow 0$ so that $y(x^*) = x^*$.

Otherwise suppose that there exists a subsequence $\{\alpha_{k'}\} \subseteq \{\tilde{\alpha}_k\}$ such that $\alpha_{k'} \longrightarrow 0$. By the line search rule we have that

$$\frac{h(x_{k'}) - h(x_{k'} + \bar{\alpha}_{k'} d(x_{k'}))}{\bar{\alpha}_{k'}} < \sigma \|d(x_{k'})\|^2, \tag{3.6}$$

where $\bar{\alpha}_{k'} = \frac{\alpha_{k'}}{\beta}$.

Taking the limit in (3.6) for $k \longrightarrow \infty$, since $\bar{\alpha}_{k'} \longrightarrow 0$ and h is continuously differentiable, we obtain

$$-\langle \nabla h(x^*), d^* \rangle \le \sigma \|d^*\|^2. \tag{3.7}$$

Recalling Proposition 3.1, we have also

$$-\langle \nabla h(x^*), d^* \rangle \ge \mu \|d^*\|^2.$$

Since $\sigma < \frac{\mu}{2}$, it must be $\|d^*\| = 0$, which implies $y(x^*) = x^*$. $\square$

4   SOME APPLICATIONS AND EXTENSIONS OF MINTY VARIATIONAL INEQUALITY

Besides the already mentioned equivalence with the classic VI, Minty variational inequality enjoys some peculiar properties that justify the interest in the development of the analysis. We will briefly recall some applications in the field of optimization problems and in the theory of dynamical systems. Finally we will outline the recently introduced extension to the vector case (Giannessi (1998)).

Consider the problem

$$\min\ f(x), \quad \text{s.t.} \quad x \in K, \tag{4.1}$$

where $f : \mathbb{R}^n \longrightarrow \mathbb{R}$ is a continuously differentiable function on the convex set K.

The following statement has been proved by Komlosi (1999).

Theorem 4.1 *Let $F := \nabla f$. If x^* is a solution of VI^* then x^* is a global minimum point for (4.1).*

In some particular cases, the previous result leads to an alternative characterization of a global minimum point of (4.1).

Corollary 4.1 *Let $F := \nabla f$ and suppose that f is a quasi–convex function on K. Then x^* is a solution of VI^* if and only if it is a global minimum point for (4.1).*

Proof: Suppose that x^* is a global minimum point of (4.1). By the equivalent characterization (1.2) of the quasi-convexity, in the differentiable case, it follows that x^* is a solution of VI^*. The converse implication follows from Theorem 4.1. $\qquad\qquad\square$

A further interesting application can be found in the field of autonomous dynamical systems:

$$\frac{dx}{dt} + F(x) = 0, \quad x \in K, \qquad\qquad (DS)$$

where $x = x(t)$, $t \geq 0$.

Suppose that ∇F is continuous on the set

$$\Omega := \{x \in K : \|x\| < A\},$$

where $A > 0$, so that there exists a unique solution $x(t)$ of DS with $x(t_0) = x_0$. Consider an equilibrium point $x^* \in \Omega$, which fulfils the relation $F(x^*) = 0$. It is obvious that

$$x(t) = x^*, \ \forall t \geq 0, \quad x(t_0) = x^*,$$

is a solution for DS. The following definition clarifies the concept of stability of the previous solution.

Definition 4.1 *The equilibrium point x^* is said stable for DS if, for every $0 \leq \epsilon < A$, there exists $0 \leq \delta \leq \epsilon$ such that if $\|x_0 - x^*\| \leq \delta$, then $\|x(t) - x^*\| \leq \epsilon, \forall t \geq 0$, where $x(t)$ is the solution of DS with the initial condition $x(t_0) = x_0$.*

Minty Variational Inequality provides a sufficient condition for the equilibrium point x^* to be stable.

Theorem 4.2 (John (1998)) *Let x^* be an equilibrium point for DS. If*

$$\langle F(y), x^* - y \rangle \leq 0, \quad \forall y \in \Omega,$$

then x^* is stable.

Giannessi (1998) has extended the analysis of VI^* to the vector case and has obtained a first order optimality condition for a Pareto solution of the vector optimization problem:

$$\min_{C\setminus\{0\}} f(x) \quad \text{s.t.} \quad x \in K, \tag{4.2}$$

where C is a convex cone in $\mathbb{R}^\ell$, $f : K \longrightarrow \mathbb{R}^\ell$ and $K \subseteq \mathbb{R}^n$.

The Minty vector variational inequality is defined by the following problem:

find $x \in K$ such that

$$F(y)(x - y) \not\geq_{C\setminus\{0\}} 0, \quad \forall y \in K, \tag{VVI^*}$$

where, $a \geq_{C\setminus\{0\}} b$ iff $a - b \in C \setminus \{0\}$, $F : \mathbb{R}^n \longrightarrow \mathbb{R}^{\ell \times n}$.

We observe that, if $C := \mathbb{R}_+$, then Minty vector variational inequality collapses into VI^*.

In the hypotheses that $C = \text{int} \mathbb{R}_+^\ell$, $F = \nabla f$ and f is a (componentwise) convex function, Giannessi (1998) proved that x is an optimal solution for (4.2) if and only if it is a solution of VVI^*.

Further developments in the analysis of VVI^* can be found in Giannessi (1998), Komlosi (1999), Mastroeni (2000).

5   CONCLUDING REMARKS

We have shown that the gap function theory developed for the classic VI, introduced by Stampacchia, can be extended, under further suitable assumptions, to the Minty Variational Inequality. These extensions are concerned not only with the theoretical point of view, but also with the algorithmic one: under strict or strong monotonicity assumptions on the operator F, exact or inexact descent methods, respectively, can be defined for VI^* following the line developed for VI.

It would be of interest to analyse the relationships between the class of gap functions associated to VI and the one associated to VI^* in the hypothesis of pseudomonotonicity of the operator F, which guarantees the equivalence of the two problems. This might allow to define a resolution method, based on the simultaneous use of both gap functions related to VI and VI^*.

6   APPENDIX

In this appendix we recall the main theorems that have been employed in the proofs of the results stated in the present paper.

Theorem 6.1 (Bank et al (1983)) is concerned with the continuity of the optimal solution map of a parametric optimization problem. Theorem 6.2 (Auslender (1976)) is a generalization of well-known results on directional differentiability of extremal-value functions. Theorem 6.3 is the Zangwill convergence theorem for a general algorithm formalized under the form of a multifunction.

Consider the following parametric optimization problem:

$$v(x) := inf\{f(x,y) \quad s.t. \quad y \in M(x)\},$$

where $f : \Lambda \times Y \longrightarrow \mathbb{R}$, $M : \Lambda \longrightarrow 2^Y$, $Y \subseteq \mathbb{R}^n$ and Λ is a metric space. Let $\psi : \Lambda \longrightarrow 2^Y$ be the optimal set mapping

$$\psi(x) = \Big\{y \in M(x) : f(x,y) = v(x)\Big\}.$$

Theorem 6.1 (Bank et al (1983), Theorem 4.3.3) *Let $Y := \mathbb{R}^n$ and $x^0 \in \Lambda$. Suppose that the following condition are fulfilled:*

1. *$\psi(x^0)$ is non-empty and bounded;*

2. *f is lower semicontinuous on $\{x^0\} \times Y$ and a point $y^0 \in \psi(x^0)$ exists such that f is upper semicontinuous at (x^0, y^0);*

3. *$f(x, \cdot)$ is quasiconvex on Y for each fixed $x \in \Lambda$;*

4. *$M(x)$ is a convex set, $\forall x \in \Lambda$;*

5. *$M(x^0)$ is closed and the mapping M is closed and lower semicontinuous, according to Berge, at x^0.*

Then ψ is upper semicontinuous according to Berge at x^0.

We observe that, if $M(x) = K$, $\forall x \in \Lambda$, where K is a nonempty convex and compact set in $\mathbb{R}^n$, then the assumptions *1,4* and *5*, of Theorem 6.1, are clearly fulfilled and it is possible to replace the assumption $Y := \mathbb{R}^n$ with $Y := K$.

Next result is well-known and can be found in many generalized versions: we report the statement of Auslender (1976). We recall that a function $h : \mathbb{R}^p \longrightarrow \mathbb{R}$ is said to be "directionally differentiable" at the point $x^* \in \mathbb{R}^p$ in the direction d, iff there exists finite:

$$\lim_{t \to 0+} \frac{h(x^* + td) - h(x^*)}{t} =: h'(x^*; d).$$

If there exists $z^* \in \mathbb{R}^p$ suct that $h'(x^*, d) = \langle z^*, d \rangle$ then h is said to be differentiable in the sense of Gateaux at x^*, and z^* is denoted by $h'(x^*)$.

Theorem 6.2 *(Auslender (1976), Theorem 1.7, Chapter 4) Let*

$$v(x) := \inf_{y \in Y} f(x, y),$$

where $f : \mathbb{R}^p \times Y \longrightarrow \mathbb{R}$. Suppose that

1. *f is continuous on $\mathbb{R}^p \times Y$;*

2. *$\nabla_x f$ exists and is continuous on $\Omega \times Y$, where Ω is an open set in $\mathbb{R}^p$;*

3. *Y is a closed set in $\mathbb{R}^n$;*

4. *For every $x \in \mathbb{R}^p$, $\psi(x) := \{y \in Y : f(x, y) = v(x)\}$ is nonempty and there exists a neighbourhood $V(x)$ of x, such that $\cup_{z \in V(x)} \psi(z)$ is bounded.*

Then, for every $x \in \Omega$, we have:

$$v'(x; d) = \inf_{y \in \psi(x)} \langle \nabla_x f(x, y), d \rangle.$$

Moreover, if for a point $x^* \in \Omega$, $\psi(x^*)$ contains exactly one element $y(x^*)$, then v is differentiable, in the sense of Gateaux, at x^* and

$$v'(x^*) = \nabla_x f(x^*, y(x^*)).$$

We observe that, when Y is a nonempty compact set, then the assumptions *3* and *4*, of Theorem 6.2, are obviously fulfilled. The reader can also refer to Hogan (1973) and references therein for similar versions of the previous theorem.

Finally, we recall the statement of Zangwill Convergence Theorem as reported in Minoux (1986). Given an optimization problem P defined on $X \subseteq \mathbb{R}^n$, let $\mathcal{M}$ be the set of the points of X that fulfil a suitable necessary optimality condition. Suppose that, in order to solve P, it is used an algorithm represented by a point to set map $A : X \longrightarrow 2^X$.

Definition 6.1 *We say that $z : X \longrightarrow \mathbb{R}$ is a descent function (related to the algorithm A) if it is continuous and has the following properties:*

1. *$x \notin \mathcal{M}$ implies $z(y) < z(x) \quad \forall y \in A(x)$,*

2. *$x \in \mathcal{M}$ implies $z(y) \leq z(x) \quad \forall y \in A(x)$.*

Theorem 6.3 (Zangwill (1969)) *Let P be an optimization problem on X and $\mathcal{M}$ be the set of the points of X that fulfil a certain necessary optimality condition.*

Let $A : X \longrightarrow 2^X$ be the algorithmic point to set mapping and consider a sequence $\{x^k\}$ generated by the algorithm, i.e. satisfying $x^{k+1} \in A(x^k)$.

Suppose that the following three conditions hold:

1. *Every point x^k is contained in a compact set $K \subset X$;*

2. *There exists a descent function z;*

3. *The point to set map A is closed on $X \setminus \mathcal{M}$ and $\forall x \in X \setminus \mathcal{M}, \ A(x) \neq \emptyset$.*

Then, for every x which is the limit of a convergent subsequence of $\{x^k\}$, we have that $x \in \mathcal{M}$.

References

Auslender A. (1976), *Optimization. Methodes Numeriques*, Masson, Paris.

Avriel M., Diewert W.E., Schaible S. and Ziemba W.T. (1981), Introduction to concave and generalized concave functions, in *"Generalized Concavity in Optimization and Economics"*, S. Schaible, W.T. Ziemba (Eds.), pp. 21-50.

Bank B., Guddat J., Klatte D., Kummer B. and Tammer K. (1983), *Nonlinear Parametric Optimization*, Birkhauser Verlag.

Fukushima M. (1992), Equivalent differentiable optimization problems and descent methods for asymmetric variational inequality problems, *Mathematical Programming*, Vol. 53, pp. 99-110.

Giannessi F. (1998), On Minty variational principle, in *"New Trends in Mathematical Programming"*, F. Giannessi, S. Komlosi, T. Rapcsak (Eds.), Kluwer Academic Publishers, Dordrecht, Boston, London.

Harker P.T., Pang J.S. (1990), Finite–dimensional variational inequalities and nonlinear complementarity problem: a survey of theory, algorithms and applications, *Mathematical Programming*, Vol. 48, pp. 161-220.

Hogan W. (1973), Directional derivatives for extremal-value functions with applications to the completely convex case, *Operations Research*, Vol. 21, N.1, pp.188-209.

John R. (1998), Variational inequalities and pseudomonotone functions: some characterizations, in " Generalized Convexity, Generalized Monotonicity",

J.P. Crouzeix, J.E. Martinez-Legaz (Eds.), Kluwer Academic Publishers, Dordrecht, Boston, London.

Karamardian S. (1976), An existence theorem for the complementary problem, *Journal of Optimization Theory and Applications*, Vol. 18, pp. 445-454.

Karamardian S. (1967), Strictly quasi–convex functions and duality in mathematical programming, *Journal of Mathematical Analysis and Applications*, Vol. 20, pp. 344-358.

Komlosi S. (1999), On the Stampacchia and Minty variational inequalities, in *"Generalized Convexity and Optimization for Economic and Financial Decisions"*, G. Giorgi and F. Rossi (Eds.), Pitagora, Bologna, Italy, pp. 231-260.

Lemarechal C., Nemirovskii A. and Nesterov Y. (1995), New variants of bundle methods, *Mathematical Programming*, Vol. 69, pp. 111-147.

Mastroeni G. (1999), Minimax and extremum problems associated to a variational inequality, *Rendiconti del Circolo Matematico di Palermo*, Vol. 58 , pp. 185-196 .

Mastroeni G. (2000), On Minty vector variational inequality, in *"Vector Variational Inequalities and Vector Equilibria. Mathematical Theories"*, F. Giannessi (Ed.), Kluwer Academic Publishers, Dordrecht, Boston, London pp. 351-361.

Minoux M. (1986), *Mathematical Programming, Theory and Algorithms*, John Wiley, New York.

Minty G. J. (1962), Monotone (non linear) operators in Hilbert space, *Duke Math. Journal*, Vol. 29, pp. 341-346.

Ortega J.M., Rheinboldt W.C. (1970), *Iterative solutions of nonlinear equations in several variables*, Academic Press, New York.

Ponstein J. (1967), Seven kind of convexity, *S.I.A.M. Rev., 9*, pp. 115-119.

Thompson W.A., Parke D.W. (1973), Some properties of generalized concave functions, *Operations Research*, Vol. 21, pp. 305-313.

Yamashita N., Taji K. and Fukushima M. (1997), Unconstrained optimization reformulations of variational inequality problems, *Journal of Optimization Theory and Applications*, Vol. 92, pp. 439-456.

Zangwill W.I. (1969), *Nonlinear Programming: a unified approach*, Prentice–Hall, Englewood Cliffs, New York.

Zhu D.L., Marcotte P. (1994), An extended descent framework for variational inequalities, *Journal of Optimization Theory and Applications*, Vol. 80, pp. 349-366.

28 A NEW CLASS OF PROXIMAL ALGORITHMS FOR THE NONLINEAR COMPLEMENTARITY PROBLEM

G.J.P. DA Silva

Universidade Federal de Goiás
Instituto de Matemática e Estatística
Campus II, CP 131
Goiânia, GO, CEP 74001-970, Brazil.
geci@cos.ufrj.br

and P.R. Oliveira

Engenharia de Sistemas e Computação
COPPE-UFRJ,CP 68511
Rio de Janeiro, RJ, CEP 21945-970, Brazil.
poliveir@cos.ufrj.br

Abstract: In this paper, we consider a new variable proximal regularization method for solving the nonlinear complementarity problem(NCP) for P_0 functions.

Key words: Nonlinear complementarity problem, P_0 function, proximal regularization.

1 INTRODUCTION

Consider the nonlinear complementarity problem $(NCP(F))$,

$$x \geq 0, \quad F(x) \geq 0, \quad x^T F(x) = 0,$$

where we assume that $F : \mathbb{R}^n \longrightarrow \mathbb{R}^n$ is a continuously differentiable P_0 function. We recall that F is a P_0 function if for any $x \neq y$ in $\mathbb{R}^n$

$$\max_{x_i \neq y_i} (x_i - y_i)(F_i(x) - F_i(y)) \geq 0.$$

Note that the class of P_0 functions includes the class of monotone functions. Applications of NCP can be found in many important fields such as mathematical programming, economics, engineering and mechanics (see, e.g.,Cottle et al (1992); Harker and Pang (1990)).

There exist several methods for the solution of the complementarity problem. In this paper are considered regularization methods, which are designed to handle ill-posed problems. Very roughly speaking, an ill-posed problem may be difficult to solve since small errors in the computations can lead to a totally wrong solution.

For the class of the P_0 functions, Facchinei and Kanzow (1999) considered the Tikhonov-regularization, this scheme consist of solving a sequence of complementarity problems $NCP(F_k)$, where $F_k(x) := F(x) + c_k x$ and c_k is a positive parameter converging to 0, and Yamashita et al (1999), considered the proximal point algorithm, proposed by Martinet (1970) and further studied by Rockafellar (1976). For the $NCP(F)$, given the current point x^k, the proximal point algorithm produces the next iterate by approximately solving the subproblem $NCP(F_k)$, where $F_k(x) := F(x) + c_k(x - x^k)$ and c_k is a positive parameter that does not necessary converge to 0. In the case above, if F is a P_0 function, then F_k is a P function, that is, for any $x, y \in \mathbb{R}^n$ with $x \neq y$,

$$\max_i \{(x_i - y_i)(F_i(x) - F_i(y))\} > 0.$$

Therefore the subproblem $NCP(F_k)$ is better tractable than $NCP(F)$, since if F is a P function, then $NCP(F)$ has at most one solution.

In this paper, we consider a variable proximal regularization algorithm. Given the current point $x^k > 0$, the variable proximal regularization algorithm produces the next iterate by approximately solving the subproblem $NCP(F^k)$, where

$$F^k(x) := F(x) + c_k(X^k)^{-r}(x - x^k),$$

$(X^k)^{-r}$ is defined by $(X^k)^{-r} = diag\{(x_1^k)^{-r}, \cdots, (x_n^k)^{-r}\}$, $r \geq 1$ and c_k is a positive parameter.

Now, some words about our motivation. It comes from the application of some tools of Riemannian geometry to the continuous optimization. This is object of research by many authors, as can be seen in Bayer and Lagarias (1989); Bayer and Lagarias (1989a); Cruz Neto, Lima and Oliveira (1998); Ferreira and Oliveira (1998); Ferreira and Oliveira (2002); Gabay (1982); Karmarkar (1990); Nesterov and Todd (2002); Rapcsák (1997); Udriste (1994), and in the bibliography therein. One of the trends is given in Cruz Neto and OLiveira (1995), where it is explored the idea that associates Riemannian metrics and descent directions. Specifically, they start from the equivalence property between a metric dependent gradient and the generator descent direction (also metric dependent). The authors had unified a large variety of primal methods, seen as gradient ones, and obtained other classes. That includes primal interior point methods such as Dikin (1967), Karmarkar (1984), and the Eggermont multiplicative algorithm, by Eggermont (1990). A general theory for such gradient methods can be seen in Cruz Neto, Lima and Oliveira (1998). Particularly, in Cruz Neto and OLiveira (1995), they considered the positive orthant $R_+{}^n$ as a manifold, associated with a class of metrics generated by the Hessian of some separable functions $p(x) = \sum_{i=1}^{n} p_i(x_i)$, $p_i : R_+{} \to R$, $p_i(x_i) > 0$, for $i = 1, 2, ..., n$. Clearly, the functions such that $p_i = x_i^{-r}$, $i = 1, 2, ..., n$, and $r \geq 1$, is contained in that class. For $r = 1$, $r = 2$, and $r = 3$, they correspond, respectively, to the Hessian of Eggermont multiplicative, log and Fiacco- McCormick barriers. Those metrics, denoted by X^{-r}, lead to projective (affine) interior point methods in Pinto, Oliveira and Cruz Neto (2002), and proximal interior point algorithms in Oliveira and Oliveira (2002). Here, we exploit those ideas in a context of NCP. Observe that in the case where NCP is equivalent to the optimality conditions of some linear or nonlinear programming problem, our method can be seen as an infeasible interior point class of algorithms.

By using the Mountain Pass Theorem, we show that the method converges globally if F is a P_0 function and the solution set of $NCP(F)$ is nonempty and bounded. Facchinei (1998) give necessary and sufficient conditions that ensure that the solution set of $NCP(F)$ is bounded.

The paper is organized as follows. In Section 2, we review some results which will be used in the following sections. In section 3, we prove that the regularized problem has a unique solution. In Section 4, we describe the proposed algorithm and we show its convergence properties.

2   PRELIMINARIES

In this section we review some basic definitions and properties which will be used in the subsequent analysis.

We first restate the basic definition.

Definition 2.1 *A matrix $M \in \mathbb{R}^{n \times n}$ is called*

1. *a P_0-matrix if, for every $x \in \mathbb{R}^n$ with $x \neq 0$, there is an index $i = i(x)$ with*

$$x_i \neq 0 \quad \text{and} \quad x_i[Mx]_i \geq 0;$$

2. *a P-matrix if, for every $x \in \mathbb{R}^n$ with $x \neq 0$, it holds that*

$$\max_i \{x_i[Mx]_i\} > 0;$$

The following proposition summarizes some useful properties that play an important role in the analysis of the uniqueness of solution of the regularized problems $NCP(F^k)$.

Proposition 2.1 *Let F be function from $\mathbb{R}^n$ in $\mathbb{R}^n$.*

(a) *(Moré and Rheinboldt (1973), Theorem 5.8) If F is a P_0 function, then the Jacobian matrix, $F'(x)$, is a P_0-matrix for every $x \in \mathbb{R}^n$;*

(b) *(Moré and Rheinboldt (1973), Theorem 5.2) If $F'(x)$ is a P-matrix for every $x \in \mathbb{R}^n$, then F is a P function;*

(c) *(Moré (74), Theorem 2.3) If F is a P function, then $NCP(F)$ has at most one solution.*

The following theorem is a version of the mountain pass theorem (see, Palais and Terng (1988)), that will be used to establish a global convergence theorem for the proposed algorithm.

Theorem 2.1 *Let $f : \mathbb{R}^n \longrightarrow \mathbb{R}$ be continuously differentiable and coercive. Let $S \subset \mathbb{R}^n$ be a nonempty and compact set and define m to be least value of f on the (compact) boundary of S:*

$$m := \min_{x \in \partial S} f(x).$$

Assume further that there are two points $a \in S$ and $b \notin S$ such that $f(a) < m$ and $f(b) < m$. Then there exists a points $c \in \mathbb{R}^n$ such that $\nabla f(c) = 0$ and $f(c) \geq m$.

3 EXISTENCE OF REGULARIZED SOLUTIONS

In this section, we prove that the regularized problem $NCP(F^k)$ has a unique solution for every k. For this, we consider the equivalent reformulation of $NCP(F)$ using the Fischer- Burmeister function (see, Fischer (1992)), $\varphi : \mathbb{R}^2 \longrightarrow \mathbb{R}$, defined by

$$\varphi(a, b) = \sqrt{a^2 + b^2} - a - b.$$

The most fundamental property of this function is that

$$\varphi(a, b) = 0 \quad \Longleftrightarrow \quad a \geq 0, \quad b \geq 0 \quad \text{and} \quad ab = 0.$$

Using this function, we obtain the following system of equations equivalent to $NCP(F)$

$$\Psi(x) = \begin{bmatrix} \varphi(x_1, F_1(x)) \\ \vdots \\ \varphi(x_n, F_n(x)) \end{bmatrix} = 0$$

With this operator, we define a merit fuction $\Phi : \mathbb{R}^n \longrightarrow \mathbb{R}$ through

$$\Phi(x) = \frac{1}{2} \|\Psi(x)\|^2.$$

For the regularized problem, we define the corresponding operator and the corresponding merit function similarly as

$$\Psi^k(x) = \begin{bmatrix} \varphi(x_1, F_1^k(x)) \\ \vdots \\ \varphi(x_n, F_n^k(x)) \end{bmatrix}$$

and

$$\Phi^k(x) = \frac{1}{2} \|\Psi^k(x)\|^2.$$

We summarize some useful properties of those functions in the following result (see, Facchinei and Kanzow (1999)).

Proposition 3.1 *The following statements hold:*

1. *$x^* \in \mathbb{R}^n$ solves $NCP(F)$ if and only if x^* solves the nonlinear system of equations $\Psi(x) = 0$.*

2. *The merit function Φ is continuosly differentiable on the wholc space $\mathbb{R}^n$.*

3. *If F is a P_0 function, then every stationary point of Φ is a solution of $NCP(F)$.*

The proof of the following lemma can be found in Kanzow (1996).

Lemma 3.1 *Let $\{a^k\}, \{b^k\} \subset \mathbb{R}$ be any two sequences such that $a^k, b^k \longrightarrow +\infty$ or $a^k \longrightarrow -\infty$ or $b^k \longrightarrow -\infty$. Then $\left|\varphi(a^k, b^k)\right| \longrightarrow +\infty$.*

The following proposition plays an important role in proving the existence of solution of the regularized problems $NCP(F^k)$.

Proposition 3.2 *Suppose that F is a P_0 function. Then the merit function Φ^k is coercive for every k, i.e.,*

$$\lim_{\|x\| \longrightarrow \infty} \Phi^k(x) = +\infty.$$

Proof: Suppose by contradiction that there exists an unbounded sequence $\{z^l\}$ such that $\{\Phi^k(z^l)\}$ is bounded. Since the sequence $\{z^l\}$ is unbounded, the index set

$$J := \{j \in \{1, \cdots, n\} | \{z_j^l\} \text{ is unbounded}\},$$

is nonempty. Subsequencing if necessary, we can assume without loss of generality that $|z_j^l| \longrightarrow \infty$ for all $j \in J$. Therefore, we consider two possible cases.

Case 1. $z_j^l \longrightarrow \infty$ Let $\{y^l\}$ denotes the bounded sequence defined through

$$y_j^l := \begin{cases} 0 & \text{if } j \in J \\ z_j^l & \text{if } j \notin J. \end{cases}$$

¿From the definition of $\{y^l\}$ and the assumption that F is a P_0 function, we get

$$
\begin{aligned}
0 \;\leq\; & \max_{1\leq i\leq n}(z_i^l - y_i^l)[F_i(z^l) - F_i(y^l)] \\
=\; & \max_{i\in J} z_i^l[F_i(z^l) - F_i(y^l)] \\
=\; & z_j^l[F_j(z^l) - F_j(y^l)],
\end{aligned}
$$

where j is one of the indices for which the max is attained, that is independent of l. Since $\{y^l\}$ is bounded, by continuity of F_j it follows that $\{F_j(y^l)\}$ is bounded. Therefore, $z_j^l[F_j(z^l) - F_j(y^l)] \geq 0$ implies that $\{F_j(z^l)\}$ does not tend to $-\infty$. This, in turn, implies $z_j^l \longrightarrow \infty$ e $F_j(z^l) + c_k(x_j^k)^{-r}(z_j^l - x_j^k) \longrightarrow \infty$. By Lemma 3.1 we have that

$$
\sqrt{(z_j^l)^2 + (F_j(z^l) + c_k(x_j^k)^{-r}(z_j^l - x_j^k))^2} \;- \\
-\; z_j^l - (F_j(z^l) + c_k(x_j^k)^{-r}(z_j^l - x_j^k)) \longrightarrow \infty,
$$

contradicting the boudedness of the sequence $\{\Phi^k(z^l)\}$.

Case 2. $z_j^l \longrightarrow -\infty$ We have immediately from Lemma 3.1 that

$$
\sqrt{(z_j^l)^2 + (F_j(z^l) + c_k(x_j^k)^{-r}(z_j^l - x_j^k))^2} \;- \\
-\; z_j^l - (F_j(z^l) + c_k(x_j^k)^{-r}(z_j^l - x_j^k)) \longrightarrow \infty,
$$

contradicting the boundedness of the sequence $\{\Phi^k(z^l)\}$.

We are now in position to prove the following existence and uniqueness result.

Theorem 3.1 *If F is a P_0 function, then the $NCP(F^k)$ has a unique solution for every k.*

Proof: Since $F'(x)$ is a P_0-matrix, $c_k > 0$ and $(X^k)^{-r}$ is positive definite, by the definition of F^k we have that $(F^k)'(x) = F'(x) + c_k(X^k)^{-r}$ is a P-matrix. Therefore, by Proposition 2.1 **(b)**it follows that F^k is a P function. This in turn implies that $NCP(F^k)$ has at most one solution, due to the Proposition 2.1 **(c)**.

We prove now the existence of solution. From the Proposition 3.2 Φ^k is coercive. Since the function Φ^k is continuos, it attains a global minimum. This, in turn, implies, also using Proposition 3.1, item 2, that the global minimum is a stationary point of Φ^k. However, Φ^k is a

P function; in particular, Φ^k itself is a P_0 function, so that the global minimum must be a solution of $NCP(F^k)$, due to Proposition 3.1, item 3.

4   ALGORITHM AND CONVERGENCE

In this section we propose a variable metric proximal algorithm and we show convergence properties.

Algorithm 4.1   Step 0: *Choose $c_0 > 0$, $\delta_0 \in (0,1)$ and $x^0 \in \mathbb{R}^n_{++}$. Set $k := 0$.*

Step 1: *Given $c_k > 0$, $\delta_k \in (0,1)$ and $x^k \in \mathbb{R}^n_{++}$, obtain $x^{k+1} \in \mathbb{R}^n_{++}$ such that $\Phi^k(x^{k+1})^{\frac{1}{2}} \leq \delta_k$.*

Step 2: *Choose $c_{k+1} \in (0, c_k)$ and $\delta_{k+1} \in (0, \delta_k)$. Set $k := k + 1$, and go to Step 1.*

The Algorithm 4.1 is well defined, since by Theorem 3.1 the $NCP(F^k)$ has a unique solution, therefore, as Φ^k is continuous, given $\delta_k > 0$, there exists $x^{k+1} \in \mathbb{R}^n_{++}$, in a neigborhood of the unique solution of the $NCP(F^k)$, such that $\Phi^k(x^{k+1})^{\frac{1}{2}} \leq \delta_k$.

Now, we give conditions under which the algorithm converges globally to a solution of $NCP(F)$. The sequence $\{c_k\}$ satisfies the following conditions:

(A) $c_k(X^k)^{-r}(x^{k+1} - x^k) \longrightarrow 0$ if $\{x^k\}$ is bounded;

(B) $c_k(X^k)^s \longrightarrow 0$ if $\{x^k\}$ is unbounded and $s \leq 0$.

Remark 4.1 *In Oliveira and Oliveira (2002), they introduce a family of variable metric interior-proximal methods which considering $F = \nabla f$ and assuming that F is a Lipschtz continuous monotone function, they showed that hold **(A)**.*

Remark 4.2 *The conditions* **(A)** *and* **(B)** *can be verified if we define* $c_k = \dfrac{\beta^k}{\|(X^k)^s\|}$, *for* $\beta \in (0,1)$. *In this case* $c_k \longrightarrow 0$.

Remark 4.3 *When* $s = 0$ *in* **(B)**, *then* $c_k \longrightarrow 0$.

Under above conditions, we have the following lemma.

Lemma 4.1 *Suppose that condition* **(B)** *holds. Let* $S \subset \mathbb{R}^n$ *be an arbitrary compact set. If* $\{x^k\}$ *is unbounded, then for any* $\varepsilon > 0$, *there exists a sufficiently large* k_0 *such that for all* $k \geq k_0$*

$$\left|\Phi^k(x) - \Phi(x)\right| \leq \varepsilon \quad \text{for all} \quad x \in S.$$

Proof: By definition of F^k we have that for any $x \in S$

$$F^k(x) - F(x) = c_k(X^k)^{-r}(x - x^k) = c_k(X^k)^{-r}x - c_k(X^k)^{1-r}e,$$

where $e = (1, \cdots, 1)^T \in \mathbb{R}^n$. Therefore by condition **(B)**, and the fact that S is compact we have $\|F^k(x) - F(x)\| \longrightarrow 0$, since $r \geq 1$ e $x \in S$. Now, for any $a, b, c \in \mathbb{R}$ we have that

$$
\begin{aligned}
|\varphi(a, b+c) - \varphi(a,b)| &= \left|\sqrt{a^2 + (b+c)^2} - \sqrt{a^2 + b^2} - c\right| \\
&\leq \left|\sqrt{a^2 + (b+c)^2} - \sqrt{a^2 + b^2}\right| + |c| \\
&= \big|\, \|(a, b+c)\| - \|(a,b)\| \,\big| + |c| \\
&\leq \|(0,c)\| + |c| = 2\,|c|.
\end{aligned}
$$

Applying the result above with $a = x_i$, $b = F_i(x)$ and $c = c_k(x_i^k)^{-r}(x_i - x_i^k)$ we have that $|\varphi(x_i, F_i^k(x)) - \varphi(x_i, F_i(x))| \leq 2|c_k(x_i^k)^{-r}(x_i - x_i^k)| \longrightarrow 0$.

Since S is compact, F^k converges uniformly to F in S; furthermore, φ, F and F^k are continuous, so we have that for all i

$$|\varphi^2(x_i, F_i^k(x)) - \varphi^2(x_i, F_i(x))| =$$
$$= |\varphi(x_i, F_i^k(x)) - \varphi(x_i, F_i(x))||\varphi(x_i, F_i^k(x)) + \varphi(x_i, F_i(x))| \longrightarrow 0.$$

Therefore, it follows that Φ^k converges uniformly to Φ in S.

The following result is our main convergence theorem for the Algorithm 4.1.

Theorem 4.1 *Suppose that F is a P_0 function and assume that the solution set S^* of $NCP(F)$ is nonempty and bounded. Suppose also that conditions* **(A)** *and* **(B)** *hold. If $\delta_k \longrightarrow 0$, then $\{x^k\}$ is bounded and any accumulation point of $\{x^k\}$ is a solution of $NCP(F)$.*

Proof: First we show that $\{x^k\}$ is bounded. Suppose that the sequence $\{x^k\}$ is not bounded. Then there exists a subsequence $\{x^k\}_{k \in K}$ such that $\|x^k\| \to \infty$ as $k \to \infty$ with $k \in K$. Since S^* is bounded, there exists a nonempty compact set $S \subset \mathbb{R}^n$ such that $S^* \subset int(S)$ and $x^k \notin S$ for all $k \in K$, sufficiently large. If $x^* \in S^*$, then we have $\Phi(x^*) = 0$. Let

$$\alpha := \min_{x \in \partial S} \Phi(x) > 0.$$

Applying Lemma 4.1 with $\varepsilon := \frac{\alpha}{4}$, there exists some k_0 such that for all $k \geq k_0$

$$\Phi^k(x^*) \leq \frac{\alpha}{4}$$

and

$$m := \min_{x \in \partial S} \Phi^k(x) \geq \frac{3\alpha}{4}.$$

Since $\Phi^k(x^{k+1}) \leq \delta_k^2$ by Step 1 of Algorithm 4.1, there exists some k_1 such that for all $k \geq k_1$,

$$\Phi^k(x^{k+1}) \leq \frac{\alpha}{4},$$

since $\delta_k \longrightarrow 0$ by our asumption.

Now, consider a fixed index $k \geq \max\{k_0, k_1\}$ with $k \in K$ and set $a = x^*$ and $b = x^{k+1}$, we have from Theorem 2.1 that there exists a vector $c \in \mathbb{R}^n$ such that

$$\nabla \Phi^k(c) = 0 \quad \text{and} \quad \Phi^k(c) \geq m \geq \frac{3\alpha}{4} > 0.$$

Therefore c is a stationary point of Φ^k, which does not minimize Φ^k globally. However this contradicts Proposition 3.1, item 3. Hence $\{x^k\}$ is bounded.

Next, we show that any accumulation point of $\{x^k\}$ is a solution of $NCP(F)$. Since $\{x^k\}$ is bounded, we have $\|F^k(x^{k+1}) - F(x^{k+1})\| \longrightarrow 0$ by condition **(A)**, and hence $|\Phi^k(x^{k+1}) - \Phi(x^{k+1})| \longrightarrow 0$. By Step 1 of algorithm and the assumption that $\delta_k \longrightarrow 0$, we have $\Phi^k(x^{k+1}) \longrightarrow 0$. Consequently it holds that $\Phi(x^{k+1}) \longrightarrow 0$, which means that every accumulation point of the sequence $\{x^k\}$ is a solution of $NCP(F)$.

5   CONCLUSIONS

We presented a new class of algorithms for NCP, with convergence results for P_0 functions. As we have seen in the introduction, a few papers were produced for that class of NCP. As a further working, we are particularly interested in the behavior of the algorithm, as depending of the $r \geq 1$ parameter, also, when applied to monotone functions. On the other hand, observe that although Fischer-Burmeister function (Fischer (1992)), was essential in our theoretical analysis, it could be used someother function in step 1 of our algorithm. Now, we are working on different choices, in order to get a easier computable c_k. As a final remark, it is worthwhile to consider the case when the solution set of $NCP(F)$ is unbounded.

Acknowledgments

The author Da Silva thanks CAPES/PICDT/UFG for support. The author Oliveira thanks CNPq for support.

References

Palais, R.S. and Terng, C.L. (1988), *Critical point theory and submanifold geometry, Lecture Note in Mathematics*, 1353, Springer Verlag, Berlin.

Cottle, R.W., Pang, J.S. and Stone, R.E. (1992), *The Linear Complementarity Problem*, Academic Press, New York.

Rapcsák, T. (1997), *Smooth Nonlinear Optimization in R^n*, Kluwer Academic Publishers, Dordrecht, Netherlands.

Udriste, C. (1994), *Convex Functions and Optimization Methods in Riemannian Geometry*, Kluwer Academic Publishers, Dordrecht, Netherlands.

Bayer, D.A. and Lagarias, J.C. (1989), The Nonlinear Geometry of Linear Programming I, Affine and Projective Scaling Trajectories, *Transactions of the American Mathematical Society*, Vol. 314, No 2, pp. 499-526.

Bayer, D.A. and Lagarias, J.C. (1989), The Nonlinear Geometry of Linear Programming II, Legendre Transform Coordinates and Central Trajectories, *Transactions of the American Mathematical Society*, Vol. 314, No 2, pp. 527-581.

Cruz Neto, J.X. and Oliveira, P.R. (1995), Geodesic Methods in Riemannian Manifolds, *Preprint, RT 95-10, PESC/COPPE - Federal University of Rio de Janeiro, BR*

Cruz Neto, J. X., Lima, L.L. and Oliveira, P. R. (1998), Geodesic Algorithm in Riemannian Manifold, *Balkan JournaL of Geometry and Applications*, Vol. 3, No 2, pp. 89-100.

Ferreira, O.P. and Oliveira, P.R. (1998), Subgradient Algorithm on Riemannian Manifold, *Journal of Optimization Theory and Applications*, Vol. 97, No 1, pp. 93-104.

Ferreira, O.P. and Oliveira, P.R. (2002), Proximal Point Algorithm on Riemannian Manifolds, *Optimization*, Vol. 51, No 2, pp. 257-270.

Gabay, D. (1982), Minimizing a Differentiable Function over a Differential Manifold, *Journal of Optimization Theory and Applications*, Vol. 37, No 2, pp. 177-219.

Karmarkar, N. (1990), Riemannian Geometry Underlying Interior-Point Methods for Linear Programming, *Contemporary Mathematics*, Vol. 114, pp. 51-75.

Karmarkar, N. (1984), A New Polynomial-Time Algorithm for Linear Programming, *Combinatorics*, Vol. 4, pp. 373-395.

Nesterov, Y.E. and Todd, M. (2002), On the Riemannian Geometry Defined by self- Concordant Barriers and Interior-Point Methods, *Preprint*

Dikin, I.I. (1967), Iterative Solution of Problems of Linear and Quadratic Programming, *Soviet Mathematics Doklady*, Vol. 8, pp. 647-675.

Eggermont, P.P.B. (1990), Multiplicative Iterative Algorithms for Convex Programming, *Linear Algebra and its Applications*, Vol. 130, pp. 25-42.

Oliveira, G.L. and Oliveira, P.R. (2002), A New Class of Interior-Point Methods for Optimization Under Positivity Constraints, *TR PESC/COPPE-FURJ, preprint,*

Pinto, A.W.M., Oliveira, P.R. and Cruz Neto, J. X. (2002), A New Class of Potential Affine Algorithms for Linear Convex Programming, *TR PESC/COPPE-FURJ, preprint,*

Moré, J.J. and Rheinboldt, W.C. (1973), On P- and S-functions and related classes of n- dimensional nonlinear mappings, *Linear Algebra Appl.*, Vol. 6, pp. 45-68.

Harker, P.T. and Pang, J.S. (1990), Finite dimensional variational inequality and nonlinear complementarity problems: A survey of theory, algorithms and applications, *Mathematical Programming*, Vol. 48, pp. 161-220.

Rockafellar, R.T. (1976), Monotone operators and the proximal point algorithm, *SIAM Journal on Control and Optimization*, Vol. 14, pp. 877-898.

Martinet, B. (1970), Regularisation d'inequations variationelles par approximations sucessives, *Revue Française d'Informatique et de Recherche Operationelle*, Vol. 4, pp. 154-159.

Facchinei, F. (1998), Strutural and stability properties of P_0 nonlinear complementarity problems, *Mathematics of Operations Research*, Vol. 23, pp. 735-745.

Facchinei, F. and Kanzow, C. (1999), Beyond Monotonicity in regularization methods for nonlinear complementarity problems, *SIAM Journal on Control and Optimization*, Vol. 37, pp. 1150-1161.

Yamashita, N., Imai, I. and Fukushima, M. (2001), The proximal point algorithm for the P_0 complementarity problem, *Complementarity: Algorithms and extensions* , Edited by Ferris, M.C., Mangasarian, O. L. and Pang, J. S., *Kluwer Academic Publishers*, pp. 361-379.

Kanzow, C. (1996), Global convergence properties of some iterative methods for linear complementarity problems, *SIAM Journal of Optimization*, Vol. 6, pp. 326-341.

Moré, J.J. (1974), Coercivity conditions in nonlinear complementarity problem, *SIAM Rev.*, Vol. 16, pp. 1-16.

Fischer, A. (1992), A special Newton-type optimization method, *Optmization*, Vol. 24, pp. 269-284.